Sedimentology and Stratigraphy

Sedimentology and Stratigraphy

Editor: Aiden Williams

Callisto Reference
www.callistoreference.com

Callisto Reference,
118-35 Queens Blvd., Suite 400,
Forest Hills, NY 11375, USA

Visit us on the World Wide Web at:
www.callistoreference.com

ISBN: 978-1-64116-075-9 (Hardback)

Cataloging-in-Publication Data

Sedimentology and stratigraphy / edited by Aiden Williams.
p. cm.
Includes bibliographical references and index.
ISBN 978-1-64116-075-9
1. Sedimentology. 2. Geology, Stratigraphic. 3. Geology. I. Williams, Aiden.
QE471 .S43 2019
551.3--dc23

Table of Contents

Preface

This book aims to highlight the current researches and provides a platform to further the scope of innovations in this area. This book is a product of the combined efforts of many researchers and scientists, after going through thorough studies and analysis from different parts of the world. The objective of this book is to provide the readers with the latest information of the field.

The study of sediments such as silt, clay and sand, and the processes that shape their formation is referred to as sedimentology. Some of these processes are weathering, erosion, deposition, transport and diagenesis. Studies of sedimentary rocks and structures are fundamental to the reconstruction of past environments and understanding of the Earth's geologic history. The principles of superposition, original horizontality, lateral continuity and cross-cutting relationships are vital to the study of sedimentology. This field is closely associated with stratigraphy. It is a branch of geology that studies rock layers and stratification. It is crucial for the study of layered volcanic rocks and sedimentology. The sub-fields of stratigraphy are biostratigraphy and lithostratigraphy. Descriptions of rock core, sequence stratigraphy and lithology of the rock are some of the focus areas of sedimentology as well as stratigraphy. This book provides comprehensive insights into the fields of sedimentology and stratigraphy. Also included in this book is a detailed explanation of the various concepts and applications of these domains. In this book, using case studies and examples, constant effort has been made to make the understanding of the difficult concepts of these disciplines as easy and informative as possible for the readers.

I would like to express my sincere thanks to the authors for their dedicated efforts in the completion of this book. I acknowledge the efforts of the publisher for providing constant support. Lastly, I would like to thank my family for their support in all academic endeavors.

Editor

1

Fault-block rotation controlling the distribution of fluvial sediments; a quantitative test on a Lower Pennsylvanian (Carboniferous) cyclothem succession

FRANK J. G. VAN DEN BELT*, POPPE L. DE BOER* and FRANK VAN BERGEN†,[1]

*Department of Earth Sciences, University of Utrecht, P.O. Box 80021, 3508 TA, Utrecht, The Netherlands (E-mail: f.j.g.vandenbelt@uu.nl)
†TNO/Geological Survey of the Netherlands, Princetonlaan 6, P.O. Box 80015, 3508 TA, Utrecht, The Netherlands

Keywords
Cyclothems, floodplain, fluvial, Pennsylvanian, subsidence.

[1]Present address: Nexen Petroleum UK Ltd., Prospect House, 97 Oxford Road, Uxbridge UB8 1LU, UK.

ABSTRACT

Depositional models of axial fluvial systems in half-grabens predict that the fluvial-sandstone percentage increases towards the downthrown side of a fault, because channel systems tend to migrate to the area of maximum subsidence. This migration is at the expense of mudstone, but floodplain deposition occurs near faults occasionally. The models assume gradual, transverse tilting and no external base-level change, and their applicability to cases involving tectonics and/or sea-level change may therefore be restricted. Here, a quantitative analysis is presented on a subsurface data set from a Lower Pennsylvanian cyclothem succession, which formed under conditions of differential subsidence and fluctuating sea-level. The studied interval is wedge-shaped and shows a systematic thickness increase from 165 to 245 m, controlled by syndepositional fault-block tilting. It comprises three depositional units, bounded by coal groups. These units display an upward change from wedge-shaped (75 to 120 m) to tabular (42 to 55 m). Despite their variable thickness, the units contain almost equal amounts of *ca* 45 m of floodplain deposits, plus *ca* 5 m of encased channel sandstones, in all boreholes. Where units are thicker, the remaining thickness comprises fluvial-braidplain sandstone. This arrangement indicates that the units represent equal time periods, during which background subsidence allowed the deposition of thin channel sands and overbank mud on a level floodplain. Occasional tilting produced additional accommodation space, which was completely filled by sand-dominated braided systems. The temporary cessation of floodplain-mud deposition suggests that aggradation of the river system could not keep up with floodplain tilting. In addition, bypass of floodplain fines may have been promoted by a basin parallel tilting component. It is shown that (i) cases in which the standard models fully apply, and (ii) cases in which differential subsidence is too strong or too abrupt, can be distinguished by analysing cross-plots of cumulative-sandstone and cumulative-mudstone thickness.

INTRODUCTION

Pennsylvanian (Upper Carboniferous) sedimentary successions in Euramerican basins are characterized by repetitive fluvio-deltaic cycles that formed in response to glacio-eustatic sea-level fluctuations (Davies, 2008; Greb *et al.*, 2008; Rygel *et al.*, 2008). These cycles, also known as 'cyclothems' (Weller, 1930), are a few metres to tens of metres thick and commonly comprise deltaic or marine shales overlain by alternations of fluvial sandstone, floodplain mud and coal-bearing coastal-plain deposits (Fielding, 1984a; Guion *et al.*, 1995). The fluvial sandstones are mostly extensive, erosively based bodies with a thickness up to 15 to 20 m, and their width is estimated to be anywhere between a few and tens of kilometres (Fielding, 1986; Aitken & Flint, 1995; Guion & Rippon, 1995; Rippon, 1996; Jones & Glover, 2005; Rygel *et al.*, 2008).

Synsedimentary tectonics may influence or control the distribution and (stacked) thickness of the fluvial sandstones. In the Pennine Basin (UK) major fluvial-sandstone bodies appear stacked on hangingwall blocks (Fielding, 1984a, 1986; Fielding & Johnson, 1987; Guion & Fielding, 1988; Rippon, 1996). This arrangement was attributed to the tendency of fluvial channels to seek low-lying areas after avulsion events (cf. Alexander & Leeder, 1987; Leeder & Gawthorpe, 1987). Collinson *et al.* (1993) described comparable sandstone stacking in the Pennsylvanian below the adjacent Southern North Sea. The laterally offset, 'diagonal' stacking of channel bodies observed by Fielding (1984a) was considered evidence of progressive downslope channel migration. In various sub-basins of the Appalachian foreland (USA), similar observations were made. For the Warrior Basin (Alabama), Weisenfluh & Ferm (1984) and Ferm & Weisenfluh (1989) described how a number of *ca* 1 km wide fluvial channels flowed parallel to a major fault on the hangingwall block over a distance of 10 to 15 km, stacking into a 30 m thick compound sandstone unit. Horne (1978) and Allen (1993) observed similar channel-stacking patterns in eastern Kentucky and West Virginia. Also, in those cases, clustering of fluvial-channel sandstone bodies is attributed to the tendency of channels to migrate to low-lying areas at the downthrown side of synsedimentary active faults (Ferm & Weisenfluh, 1989).

The tendency of fluvial channels to preferentially occupy areas of maximum subsidence, and associated sandstone-body stacking near faults, was first described by Alexander & Leeder (1987) and Leeder & Gawthorpe (1987) and has been incorporated in modelling studies since the late 70s. Pioneering theoretical work by Allen (1978) and Bridge & Leeder (1979) was followed by numerous quantitative studies investigating various aspects of alluvial sedimentation (Hajek *et al.*, 2010; Straub *et al.*, 2013; Kopp & Kim, 2015), including the response of fluvial systems to half-graben tilting (Bridge & Mackey, 1993; Mackey & Bridge, 1995; Kim *et al.*, 2010). Here, we refer to these studies collectively as LAB-models (Leeder, Allen and Bridge), following Bryant *et al.* (1995). These theoretical models are supported by a few (semi-)quantitative field tests only (Mack & James, 1993; Leeder *et al.*, 1996; Peakall, 1998) and the rarity of high-quality quantitative data sets has hampered thorough testing of the models (Paola, 2000).

An element of some of the LAB-models is the prediction of the response of fluvial systems to tilting of a floodplain and how this controls alluvial stratigraphy. For example, cross-sections from Bridge & Leeder (1979) and Bridge & Mackey (1993) show slightly different results, but overall the model output displays an increase of cumulative-sandstone thickness (Fig. 1). This increase is non-linear, because at the same time the sandstone percentage increases. Despite the related decrease in the

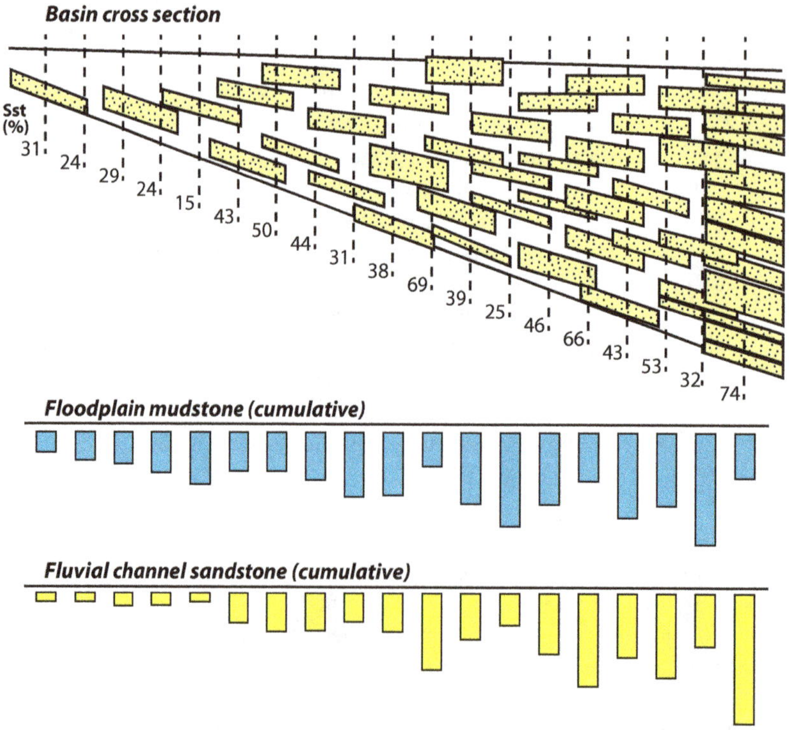

Fig. 1. Half-graben cross-section showing the distribution of fluvial channels (redrawn after Bridge & Mackey, 1993). Sandstone percentage, cumulative-sandstone and cumulative-mudstone thickness are indicated.

mudstone percentage, the predictions show that the cumulative-mudstone thickness does still increase towards the fault.

The LAB-models are based on gradual subsidence and do not account for drastic changes in base level, i.e. due to glacio-eustasy, which may put limitations on their general applicability. Furthermore, the models are based on situations in which a floodplain is tilted exactly perpendicular to the block-margin fault; in actual situations, however, there may be a longitudinal tilting component. Longitudinal tilting has been dealt with in a number of studies (Mackey & Bridge, 1995; Heller & Paola, 1996), although for symmetrical basins only, and it is found to result in downstream changes of alluvial architecture by promoting channel avulsions.

A number of recent studies have pointed out that floodplain tilting does not automatically result in channel migration towards the area of maximum subsidence. Kim *et al.* (2010) have shown, based on laboratory simulations, that axial channels are only forced to migrate towards a fault when lateral channel mobility and/or tilting rates are high. Peakall (1998) has determined for the Holocene Carson River (U.S.A.) that channel systems indeed migrated towards the marginal fault shortly after faulting events, but that they migrated away from the fault during periods of tectonic quiescence. Furthermore, channel systems may be forced to migrate by means of avulsion when the lateral tilt rate is high, or gradually when low (Peakall *et al.*, 2000).

In this paper, results are presented of a detailed quantitative analysis of the control of subsidence on the distribution of fluvial sandstones under strong differential subsidence in a rotational fault-block setting with a fault-parallel tilting component. A three-dimensional borehole data set was analysed, from a Lower Pennsylvanian coal-bearing sequence in a coal-mining concession area in the Upper Silesian Coal Basin (USCB) in Poland (Van Bergen *et al.*, 2006, 2009; Van den Belt, 2012). Only rudimentary sedimentary descriptions were available for the boreholes, which did not allow accurate interpretation of facies. However, a high borehole density in an area of considerable differential subsidence, and a well-established coal-bed framework, makes this an interesting data set to quantitatively test how differential subsidence may control the distribution of fluvial sands.

GEOLOGICAL BACKGROUND

The USCB is a narrow, north–south aligned foreland basin (Fig. 2) in Poland and the Czech Republic (Ziegler, 1990; Zdanowski & Zakowa, 1995). It is bordered by the Moldanubian thrust zone in the west and the East Silesian High in the east, and is strongly asymmetrical. The basement consists of strongly folded metamorphic and plutonic Precambrian and Cambrian rocks. In many parts of the basin, including the study area, the basement is block-faulted, and normal faulting and fault-block rotation associated with this fault system influenced sedimentation during the Devonian (Ziegler, 1990) and Carboniferous (Jureczka & Kotas, 1995). Ziegler (1990) attributed subsidence to tectonic loading of foreland crust and noted that a dextral shear component is required to explain crustal shortening in the Rhenohercynian Basin.

During the Mississippian (Early Carboniferous), the basin was characterized by carbonate-platform deposition throughout, and at the onset of the Namurian deep-water clastics accumulated in the narrow foredeep east of the Moldanubian thrust zone, indicating the onset of regional compression and accelerated subsidence (Zdanowski & Zakowa, 1995). Thrusting started in the Namurian and resulted in a Namurian-Westphalian clastic sediment wedge that thickens westwards (Fig. 2C). In the north-western part of the basin the stratigraphic column is up to 8·5 km thick locally (Kotas, 1994; Zdanowski & Zakowa, 1995; Doktor, 2007). Alternations of fluvio-deltaic sandstone, mudstone and coal characterize the Namurian-Westphalian succession. Fluvial systems originated in the Moldanubian thrust zone and drained to the north and north-east (Fig. 2C; Kotas, 1994).

The Carboniferous stratigraphy of the basin is illustrated in Fig. 3. The studied interval spans the Upper Silesian Sandstone Series (USSS; Namurian C) and the lower part of the Mudstone Series (Westphalian A-B). The USSS is a fluvial sandstone-rich unit with a thickness of up to 1·1 km close to the thrust zone that rapidly thins eastwards (Kotas, 1994). In the study area, it is between 125 and 200 m thick. It unconformably overlies the mudstone-dominated sediments from the Namurian A (Paralic Series) and across most of the basin, its base is marked by regional coal bed 510. The top of the USSS is a goniatite-bearing 'marine band' that marks the Namurian-Westphalian boundary.

The USSS contains nearly 10% coal and many of the coal beds are between 4 and 8 m thick (Kotas, 1994). The abundance of fluvial-sandstone bodies is attributed to the Late-Namurian 'Erzgebirgian' thrusting event taking place at the western basin margin (Kotas, 1994; Zdanowski & Zakowa, 1995). Sediment sourced from the upthrusted area was transported in an overall north–north-eastward direction by basin-parallel fluvial systems, close to the axis of maximum subsidence (Kotas, 1994; Doktor, 2007). Kotas (1995) reports eastward and north-eastward fluvial transport directions for the USSS.

In the eastern parts of the basin subsidence rates were low during deposition of the USSS, thus shielding the area from sediment input and promoting the formation

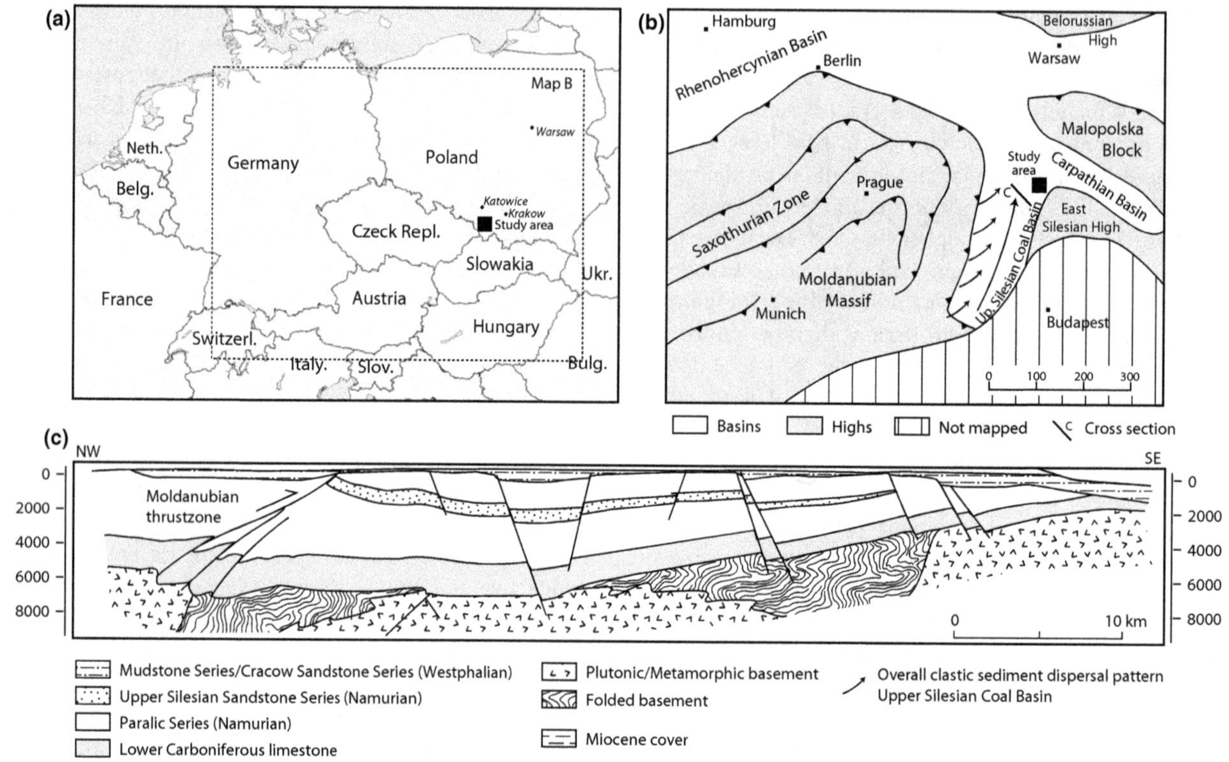

Fig. 2. Palaeogeographical and structural setting of the Upper Silesian Coal Basin and the study area. (A) Map of Central Europe showing location of the study area, (B) Structural organization of north-central Europe during the Namurian (after Ziegler, 1990), (C) NW-SE cross-section through the northern part of the Upper Silesian Coal Basin close to the study area (after Kotas, 1994).

of thick peat bodies. For instance, the thickness of coal bed 510, which constitutes the base of the USSS, increases from 6 m in the study area to 24 m in the east.

The overlying Mudstone Series is dominated by meandering channel and floodplain deposits (Doktor & Gradzinski, 1985; Gradzinski *et al.*, 1995). It is unconformably overlain by the Cracow Sandstone Series (Westphalian C-D), a unit dominated by sandy and conglomeratic braided-fluvial-channel deposits (Gradzinski *et al.*, 1995; Doktor, 2007) that displays overall north–north-eastward sediment transport as well (Kotas, 1995). In the east, the Cracow Sandstone Series is unconformably overlain by red and variegated sediments without any intercalated coal beds; these are probably of Stephanian age (Zdanowski & Zakowa, 1995). The Pennsylvanian section is truncated by an Alpine unconformity and is overlain by Miocene marls, with the depth of truncation increasing westwards (Fig. 2).

Study area

This study is based on sedimentary records of eight boreholes and fault maps from a 5 km^2 area within the 'Silesia' coal-mining concession. The area is located approximately 40 km south of the city of Katowice in southern Poland (Fig. 4). The youngest Pennsylvanian deposits in the area are Westphalian C sandstones of the Cracow Sandstone Series; these are buried below a *ca* 250 m Miocene cover (Van Bergen *et al.*, 2006). The stratigraphic interval under investigation comprises the USSS and the basal section of the conformably overlying Mudstone Series. In the study area, this interval is present at depths between 950 and 1250 m.

The concession area is dissected by two sets of steep faults that strike NE–SW and NW–SE (Fig. 4). The faults intersect at *ca* 60° angles and the NW–SE striking faults abutt the NE–SW striking faults. Offset of coal beds indicates normal displacement. Both the steepness of the faults and the 60° to 70° intersection angles of the two sets point to normal reactivation of an original strike-slip fault system, possibly related to Late Devonian dextral shear (Ziegler, 1990). The E–W trending fault that defines the southern margin of the coal-mining concession area is of Alpine origin (Van Bergen *et al.*, 2006).

The study area is located at the eastern end of the coal-mining concession area (Fig. 4). It is characterized by a major NE–SW trending normal fault (F1) that separates a footwall block in the south-west (fault block I) and a

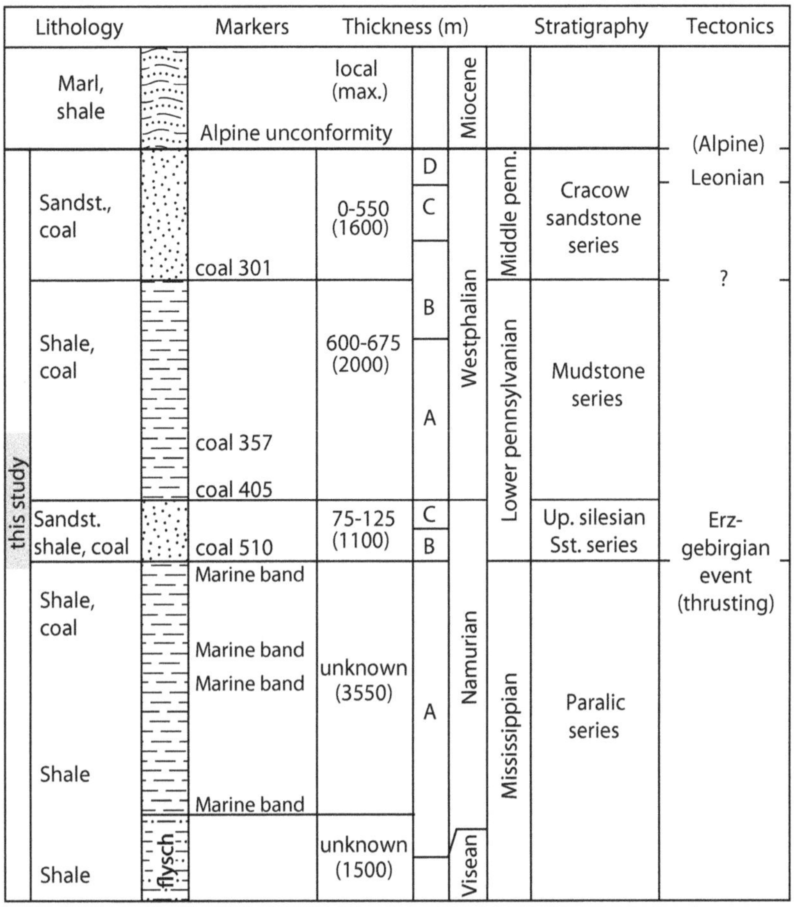

Lithology
Markers
Thickness (m)
Stratigraphy
Tectonics
Marl, shale
local (max.)
Alpine unconformity
Miocene
(Alpine)
Leonian
Sandst., coal
coal 301
0-550 (1600)
D
C
Middle penn.
Cracow sandstone series
?
Shale, coal
coal 357
coal 405
600-675 (2000)
B
A
Westphalian
Lower pennsylvanian
Mudstone series
this study
Sandst. shale, coal
coal 510
75-125 (1100)
C
B
Up. silesian Sst. series
Erz-gebirgian event (thrusting)
Shale, coal
Marine band
Marine band
Marine band
unknown (3550)
A
Namurian
Mississippian
Paralic series
Shale
Marine band
Shale
flysch
unknown (1500)
Visean

Fig. 3. Stratigraphic column for the Upper Silesian Coal Basin. Major coal beds/zones and marine bands are after Kotas (1994); thickness values are for the study area, maximum thickness in the basin are given between brackets.

composite hangingwall block in the north-west. This hangingwall block consists of two higher order fault blocks (II and III) that are separated by fault F2. Note that fault block II is part of the hangingwall block of fault block I, but it also serves as a (higher order) footwall block to fault block III. Extrapolation of coal-bed depth trends, observed on mining company cross-sections, has indicated the presence of a fault between well Si-18 and MB-90 (Fig. 4), the exact orientation of which is not known.

SEDIMENTARY FRAMEWORK

The sedimentary framework is based on core descriptions, wireline logs and coal-bed depth/thickness maps. The available core descriptions were drafted for coal-exploration purposes, with a lithology record but lacking descriptions of sedimentary structures. From regional work it is known that the USSS generally comprises alternating floodplain and coastal-plain mudstones, coal beds and braided fluvial sandstones; marginal marine shales are restricted to the interval below coal bed 405 (Kotas, 1994; Zdanowski & Zakowa, 1995). The elementary character of the sedimentary descriptions did not permit the recognition of, for example, lacustrine intercalations within floodplain shales, or the distinction between small fluvial-channel and thin crevasse delta deposits. Some of the interpretations may therefore be slightly simplified.

Figure 5 shows N-S and E-W correlation panels based on coal-bed interpretations of the mining company, using regional coal-bed terminology. The studied interval, comprising the USSS and the lower 50 m of the Westphalian Mudstone Series is characterized by four relatively thick, correlatable coal beds (510, 405, 401, 354). These main coal beds define three (major) sedimentary units (Unit 1 to 3) with an average thickness that decreases upwards from 115 m to 50 m. Units 1 and 2 have a distinct wedge shape and thicken to the north-west. Unit 3 has a more tabular shape. Internally, each unit is composed of three to four (preserved) sedimentary cycles, which are typical cyclic alternations of fluvial sandstone, grey floodplain or delta-plain mudstones and coal beds (cyclothems). Units 1 and 2 are dominated by thick, laterally extensive

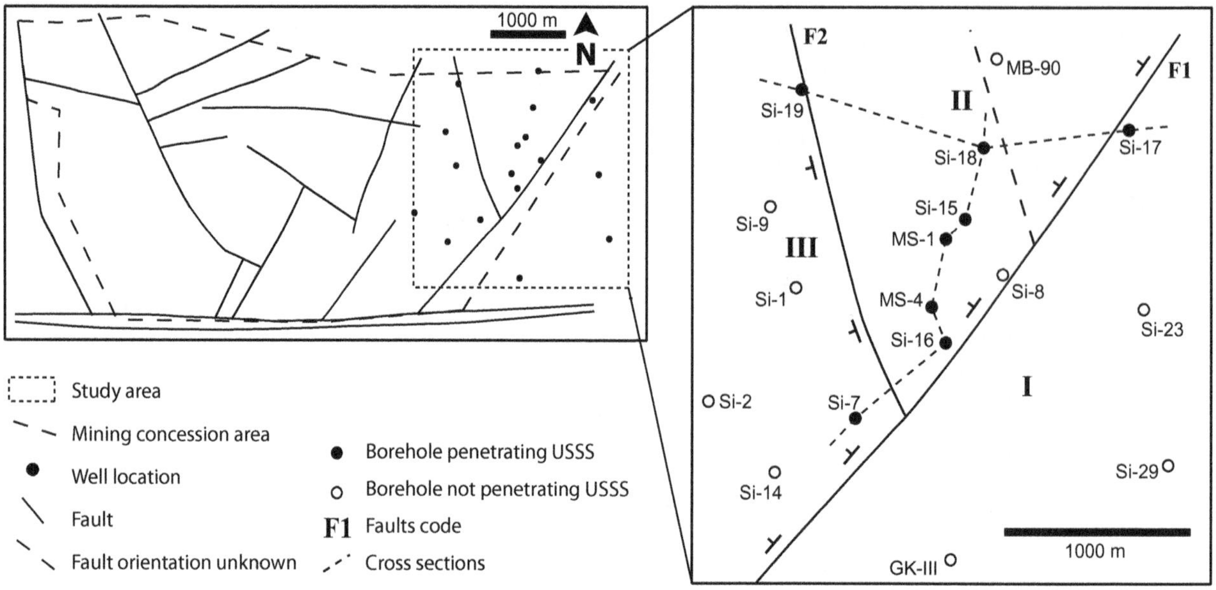

Fig. 4. Map of the Silesia mining concession showing major faults and borehole locations for the study area. The study area covers 3 fault blocks (numbered I to III) with densely spaced boreholes. Wells not incorporated in the cross-sections do not penetrate the studied interval or only partly and were therefore not included in the quantitative analysis (open circles).

sandstone bodies, whereas the sandstone bodies in Unit 3 are thinner and more isolated.

Thick sandstone bodies are numerous in Units 1 and 2; they alternate with laminated and rooted grey mudstone intervals and coal beds. Their number decreases upwards from Unit 1 (average sandstone content *ca* 55%) to Unit 2 (*ca* 30%). Thick sandstone bodies have not been encountered in Unit 3. The bodies are between 10 and 20 m thick and most of them appear laterally extensive. Occasionally, sandstone bodies in the vicinity of faults wedge out towards adjacent boreholes (e.g. borehole Silesia-18/19). Towards the downthrown sides of faults, thick sandstone bodies are more numerous and occur stacked into compound bodies, and their cumulative thickness gradually increases. This suggests that channels preferentially followed fault trajectories (Fig. 5, inset).

Thin, isolated sandstone bodies are observed in all three units. They are most pronounced in Unit 3, where they make up *ca* 10% of the succession. These sandstone bodies are usually only a few metres thick and cannot be traced to neighbouring boreholes. In contrast to the thick fluvial sandstones that dominate Units 1 and 2, these channel sands are encased in floodplain fines.

Coal beds are present throughout the sequence and occur in distinct bundles. Two types of coal bed characterize the study area. The first type comprises thick coal beds (1 to 6 m) that are present across the entire study area (Fig. 5). The other type comprises non-continuous thin coals (<1 m); these wedge out laterally or are possibly truncated by fluvial-sandstone bodies. Coal beds show pronounced thickness variation across the study area. For example, the thickness of coal bed 510 increases from 0·6 m (Silesia-17) to 6 m (Silesia-19) over a distance of *ca* 3 km.

Interpretation

The thick, laterally extensive sandstone bodies that dominate Units 1 and 2 have dimensions in the range of major fluvial-sandstone bodies in the Pennsylvanian of Europe and North America (Fielding, 1984a; Aitken & Flint, 1995; Jones & Glover, 2005; Greb *et al.*, 2008). In general, Pennsylvanian fluvial systems are interpreted as sheet-like fluvial-braidplain deposits with erosional basal surfaces (Haszeldine & Anderton, 1980; Jones & Hartley, 1993; Jones & Glover, 2005) or as incised valley deposits (Aitken & Flint, 1995; Hampson *et al.*, 1999). Based on their sheet character and the regular alternations with mudstone and coal, the majority of the major sandstones in the study area are interpreted as the deposits of laterally extensive fluvial-braidplain systems. Some of the sandstone bodies in the vicinity of faults, however, wedge out rapidly away from these fault, which could point to channels incising into the substrate where subsidence is highest. The thin, isolated sandstone bodies encased in floodplain mud are interpreted as the deposits of small channels meandering across the floodplain. Other types of floodplain sandstones, such as crevasse splays, may be represented as well

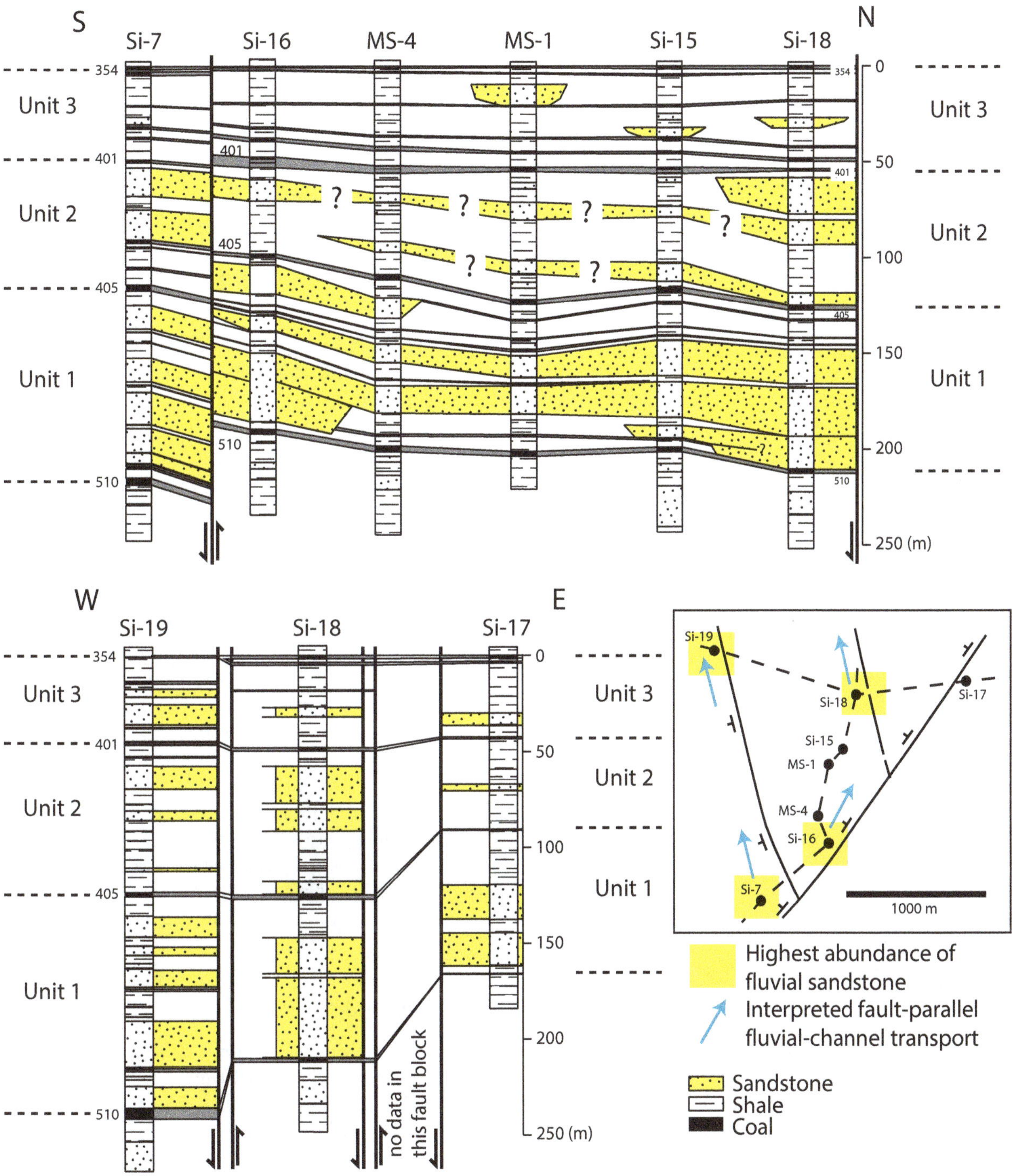

Fig. 5. Correlation panels showing north–south and east–west transects through the study area. Correlations based on coal-bed stratigraphy. Sandstone bodies seem well correlatable in the lower part of the studied interval; correlations in the middle part are more tentative. Numbers 510, 405, 401 and 354 indicate major coal beds.

(Fielding, 1986), but could not be confidently distinguished due to the rudimentary nature of the data set.

Thick upper Carboniferous coal beds are commonly interpreted as coastal-plain and floodplain peat accumulations (Fielding, 1984b; McCabe, 1984) and, based on the associated facies, the coals in the study area are probably floodplain coals. The great thickness and lateral extent of some of the coals indicate that peat swamps were exten-

Table 1. Thickness and lithological data for the studied interval (between coal beds 354 to 510).

Borehole	Sediment column (m)	Lithology (cumulative thickness) Sandstone (m)	Coal (m)	Mudstone (m)	Mudstone + coal (m)	Lithology Sandstone (%)	Coal (%)	Mudstone (%)	Mudstone + coal (%)	Deviation from mean Sediment column (%)	Sandstone (m) (%)	Coal (m) (%)	Mudstone (m) (%)	Mudstone + coal (m) (%)
Silesia-07	218	94	1	107	124	43·1	7·8	49·1	56·9	6·2	13·9	37·8	−3·2	−3·9
Silesia-15	202	69	15	118	133	34·2	7·4	58·4	65·8	−1·6	−16·4	21·6	6·8	3·1
Silesia-16	192	70	11	111	122	36·5	5·7	57·8	63·5	−6·5	−15·2	−10·8	0·5	−5·5
Silesia-17	166	44·5	4·5	117	121·5	26·8	2·7	70·5	73·2	−19·2	−46·1	−63·5	5·9	−5·9
Silesia-18	213	104	12·5	96	108·5	48·9	5·9	45·2	51·1	3·5	26·1	1·4	−13·1	−15·9
Silesia-19	242	114	14	114	128	47·0	5·8	47·2	53·0	17·6	37·6	13·5	3·2	−0·8
MS-1	204	62·5	13·5	128	141·5	30·6	6·6	62·7	69·4	−0·6	−24·2	9·5	15·8	9·6
MS-4	201	47	10·5	143·5	154	23·4	5·2	71·4	76·6	−2·1	−43·0	−14·9	29·9	19·3
Mean	205	83	12	111	129	39	6	55	55					
Correlation coefficient (with sediment column thickness)		0·84	0·78	−0·20	0·00									

sive and built considerable peat accumulations over prolonged periods. The thin, non-continuous coal beds probably formed in local depressions, such as lakes and abandoned channels (cf. Greb & Chesnut, 1992). The change in coal thickness over short distances may be related to infilling of inherited floodplain topography (Greb & Chesnut, 1992; Greb *et al.*, 1999), but overall coal thickness increases towards the downthrown sides of faults, which indicates that synsedimentary faulting and related differential subsidence played a role (cf. Guion & Fielding, 1988; Greb *et al.*, 2002, 2005).

INFLUENCE OF FAULT-BLOCK ROTATION ON SEDIMENTATION

Regional subsidence variations were approximated by measuring the thickness of the studied interval (coal-bed interval 354 to 510) and the thickness of the three units at the borehole locations (Table 1). This approach provides a reasonable approximation because the occurrence of coal beds throughout the sequence indicates deposition near base level (Bohacs & Suter, 1997), i.e. at or close to sea-level, and also that sedimentation overall kept up with subsidence. The studied interval contains relatively equal amounts of mudstone at the various borehole locations (Table 1), so that the thickness variations cannot be due to differential compaction of mud. Because coal is more abundant in the thicker sequences, the thickness variations of the studied interval cannot be attributed to preferential compaction of peat. Also note that Nadon (1998) has shown that the deep burial compaction of peat, long considered to be around 10 : 1 (McCabe, 1984), is not higher than *ca* 2·5 : 1 which equals the compaction of mud. This ratio applies because most compaction occurs in the plant-to-peat rather than the peat-to-coal stage (Van Asselen *et al.*, 2010).

Thickness maps

In Fig. 6 thickness maps are shown for the studied interval and for the individual Units 1 to 3. Map *a* shows the variation in thickness of the entire sediment column (between coals 510 to 354); it ranges from 166 m (Silesia-17) on fault block I to 242 m on fault block III (Silesia-19). With a mean thickness of 205 m these extremes reflect a differential-subsidence range between −19% and +18% (Table 1). The data further show that the thickness increases northwards across fault blocks II and III.

Figure 6B shows the variation in interval thickness for Unit 1. The unit has a maximum thickness of 119 m in borehole Silesia-19 and minimum thickness of 76 m in borehole Silesia-17. With an average thickness of 94 m, differential subsidence ranges between −18% and +28%.

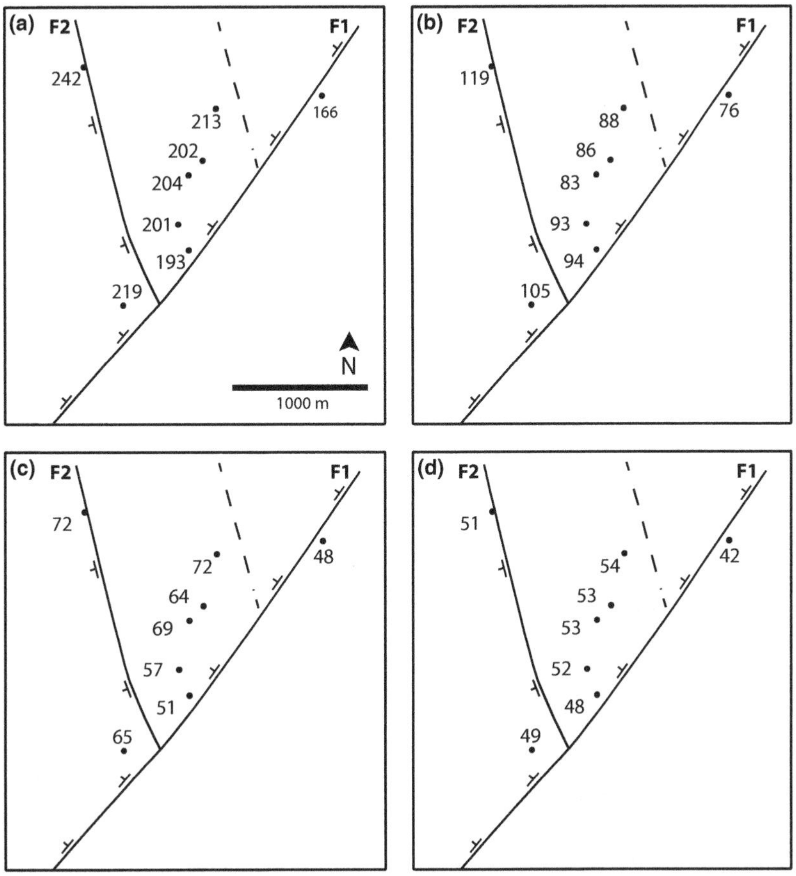

Fig. 6. Thickness maps for the studied stratigraphic interval. (A) complete interval, (B) Unit 1, (C) Unit 2 and (D) Unit 3.

The thickness of Unit 1 on fault block II shows a slight southward increase from 88 to 94 m; on block II the unit thickness increases from 105 to 119 m northward.

In Fig. 6C the thickness variation for Unit 2 is shown. In borehole Silesia-19, the unit has a maximum thickness of 72 m; it has a minimum thickness of 48 m in borehole Silesia-17. The mean thickness is 62 m, giving a differential-subsidence range from −23% to +16%. On fault block II the thickness increases northwards from 51 to 72 m and on fault block III the thickness increases northwards as well, from 65 to 79 m.

Figure 6D shows the thickness variation for Unit 3. It is characterized by a fairly constant thickness of 48 m to 54 m across most of the study area, indicating little differential subsidence, ranging between −5% and +8%. Borehole Si-17 on fault block I shows a more reduced thickness of 42 m (−17%).

Subsidence history

The maps of Fig. 6 indicate that the thickness variations are not random, but follow directional trends. The data indicate that much of the difference in thickness can be attributed to relative movements between fault blocks, but thickness changes on individual fault blocks indicate that rotation of fault blocks contributed considerably.

Subsidence rates were lowest in the south-east and increased overall to the north-west. A reduced thickness for all units in borehole Silesia-17 indicates that fault block I served as a footwall block to fault blocks II and III. Fault block III experienced the highest subsidence rates; during the deposition of Units 1 and 2 it subsided more rapidly than the fault blocks to the east and south-east. The rather constant thickness of Unit 3 north of fault F1 indicates that differential subsidence between fault blocks II and III came to a halt after deposition of Unit 2, and that they continued to subside as a compound block, only slightly faster than fault block I.

The thickness differences indicate that differential subsidence was greatest during and shortly after deposition of coal bed 510 and had mostly ceased after deposition of Unit 2, the top of which marks the Namurian-Westphalian boundary. This boundary coincides with the end-

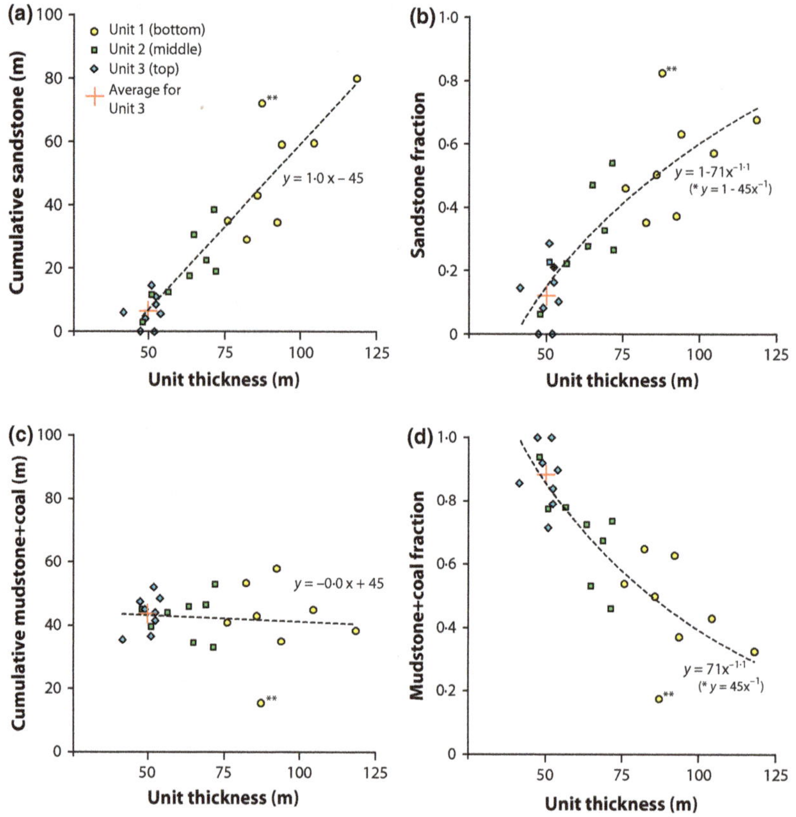

Fig. 7. Cross-plots of lithological thickness data versus cycle thickness. (A) cumulative-sandstone thickness, (B) sandstone fraction, (C) cumulative-mudstone/coal thickness and (D) mudstone/coal fraction. *Regression equation with outlier** excluded.

Namurian termination of the 'Erzgebirgian' thrusting event (Kotas, 1994), possibly indicating that thrusting events caused immediate loading-induced subsidence, which was accommodated by normal displacement of the block-faulted basement. Then, when thrusting came to a halt subsidence was dominated again by background subsidence.

Thickness variations on individual fault blocks point to superimposed rotation, which was consistently northwards for fault block III, whereas fault block II rotated northwards during the deposition of Unit 2 and shows a westward or south-westward rotational component during the deposition of Unit 1. For the other fault blocks, the rotation history could not be reconstructed due to limited borehole control.

Fluvial transport directions

Due to data limitations, there is no detailed information on sandstone-body orientation or palaeocurrent directions, other than regional information indicating overall north to north-eastward fluvial transport. However, the distribution of sandstone bodies may give some general information about their orientation. The correlation panels (Fig. 5) show that fluvial-sandstone bodies are more numerous and have a greater combined thickness at the downthrown sides of faults, i.e. in wells Si-7, Si-16, Si-18 and Si-19. This is taken as an indication that the channels flowed primarily parallel to these faults, as indicated on the inset of Fig. 5.

Note that the orientation of faults in the study area is such that the overall transport direction of the fluvial systems is consistent with the large-scale, north–north-westward direction observed for the basin during deposition of the USSS (Kotas, 1994; Zdanowski & Zakowa, 1995).

QUANTITATIVE ANALYSIS OF SEDIMENT DISTRIBUTION

To assess the mechanism that controlled sediment distribution, a quantitative analysis of the influence of subsidence on the distribution of sandstone, mudstone and coal was carried out for the entire studied interval (between coal beds 510 and 354) and for the three units separately.

Cumulative thicknesses and proportions of the different lithologies are presented in Table 1. A positive correlation of 0·84 exists between the thickness of the studied interval

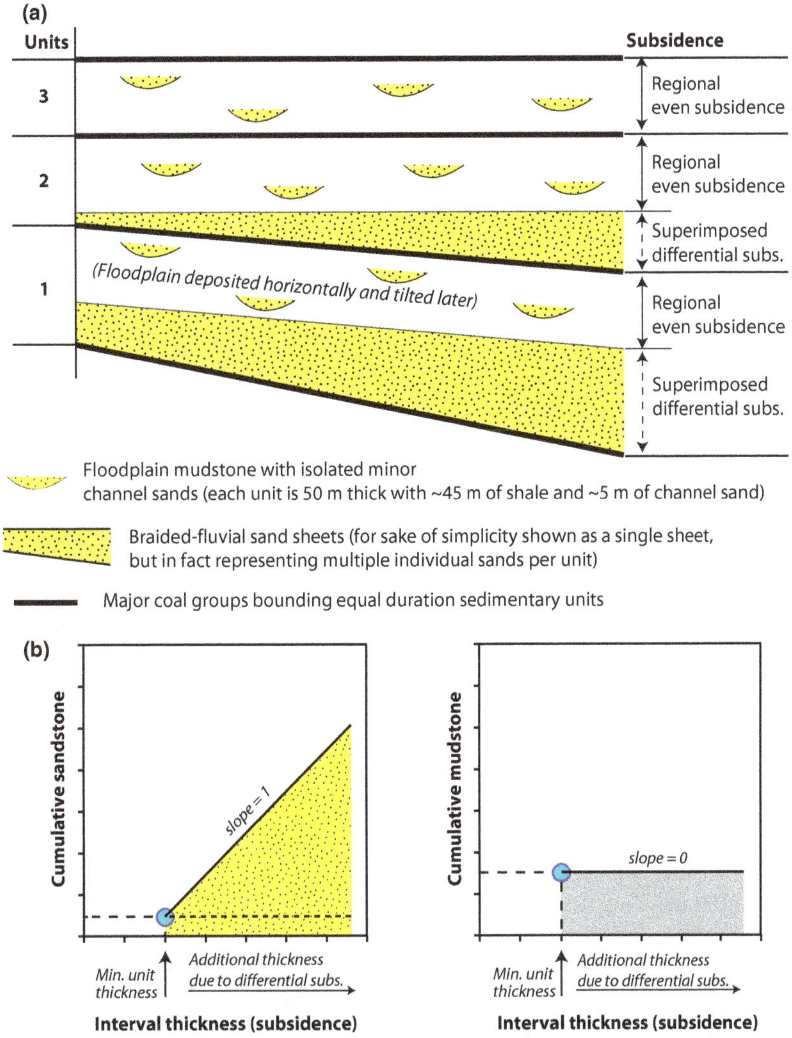

Fig. 8. Diagram showing (A) the overall distribution of braided-fluvial sands and coal-bearing floodplain deposits in Units 1, 2 and 3; and (B) cross-plot trends for cumulative-sandstone and cumulative-mudstone thickness.

and the cumulative-sandstone thickness (Table 1, Fig. 7A). This correlation is to be expected, because an overall thicker sequence is likely to contain more sandstone than a thin sequence, but the percentage of sandstone increases with increasing interval thickness as well. A similar relationship is observed for coal, with higher values for cumulative coal thickness and coal percentage where the thickness of the studied interval is greater (correlation coefficient: 0·78). Conversely, the cumulative-mudstone content shows a slightly negative correlation (−0·20) with interval thickness. Note, however, that the cumulative thickness of mudstone and coal (those lithologies added) shows a correlation coefficient of zero with interval thickness.

The cross-plots in Fig. 7 show that the strong positive correlation between interval thickness and cumulative-sandstone thickness is maintained at the scale of individual units. Note that the cumulative-sandstone-thickness data for each unit constitute a well-defined data cluster in the cross-plot, and that only the data for Units 1 and 2 overlap slightly. The plots for the cumulative mudstone/coal thickness and the mudstone/coal percentage (Fig. 7C) show that the three units contain approximately equal amounts of mudstone/coal (*ca* 45 m), and that no relationship exists between unit thickness and mudstone/coal content. Figure 7A further shows that a single regression line fits the sandstone values for all three units, and that this line originates from the mean of the data cluster for Unit 3, which is the unit that hardly experienced superimposed differential subsidence. This mean value represents a thickness of 50 m, comprising *ca* 45 m of mudstone/coal and *ca* 5 m of minor-channel sandstone on average. The fact that the regression line originates from this mean value indicates that each other unit is essentially a Unit-3 unit complemented by a differential-subsidence dependent amount of sandstone. The regression-line equation for mudstone/coal ($y = 0{\cdot}0x + 45$) reflects that the amount of mudstone/coal equals *ca* 45 m at all borehole locations and its slope of 0 stresses the independence of differential subsidence. The regression-line equation for sandstone ($y = 1{\cdot}0x - 45$) with a slope of 1 shows that all accommodation space, minus the 45 m for mudstone and coal, comprises sandstone.

In Fig. 8, the above interpretation is shown diagrammatically. The fact that amounts of mudstone and coal in each cycle are regionally constant indicates that all mudstone and coal was deposited when the depositional plain experienced laterally even subsidence. If mudstone and coal would have been deposited during phases of differential subsidence as well, then some additional floodplain mudstone/coal deposition would be expected in areas of higher subsidence, as predicted by the LAB-models (see Fig. 1; Bridge & Mackey, 1993). That each cycle contains the same cumulative thickness of mudstone and coal (*ca* 45 m) further suggest that the three coal group-bounded units represent equal time periods, during which long-term subsidence resulted in regionally constant, relatively slow creation of accommodation space. That thick coal-bounded units represent equal time periods is a common interpretation for Pennsylvanian sequences, attributed to strong glacio-eustatic control (Klein & Willard, 1989; Greb *et al.*, 2008), and such cycles are typically thought to be short eccentricity (100 kyr) or long-eccentricity (400 kyr) cycles (Heckel, 1986, 2008).

That the cumulative-sandstone thickness is much higher in areas of stronger subsidence while equal amounts of mudstone/coal are present at each borehole location indicates that fault-block tilting was episodic, only shortly interrupting background subsidence. When differential subsidence commenced, it resulted immediately in an overall change from accumulation of floodplain mud and associated minor-channel sands to bypassing of mud and accommodation of sand-sized sediment only.

SYNTECTONIC DEPOSITIONAL MODEL

As explained above, the alternation of tabular mudstone/coal deposits, with encased small-scale fluvial-channel sandstone, and wedge-shaped fluvial-braidplain deposits reflects that continuous, regionally constant subsidence was at times overprinted by pulses of fault-block rotation. It is envisaged that thrusting-induced loading was accommodated by reactivation of the block-faulted basement.

In Fig. 9, a depositional model is depicted that shows how periodic fault-block tilting superimposed on regional subsidence, explains the sediment distribution. An overall

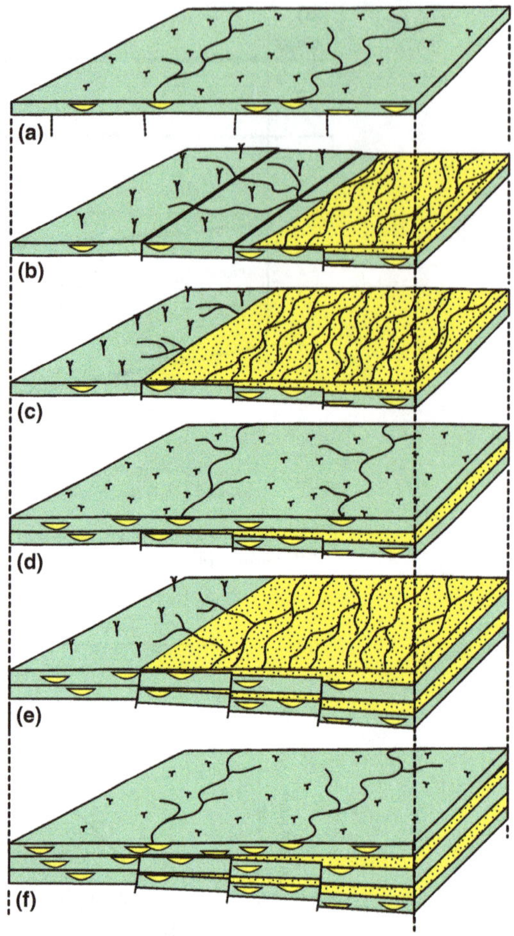

Fig. 9. Depositional model for fluvial deposition on top of faulted topography where tilting has a downstream component. (A) Under conditions of regionally constant subsidence a level floodplain should exist. Floodplain deposition is dominated by overbank mudstone, with small channels, contained within stable vegetated levees, distributed randomly. Dashed vertical lines below the floodplain indicate hidden, temporarily inactive faults; (B) When faulting occurs, additional accommodation space is created which becomes occupied by a high-energy braidplain system, eventually filling up the faulted topography (C); note that laterally the floodplain is a surface of non-deposition. The existence of a gradient in the downstream direction results in bypassing of the mud fraction, and allowing deposition of sand only. (D) Once faults become inactive and the available accommodation space has been filled by the braidplain system, a level depositional plain is re-established, and floodplain deposition takes over again. (E & F) Alternating periods of floodplain deposition and braidplain deposition result in a stacking of (i) tabular units consisting of mudstone, coal and rare isolated sandstone bodies and (ii) wedge-shaped sandstone bodies.

level depositional plain, characterized by depositional topography only, existed when regionally constant subsidence prevailed. The absence of a strong depositional gradient permitted standing water, thus promoting the settling of fine-grained sediment and the accumulation of

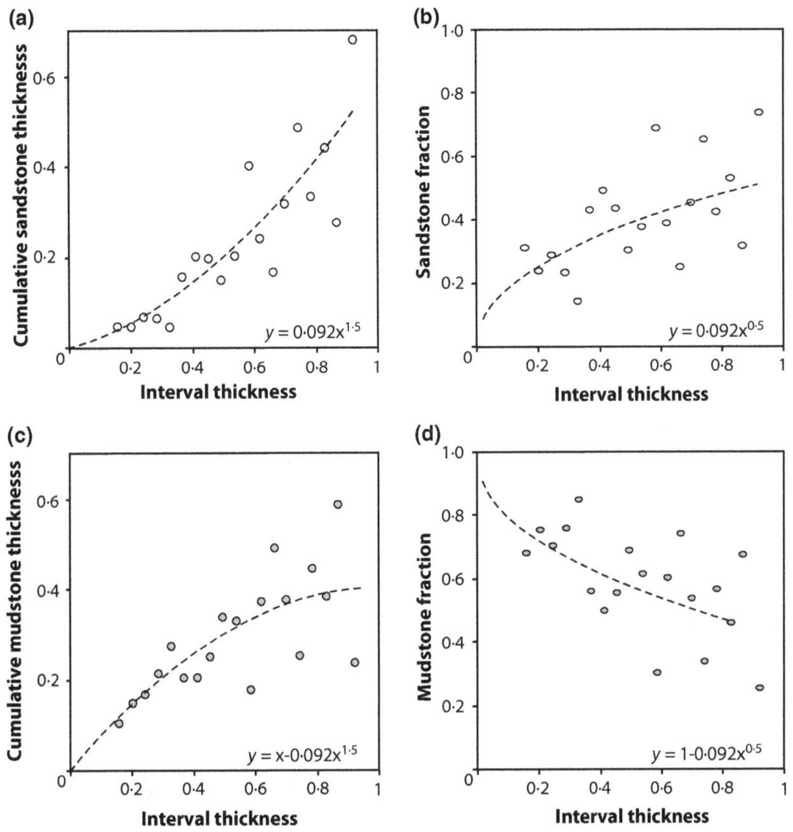

Fig. 10. Cross-plots determined for the LAB-model sequence in Figure 1.

peat (Fig. 9A). Channels flowed across the floodplain and experienced regular avulsions, resulting in a more or less random sandstone distribution. This random distribution is considered the direct result of constant subsidence rates throughout the area, i.e. there were no low-lying areas that preferentially attracted channel systems (cf. Alexander & Leeder, 1987).

Thrusting events caused periodic rotation of basement fault blocks and tilting of the depositional plain in various directions, but mostly with a northward-tilting component, i.e. in the direction of regional downbasin fluvial transport, which led to the interruption of floodplain conditions. It is thought that tilting was too fast for the fluvial system to compensate by means of aggradation and thus developed a depositional gradient (cf. Heller & Paola, 1996; Connell *et al.*, 2012). This gradient then resulted in bypass of fines downbasin. Fluvial sand accumulated in the low-lying areas, probably in braidplain systems, adapting and building to the new base-level profile (Fig. 9B). The braidplain systems quickly filled the newly available accommodation space (Fig. 9C). Once the fault activity had stopped, and the faulted topography had been levelled, a horizontal floodplain depositional system was re-established (Fig. 9D).

Coal beds are slightly thicker in areas of higher subsidence, although they are part of the tabular floodplain layers. This may be because peat swamps were formed where the groundwater level was highest during the transition from braidplain to floodplain, i.e. when the area was not completely level. Such a floodplain arrangement, with swamps preferentially on the hangingwall blocks, may have been maintained during the subsequent floodplain phase; also because peat lands may have grown well above the ground-water table.

Repeated alternations of floodplain and braidplain conditions resulted in the architecture depicted in Fig. 9E, with wedge-shape sandstone layers alternating with more tabular, mudstone-dominated layers. The thickness of these layers is variable, depending on the frequency and intensity of thrusting and fault-block rotation events. Note that the model presented here applies to low base-level situations only, otherwise faulting events would have resulted in the formation of lake bodies on top of the faulted topography (cf. Blair & Bilodeau, 1988).

Because the Pennsylvanian period was characterized by high-frequency sea-level fluctuations (Heckel, 1986; Rygel *et al.*, 2008), it is possible that infilling of the topography took place during transgressions, along the lines as described for Pennsylvanian major fluvial-channel systems in other basins (Davies *et al.*, 1992; Aitken & Flint, 1995; Hampson *et al.*, 1999). The formation of those bodies is commonly explained in terms of aggradational infill of lowstand valleys in response to transgressions (Davies *et al.*, 1992; Shanley & McCabe, 1993).

The final situation (Fig. 9F) shows how two wedge-shape fluvial sandstone units are sandwiched between three tabular units of floodplain mudstone and randomly distributed minor-channel sands.

Lateral extent of fluvial-sandstone bodies

Within the study area there is no evidence of uplift of fault blocks during rotation, as the wedge-shaped units are thicker than the 50 m of the baseline unit at all locations. The major sandstone bodies are extensive and mostly run from fault to fault, their thickness gradually increasing towards areas of higher subsidence. Their regular, predictable distribution implies that the sandstone bodies are largely aggradational in nature and are hardly incised into the substrate, although minor erosion is bound to be associated with their basal scour surfaces.

The model predicts that the major sandstone bodies are continuous throughout the study area. Local deviations may result from, for example, the random distribution of small channels, the presence of mud-filled channels or differential compaction, but in a few boreholes the sandstone content deviates strongly. For instance, borehole Silesia-16 contains thick sandstone bodies at the base of Unit 1 that wedge-out over a few 100s of metres (Fig. 5). This borehole is located very close to a fault, where the subsidence rate was maximal. Such locations must have attracted most of the run-off and may have been more sensitive to erosion, resulting in local incision, either as isolated bodies or as localized deeper basal incisions at the base of aggradational sandstone sheets (Fig. 8).

Cyclothems and time

The above analysis indicates that the accumulation of the major sandstones occurred after faulting events. Hence, the apparent 'cyclothemic' alternation of fluvial sandstone, floodplain mudstone and coal beds, observed within the individual depositional units, is primarily of tectonic origin, rather than being controlled by glacio-eustasy. This conclusion is consistent with interpretations by others for sandstone-rich Pennsylvanian successions, that cyclothemic arrangement has a strong tectonic overprint or is tectonically controlled (Klein & Willard, 1989; Klein & Kupperman, 1992; Jones & Glover, 2005; Greb *et al.*, 2008). As discussed in more detail above, however, the actual infilling of the accommodation space generated by tilting may have taken place during transgressions.

The three coal-bounded units that constitute the main sedimentary framework were likely eustatically controlled, because the equal amounts of mudstone/coal in the units, deposited under conditions of regionally constant subsidence, point to an equal duration. On a large scale, cyclothem sequences are generally believed to be built of sedimentary cycles representing short and long eccentricity (Greb *et al.*, 2008; Heckel, 2008). Based on a minimum thickness of 50 m for each of the three units, and considering the fact that the study area is located at the low-subsidence, eastern end of the basin, long eccentricity is the more likely candidate. It requires a net, post-compaction subsidence rate of *ca* 12 cm kyr^{-1}, compared to more than 50 cm kyr^{-1} for short eccentricity. Based on local and maximum Westphalian A-B stratigraphic thickness values (650 to 2000 m; Fig. 2), and a duration for that interval of about 3 to 4 Ma (Gradstein *et al.*, 2004), subsidence must have been in the range of max. *ca* 65 cm kyr^{-1} near the thrust belt to min. *ca* 15 cm kyr^{-1} in the study area, which further supports the long-eccentricity interpretation.

DISCUSSION AND CONCLUSIONS

This study demonstrates that differential subsidence had a strong control over the lateral and vertical distribution of sandstone, mudstone and coal. Sandstone content increases more than linearly towards the areas of higher subsidence. This outcome is close to LAB-model predictions (Allen, 1978; Bridge & Leeder, 1979; Bridge & Mackey, 1993; Mackey & Bridge, 1995), but the lateral distribution of mudstone is intrinsically different. The work by Alexander & Leeder (1987), Leeder & Gawthorpe (1987) and subsequent LAB-model results predict that, despite a decrease in mudstone percentage, the cumulative floodplain-mudstone thickness increases towards the downthrown side of a fault. In the study area, the amount of floodplain mudstone is constant, hence independent of subsidence.

A number of geological circumstances may explain why the observed alluvial architecture does not fully comply with the standard model. Faulting and associated differential subsidence were intense and probably episodic, thus causing relatively steep gradients, also in the downbasin direction. This condition may have prevented the fluvial systems from maintaining a level floodplain by means of aggradation (Heller & Paola, 1996; Kim *et al.*, 2010; Connell *et al.*, 2012) and causing bypass of fine-grained sediment further downstream. In addition, the studied sediments are from a time period when glacio-eustatic sea-level fluctuations were prominent. It is possible that lowstands of sea-level, concurrent with faulting episodes, further enhanced gradients, and during subsequent transgression promoted infilling of tectonic accommodation space with the sandy sediments of braided-fluvial systems, in the same way that lowstand valleys are filled by aggrading fluvial systems under conditions of rising sea-level (Shanley & McCabe, 1993).

For the sake of comparison, a quantitative analysis was performed on the LAB-model half-graben cross-section of Bridge & Mackey (1993, shown here as Fig. 1), the results of which are shown in Fig. 10. There is quite a bit of data scatter, because of the low number of channels in the sediment volume in combination with their random distribution. However, the cross-plots clearly show that the increase of cumulative-sandstone with increasing interval thickness (subsidence) is not linear, but probably follow a power law ($y = ax^b$), i.e. the sandstone thickness increases more than linearly with subsidence. At the same time cumulative-mudstone increases less than linearly with subsidence ($y = x - ax^b$). At first glance, the data clouds may actually suggest a linear relationship, but it should be taken into account that a linear trend line must intersect the origin of the graph (0,0), because the standard model is based on rotational subsidence only (Fig. 1). Hence, where subsidence is zero, or approaches it, cumulative-sandstone and cumulative-mudstone thickness must be zero as well. The studied sequence, by comparison, experienced rotational subsidence superimposed on regional background subsidence, which resulted in the creation of accommodation space across the entire area (at least 50 m per unit). Therefore, the linear trend line does not intersect the origin of the graph, but originates from the point for which $x = 50$ and $y = 5$ (Fig. 7A). This point represents the minimal unit thickness of 50 m and the average cumulative-sandstone thickness of *ca* 5 m for such a unit.

Both mechanisms result in a high sandstone fraction in areas of high subsidence, but sandstone-body width and interconnectedness are likely to be different. The LAB-models predict numerous sandstone bodies of limited lateral extent, which are likely to be vertically connected. The model described here predicts laterally extensive sandstone bodies that in most cases are vertically disconnected by intervening floodplain mudstones. In addition to the above-described local variations in sandstone percentage, such architectural differences should be taken into account when these models are used as a predictive tool in hydrocarbon exploration and production studies.

The study area is small, but the straightforward relationship with differential subsidence suggests that the mechanism at work is more widely applicable. This approach requires further quantitative testing, on other Pennsylvanian cyclothem sequences and on fluvial sequences from geological periods when high-amplitude sea-level fluctuations did not interfere.

ACKNOWLEDGEMENTS

This study originates from the RECOPOL field experiment for underground storage of carbon dioxide in coal beds in southern Poland, a project funded by the 5th framework of the European Commission. We are grateful to Chris Paola, an anonymous reviewer and journal editor Paul Carling for constructive comments and suggestions. Discussions with K. Geel, H. Pagnier and E. van der Most helped shape our ideas. The Central Mining Institute of Poland is thanked for providing access to the data.

References

Aitken, J.F. and **Flint, S.S.** (1995) The application of high-resolution sequence stratigraphy to fluvial systems: a case study from the Upper Carboniferous Breathitt Group, eastern Kentucky, USA. *Sedimentology*, **42**, 3–30.

Alexander, J. and **Leeder, M.R.** (1987) Active tectonic control of alluvial architecture. In: *Recent Developments in Fluvial Sedimentology* (Eds F.G. Ethridge, R.M. Flores and M.D. Harvey), *SEPM Spec. Publ.*, **39**, 243–252.

Allen, J.R.L. (1978) Studies in fluvial sedimentation: an exploratory quantitative model for the architecture of avulsion-controlled alluvial suites. *Sed. Geol.*, **21**, 129–147.

Allen, J.L. (1993) Lithofacies relations and controls on deposition of fluvial-deltaic rocks of the Upper Pocahontas Formation in southern West Virginia. *Southeast. Geol.*, **33**, 131–147.

Blair, T.C. and **Bilodeau, W.L.** (1988) Development of tectonic cyclothems in rift, pull-apart, and foreland basins: sedimentary response to episodic tectonism. *Geology*, **16**, 517–520.

Bohacs, K.M. and **Suter, J.** (1997) Sequence stratigraphic distribution of coaly rocks; fundamental controls and paralic examples. *AAPG Bull.*, **81**, 1612–1639.

Bridge, J.S. and **Leeder, M.R.** (1979) A simulation model of alluvial stratigraphy. *Sedimentology*, **26**, 617–644.

Bridge, J.S. and **Mackey, S.D.** (1993) A revised alluvial stratigraphy model. In: *Alluvial Sedimentation* (Eds M. Marzo and C. Puigdefabregas), *IAS Spec. Publ.*, **17**, 319–336.

Bryant, M., Falk, P. and **Paola, C.** (1995) Experimental study of avulsion frequency and rate of deposition. *Geology*, **23**, 365–368.

Collinson, J.D., Jones, C.M., Blackbourn, G.A., Besly, B.M., Archard, G.M. and **McMahon, A.H.** (1993) Carboniferous depositional systems of the Southern North Sea. In: *Petroleum Geology of Northwest Europe: Proceedings of the 4th Conference* (Ed. J.R. Parker), *Geol. Soc. London*, **4**, 677–687.

Connell, S.D., Wonsuck, K., Paola, C. and **Smith, G.A.** (2012) Fluvial morphology and sediment-flux steering of axial–transverse boundaries in an experimental basin. *J. Sed. Res.*, **82**, 310–325.

Davies, S.J. (2008) The record of carboniferous sea-level change in low-latitude sedimentary successions from Britain and Ireland during the onset of the late paleozoic ice age. In: *Resolving the Late Paleozoic Ice Age in Time and Space* (Eds C.R. Fielding, T.D. Frank and J.L. Isbell), *GSA Spec. Pap.*, **441**, 187–204.

Davies, H., **Burn, M.**, **Budding, M.** and **Williams, H.** (1992) High-resolution sequence stratigraphic analysis of fluvio-deltaic cyclothems: the Pennsylvanian Breathitt Group, east Kentucky. AAPG 1992 Annual Convention Programme, pp.27, *AAPG Search and Discovery Article #91012.*

Doktor, M. (2007) Conditions of accumulation and sedimentary architecture of the Upper Westphalian Cracow Sandstone Series (Upper Silesia Coal Basin, Poland). *Ann. Soc. Geol. Pol.*, **77**, 219–268.

Doktor, M. and **Gradzinski, R.** (1985) Alluvial depositional environment of coal-bearing "mudstone series" (Upper Carboniferous, Upper Silesian Coal Basin). *Stud. Geol. Pol.*, **82**, 5–67.

Ferm, J.C. and **Weisenfluh, G.A.** (1989) Evolution of some depositional models in Late Carboniferous rocks of the Appalachian coal fields. *Int. J. Coal Geol.*, **12**, 259–292.

Fielding, C.R. (1984a) Upper delta plain lacustrine and fluviolacustrine facies from the Westphalian of the Durham coalfield, NE England. *Sedimentology*, **31**, 547–567.

Fielding, C.R. (1984b) A coal depositional model for the Durham Coal Measures of NE England. *J. Geol. Soc. London*, **141**, 919–931.

Fielding, C.R. (1986) Fluvial channel and overbank deposits from the Westphalian of the Durham coalfield, NE England. *Sedimentology*, **33**, 119–140.

Fielding, C.R. and **Johnson, G.A.L.** (1987) Sedimentary structures associated with extensional fault movement from the Westphalian of NE England. In: *Continental Extensional Tectonics* (Eds M.P. Coward, J.F. Dewey and P.L. Hancock), *Geol. Soc. London. Spec. Publ.*, **28**, 511–516.

Gradstein, F.M., **Ogg, J.G.** and **Smith, A.G.** (2004) *A Geologic Time Scale 2004.* Cambridge University Press, Cambridge, 610 pp.

Gradzinski, R., **Doktor, M.** and **Slomka, T.** (1995) Depositional environments of the coal-bearing Cracow Sandstone Series (Upper Westphalian), Upper Silesia, Poland. *Stud. Geol. Pol.*, **108**, 149–170.

Greb, S.F. and **Chesnut, D.R.** (1992) Transgressive channel filling in the Breathitt Formation (Upper Carboniferous), eastern Kentucky coal field, USA. *Sed. Geol.*, **75**, 209–221.

Greb, S.F., **Eble, C.F.** and **Hower, J.C.** (1999) Depositional history of the Fire Clay coal bed (Late Duckmantian), Eastern Kentucky, USA. *Int. J. Coal Geol.*, **40**, 255–280.

Greb, S.F., **Eble, C.F.**, **Hower, J.C.** and **Andrews, W.M.** (2002) Multiple-bench architecture and interpretations of original mire phases—Examples from the Middle Pennsylvanian of the Central Appalachian Basin, USA. *Int. J. Coal Geol.*, **49**, 147–175.

Greb, S.F., **Eble, C.F.** and **Hower, J.C.** (2005) Subtle structural influences on coal thickness and distribution: examples from the Lower Broas-Stockton coal (Middle Pennsylvanian), Eastern Kentucky Coal Field, USA. In: *Coal Systems Analysis* (Ed. P.D. Warwick), *Geol. Soc. Am. Spec. Pap.*, **387**, 31–50.

Greb, S.F., **Pashin, J.C.**, **Martino, R.L.** and **Eble, C.F.** (2008) Appalachian sedimentary cycles during the Pennsylvanian: changing influences of sea level, climate, and tectonics. *Geol. Soc. Am. Spec. Pap.*, **441**, 235–248.

Guion, P.D. and **Fielding, C.R.** (1988) Westphalian A and B Sedimentation in the Pennine Basin, UK. In: *Sedimentation in a Syn-orogenic Basin Compex: The Upper Carboniferous of Northwest Europe* (Eds B.M. Besly and G. Kelling), pp. 153–177. Blackie, Glasgow.

Guion, P.D. and **Rippon, J.H.** (1995) The Silkstone Rock (Westphalian A) from the east Pennines, England: implications for sand body genesis. *J. Geol. Soc. London*, **152**, 819–832.

Guion, P.D., **Fulton, I.M.** and **Jones, N.S.** (1995) Sedimentary facies of the coal-bearing Westphalian A and B north of the Wales-Brabant High. *Geol. Soc. London. Spec. Publ.*, **82**, 45–78.

Hajek, E.A., **Heller, P.L.** and **Sheets, B.A.** (2010) Significance of channel-belt clustering in alluvial basins. *Geology*, **38**, 535–536.

Hampson, G.J., **Davies, S.J.**, **Elliott, T.**, **Flint, S.S.** and **Stollhofen, H.** (1999) Incised valley fill sandstone bodies in Upper Carboniferous fluvio–deltaic strata: recognition and reservoir characterization of Southern North Sea analogues. *Geol. Soc. London Petrol. Geol. Conf. Ser.*, **5**, 771–788.

Haszeldine, R.S. and **Anderton, R.** (1980) A braidplain facies model for the Westphalian B Coal Measures of north-east England. *Nature*, **284**, 51–53.

Heckel, P.H. (1986) Sea-level curve for Pennsylvanian eustatic marine transgressive-regressive depositional cycles along midcontinent outcrop belt, North America. *Geology*, **14**, 330–334.

Heckel, P.H. (2008) Pennsylvanian cyclothems in Midcontinent North America as far-field effects of waxing and waning of Gondwana ice sheets. *Geol. Soc. Am. Spec. Pap.*, **441**, 275–289.

Heller, P.L. and **Paola, C.** (1996) Downstream changes in alluvial architecture: an exploration of controls on channel stacking patterns. *J. Sed. Res.*, **66**, 297–306.

Horne, J.C. (1978) Sedimentary responses to contemporaneous tectonism. In: *Carboniferous Depositional Environments: Eastern Kentucky and West Virginia* (Eds J.C. Horne and J.C. Ferm), pp. 27–34. University of South Carolina, Field Guide, Columbia, SC.

Jones, N.S. and **Glover, B.W.** (2005) Fluvial sandbody architecture, cyclicity and sequence stratigraphical setting – implications for hydrocarbon reservoirs: the Westphalian C and D of the Osnabrück and Ibbenbüren area, Northwest

Germany. In: *Carboniferous Hydrocarbon Geology – The Southern North Sea and Surrounding Onshore Areas* (Eds J.D. Collinson, D.J. Evans, D.W. Holliday and N.S. Jones), *Yorks. Geol. Soc. Occ. Publ.*, **7**, 57–74.

Jones, J.A. and **Hartley, A.J.** (1993) Reservoir characteristics of a braid-plain depositional system: the Upper Carboniferous Pennant Sandstone of South Wales. In: *Characterization of Fluvial and Aeolian Reservoirs* (Eds C.P. North and D.J. Prosser), *Geol. Soc. London. Spec. Publ.*, **73**, 143–156.

Jureczka, J. and **Kotas, A.** (1995) Coal Deposits – Upper Silesian Coal Basin. In: *The Carboniferous System in Poland* (Eds A. Zdanowski and H. Zakowa), pp. 164–172. Prace Panstwowego Instytutu Geologicznego, Warsaw.

Kim, W., **Sheetsz, B.A.** and **Paola, C.** (2010) Steering of experimental channels by lateral basin tilting. *Basin Res.*, **22**, 286–302.

Klein, G. and **Kupperman, J.B.** (1992) Pennsylvanian cyclothems: methods of distinguishing tectonically induced changes in sea level from climatically induced changes. *Geol. Soc. Am. Bull.*, **104**, 166–175.

Klein, G. and **Willard, D.A.** (1989) Origin of Pennsylvanian coal-bearing cyclothems of North America. *Geology*, **17**, 152–155.

Kopp, J. and **Kim, W.** (2015) The effect of lateral tectonic tilting on fluviodeltaic surficial and stratal asymmetries: experiment and theory. *Basin Res.*, 14. Online early view.

Kotas, A. (1994) Coal-bed methane potential of the Upper Silesian Coal Basin, Poland. *Pr. Panstw. Inst. Geol. Warsaw*, **142**, 81.

Kotas, A. (1995) Lithostratigraphy and sedimentologic-paleogeographic development: Upper Silesian Coal Basin. In: *The Carboniferous System of Poland* (Eds A. Zdanowski and H. Zakowa), *Pr. Panstw. Inst. Geol.*, **148**, 124–134.

Leeder, M.R. and **Gawthorpe, R.L.** (1987) Sedimentary models for extensional tilt-block/half-graben basins. In: *Continental Extensional Tectonics* (Eds M.P. Coward, J.F. Dewey and P.L. Hancock), *Geol. Soc. London. Spec. Publ.*, **28**, 139–152.

Leeder, M.R., **Mack, G.H.**, **Peakall, J.** and **Salyards, S.L.** (1996) First quantitative test of alluvial stratigraphic models: Southern Rio Grande rift, New Mexico. *Geology*, **24**, 87–90.

Mack, G.H. and **James, W.C.** (1993) Control of basin symmetry on fluvial lithofacies, Camp Rice and Palomas Formations (Plio-Pleistocene), southern Rio Grande rift, USA. In: *Alluvial Sedimentation* (Eds M. Marzo and C. Puigdefabregas), *IAS Spec. Publ.*, **17**, 439–449.

Mackey, S.D. and **Bridge, J.S.** (1995) Three-dimensional model of alluvial stratigraphy; theory and applications. *J. Sed. Res.*, **65**, 7–31.

McCabe, P.J. (1984) Depositional environments of coal and coal-bearing strata. In: *Sedimentology of Coal and Coal-bearing Sequences* (Eds R.A. Rahmani and R.M. Flores), *IAS Spec. Publ.*, **7**, 11–42.

Nadon, G.C. (1998) Magnitude and timing of peat-to-coal compaction. *Geology*, **26**, 727–730.

Paola, C. (2000) Quantitative models of sedimentary basin filling. *Sedimentology*, **47**, 121–178.

Peakall, J. (1998) Axial river evolution in response to half-graben faulting: Carson River, Nevada, U.S.A. *J. Sed. Res.*, **68**, 788–799.

Peakall, J., **Leeder, M.**, **Best, J.** and **Ashworth, P.** (2000) River response to lateral ground tilting: a synthesis and some implications for the modeling of alluvial architecture in extensional basins. *Basin Res.*, **12**, 413–424.

Rippon, J.H. (1996) Sand body orientation, palaeoslope analysis and basin-fill implications in the Westphalian A-C of Great Britain. *J. Geol. Soc. London*, **153**, 881–900.

Rygel, M.C., **Fielding, C.R.**, **Frank, T.D.** and **Birgenheier, L.P.** (2008) The magnitude of Late Paleozoic glacioeustatic fluctuations: a synthesis. *J. Sed. Res.*, **78**, 500–511.

Shanley, K.W. and **McCabe, P.J.** (1993) Alluvial architecture in a sequence stratigraphic framework: a case study from the Upper Cretaceous of southern Utah, USA. In: *The Geological Modelling of Hydrocarbon Reservoirs and Outcrop Analogues* (Eds S.S. Flint and I.D. Bryant), *IAS Spec. Publ.*, **15**, 21–56.

Straub, K.M., **Paola, C.**, **Kim, W.** and **Sheets, B.** (2013) Experimental investigation of sediment-dominated vs. tectonics-dominated sediment transport systems in subsiding basins. *J. Sed. Res.*, **83**, 1162–1180.

Van Asselen, S., **Stouthamer, E.** and **Smith, N.D.** (2010) Factors controlling peat compaction in alluvial floodplains: a case study in the Cold-Temperate Cumberland Marshes, Canada. *J. Sed. Res.*, **80**, 155–166.

Van Bergen, F., **Pagnier, H.** and **Krzystolik, P.** (2006) Field experiment of enhanced coalbed methane-CO_2 in the upper Silesian basin of Poland. *Environ. Geosci.*, **13**, 201–224.

Van Bergen, F., **Krzystolik, P.**, **van Wageningen, N.**, **Pagnier, H.**, **Jura, B.**, **Skiba, J.**, **Winthaegen, P.** and **Kobiela, Z.** (2009) Production of gas from coal seams in the Upper Silesian Coal Basin in Poland in the post-injection period of an ECBM pilot site. *Int. J. Coal Geol.*, **77**, 175–187.

Van den Belt, F.J.G. (2012) *Sedimentary Cycles in Coal and Evaporite Basins and the Reconstruction of Palaeozoic Climate.* Unpublished PhD Dissertation, University of Utrecht, 100 pp.

Weisenfluh, G.A. and **Ferm, J.C.** (1984) Geologic controls on deposition of the Pratt Seam, Black Warrior Basin, Alabama, U.S.A. In: *Sedimentology of Coal and Coal-bearing Sequences* (Eds R.A. Rahmani and R.M. Flores), *IAS Spec. Publ.*, **17**, 317–330.

Weller, J.M. (1930) Cyclical sedimentation of the Pennsylvanian period and its significance. *J. Geol.*, **38**, 97–135.

Zdanowski, A. and **Zakowa, H.** (1995) The Carboniferous System of Poland. *Pr. Panstw. Inst. Geol. Warsaw*, **148**, 215.

Ziegler, P.A. (1990) *Geological Atlas of Western and Central Europe.* Shell Internationale Petroleum Maatschappij, Den Haag, 239 pp.

2

Early diagenetic evolution of Chalk in Eastern Denmark

JULIEN MOREAU*, MYRIAM BOUSSAHA*, LARS NIELSEN*, NICOLAS THIBAULT*, CLEMENS V. ULLMANN† and LARS STEMMERIK‡

*Department of Geosciences and Natural Resource Management, University of Copenhagen, Øster Voldgade 10, 1350 Copenhagen, Denmark (E-mail: moreau.juli1@gmail.com)

†Camborne School of Mines, University of Exeter, Penryn Campus, Penryn TR10 9FE, UK

‡Natural History Museum, University of Copenhagen, Øster Voldgade 10, 1350 Copenhagen, Denmark

Keywords
Chalk, deformation bands, Denmark, diagenesis, polygonal faults, stylolites.

ABSTRACT

The genesis of polygonal faults is an intriguing diagenetic phenomenon. This study discusses their origin in carbonate mudstones together with other associated diagenetic features. In the eastern Danish Basin, at the fringe of the Baltic Sea, the Stevns peninsula offers a unique opportunity to study the early diagenesis of Upper Cretaceous Chalk deposits, buried between 500 m and 1400 m. This paper combines data from onshore and offshore high-resolution seismic reflection profiles, a fully cored borehole with high-resolution wireline logs and quarry and coastal cliff outcrops to study early diagenetic features at different scales. Chalk is affected by an extensive polygonal fault system that is detected in onshore and offshore seismic data. Outcrop and core data provide a better understanding of the distribution of contraction-related features like deformation bands (hairline fractures), stylolites and fluid escape structures. An original model of genetic relationships between these different diagenetic processes is documented for Chalk. The spatial relationships between stylolites and fractures suggest that pressure-solution processes triggered shear failure that initiated the polygonal fault systems. The early diagenetic processes affect the reservoir properties of Chalk by creating compartments and vertical connections. Taking these features into account will allow for a more detailed understanding of early diagenesis and better models for exploiting drinking water or hydrocarbons hosted in Chalk.

INTRODUCTION

Chalk is a singular material consisting principally of an accumulation of micron-sized calcareous nannofossil grains. It is similar to both fine-grained siliciclastic sediments and carbonate bio-accumulations and mechanisms common to both realms can be invoked during early diagenesis (Cartwright *et al.*, 2003; Hibsch *et al.*, 2003; Fabricius & Borre, 2007; Goulty, 2008; Gaviglio *et al.*, 2009; Wennberg *et al.*, 2013). This paper aimed to provide new insights into the first phases of Chalk diagenesis by untangling the complex relationship between centimetre to metre-scale observations made on a fully cored borehole with high-resolution wireline logs, on quarry exposures and on coastal cliff outcrops of the Stevns peninsula (Denmark), and 100 m-scale observations from onshore and offshore high-resolution seismic reflection profiles in the same area.

For this study, the focus is on Chalk from the eastern Danish Basin which reputedly underwent very shallow burial preceding its modern exposure on the Stevns peninsula (Fig. 1; Nielsen *et al.*, 2011). Previous studies of the area have documented the stratigraphy, depositional facies, porosity variations with depth and the associated seismic velocities (Frykman, 2001; Lykke-Andersen & Surlyk, 2004; Stemmerik *et al.*, 2006; Surlyk *et al.*, 2006, 2013; Anderskouv *et al.*, 2007; Esmerode *et al.*, 2007; Nielsen *et al.*, 2011). These studies allow the seismic response of the Chalk to be assessed according to stratigraphic variations and compositional changes, and in particular, the impact on porosity and seismic velocities (Nielsen *et al.*, 2011). Porosity loss, measured directly on core plugs or indirectly with borehole geophysics and seismic refraction data, increases with both burial depth and clay content (Japsen, 2000; Nielsen *et al.*, 2011). According to Wennberg *et al.* (2013), local porosity loss

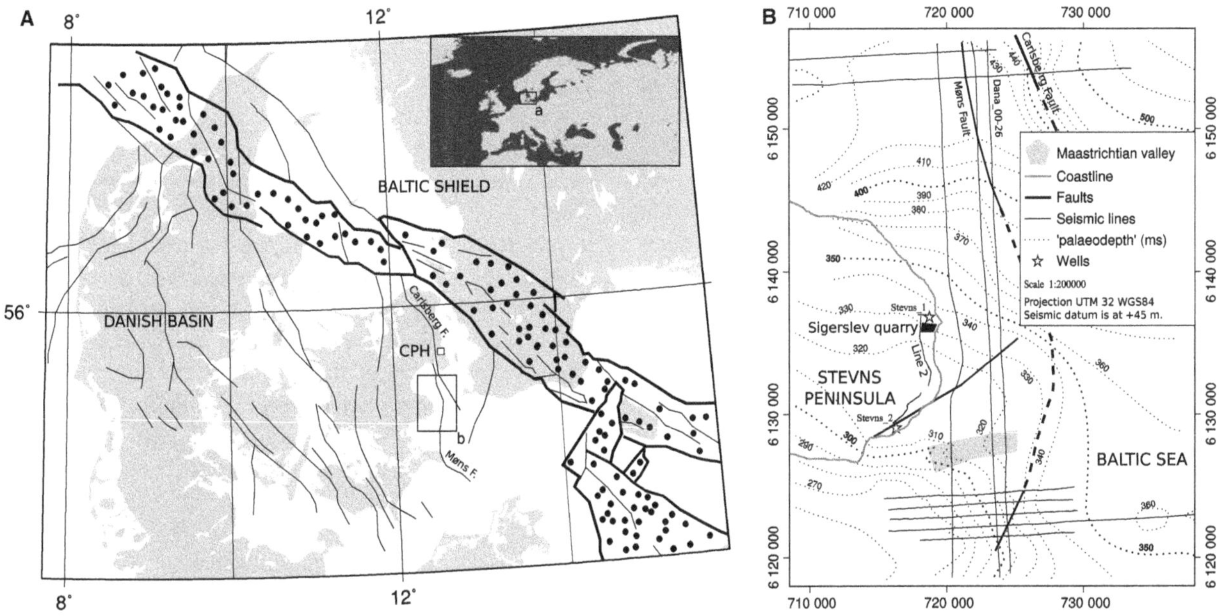

Fig. 1. Location map of the study area. (A) Main faults affecting the top Zechstein, modified from Graversen (2009). The dotted area is the Sorgenfrei-Tornquist zone, the suture where inversion occurs during the Maastrichtian. The Møns Fault is also inverted during that time (this study). CPH = Copenhagen, F. = Fault. Projection UTM 32, coastline from GSHHS database. (B) Location of the seismic lines and wells. Assuming the upper part of the Late Cretaceous in the Stevns peninsula has not been tectonically deformed, one horizon close to the Campanian–Maastrichtian boundary has been picked to represent the palaeo-sea floor depth. The horizon represents the slope of the Late Cretaceous ramp, each dotted line is separated by *ca* 15 m.

also occurs within Chalk deformation bands (hairline fractures). These deformation bands are compaction features corresponding to jointing and local pore-space collapse.

Depending on the methods used, burial depths between 500 m and 750 m have been estimated for the Stevns area based on seismic velocities of Chalk (Japsen & Bidstrup, 1999; Japsen, 2000; Nielsen *et al.*, 2011). The 750 m estimate in Japsen (2000) is based on an extensive data set from Danish offshore and onshore areas showing a direct link between seismic velocity and porosity of Chalk, and assuming a linear porosity reduction during burial. The 500 m estimate in Nielsen *et al.* (2011) is based on the recognition of three discrete steps with increasing velocity at *ca* 100 m, 300 m and 600 m most likely linked to significant diagenetic boundaries and thus implying a stepwise transformation of Chalk.

Polygonal fault systems are regionally extensive normal faults restricted to a specific stratigraphic interval (Cartwright, 2011 and reference therein). The origin of these faults is known to be diagenetic and disconnected from tectonic processes (Henriet *et al.*, 1991), except in one documented case (anticline formation, Petracchini *et al.*, 2015). Several genetic processes have been discussed in the literature and an extensive review can be found in Cartwright (2011). The latest and most accepted model invokes an overall contraction (in opposition to loading-related strain) of the host-rock, where focused grain dissolution weakens high porosity fine-grained sediments and generates shear failure which then propagates into slip and the formation of a fault plane (Shin *et al.*, 2008). The model of Shin *et al.* (2008) was numerically and experimentally conducted and upscaled and compared with 3D seismic data (Shin *et al.*, 2010). Such numerical models have only considered sedimentary rocks dominated by siliciclastic rock dissolution. As documented in this study, however, polygonal fault systems also affect carbonate mudstones and have been documented in multiple basins and settings (Cartwright *et al.*, 2003; Hibsch *et al.*, 2003; Hansen *et al.*, 2004; Cartwright, 2011; Sandrin *et al.*, 2012; Tewksbury *et al.*, 2014).

In this study, seismic reflection profiles were used to describe faults. By combining the seismic interpretation with studies of diagenetic features like hairline fractures, stylolites and flint found in the Stevns-2 core as well as in outcrop, the links between early diagenetic transformations are explored at different scales (Fig. 1). For the first time, it is possible to link polygonal fault systems and deformation bands to the occurrence of stylolites. By including data from a Chalk succession and diagenetic suite, the observations may help to consolidate the realization of a common theory for the formation of polygonal fault systems by grain dissolution during early diagenesis (Shin *et al.*, 2008, 2010; Cartwright, 2011).

GEOLOGICAL SETTING

The Stevns peninsula is located at the transition between the eastern edge of the Danish Basin and the western edge of the Baltic Sea (Fig. 1). Below the Stevns peninsula, a full Mesozoic sedimentary succession is supposedly present (from the Triassic to the Early Paleogene; Erlström *et al.*, 1997). Broadly, the Mesozoic and Cenozoic sedimentary rocks draw a large wedge, opening towards the west in the Central Graben (North Sea) and pinching out in the study area constituting the eastern edge of the subsiding intracratonic area (Lassen & Thybo, 2012). Eastward and southward, the basin progressively deepens again into the Polish Trough and the Sorgenfrei-Tornquist zone (Graversen, 2004; Lassen & Thybo, 2012; Sopher & Juhlin, 2013).

The sedimentary rocks are deformed in narrow zones along WNW-ESE-trending tectonic lineaments, the Ringkøbing-Fyn High in the south and the Tornquist suture in the north (Lassen & Thybo, 2012). Additionally, important NNW-SSE oriented tectonic lineaments, some apparently inverted during the Maastrichtian, have been identified in our data set (Møns Fault; Fig. 1; Graversen, 2009). The latter lineaments have not been previously studied individually. They are parallel to structures characterizing the area such as the Sorgenfrei-Tornquist zone and the Carlsberg Fault, and consequently are believed to have similar, mainly strike-slip, motions (Rosenbom & Jakobsen, 2005; Graversen, 2009). These structures have been active since at least the Palaeozoic and still generate earthquakes (Graversen, 2009; GEUS, 2015; Kammann *et al.*, 2016). Despite the presence of these large tectonic structures in the offshore areas east of the Stevns peninsula, the coastal cliff of the Stevns peninsula is without major fault or internal offsets within strata and the upper Maastrichtian – lower Danian sedimentary rocks are subhorizontal (Surlyk *et al.*, 2006). Analysis of the outcrops highlights the dominance of recent fractures and subhorizontal joints associated with relaxation after ice sheet unloading combined with the local stress field, resulting from the deglaciation of the area (11 to 20 ka; Frykman, 2001; Rosenbom & Jakobsen, 2005). The sediments formerly covering Chalk have been removed after the area was exhumed (Japsen *et al.*, 2007; Nielsen *et al.*, 2011).

This study mainly focuses on the Campanian and Maastrichtian Chalk which has been cored in the Stevns-1 and Stevns-2 boreholes (Figs 1 and 2; Stemmerik *et al.*, 2006). The overall stratigraphy and depositional evolution of the cored succession is presented by Surlyk *et al.* (2013) and a detailed description of the Stevns-2 core is given by Boussaha *et al.* (2016). The Stevns-2 core consists of a lower, *ca* 100 m thick interval of marl-Chalk alternations of Campanian – earliest Maastrichtian age overlain by 70 m of pure white Chalk and 170 m of Chalk with flint bands (Fig. 2C). At the top of the Maastrichtian, bryozoans become more abundant and the flint bands undulate, highlighting the presence of mounds and Chalk waves forming in response to contour currents (Anderskouv *et al.*, 2007). The Maastrichtian is locally terminated by the famous Iridium-rich clay layer characterizing the boundary between the Mesozoic and the Cenozoic although, like in the Stevns-2 core, this may be locally eroded (Surlyk *et al.*, 2006; Boussaha *et al.*, 2016). The last few metres of the Stevns succession are formed by spectacular bryozoan mounds of Danian age (Fig. 2; Bjerager & Surlyk, 2007; Boussaha *et al.*, 2016).

Stylolites have formed through the succession, and carbonate remobilization is observed in some fractures (Fig. 2; Rasmussen & Surlyk, 2012; Surlyk *et al.*, 2013). In the Stevns-1 core, suboptimal preservation of the calcareous nannofossils forming Chalk has been related to partial dissolution caused by fluid migration (Thibault *et al.*, 2012). Such fluid flow within Chalk has also been interpreted as being the cause of brecciation in the Stevns-1 well (Rasmussen & Surlyk, 2012). Although these indicators of chemical and mechanical compaction have been studied (Fabricius & Borre, 2007; Fabricius *et al.*, 2010), relatively little has yet been done to understand their implications for the structural architecture of eastern Denmark Chalk at the regional scale.

Data collection and reporting

This study is based on analyses of seismic reflection profiles, well logs, core and outcrop data (Fig. 1). The seismic sections are high-resolution onshore and offshore data imaging the area of the Stevns peninsula (Fig. 1). The onshore seismic data have a central frequency of 85 Hz. The bin spacing is 2·5 m, and the sampling of the 800 ms (*ca* 1·2 km) long record is done every millisecond. Locally, some acquisition problems disturbed the record (strong winds). In consequence, the stratigraphic patterns are not properly resolved in the east–west sections as well as certain parts of the north–south sections of the onshore survey.

The offshore data have a central frequency of 80 Hz, the bin spacing is 10 m, the traces are 2 s TWT long and sampled every millisecond (Lykke-Andersen & Surlyk, 2004). Ties with the wells Stevns-1 and Stevns-2 are performed using velocities from seismic refraction data, corrected with wireline logs of density and sonic (Nielsen *et al.*, 2011; Figs 1 and 2). The theoretical vertical resolution of the reflection seismic data using the mean velocity of the studied interval is *ca* 8 m (¼ of the wavelength, 2·6 km s^{-1}; Nielsen *et al.*, 2011).

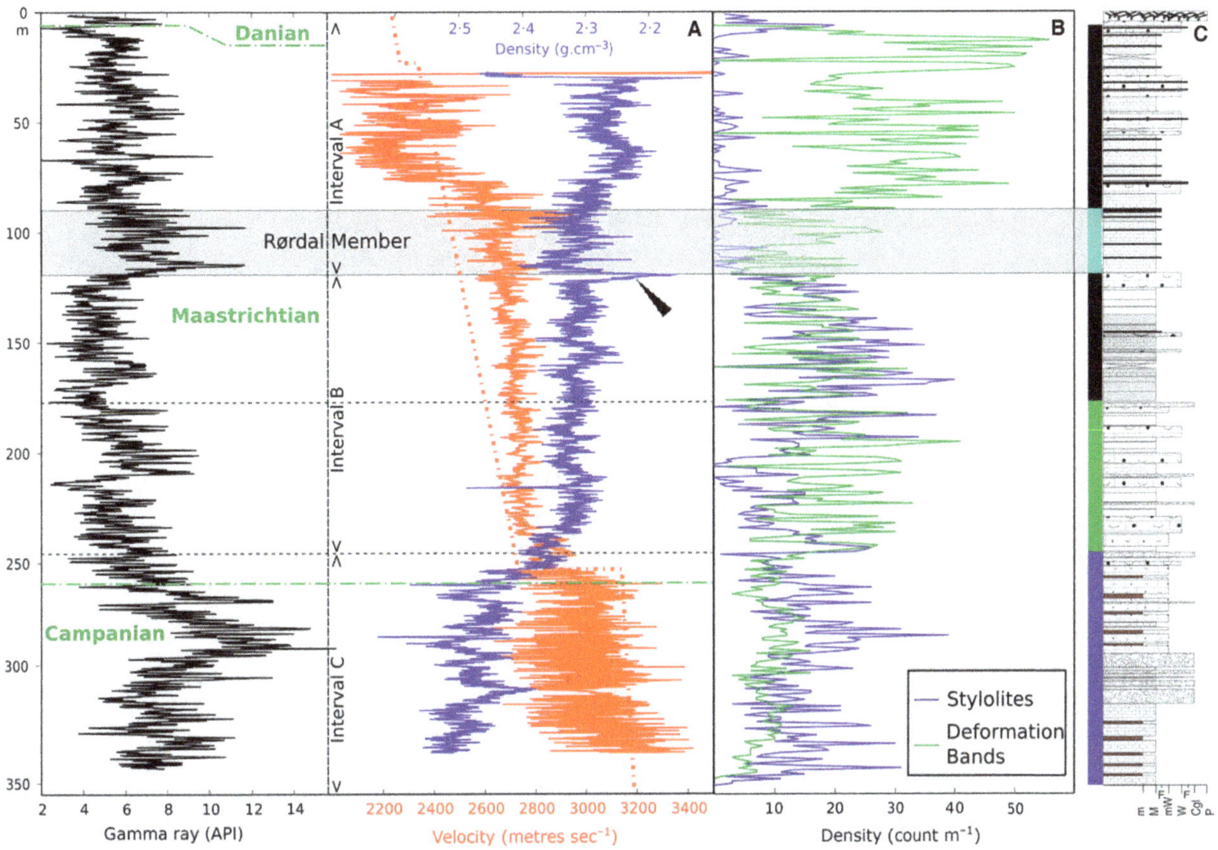

Fig. 2. Petrophysical (A), structural (B) and sedimentological (C) data from the Stevns-2 core. In C, the facies scale is modified from Boussaha *et al.* (2016) with: m = marls, M = mudstones, F = flint, mW = micro-wackestones, W = wackestones, Cgl = Conglomerate, P = packstones. The patterns used in the sedimentary log follow this scale. The sudden rise below a clay-rich interval (grey interval, Rørdal Member) in stylolites is attributed to critical burial depth reached (1 km) and local compartmentalization. Deformation bands (hairlines) may be disappearing with stylolite formation. The drop in density (black arrow) is considered to be the result of the local dissolution of carbonates. Colour codes in C correspond to the colour used in the cross-plots of Fig. 6. [Colour figure can be viewed at wileyonlinelibrary.com]

In the offshore data, a dip-steering median filter has been applied on the lines to favour continuous reflective events along the structural dip but preserving edges (details of the technique in Torvela *et al.*, 2013; www.opendtect.org). After careful picking and filtering, the overall quality of the image is locally good to very good for stratigraphic analysis (Fig. 3). In order to image subtle fracture patterns with small offsets and reflectivity lows in the onshore data, a fault enhancement filter has been used (www.opendtect.org). It is a combination of a (i) dip-steered median filter and a (ii) dip-steered diffusion filter. Broadly, when evaluating data in a dip-steered ellipsoid, if the similarity is high, filter (i) is applied, raising the continuity; whereas if the similarity is low (near a fault), filter (ii) is applied, making a sharp fault break. In order to compare the seismic imaging and the core observations, a linear velocity function is applied to project the well data and convert times measured on the profiles to depth. The function is extracted from the refraction study of Nielsen *et al.* (2011) and corresponds to a starting velocity of 2200 m s^{-1} with an acceleration of 1000 m s^{-2}. The first 50 ms of the onshore seismic profiles contain very intricate reflections which seem to cross-cut and are not present in the lower resolution offshore line. Potential Quaternary incisions/deformations or undetermined artefacts might be the cause of this architecture and the interval is consequently not considered in the final models. Similarly, the presence of sea-bed multiples and multiples of the quaternary sediments overprint the signal offshore and make it difficult to interpret the first 100 ms of the profiles (Fig. 3). A standard seismic stratigraphic technique based on reflection terminations has been applied to visualize seismic sequences in the profiles (Figs 3 and 4). The sequences are constructed based on onlaps onto horizons that truncate the underlying reflections. These horizons highlight the base of seismic sequences within Chalk (alternating colours in Figs 3 and 4). In addition, in order to characterize throw distribution patterns, 125

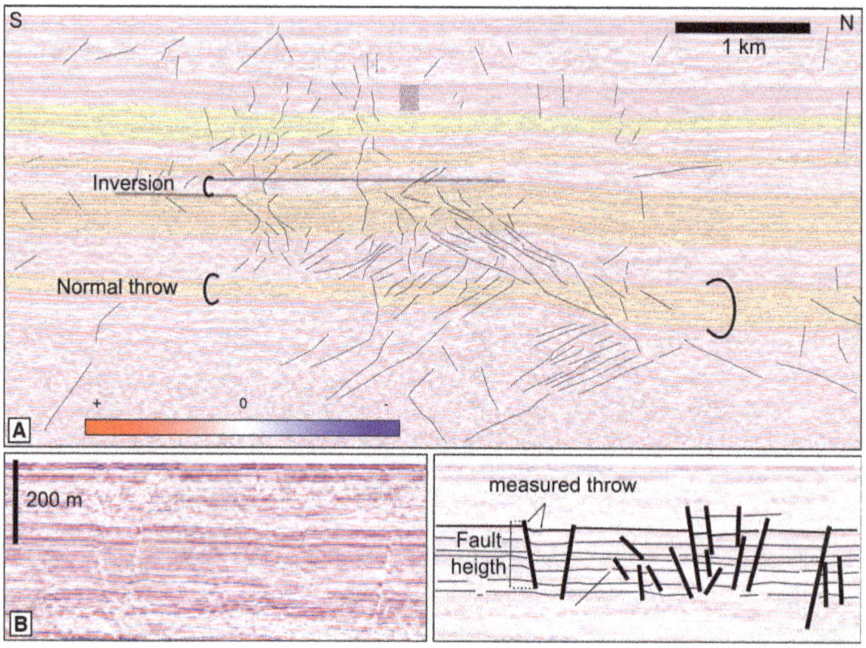

Fig. 3. Details of features and how they were measured from offshore seismic reflection data at a 1 : 1 scale. (A) Example of a section through the Møns Fault (location in Fig. 1) showing its extent across the whole imaged sedimentary succession. Note its complex history with extension at base, followed by inversion and formation of a broad anticline affecting the depositional sequences (marked by the alternating colours). The fault forms a cluster which draws diamond shapes. (B) Example of strata-bound faults (polygonal faults). This figure illustrates the geometric characteristics of the faults (high angle, strata bound), their composite nature and the way the throw and the height are measured. [Colour figure can be viewed at wileyonlinelibrary.com]

fault segments have been measured in the offshore seismic profiles (Fig. 5). These measures are standard for blind-fault analysis with their apparent dips (Fig. 5A), the ratio between the maximum throw and the height (Fig. 5B; Baudon & Cartwright, 2008; Shin *et al.*, 2010; Cartwright, 2011). The throw measured on individual offset reflections as well as the location of maximum observed throw have been plotted against normalized fault plane intersection (with the seismic profile; Fig. 5C). To compensate for the lack of 3D data, Gaussian kernel density estimations are calculated for the values repartition of the ratio between fault height and the maximum throw as well as the throw repartition against the normalized fault plane intersection, giving a statistical vision of the measurements as if they were continuously sampled (colour mapping in Fig. 5). Since the seismic profiles form a suborthogonal grid and because apparent dip angles have a large variability, the seismic grid can be considered as intersected by faults with a large variability of strikes (producing various apparent dip angles). Therefore, the obtained density functions are considered representative of the probability of occurrences in 3D.

The Stevns-1 and -2 boreholes are entirely cored with excellent recovery (>98%; Stemmerik *et al.*, 2006). A suite of geophysical measurements have been logged in the boreholes (spectral gamma ray, neutron density, sonic) and on the core (gamma ray, density). During investigations of the Stevns-2 core particular attention was given to the quantification of structural and diagenetic features and the abundance (i.e. count of occurrence per 1 m interval) of stylolites, marl layers, fractures and joints (deformation bands) has been precisely mapped (Fig. 2). This allows precise correlation of petrophysical data with core-based observations of density of structural and diagenetic features (Figs 2 and 6).

Field work in the Sigerslev Chalk quarry and the nearby coastal sections (Fig. 1) focussed on structural observations and mapping of flint (Supplementary material 1). Since flint bands mostly nucleate very close to the seabed, they mimic the palaeo-sea floor and are the only markers which allow for approximation of the sedimentary architecture in pure Chalk successions (Surlyk *et al.*, 2006; Anderskouv *et al.*, 2007; Madsen & Stemmerik, 2010). In addition, in the otherwise homogeneous Chalk of the study area flint has precipitated in faults and 16 segments exhibiting this have been measured and mapped (Fig. 7). Delineation of the structures in Fig. 8 is made by drawing the flint nodules on a high-resolution panoramic photograph stitched using Hugin software and Panini projections to reduce parallax distortions (http://hugin.-sourceforge.net/; Supplementary material 1). However, since the picture is taken from the bottom of the quarry

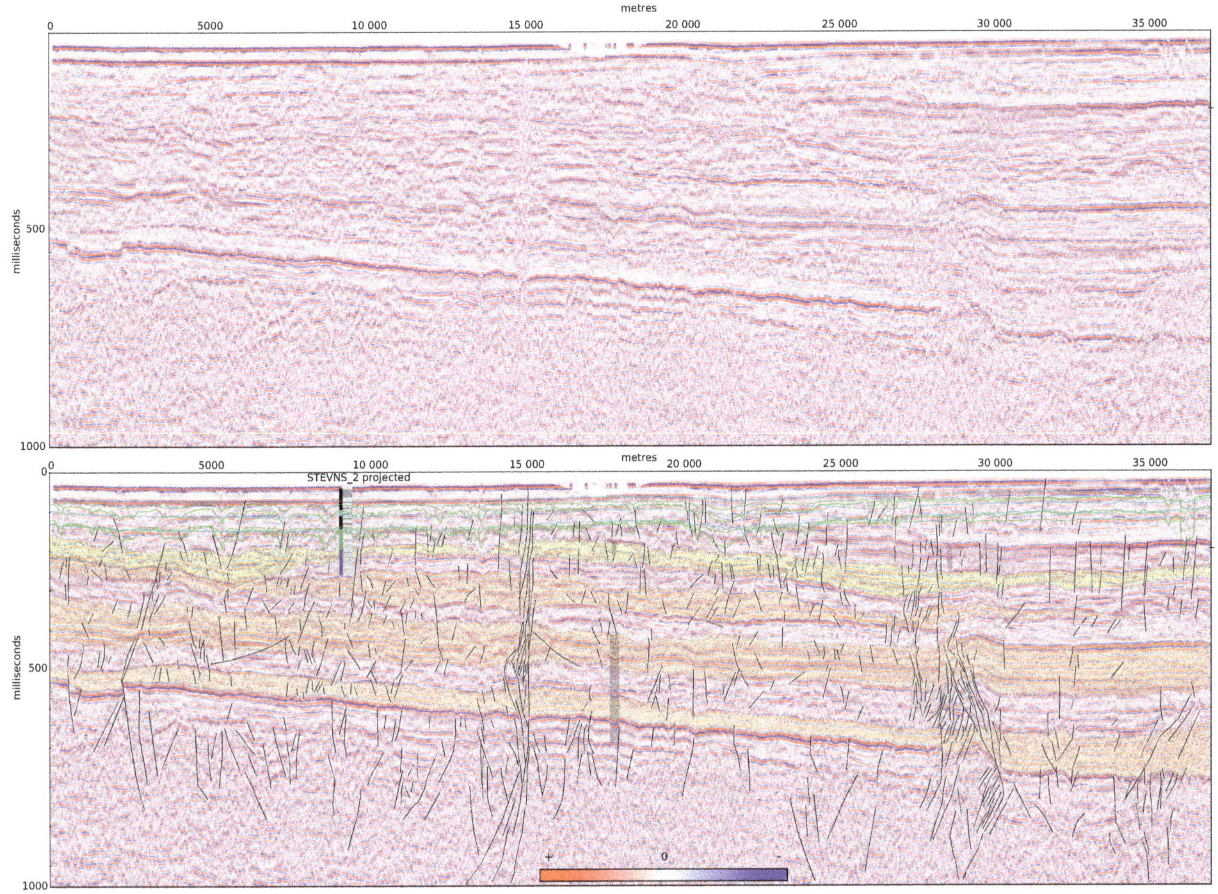

Fig. 4. Seismic interpretation of line Dana 00-26 offshore Stevns peninsula, location in Fig. 1. Blue lines represent the sea-bed multiples with the water–air interface. Green lines are multiples made by the reflection between the quaternary – Chalk and the sea-bed – Chalk interfaces. Thick quaternary sequences are generating push down artefacts locally. The colours are the main seismic sequences observed. The basal sequence probably starts in the Turonian, just above the most pronounced seismic event, marking the base of Chalk deposition in the area (Lykke-Andersen & Surlyk, 2004). The yellow sequence shows Upper Campanian deposition which in the Stevns-2 well corresponds to marl-Chalk alternations. The grey shadings are potential pock mark conduit sections. By comparison with the local stratigraphy some could be traced to the base of the Cretaceous (Erlström *et al.*, 1997). The top of the section has not been interpreted because of the predominance of the multiples over the direct reflections. [Colour figure can be viewed at wileyonlinelibrary.com]

wall, the top of the image remains distorted, and horizontal beds appear convex upward.

Data presentation and interpretation

Seismic observations

The seismic data vary in quality throughout the study area but the stratigraphy can be interpreted and the main reflections can be correlated between offshore and onshore lines (Lykke-Andersen & Surlyk, 2004; Esmerode *et al.*, 2007). However, due to lack of a proper tie, all age attributions older than the upper Campanian should be considered with caution.

Based on the drilled Maastrichtian sedimentary rocks, the seismic reflection pattern on the high-resolution onshore profiles can tentatively be correlated with the broad facies subdivision identified in the nearby Stevns-1 and Stevns-2 cores. There are two end-members in the reflection patterns attributed to (i) high energy (relatively) reflections that are continuous on all seismic profiles and (ii) low reflectivity, chaotic to disrupted reflections (Figs 3, 4 and 9). Onshore, the patterns are correlated with the borehole data and correspond to (i) marl-Chalk alternations and (ii) the 'white Chalk', which is a very pure mud-sized Chalk (Figs 3, 4 and 9). The picked reflections show numerous disruptions in the form of low- to high-angle, transparent and continuous segments from 10 to 100 ms in height (Figs 3, 4 and 9). The segments are completely nonreflective, highlighting fractures that are organized in clusters (e.g. Fig. 9). The highly reflective events (marl-

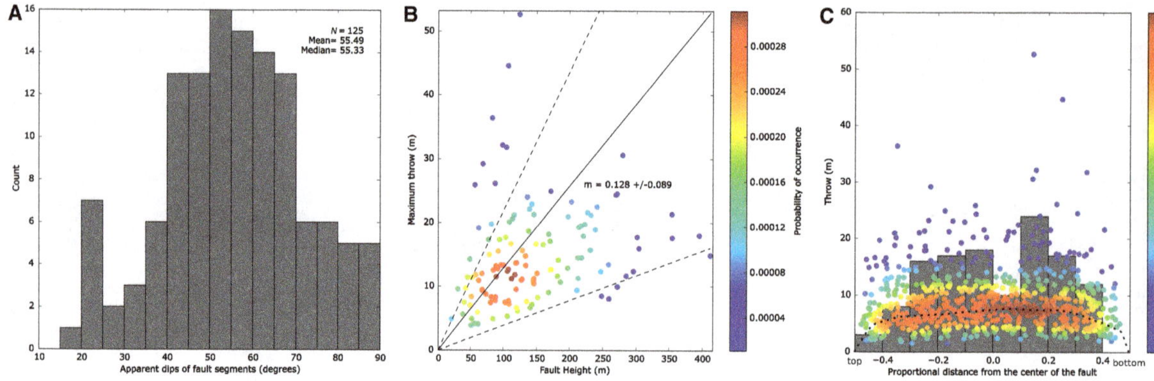

Fig. 5. Geometric characteristics of the 125 measured strata-bound faults. (A) Apparent dip of the fault segments on the seismic profiles. (B) Maximum displacement (throw) versus fault height and the density function of the data point probability of occurrence. The ratio between the two values is very high and indicates probably a high degree of lateral connectivity with other polygonal faults (Shin *et al.*, 2010). (C) The density function in colour shows the distribution of measured throws against the normalized fault height. 0 is at the centre of the faults, 0·5 the lower tip and −0·5 the upper tip. The histogram in the background indicates the position of maximum throws on the normalized fault planes (in count). Both distributions are slightly skewed towards the lower parts, but very mildly. Overall the throws seem equally distributed with a M-type distribution (Baudon & Cartwright, 2008). [Colour figure can be viewed at wileyonlinelibrary.com]

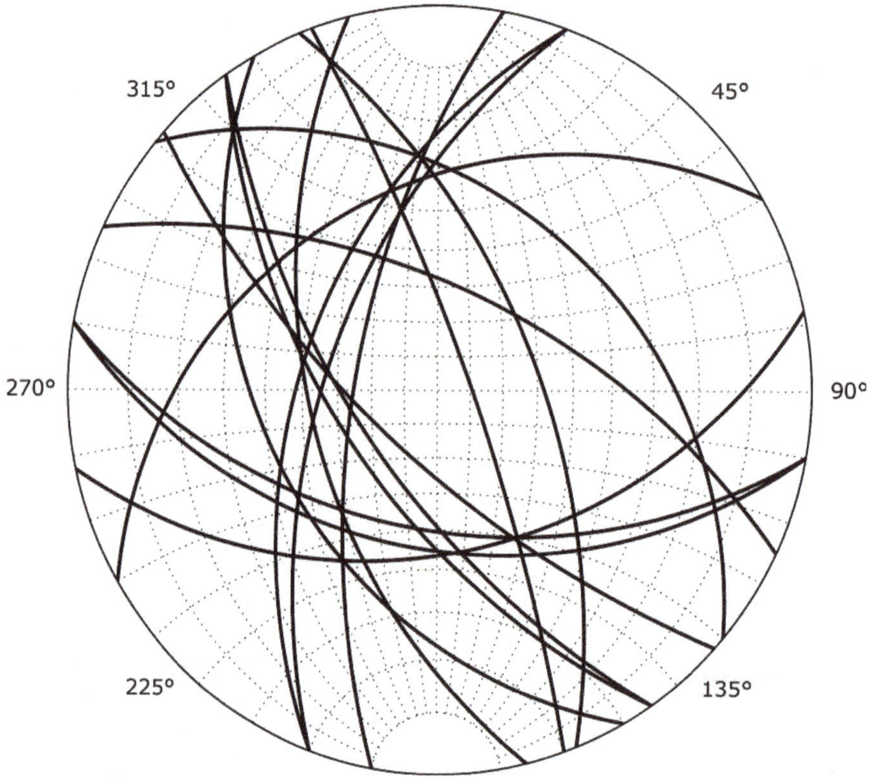

Fig. 6. Stereonet projection of the exposed fault planes marked by flint bands. No preferential orientation is distinguished on these 16 measured planes. [Colour figure can be viewed at wileyonlinelibrary.com] [Correction added on 13th Feb, after first online publication: Figure captions of figures 6,7 and 9 switched]

Chalk alternations) have very few disruptions but larger offsets reaching *ca* 60 m (*ca* 20 ms; Figs 3, 4, 5 and 9). The weakly reflective parts (white Chalk) contain most of the clusters, but contain subtle fault throws (Figs 3, 4 and 9). The faults have an apparent dip distribution which has almost a Gaussian distribution with a mean and a median of *ca* 55° and a standard deviation of 16° (Fig. 5A). The ratio of the fault heights against their observed maximum

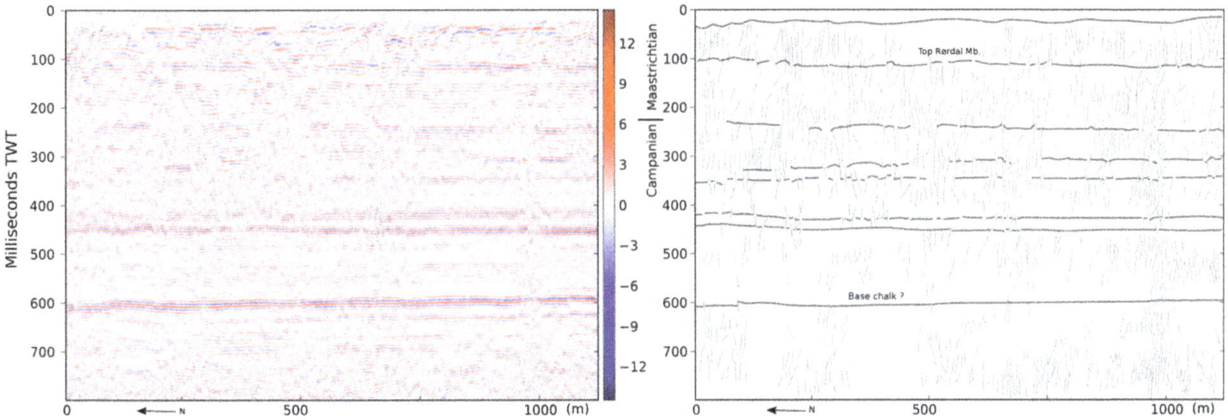

Fig. 7. Seismic interpretation of the line 2 onshore Stevns peninsula, location in Fig. 1. Notice the abundance of small faults between the low reflectivity intervals (e.g. base Maastrichtian) compared to the high reflectivity intervals (e.g. upper Campanian). The high reflectivity layers show drag folds and conical patterns of fault clusters. [Colour figure can be viewed at wileyonlinelibrary.com] [Correction added on 13th Feb, after first online publication: Figure captions of figures 6,7 and 9 switched]

throw is 0·13 ± 0·09 (Fig. 5B). The throw distribution along the faults is typical M-type (trapezoid-shape) with no clear asymmetry with depth (Fig. 5C; Muraoka & Kamata,

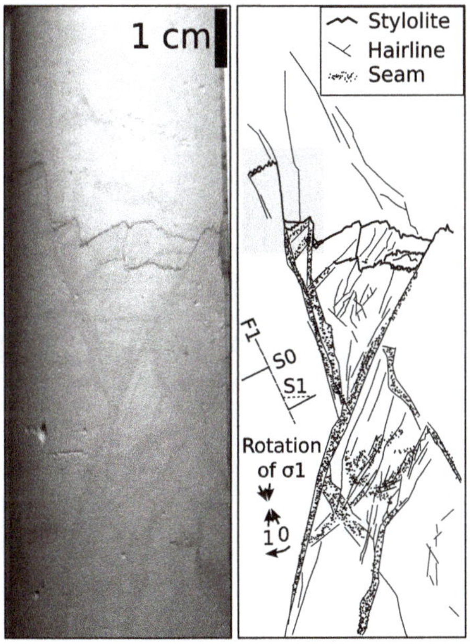

Fig. 8. Relationship between hairlines, stylolites and carbonate seams. The core picture is taken from the Stevns-2 core at 183 m MD. It shows that the stylolites are genetically associated with the hairline fractures. Here, the progressive normal throw is followed by several subtle re-equilibration of the stylolite profile which tries to stay perpendicular to the main stress vectors while the footwall of the fracture rotates (shaded area illustrates the rotation). The carbonate seams start directly under the stylolites and fill the space surrounding the underlying fractures, illustrating the downward carbonate mass transfer associated with the pressure-solution phenomenon. Note that the core is not slabbed, so the surface here is half a cylinder and distortion of the perspective occurs on the side of the picture.

1983; Baudon & Cartwright, 2008). Maximum throws are not necessarily centred on the fault planes and seem more common in the lower part of the faults (histogram, Fig. 5C). The throw distribution is relatively smooth with small throws close to the tip of the faults and relatively constant throws in the main part of the faults (local averages ranging from 5 to 7 m; Fig. 5C). Throws seem to be only slightly larger in the lower half of the faults, making their distribution slightly skewed at the bottom (Fig. 5C). A large part of the clusters have their faults terminating on the relatively strong reflective events without offsetting them (Figs 3B, 4 and 9). On the contrary, some of the clusters have a continuous vertical extension through the whole sedimentary succession and form regional lineaments routed deep below the Chalk Group and imaged in several adjacent profiles (Figs 3A, 4 and 9). The latter clusters have typical diamond-shaped sections (Figs 3A and 4). One cluster shows a complex history during Chalk deposition with a normal throw at the base followed by inversion of the structure and creation of an anticline (fault propagation fold; Figs 3A and 4). Vertically stacked depressions are observed, some of them as deep as the Lower Cretaceous (grey, Figs 3A and 4).

Borehole observations

The wireline log data show an increase in sonic velocity with depth and a corresponding increase in density (equals decrease in porosity) confirming the link between seismic velocity and porosity of the Chalk suggested by Japsen (2000; Figs 2B and 3A). However, the increase in sonic velocity is stepwise – as is the increase in density – with major jumps at 75 m MD and 240 m MD (Measured Depth; Fig. 2A). The vertical evolution of the

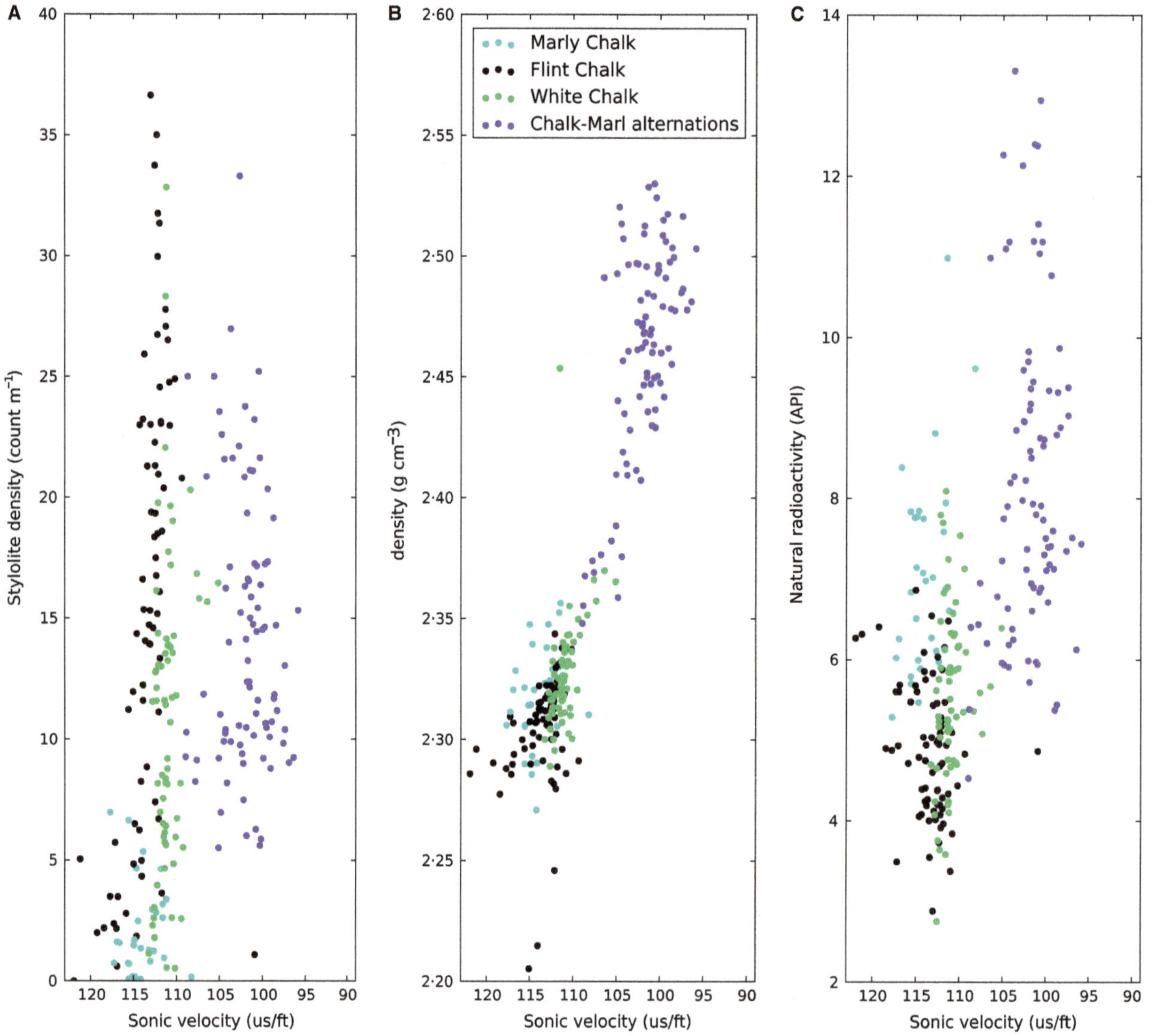

Fig. 9. Cross-plot of Chalk characteristics and features versus sonic velocity in the Stevns-2 well. Colours represent the main Chalk facies from core analysis (cf. Fig. 2). Sonic velocities are from wireline logging. (A) Cross-plot between sonic velocity versus the number of stylolites per metres counted on the core. The plot shows no clear correlation between the amount of (diagenetic) pressure-solution and the sonic velocity. (B) Plot of the sonic velocities against density. The diagram shows a clear trend, however, except when the Chalk contains marls, distinction between the main Chalk facies is not possible based on these two standard wireline tools. (C) Plot of the sonic velocities versus the natural radioactivity wireline log. No clear trend between the clay content and the velocity can be evidenced here. [Correction added on 13th Feb, after first online publication: Figure captions of figures 6,7 and 9 switched]

measured velocities indicates a partitioning of the section in three intervals (A, B and C). Interval A is defined by acceleration with depth until 120 m MD (Fig. 2A). Within Interval A, the velocity is very variable with ± 200 m s^{-1} high-frequency variations. The base of Interval A is formed by the Rørdal Member, a marly Chalk unit, whereas the top is essentially composed of Chalk with occasional flint bands (Fig. 2). Interval B comprises white Chalk and Chalk with flint layers between 120 m and to 240 m MD (Fig. 2). In this interval, the velocity is stable around 2700 ± 100 m s^{-1}. Density shows a consequent drop in the first 10 m at the top of Interval B (directly below the Rørdal Mb.; black arrow, Fig. 2A). The interval below 240 m MD forms Interval C and is characterized by higher velocities and densities and by very high-frequency variations (*ca* 3000 ± 200 m s^{-1} and $+0{\cdot}2$ g cm^{-3}; Fig. 2A). Interval C is composed of interbedded marls and Chalk Boussaha *et al.* (2016).

Diagenetic compaction features, consisting mainly of stylolites and deformation bands, have been quantified.

Stylolites have classical saw-teeth structures with stylus-shaped teeth reaching maximum amplitudes of a couple of centimetres. They resemble those described by Lind (1993) from ODP leg 130 where the dark colour is probably due to the accumulation of insoluble material (Fig. 8). The deformation bands, also called 'hairline fractures' in the literature, are also typical of Chalk (Wennberg *et al.*, 2013 and references therein). In the Stevns-2 core, the individual deformation bands are extremely thin and barely visible with the naked eye (consequently probably in the order of 0·1 mm in width; Fig. 8). They can extend over a few millimetres to 10 cm in height (Fig. 8). Some organize in swarms of coalescent parallel fractures making them more visible (Fig. 8), and can also present synthetic/antithetic organizations (Fig. 8). Some of the swarms are surrounded by a diffuse area of darker Chalk attributed to carbonate reprecipitation (carbonate seams, dotted area in Fig. 10). Detailed core observations have been used to understand the geometrical relationships between stylolites, hairlines and carbonate seams (Fig. 8). The hairline fractures (F1; Fig. 8) have a normal throw, offsetting a first generation of stylolites (S0; Fig. 8). A second generation of stylolites (S1) is in today's 'horizontal' position and is developed within the footwall of the fracture plane (Fig. 8). S0 and S1 merge at different angles, S0 being tilted due to the fracture's normal throw. Downward, the pressure-solution structures (stylolites) and the darker Chalk along the hairline fractures indicates the presence of carbonate seams (Fig. 8). There are no seams above the solution structures, they only appear impregnating the fractures downward of the stylolites (Fig. 8).

The quantification of structural and diagenetic features exhibited by the core show that hairline fractures are generally less abundant with depth and have a local minimum at the base of Interval A (Fig. 2B). Interval B (from 120 to 240 m) contains two parts with many fractures that are separated by an interval with very few fractures at 175 m MD (Fig. 2B). This lower fracture density interval corresponds to Chalk with flint/white Chalk transition (black to green; Fig. 2B and C). There is a noticeable drop in the density of deformation bands in Interval C.

The trend of stylolite density with depth contrasts with the distribution of hairline fractures, showing a stepwise overall increase in concentration (Fig. 2B). The stylolite concentration in the uppermost 10 m of the core is higher than in the rest of Interval A (Fig. 2B). The stylolite density increases rapidly at 120 m and shows an overall increase

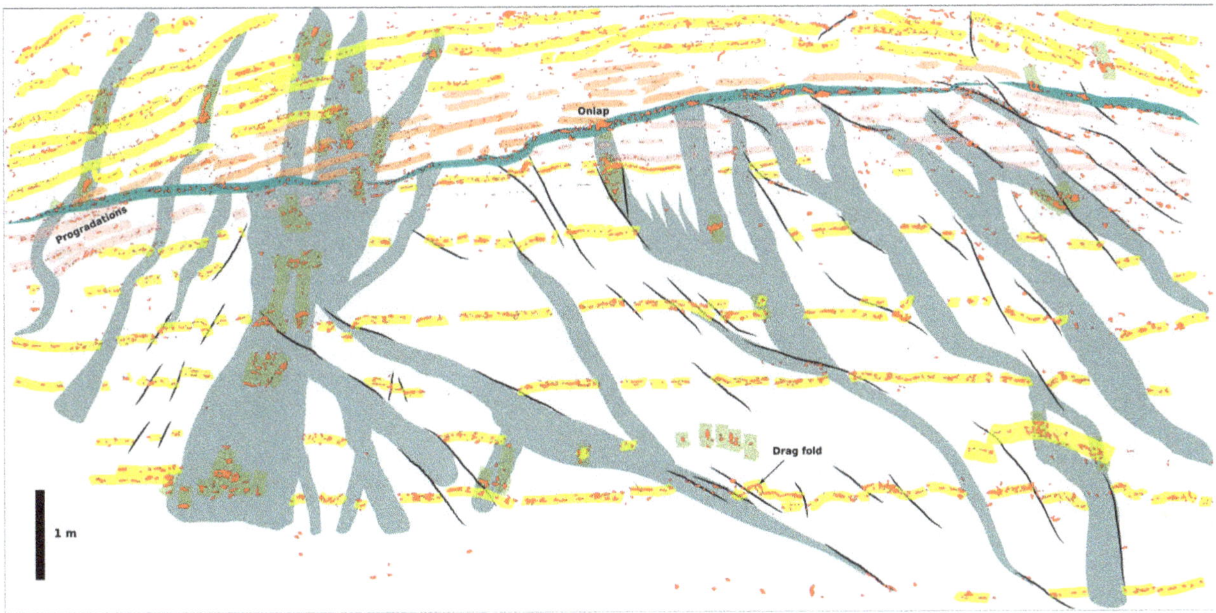

Fig. 10. Interpretation of panoramic pictures taken in the Sigerslev Quarry illustrating the relationship between polygonal faults and fluid expulsion conduits terminating in a pockmark (Fig. 1). All the flint nodules of this quarry face have been drawn in red. Original picture is in the Supplementary material 1. Flint bands showing the regular stratigraphy are in yellow. Areas where paramoudra-like flint structures (green) are present and showing gaps in the (stratigraphic) flint bands are considered as former fluid conduits (blue-grey). Steeply dipping, elongated flint nodules (Fig. 11A and B) and drag folds in flint bed indicate the presence of faults (polygonal, in black). The constructive part of the pockmark spreading out laterally is highlighted by the pink prograding flint bands. The depression of the pockmark is marked by a thick flint band (in turquoise). The filling of the pockmark is highlighted in orange. The filling of the pockmark and the main conduit have a slightly darker colour on the quarry wall (Supplementary material 1). [Colour figure can be viewed at wileyonlinelibrary.com]

down to 175 m (Fig. 2B). There is a clear distinction between Chalk with flint from 120 to 175 m (more stylolites) and pure white Chalk from 175 to 240 m (less stylolites, Fig. 2B and C). The bottom of the core (>240 m MD) has a variable density of stylolites with around 15 per metre.

Cross-plotting sonic velocities and density shows a good correlation and a clear separation between marls and purer Chalk (Fig. 6B). The cross-plot between stylolite density and sonic velocity in Fig. 6A shows that there is no correlation between these two parameters. This lack of relationship is also well illustrated in the interval from 120 to 240 m in Fig. 2 where the almost stable density and sonic velocity of Chalk (Fig. 2A) contrasts with the pronounced variations in stylolite density (Fig. 2B). The relationship between the sonic velocity and the natural radioactivity, which should represent the clay content, is also poorly developed (Fig. 9C). Overall, except when Chalk contains a high concentration of marl (deep blue Fig. 6), Chalk facies do not show any trend or correlation with the number of stylolites (i.e. intensity of diagenesis) or clay content (detrital input; Fig. 6).

Field observations

Chalk exposed in Sigerslev Quarry and the adjacent coastal cliffs shows spectacular palaeo-sea-bed topographies mimicked by undulating flint bands within Chalk (Anderskouv *et al.*, 2007). Locally, the flint bands are cross-cut at a high angle by flint aligned within planes which locally offset the strata (Figs 10 and 11). The stratigraphic flint bands are locally absent along these planes which occasionally exhibit small drag folds in their immediate vicinity (Fig. 10). The subvertical zones lacking stratiform flint bands (blue-grey in Fig. 10) contain cylindrical hollow flints resembling *Batichnus paramoudrae* (green in Fig. 10; Bromley *et al.*, 1975). Most of the subvertical zones end at a level marked by a particularly thick flint band (turquoise in Fig. 10). This thick band is topped by an interval containing low-angle dipping flint bands (pink in Fig. 10). The pink interval depicted in Fig. 10 seems to be constructive and grows laterally away from the biggest subvertical zone (small progradations; Fig. 8). The thick flint band delineates the base of a specific infill which is slightly greyish in colour (in orange and onlapping on the thick turquoise flint band in Fig. 8; supplementary material 1). The flint bands cross-cut the stratigraphy at high angles highlighting planes that are rarely continuously visible over more than 2 m (Fig. 11A). The planes themselves are composite and show different orientations of segments (Fig. 11B). They are difficult to measure and access, so no statistical measurement could be done but it seems that the structures have no clear preferred orientation (Fig. 7).

At the bottom of the quarry another important structure intersects the stratigraphy: a 10 cm thick cluster of stylolites can be followed continuously on the quarry walls (Fig. 11C and D). This first (in burial depth) continuous stylolite is located just below the structures observed in Fig. 10. Other stylolites which are discontinuous and not subparallel to the stratigraphy are observed at the top of the quarry (Fig. 11E).

Careful observations and cleaning of the quarry wall highlight numerous fractures and deformation structures (Fig. 11F, G, and H). Subhorizontal decollement planes have been observed, separating intervals with differential horizontal motion (Fig. 11F). Hairline fractures are observed all over the exposures as long as Chalk has enough colour contrast and markers to show the offsets (Fig. 11G and H). Fractures organized in conical sets are observed with kinematic indicators showing important volume changes (Fig. 11G). Small faults in the prolongation of the flint bands signify normal throws but also folding in the vicinity of the fault (Fig. 11H). Both of these features underline the importance of contractional volume changes (Fig. 11H).

DISCUSSION

The seismic data show the presence of two types of faults. Regionally extensive fault clusters affecting the whole imaged succession, being subparallel to the Sorgenfrei-Tornquist and the Carlsberg fault zones, are considered to be similar strike-slip lineaments (e.g. at 15 000 and 30 000 m in Fig. 4). The presence of an anticline and the differential sedimentation associated with it indicate an originally normal throw followed by inversion during Campanian times (Fig. 4). It is assumed that the regional fault system is not associated with the diagenesis of Chalk and follows the general history of the North Sea Basin (Erlström *et al.*, 1997; Japsen *et al.*, 2007; Sopher & Juhlin, 2013), and its analysis is considered out of the scope of this study.

Insights from strata-bound fault geometries on seismic data

Strata-bound faults (Figs 4 and 9), characteristic of polygonal fault systems (Cartwright *et al.*, 2003), are the second type encountered. The throw distribution analysis shows that fault geometry is very similar to other polygonal fault systems or more broadly to blind faults (Baudon & Cartwright, 2008). The dips of most measured faults range from 40° to 70° (Fig. 5A). This dip range is slightly less steep than faults forming close to the sea-bed (50° to 80°) but still much steeper than the faults forming at depth (20 to 50°; Cartwright, 2011). Knowing that Chalk

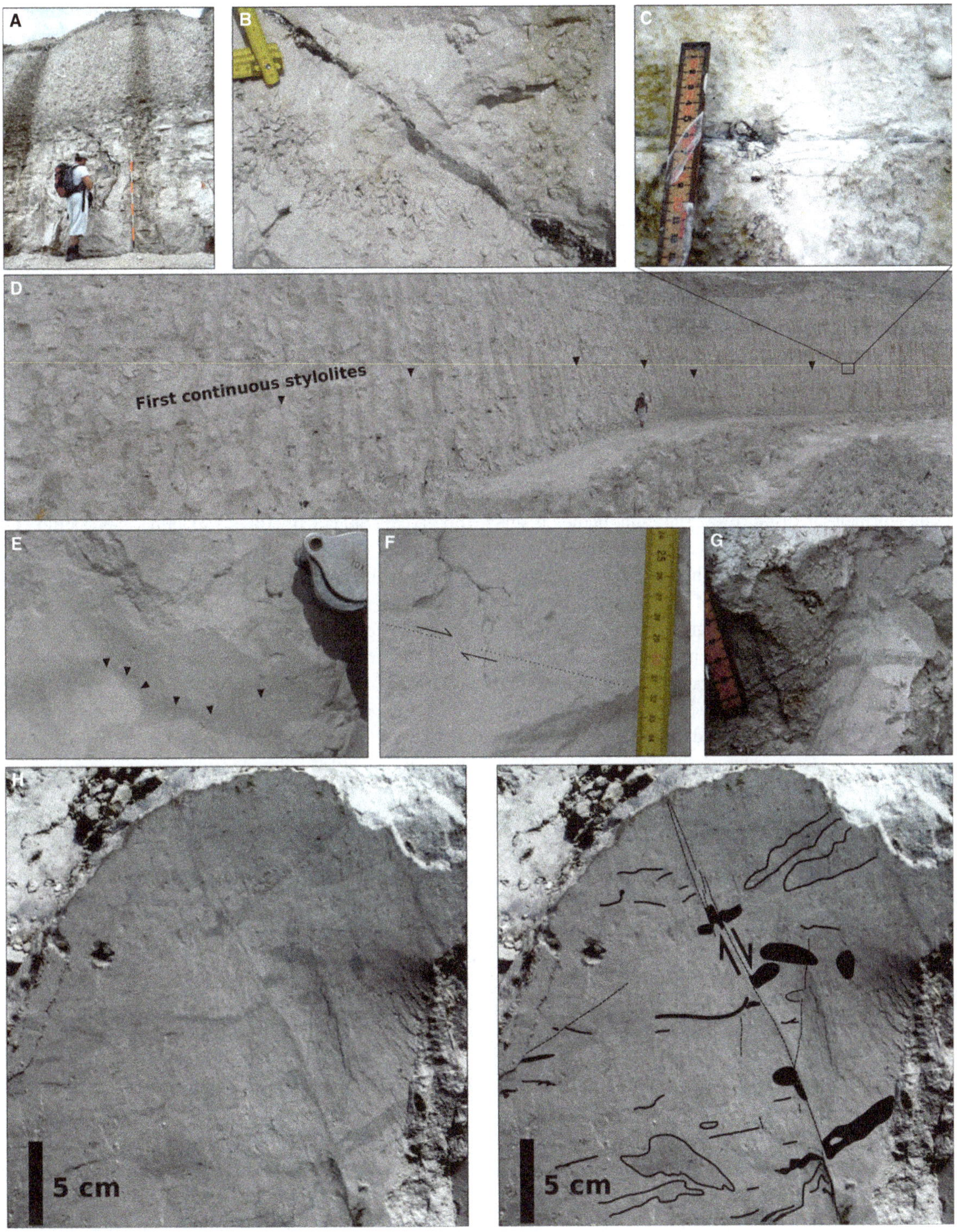

Fig. 11. Small-scale diagenetic features at the Sigerslev Quarry and surroundings. (A) 2 m long flint band within a polygonal fault at the coastal cliff. (B) Flint precipitation within a polygonal fault. Notice the discontinuity of the nodules and the absence of internal fractures. Scale in cm. (C) Close up view of the cluster of continuous stylolites situated at the very bottom of the quarry. Scale in cm. (D) Panoramic view of the first continuous stylolites. Note that they are not parallel with the flint bands (original stratigraphy). (E) Small-scale stylolite from the upper part of the quarry. (F) Decollement plane between two intervals with small subhorizontal offset in the deepest part of the quarry. (G) Hairline fractures organized in conical shape illustrating the contraction of the sediment. Scale in cm. (H) Normal fault within Chalk. The footwall is contracted at the contact with the fault. [Colour figure can be viewed at wileyonlinelibrary.com]

succession has probably reached burial depths of *ca* 500 m in the study area (Nielsen *et al.*, 2011), it can be assumed that compactional flattening with burial has made the dips shallower than their original formation geometry (Cartwright, 2011; Nielsen *et al.*, 2011). Therefore, it is possible to conclude that the faults were originally steeper, formed close to the sea-bed and were consecutively flattened by compaction. Since the maximum displacement of polygonal faults crudely scales with the fault height, this ratio is established to compare with other systems by plotting the maximum throw versus the height of individual faults (Fig. 5B; Shin *et al.*, 2010; Cartwright, 2011). The measured faults have a very large ratio of fault height to maximum throw of $0{\cdot}128 \pm 0{\cdot}089$ which is significantly higher than the standard ratio for polygonal faults of $0{\cdot}045 \pm 0{\cdot}016$ (Shin *et al.*, 2010). Since listric geometries associated with a weak basal layer are lacking, the faults probably have a high degree of lateral intersection (Shin *et al.*, 2010). The throw distribution along normalized distance on the fault planes is an indicator of fault displacement (Fig. 5C; Baudon & Cartwright, 2008; Cartwright, 2011). The displacement gradient starts abruptly from the tip of the faults, is broadly smooth in the central part and slightly more accentuated in the lower half of the faults, and in this sense similar to other studied polygonal fault systems (Fig. 5C; Cartwright, 2011). The M-type displacement pattern characterizes blind-faults and is found in polygonal fault systems (Baudon & Cartwright, 2008; Shin *et al.*, 2010; Cartwright, 2011). The maximum displacement occurs more frequently at the base of the faults; however, it seems to be highly variable (Fig. 5C). This is interpreted as being driven by heterogeneity of the mechanical properties in the sediment pile and a complex growth history (Baudon & Cartwright, 2008). The finite heterogeneous vertical displacement is most probably the result of several 'simultaneous' fault nucleations at different stratigraphic levels that link together vertically and also laterally since the faults have a high degree of intersection (large throw to height ratio). Lateral hard links between polygonal faults will limit their propagations and might be responsible for some large differences between the observed lateral displacement in Fig. 5C (Lonergan *et al.*, 1998). Therefore, it can be suggested that the largest displacements are close to the nucleation points of the faults, where the shear failure occurred. Polygonal faults are interpreted as being associated with contraction of the sediments during burial. Their initiation can be very early in the diagenetic sequence since some are considered active today at the sea-bed (Cartwright *et al.*, 2003; Goulty, 2008). The strata-bound faults are best developed in intervals, which – from correlation to the Stevns boreholes – are known to consist of relatively pure Chalk. In contrast, faulting is rare in intervals composed of Chalk containing marl layers (Fig. 4). The contrasting intensity of faulting in these two rock types indicates different diagenetic response during burial as discussed further below. This intense faulting in the white Chalk facies may be the reason for its poor reflectivity, the seismic waves being scattered by subvertical fault planes.

Relationships between faults, fractures, stylolites and fluid flows

Strata-bound faults are also observed in outcrop (Figs 10 and 11). Measurement of the faults in the field did not reveal a preferred orientation (Fig. 7). Observations of the Chalk of the eastern Danish Basin at all scales, from core and field observations to seismic profiles, highlight the link between the different diagenetic features and illustrate the transformation of the Chalk during its early burial history. Four main controls on the diagenesis can be observed: faults, fractures, stylolites and fluid escape structures.

The faults contain flint which indicates that they were associated with migration of the silica-loaded fluids and therefore formed at relatively shallow depths before opal-CT was transformed to α-quartz (Hibsch *et al.*, 2003; Madsen & Stemmerik, 2010).

The hairline fractures observed in cores and outcrop are similar to the deformation bands described by Wennberg *et al.* (2013). Deformation bands are joints in the Chalk which correspond to local pore-space collapse and are associated with the progressive burial of the sediment (Wennberg *et al.*, 2013). Surprisingly, they reduce in frequency with depth and drop considerably in abundance as stylolites become prominent (Fig. 2). This could indicate that there is a genetic link between hairline fractures and stylolites and that the joints are precursors to the stylolites. When the joints form, the pore space collapses between Chalk grains. Such collapses increase the contact area between the associated grains, facilitating the later pressure-solution processes associated with stylolite formation. While progressing through Chalk, the stylolites assimilate and dissolve the joints so that the higher the number of stylolites observed, the less precursory joints have been preserved (Fig. 2). The link between fractures and stylolites is illustrated in Fig. 8 where there is a discrepancy between the orientation of the fractures (joints) above and below the stylolite. Since it is improbable that stylolites accommodated horizontal displacement, the joints can be considered to be originally organized in a coherent synthetic-antithetic compactional configuration. Only the activity of the stylolite has modified the volumes, dissolving Chalk downward to obtain the observed structural discrepancy above and below the stylolite

(Fig. 8). In addition, the presence of subhorizontal S1 and tilted S0 stylolites along a fracture indicates that the fracture developed in conjunction with S0 (Fig. 8). S0 was originally horizontal and formed perpendicular to the main stress, the lithostatic pressure (vertical). Since the footwall of the fractures became tilted (including S0), the stylolites were not in equilibrium with the main stress, causing a new generation of horizontal stylolites (S1) to form. The rotation of the palaeo-stress indicator (stylolite) is seen as the result of the time-transgressive formation of stylolites in contracting sediment.

The compaction of the solid parts of Chalk and the increasing stylolitization during burial do not result in cementation of the sediment, indicating that the carbonate enriched solutions and the pore water were able to move. In a closed system, the density should rise because pore-space progressively disappears and the dissolved carbonate will re-precipitate as cement. Open-system carbonate mass-transfers match the observations better. This is supported by the absence of correlation between stylolite abundance and density of Chalk in the interval from 120 to 240 m (Figs 2 and 6). It is assumed that the pore water and the carbonate-rich fluids expressed during the formation of stylolites, built up pressure in the sediment which was then released by venting at a given threshold. This was not a continuous process and was most probably focussed over specific stratigraphic intervals. The difference in fault activity between pure Chalk intervals and intervals of interbedded Chalk and marl, together with the pronounced stepwise velocity jumps across these facies transitions, indicate that primary facies had a major influence on diagenesis and resulting variations in fluid pressures during early burial. However, the localization of these intervals is not completely resolved in our data set. Discrete zones characterized by variable preservation of the calcareous nannofossil assemblages have been observed in the Stevns-1 core (Thibault *et al.*, 2012), and may be candidates for localized fluid flow zones.

The significance of cold seeps

In outcrop, evidence of venting is seen as subvertical conduits and strata-bound faults capped by a thick layer of flint (Fig. 10). The flint most likely was precipitated close to the sea floor at a stable redox boundary (Madsen & Stemmerik, 2010), and judging from the flint band orientations above and below, it formed at a time of shifting depositional conditions (Fig. 10; Supplementary material 1). Based on the architectural data, it is proposed that the plumbing system is likely to have been via cold seep. Cold seeps or so-called pockmarks are structures commonly observed in seismic images in association with polygonal fault systems (Fig. 10; Cartwright *et al.*, 2003; Gay *et al.*, 2006) or exceptionally preserved in desert exposures (Tewksbury *et al.*, 2014). The conduits may be the results of fluid expulsions which locally promote the formation of flint in the form of paramoudra-like structures (the vertically stacked hollow flint cylinders; Fig. 10). Polygonal faults generally help drain fluids from depth to the sea-bed (Gay *et al.*, 2006; Tewksbury *et al.*, 2014). Most of the conduits stop within depressions or polygonal faults (Fig. 8). This indicates that the polygonal faults have reached the sea-bed and may have promoted the drainage of fluids through the low permeability Chalk. Similar-sized vertically superimposed depressions are observed on the seismic profiles and are interpreted as being fluid escape structures associated with cold seeps (Figs 3 and 4). The formation of such pockmarks may be quite common in Chalk although not easily observed in the field. Some examples of strata-bound/polygonal faults as well as pockmarks have been documented in seismic data of the North Sea and other basins with thick Chalk accumulations (Gemmer *et al.*, 2002; Hansen *et al.*, 2004; van Gent *et al.*, 2010; Sandrin *et al.*, 2012) and once in the field (Tewksbury *et al.*, 2014). In addition, igneous intrusions and ash layers are documented within the Upper Cretaceous of the nearby Scania area as well as in the Kattegat (Norling & Bergström, 1987; Ziegler, 1987). While cooling down, a volcanic intrusion at depth could also be a potential source of fluids which would have flowed upward through the upper Cretaceous sediments and stimulated the formation of vents, diagenetic reactions and polygonal fault networks (Planke *et al.*, 2005; Gay *et al.*, 2012).

From contraction to shear failure

The data presented here indicate that the hairline fractures are associated with volume change and contraction of Chalk. Observation of millimetre-scale horizontal movements additionally suggests separation in intervals even at small scale (Fig. 11F, G, and H). These characteristics are shared with polygonal fault systems (Cartwright *et al.*, 2003). It is therefore tempting to see deformation bands and polygonal faults as the results of the same processes of contraction of Chalk, just reflecting different scales and locations. These deformation bands reflect a more pervasive deformation than the polygonal faults.

Recently, it has been suggested that diagenetic dissolution of grains is the driving process for polygonal fault system nucleation (Shin *et al.*, 2010; Cartwright, 2011). The sediments involved have low post-peak shear strength (Shin *et al.*, 2010). To achieve this process, the sediments should have a high porosity and be very fine-grained while undergoing mineral dissolution (Shin *et al.*, 2010), conditions which are met in the studied Chalk; the grain size being less than 20 μm, porosities between 35% and 50% (Frykman, 2001; Nielsen *et al.*, 2011) and the presence of stylolites (pressure-solution structures) attesting to mineral dissolution. In addition, numerical experiments based on this dissolution/shear failure model predict displacement patterns which are strikingly similar to our measured displacement (Fig. 5C; Shin *et al.*, 2010). Therefore, this process is considered the main trigger of contraction-driven faults in the area.

Synthesis

Chalk, like other low permeability, high porosity and very fine-grained sediments, undergoes intense diagenetic transformations in the first stage of its burial history (Fig. 12; Cartwright *et al.*, 2003; Hibsch *et al.*, 2003). A scenario for these transformations can be established from coupled observations of core material, outcrops and seismic profiles in eastern Denmark. Pure white Chalk appears massive but is pervasively disturbed by deformation bands during shallow burial (Fig. 12). At depth, the deformation bands will progressively merge to become stylolites. The stylolites promote mass-transfers of the carbonates, trigger shear failure and the formation of polygonal fault systems (Fig. 12). The polygonal faults propagate to eventually reach the sea-bed, allowing for drainage of overpressured pore water from Chalk and eventually forming pockmarks (Fig. 12). All these processes contribute to the shrinkage of Chalk and can trigger positive feedback cycles. The initiation, temporal progression and termination of this phenomenon, however, are at present not understood.

As also emphasized by Hibsch *et al.* (2003), some faults contain flint nodules, indicating that fluid migration occurred before transformation of opal-CT to α-quartz (Madsen & Stemmerik, 2010). The formation of flint nodules is thought to be a relatively shallow and early

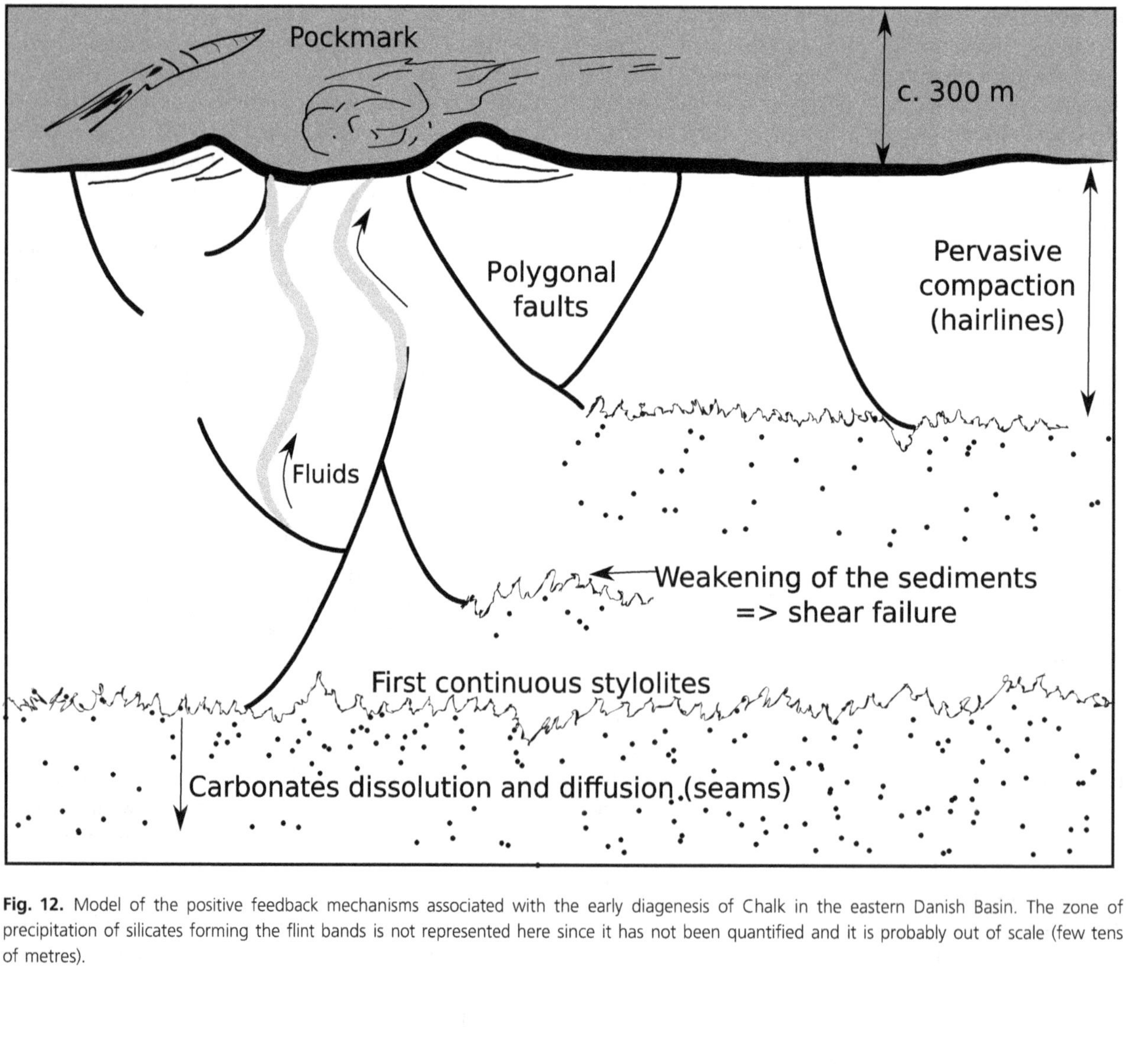

Fig. 12. Model of the positive feedback mechanisms associated with the early diagenesis of Chalk in the eastern Danish Basin. The zone of precipitation of silicates forming the flint bands is not represented here since it has not been quantified and it is probably out of scale (few tens of metres).

diagenetic process although no quantitative estimation has yet been provided for the Stevns peninsula setting (Madsen & Stemmerik, 2010).

In modern Chalk, the formation of 'large' stylolites starts at *ca* 830 m of burial, whereas the first 'small' stylolites occur at 490 m (Lind, 1993). In Sigerslev Quarry, the first large continuous stylolite occurs near the base of the quarry, corresponding to a burial depth of *ca* 580 m, while in Stevns-2 a sudden increase in stylolite concentration occurs at 120 m MD corresponding to a burial depth of 620 m using data in Nielsen *et al.* (2011). It is therefore tempting to assume that a burial depth of *ca* 600 m was needed to trigger the formation of faults and their propagation in the Upper Cretaceous Chalk. This is somewhat shallower than the depth at which modern large stylolites start to form. However, the presence of compartments of different permeability, the heterogeneity of Chalk itself as well as potential fluid supply from deeper parts of the succession may change the initial conditions of stylolite formation in the study area compared to the present-day systems (Lind, 1993). If the estimate of 600 m overburden is applied, nucleation of polygonal faults may have started in the deepest parts of the Chalk Group already during the late Campanian, whereas a late Maastrichtian age is obtained if 800 m of overburden is used as a critical value. In both cases the outlined venting system was in place during the latest stages of Chalk deposition at Stevns (Fig. 12).

The integration of observations from different sources and at different scales have helped to better understand the timing and linking of the processes involved in the transformation of oceanic ooze into Chalk. Better quantitative estimates of the palaeo-burial depths and conditions needed to achieve the observed diagenetic processes could possibly be acquired using a high-resolution 3D seismic data set from an area where Chalk has not been buried more than 1 to 2 km. Another approach would be to use numerical modelling to simulate the dynamics of polygonal faulting in conditions similar to those of the Late Cretaceous of the eastern Danish Basin in order to better constrain the diagenetic sequence affecting Chalk. The velocity layering of Chalk (Fig. 2; Nielsen *et al.*, 2011) indicates that the diagenetic transformations took place at critical depths. This compares to studies of modern ooze from the Java Plateau where the first stylolites occur at 490 m, large stylolites appear at 830 m and cementation starts around 1100 m (Fabricius & Borre, 2007). Evidently, only the deepest parts of the Chalk Group at Stevns have experienced overburdens of more than 1100 m. Considering seismic velocities, Nielsen *et al.* (2011) similarly suggested that an equivalent to the Java Plateau 1100 m event could correspond to a velocity jump identified *ca* 250 m below TD of Stevns-2 (at 600 m today's depth). The rapid increase in velocity at *ca* 240 m and the associated decrease in porosity may be a candidate for such a transformation of intensified dissolution and eventual nucleation of polygonal faults. Most probably, the upper part of Chalk at Stevns had only been affected by one generation of the contraction process associated with the faults, whereas the deeper intervals may have been subject to several generations of faults and consequently, more intense compaction. Is the velocity layering the result of multiple generations of polygonal fault systems or does it represent a stepwise diagenetic evolution of Chalk? This question could be elucidated with a thorough study of deeper buried Chalk successions.

CONCLUSIONS

This study illustrates the diagenetic processes affecting nannofossil ooze in the first kilometre of burial. The diagenetic reactions involve a first stage of pervasive contraction marked by the formation of deformation bands also commonly called hairline fractures. This stage is followed by the start of stylolite formation at *ca* 600 m of burial. For the first time, it can be shown that Chalk behaves like its fine-grained siliciclastic counterparts, and that grain dissolution triggers the formation of polygonal faults. The polygonal faults in the study area have a high degree of lateral connection and form close to the surface. In addition to mimicking the sea-bed, flint bands are fossilized vertical conduits of fluid escape and fault planes, in connection with the polygonal fault system. Like in other polygonal fault systems, the contraction of Chalk was probably associated with cold seeps. Our detailed observations of the diagenetic structures in Chalk of the Stevns Peninsula may have been overlooked elsewhere and should be visible on other good Chalk exposures.

Chalk is a major reservoir for hydrocarbons and drinking water in the North Sea and its coastal areas. Its reservoir properties are influenced by the diagenetic features which are observed in the Stevns peninsula (pore-space collapse, compartmentalization, vertical conduits). However, more quantitative data are needed to better constrain the diagenetic system. Therefore, illustrating the complex interaction of diagenetic phenomena affecting Chalk during burial is expected to stimulate more research on this outstanding sedimentary system.

ACKNOWLEDGEMENTS

We acknowledge Kresten Anderskouv for his pre-review work. Finn Surlyk is thanked for the stimulating discussions on Chalk depositional system and its evolution in Denmark. We would like to thank Lars Ole Boldreel for giving access to the original seismic data repository. Lise

Boulicault is also thanked for her help during field acquisition. We are grateful to Maersk Oil for having sponsored this research in the C-cubed project framework. We thank J. Cartwright, C. Jackson, L. Lonergan and A. Gay for their revision of a former version of the manuscript. We also would like to thank J. D'Arcy for the English-language proofing of the manuscript.

CONFLICT OF INTEREST

No conflict of interest declared.

References

Anderskouv, K., **Damholt, T.** and **Surlyk, F.** (2007) Late Maastrichtian chalk mounds, Stevns Klint, Denmark – combined physical and biogenic structures. *Sed. Geol.*, **200**, 57–72. doi:10.1016/j.sedgeo.2007.03.005.

Baudon, C. and **Cartwright, J.A.** (2008) 3D seismic characterisation of an array of blind normal faults in the Levant Basin, Eastern Mediterranean. *J. Struct. Geol.*, **30**, 746–760. doi:10.1016/j.jgs.2007.12.008.

Bjerager, M. and **Surlyk, F.** (2007) Danian Cool-Water Bryozoan Mounds at Stevns Klint, Denmark – a new class of non-cemented skeletal mounds. *J. Sed. Res.*, **77**, 634–660. doi:10.2110/jsr.2007.064.

Boussaha, M., **Thibault, N.** and **Stemmerik, L.** (2016) Integrated stratigraphy of the late Campanian – Maastrichtian in the Danish Basin: revision of the Boreal calcareous nannofossil zonation. *Newsl. Stratigr.*, **49**, 337–360. doi:10.1127/nos/2016/0075.

Bromley, R.G., **Schulz, M.-G.** and **Peake, N.B.** (1975) *Paramoudras: Giant Flints, Long Burrows and the Early Diagenesis of Chalks.* Kommissionær Munksgaard, København, 42 pp.

Cartwright, J. (2011) Diagenetically induced shear failure of fine-grained sediments and the development of polygonal fault systems. *Mar. Pet. Geol.*, **28**, 1593–1610. doi:10.1016/j.marpetgeo.2011.06.004.

Cartwright, J., **James, D.** and **Bolton, A.** (2003) The genesis of polygonal fault systems: a review. *Geol. Soc. London. Spec. Publ.*, **216**, 223–243. doi:10.1144/GSL.SP.2003.216.01.15.

Erlström, M., **Thomas, S.A.**, **Deeks, N.** and **Sivhed, U.** (1997) Structure and tectonic evolution of the Tornquist Zone and adjacent sedimentary basins in Scania and the southern Baltic Sea area. *Tectonophysics*, **271**, 191–215.

Esmerode, E.V., **Lykke-Andersen, H.** and **Surlyk, F.** (2007) Ridge and valley systems in the Upper Cretaceous chalk of the Danish Basin: contourites in an epeiric sea. *Geol. Soc. London. Spec. Publ.*, **276**, 265–282. doi:10.1144/GSL.SP.2007.276.01.13.

Fabricius, I.L. and **Borre, M.K.** (2007) Stylolites, porosity, depositional texture, and silicates in chalk facies sediments. Ontong Java Plateau, Gorm and Tyra fields, North Sea. *Sedimentology*, **54**, 183–205. doi:10.1111/j.1365-3091.2006.00828.x.

Fabricius, I.L., **Bächle, G.T.** and **Eberli, G.P.** (2010) Elastic moduli of dry and water-saturated carbonates – effect of depositional texture, porosity, and permeability. *Geophysics*, **75**, 65–78. doi:10.1190/1.3374690.

Frykman, P. (2001) Spatial variability in petrophysical properties in Upper Maastrichtian chalk outcrops at Stevns Klint, Denmark. *Mar. Petrol. Geol.*, **18**, 1041–1062. doi:10.1016/S0264-8172(01)00043-5.

Gaviglio, P., **Bekri, S.**, **Vandycke, S.**, **Adler, P.M.**, **Schroeder, C.**, **Bergerat, F.**, **Darquennes, A.** and **Coulon, M.** (2009) Faulting and deformation in chalk. *J. Struct. Geol.*, **31**, 194–207. doi:10.1016/j.jsg.2008.11.011.

Gay, A., **Lopez, M.**, **Cochonat, P.**, **Séranne, M.**, **Levaché, D.** and **Sermondadaz, G.** (2006) Isolated seafloor pockmarks linked to BSRs, fluid chimneys, polygonal faults and stacked Oligocene-Miocene turbiditic palaeochannels in the Lower Congo Basin. *Mar. Geol.*, **226**, 25–40. doi:10.1016/j.margeo.2005.09.018.

Gay, A., **Mourgues, R.**, **Berndt, C.**, **Bureau, D.**, **Planke, S.**, **Laurent, D.**, **Gautier, S.**, **Lauer, C.** and **Loggia, D.** (2012) Anatomy of a fluid pipe in the Norway Basin: initiation, propagation and 3D shape. *Mar. Geol.*, **332–334**, 75–88. doi:10.1016/j.margeo.2012.08.010.

Gemmer, L., **Huuse, M.**, **Clausen, O.R.** and **Nielsen, S.B.** (2002) Mid-Palaeocene palaeogeography of the eastern North Sea basin: integrating geological evidence and 3D geodynamic modelling. *Basin Res.*, **14**, 329–346. doi:10.1046/j.1365-2117.2002.00182.x.

van Gent, H., **Back, S.**, **Urai, J.L.** and **Kukla, P.** (2010) Small-scale faulting in the Upper Cretaceous of the Groningen block (The Netherlands): 3D seismic interpretation, fault plane analysis and regional paleostress. *J. Struct. Geol.*, **32**, 537–553. doi:10.1016/j.jsg.2010.03.003.

GEUS (2015) Tabel med fakta om registrerede danske jordskælv. Registrerede jordskælv. Available at: http://www.geus.dk/DK/nature-climate/natural-disasters/seismology/Sider/seismo_reg-dk.aspx.

Goulty, N.R. (2008) Geomechanics of polygonal fault systems: a review. *Petrol. Geosci.*, **14**, 389–397. doi:10.1144/1354-079308-781.

Graversen, O. (2004) Upper Triassic-Lower Cretaceous seismic sequence stratigraphy and basin tectonics at Bornholm, Denmark, Tornquist Zone, NW Europe. *Mar. Pet. Geol.*, **21**, 579–612. doi:10.1016/j.marpetgeo.2003.12.001.

Graversen, O. (2009) Structural analysis of superposed fault systems of the Bornholm horst block, Tornquist Zone, Denmark. *Bull. Geol. Soc. Den.*, **57**, 25–49.

Hansen, D.M., **Shimeld, J.W.**, **Williamson, M.A.** and **Lykke-Andersen, H.** (2004) Development of a major polygonal fault system in Upper Cretaceous chalk and Cenozoic mudrocks of the Sable Subbasin, Canadian Atlantic margin. *Mar. Pet. Geol.*, **21**, 1205–1219. doi:10.1016/j.marpetgeo.2004.07.004.

Henriet, J.P., **De Batist, M.D.** and **Verschuren, M.** (1991) Early fracturing of Paleogene clays, southernmost North Sea: relevance to mechanisms of primary hydrocarbon migration. In: *Generation, Accumulation and Production of Europe's Hydrocarbons* (Ed. A.M. Spenser), *Spec. Publ. Eur. Assoc. Petrol. Geol.*, **1**, 217–227.

Hibsch, C., **Cartwright, J.**, **Hansen, D.M.**, **Gaviglio, P.**, **Andre, G.**, **Cushing, M.**, **Bracq, P.**, **Juignet, P.**, **Benoit, P.** and **Allouc, J.** (2003) Normal faulting in chalk: tectonic stresses vs. compaction-related polygonal faulting. *Geol. Soc. London. Spec. Publ.*, **216**, 291–308. doi:10.1144/GSL.SP.2003.216.01.19.

Japsen, P. (2000) Investigation of multi-phase erosion using reconstructed shale trends based on sonic data. Sole Pit axis, North Sea. *Global Planet. Change*, **24**, 189–210. doi:10.1016/S0921-8181(00)00008-4.

Japsen, P. and **Bidstrup, T.** (1999) Quantification of late Cenozoic erosion in Denmark based on sonic data and basin modelling. *Bull. Geol. Soc. Den.*, **46**, 79–99.

Japsen, P., **Green, P.F.**, **Nielsen, L.H.**, **Rasmussen, E.S.** and **Bidstrup, T.** (2007) Mesozoic-Cenozoic exhumation events in the eastern North Sea Basin: a multi-disciplinary study based on palaeothermal, palaeoburial, stratigraphic and seismic data. *Basin Res.*, **19**, 451–490. doi:10.1111/j.1365-2117.2007.00329.x.

Kammann, J., **Hübscher, C.**, **Boldreel, L.O.** and **Nielsen, L.** (2016) High-resolution shear-wave seismics across the Carlsberg Fault zone south of Copenhagen – implications for linking Mesozoic and late Pleistocene structures. *Tectonophysics*, **682**, 56–64. doi:10.1016/j.tecto.2016.05.043.

Lassen, A. and **Thybo, H.** (2012) Neoproterozoic and Palaeozoic evolution of SW Scandinavia based on integrated seismic interpretation. *Precambr. Res.*, **204–205**, 75–104. doi:10.1016/j.precamres.2012.01.008.

Lind, I.L. (1993) Stylolites in chalk from Leg 130, Ontong Java Plateau. In: *Proceedings of the Ocean Drilling Program, Scientific Results* (Eds W. H. Berger, L. W. Kroenke, T. R. Janecek, and W. V. Sliter), pp. 445–451. Ocean Drilling Program, College Station, TX.

Lonergan, L., **Cartwright, J.** and **Jolly, R.** (1998) The geometry of polygonal fault systems in Tertiary mudrocks of the North Sea. *J. Struct. Geol.*, **20**, 529–548.

Lykke-Andersen, H. and **Surlyk, F.** (2004) The Cretaceous-Palaeogene boundary at Stevns Klint, Denmark: inversion tectonics or sea-floor topography? *J. Geol. Soc. London*, **161**, 343–352. doi:10.1144/0016-764903-021.

Madsen, H.B. and **Stemmerik, L.** (2010) Diagenesis of Flint and Porcellanite in the Maastrichtian Chalk at Stevns Klint, Denmark. *J. Sed. Res.*, **80**, 578–588. doi:10.2110/jsr.2010.052.

Muraoka, F. and **Kamata, H.** (1983) Displacement distribution along minor fault traces. *J. Struct. Geol.*, **5**, 483–495.

Nielsen, L., **Boldreel, L.O.**, **Hansen, T.M.**, **Lykke-Andersen, H.**, **Stemmerik, L.**, **Surlyk, F.** and **Thybo, H.** (2011) Integrated seismic analysis of the Chalk Group in eastern Denmark – implications for estimates of maximum palaeo-burial in southwest Scandinavia. *Tectonophysics*, **511**, 14–26. doi:10.1016/j.tecto.2011.08.010.

Norling, E. and **Bergström, J.** (1987) Mesozoic and Cenozoic tectonic evolution of Scania, southern Sweden. In: *Compressional Intra-Plate Deformations in the Alpine Foreland* (Ed. P.A. Ziegler), *Tectonophysics*, **137**, 7–19.

Petracchini, L., **Antonellini, M.**, **Billi, A.** and **Scrocca, D.** (2015) Syn-thrusting polygonal normal faults exposed in the hinge of the Cingoli anticline, northern Apennine, Italy. *Front. Earth Sci.*, **3**, 1–24.

Planke, S., **Rasmussen, T.**, **Rey, S.S.** and **Myklebust, R.** (2005) Seismic characteristics and distribution of volcanic intrusions and hydrothermal vent complexes in the Vøring and Møre basins. In: *Petroleum Geology: North-West Europe and Global Perspectives Proceedings of the 6th Petroleum Geology Conference* (Eds A.G. Doré and B.A. Vining), *Geol. Soc. London*, 833–844.

Rasmussen, S.L. and **Surlyk, F.** (2012) Facies and ichnology of an Upper Cretaceous chalk contourite drift complex, eastern Denmark, and the validity of contourite facies models. *J. Geol. Soc.*, **169**, 435–447. doi:10.1144/0016-76492011-136.

Rosenbom, A.E. and **Jakobsen, P.R.** (2005) Infrared thermography and fracture analysis of preferential flow in Chalk. *Vadose Zone J.*, **4**, 271–280. doi:10.2136/vzj2004.0074.

Sandrin, A., **Fehmers, G.**, **Printz, B.**, **van Buchem, F.**, **Uldall, A.** and **Hoffmann, U.** (2012) Polygonal Faulting in Chalk – an example at the Tyra Field, Danish North Sea. EAGE, Extended Abstracts, 74th EAGE Conference & Exhibition incorporating SPE EUROPEC 2012, Copenhagen, Denmark, 4–7 June 2012.

Shin, H., **Santamarina, J.C.** and **Cartwright, J.A.** (2008) Contraction-driven shear failure in compacting uncemented sediments. *Geology*, **36**, 931–935. doi:10.1130/G24951A4.

Shin, H., **Santamarina, J.C.** and **Cartwright, J.A.** (2010) Displacement field in contraction-driven faults. *J. Geophys. Res.*, **115**, 2156–2202. doi:10.1029/2009JB006572.

Sopher, D. and **Juhlin, C.** (2013) Processing and interpretation of vintage 2D marine seismic data from the outer Hanö Bay area, Baltic Sea. *J. Appl. Geophys.*, **95**, 1–15. doi:10.1016/j.jappgeo.2013.04.011.

Stemmerik, L., **Surlyk, F.**, **Klitten, K.**, **Rasmussen, S.L.** and **Schovsbo, N.H.** (2006) Shallow core drilling of the Upper Cretaceous Chalk at Stevns Klint, Denmark. *Geol. Surv. Denmark Greenland Bull.*, **10**, 13–16.

Surlyk, F., **Damholt, T.** and **Bjerager, M.** (2006) Stevns Klint, Denmark: Uppermost Maastrichtian chalk, Cretaceous-Tertiary boundary, and lower Danian bryozoan mound complex. *Bull. Geol. Soc. Den.*, **54**, 1–48.

Surlyk, F., **Rasmussen, S.L.**, **Boussaha, M.**, **Schiøler, P.**, **Schovsbo, N.H.**, **Sheldon, E.**, **Stemmerik, L.** and **Thibault, N.** (2013) Upper Campanian-Maastrichtian holostratigraphy of the eastern Danish Basin. *Cretac. Res.*, **46**, 232–256. doi:10.1016/j.cretres.2013.08.006.

Tewksbury, B.J., **Hogan, J.P.**, **Kattenhorn, S.A.**, **Mehrtens, C.J.** and **Tarabees, E.A.** (2014) Polygonal faults in chalk: insights from extensive exposures of the Khoman Formation, Western Desert, Egypt. *Geology*, **42**, 479–482.

Thibault, N., **Harlou, R.**, **Schovsbo, N.**, **Schiøler, P.**, **Minoletti, F.**, **Galbrun, B.**, **Lauridsen, B.W.**, **Sheldon, E.**, **Stemmerik, L.** and **Surlyk, F.** (2012) Upper Campanian-Maastrichtian nannofossil biostratigraphy and high-resolution carbon-isotope stratigraphy of the Danish Basin: towards a standard $\delta^{13}C$ curve for the Boreal Realm. *Cretac. Res.*, **33**, 72–90. doi:10.1016/j.cretres.2011.09.001.

Torvela, T., **Moreau, J.**, **Butler, R.W.H.**, **Korja, A.** and **Heikkinen, P.** (2013) The mode of deformation in the orogenic mid-crust revealed by seismic attribute analysis. *Geochem. Geophys. Geosyst.*, **14**, 1069–1086. doi:10.1002/ggge.20050.

Wennberg, O.P., **Casini, G.**, **Jahanpanah, A.**, **Lapponi, F.**, **Ineson, J.**, **Wall, B.G.** and **Gillespie, P.** (2013) Deformation bands in chalk, examples from the Shetland Group of the Oseberg Field, North Sea, Norway. *J. Struct. Geol.*, **56**, 103–117. doi:10.1016/j.jsg.2013.09.005.

Ziegler, P.A. (1987) Late Cretaceous and Cenozoic intra-plate Deformations in the Alpine foreland – a geodynamic model. In: *Compressional Intra-Plate Deformations in the Alpine Foreland* (Ed. P.A. Ziegler), *Tectonophysics*, **137**, 389–420.

3

Hiatuses and condensation: an estimation of time lost on a shallow carbonate platform

ANDRÉ STRASSER

Department of Geosciences, University of Fribourg, Chemin du Musée 6, 1700, Fribourg, Switzerland (Eail: andreas.strasser@unifr.ch)

Keywords
Carbonate platform, condensation, cyclostratigraphy, hiatus, sedimentation rate.

ABSTRACT

On shallow carbonate platforms, the sedimentary record is highly fragmentary because low accommodation commonly leads to non-deposition, erosion, reworking and condensation. Consequently, it is difficult to quantify the time that is actually recorded and to estimate sedimentation rates. The Berriasian platform in the Swiss and French Jura Mountains offers an opportunity to address this problem. Facies evolution through time displays deepening-shallowing trends of different orders, resulting in a hierarchical stacking of sequences. The sequence- and cyclostratigraphic analysis relates these sequences to orbital cycles with durations of 20, 100, and 400 kyr. Many surfaces in the record of shallow-water carbonates can be interpreted as hiatuses due to non-deposition in the supratidal realm, reactivation of shoals or scouring by currents. Strongly bioturbated intervals and hardgrounds commonly result from sediment starvation. These same processes and products can be observed on modern carbonate platforms, and the duration of the hiatuses or condensed intervals can be inferred (a few hours to a few hundreds of years). Amalgamations and narrow stacking of such surfaces create sequence-boundary zones that can be followed over the entire platform. Time lost or condensed in these zones may correspond to hundreds of thousands of years. Time distribution in the studied sections thus is highly irregular: short periods of sedimentation alternated with long periods of non-deposition, erosion, reworking, and condensation. Furthermore, due to substrate morphology and laterally variable depositional environments, the depositional record of a given time interval varied significantly across the platform. In one example, it can be estimated that – applying modern sedimentation rates – a sedimentary record spanning 800 kyr could have been deposited within 44 kyr if there were no hiatuses or condensations. It becomes clear that, when estimating carbonate production and sediment accumulation rates in ancient records, the shortest possible time spans must be used to minimize the effect of time lost in hiatal surfaces and condensed intervals.

INTRODUCTION

Faithful reconstruction of Earth's history depends on the completeness of the rock record. Chronostratigraphic charts are established by radiometrically and/or astrochronologically dating rocks that represent this record. No place on Earth contains the entire record from the Hadean to the Holocene, and completeness is attempted by patching fragmentary records together at a global scale. One such example is the IUGS International Chronostratigraphic Chart that is constantly updated following new findings and improved dating methods (www.stratigraphy.org).

Relatively complete records covering parts of Earth history are obtained from sediments that have been deposited at high rates in regularly subsiding oceanic or lacustrine basins below wave base and out of the influence of bottom currents, and where bioturbation is minimal due to low oxygen levels. Thus, time windows are opened in which the sedimentary, hydrological, climatic and biological evolution in a specific basin can be reconstructed. An example for a well-preserved marine record is the Miocene to Holocene sediment stack of the Mediterranean basin, which reflects orbital cyclicity (Fischer *et al.*, 2009) and where specific intervals can be interpreted with a millennial- to centennial-scale resolution (Rodrigo-Gámiz *et al.*, 2014). For lake sediments, a good example is the 600 kyr Pleistocene to Holocene record drilled and interpreted in Lake Van, Turkey (Stockhecke *et al.*, 2014).

When it comes to shallow sedimentary environments, gaps in the depositional record are the norm. If such an environment lies on land above sea or lake level, no sediment at all may be deposited and the corresponding time interval leaves no trace, or else erosion of the terrestrial sediment may partly or completely destroy the record. In the subaqueous zone, the sediment may be scoured and reworked by currents or overturned by biogenic activity, and the final record does not anymore reflect the original depositional history. This incompleteness of the depositional record has been recognized and discussed by many authors. For example, Barrell (1917) wrote that most stratigraphic records are characterized by diastems at bedding planes. In the case of tide-influenced shallow siliciclastic sediments, Reineck (1960) estimated that only 10^{-4} to 10^{-5} of geologic time is actually preserved in the sedimentary column. Ager (1980) coined the phrase that the 'stratigraphical record is a lot of holes tied together with sediment'. Sadler (1981) wrote that 'sedimentation is an essentially discontinuous process' and compared sediment accumulation rates with the time span over which this accumulation took place. Dott (1983) discussed the episodic nature of sedimentation, and Miall (2014) wrote about the 'emptiness of the stratigraphic record'. Different types of condensation processes on shallow platforms were illustrated by Gómez & Fernández-López (1994). Kemp (2012) modelled the distribution of hiatuses in the stratigraphic record, and Hill *et al.* (2012) focused on the preservation potential of shallow-water carbonate sediments exposed to high-frequency, orbitally controlled sea-level fluctuations. Good examples of complex discontinuity surfaces have been published by Sattler *et al.* (2005) and Rameil *et al.* (2012) from the Cretaceous carbonate platforms in Oman, Waite *et al.* (2013) from the Late Jurassic in the Swiss Jura Mountains, and Brlek *et al.* (2014) from the Cretaceous-Palaeogene boundary in Croatia.

Figure 1 schematically illustrates how different processes on a shallow carbonate platform lead to the sedimentary record. After some marine, subtidal sediment accumulation, a drop in relative sea-level leads to erosion that removes part of the originally deposited sediment. The sediment bed is then cemented in the freshwater vadose zone, and pedogenesis sets in. During subaerial exposure, some biological and mechanical erosion goes on but is counterbalanced by deposition, so that a thin bed of soil with limestone clasts is formed. Rapid marine transgression then picks up some of these clasts and incorporates them at the base of a bed that shows cross-bedding of an inter- to subtidal dune. Incipient hardgrounds partly consolidate the sediment, which is reworked by currents and redeposited in a marly matrix. A phase of subtidal carbonate deposition sets in, incorporating some reworked mud-clasts. The centre of the bed is characterized by intense bioturbation, implying a lowered sedimentation rate (i.e. condensation of time in a thin stratigraphic interval). The top of the bed is a hardground that is bored and encrusted by microorganisms, and mineralized. There is a balance between biological erosion and accumulation. This hypothetical history demonstrates how unequally time is distributed in the depositional record, but also that there is a potential to decipher some of this history by careful analysis.

The purpose of the present paper is to document and interpret the incompleteness of the depositional record of a shallow, subtropical carbonate platform of Berriasian (Early Cretaceous) age. It will be attempted not only to estimate the amount of time lost in non-deposition, erosion, and condensation but also to seek for the causes of the gaps and the processes that led to the final rock record as observed today in the studied outcrops. Furthermore, sedimentation rates will be discussed. The goal is to evaluate to what extent a fragmentary depositional record can be used to interpret the full history of a carbonate platform.

GEOGRAPHIC, PALAEOGEOGRAPHIC AND STRATIGRAPHIC SETTING

The present study is based on the analysis of 35 sections in the Swiss and French Jura Mountains that have been logged at centimetre-scale (Strasser, 1988; Pasquier, 1995; Hillgärtner, 1999; Tresch, 2007) (Fig. 2). They represent the shallow-water carbonate realm during the Berriasian. The Jura platform was situated between the Paris Basin and the Helvetic Shelf, at a palaeolatitude of 27 to 28°N (Dercourt *et al.*, 2000) (Fig. 3). The climate was subtropical and the potential for organic carbonate production was high. Episodically, quartz grains, clays, and nutrients were washed in from the emergent Hercynian massifs. The platform was block-faulted and structured into highs and shallow depressions, creating a complex substrate morphology (Allenbach, 2002).

The studied interval covers three formations (Fig. 4): the upper part of the Twannbach Formation (Vouglans Member; Bernier, 1984), the Goldberg Formation (Häfeli,

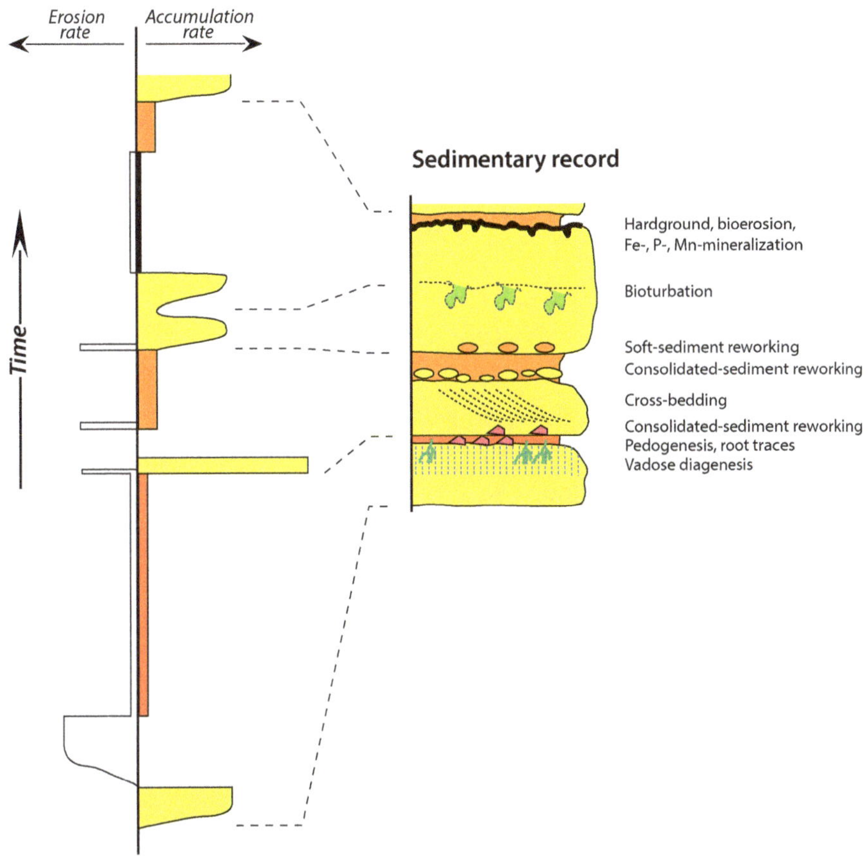

Fig. 1. Hypothetical sedimentary record on a shallow, carbonate-dominated platform, displaying typical features such as palaeosol, reworked lithoclasts, bioturbation and hardground. Hypothetical sediment accumulation and erosion rates are plotted along a time axis that represents a few thousand years. Note that erosion and sedimentation may occur simultaneously during soil and hardground formation, albeit at low rates. Erosion includes mechanical, chemical and biological processes. Sediment accumulation includes in situ carbonate production, import of particles by currents or wind, and biogenic encrustation on the hardground surface. For discussion refer to text.

1966) and the Pierre-Châtel Formation (Steinhauser & Lombard, 1969). Biostratigraphic dating is given by ammonites and benthic foraminifera (Clavel *et al.*, 1986), and by charophytes and charophyte-ostracode assemblages (Mojon, 2002). Within the frame of this biostratigraphy, the large-scale sequence boundaries recognized in the studied outcrops can be correlated with those of the sequence-chronostratigraphic chart of Hardenbol *et al.* (1998).

A sequence-stratigraphic and cyclostratigraphic interpretation has been proposed by Pasquier & Strasser (1997), Strasser & Hillgärtner (1998), and Strasser *et al.* (2004). According to the chart of Hardenbol *et al.* (1998), the time span between sequence boundaries Be 1 and Be 4 is 3·2 Myr (Fig. 4). In the same interval, 32 small-scale depositional sequences are counted (for definition of such sequences see below), suggesting that one small-scale sequence corresponds to the short eccentricity cycle of 100 kyr (Berger *et al.*, 1989). The fact that the ages of sequence boundaries Be 2 and Be 3 do not correspond to our cyclostratigraphic interpretation can be explained by the physical expression of prominent large-scale boundaries that does not necessarily happen at the same time on the platform and in the basin on which the chart of Hardenbol *et al.* (1998) is based (Strasser *et al.*, 2000). Between sequence boundary Be 4 and the base of the Pierre-Châtel Formation, sedimentation on the shallow platform was much reduced but continuous in the Vocontian Basin, where the cyclostratigraphic analysis of hemipelagic limestone-marl alternations suggests that this interval lasted about 900 kyr (Strasser *et al.*, 2004). The Pierre-Châtel Formation itself is composed of nine small-scale sequences where it is fully developed, but non-deposition and/or erosion at sequence boundary Be 5 locally cut off the topmost sequences (Pasquier, 1995; Strasser *et al.*, 2004). The total time interval between Be 4 and Be 5 would thus have lasted about 1·8 Myr, which is close to the 1·7 Myr mentioned in the chart of Hardenbol

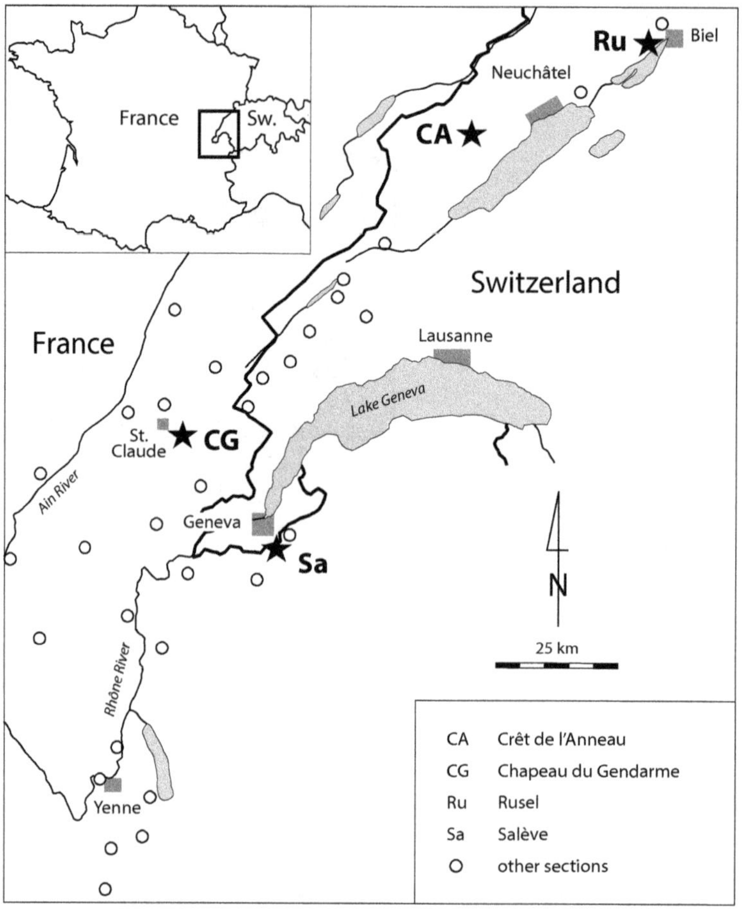

Fig. 2. Location of studied sections in the Swiss and French Jura Mountains. Stars: sections mentioned in this paper; empty circles: other sections of the Goldberg and/or Pierre-Châtel formations.

et al. (1998). It has to be noted that the ages of the sequence boundaries published by Hardenbol *et al.* (1998) derive from interpolation between the ages of the lower and upper boundaries of the Berriasian stage as given by Gradstein *et al.* (1995: 144·2 ± 2·6 and 137 ± 2·2 Ma, respectively). The newest chronostratigraphic chart (www.stratigraphy.org, 2015/01) indicates 145·0 ± 4·0 and 139·8 ± 3·0 Ma for these limits, thus reducing the duration of the Berriasian stage from ≈7·2 to ≈5·2 Myr (but increasing the error margins). Nevertheless, because the chart of Hardenbol *et al.* (1998) also compiles the biostratigraphy by which the studied formations are calibrated, the values of this older chart are retained.

METHODS

The sections (Fig. 2) have been logged at centimetre-scale and densely sampled. Thin-sections were prepared for the rock samples; marls were washed and the residue picked for microfossils. Under the optical microscope or the binocular, microfacies have been analysed using the Dunham (1962) classification and a semi-quantitative estimation of the abundance of rock constituents. Special attention has been paid to sedimentary structures and to discontinuity surfaces (Clari *et al.*, 1995; Hillgärtner, 1998). The sum of this sedimentological information was then used to interpret the depositional environments.

For the stable-isotope analyses (O^{18}/O^{16} and C^{13}/C^{12}), 5 to 10 mg of powdered bulk rock were reacted with 100% H_3PO_4 at 75°C and analysed in a Finnigan Mat 252 mass spectrometer at the University of Erlangen. The ratios are reported in ‰ relative to the Vienna Pee Dee Belemnite (PDB) standard. Mean standard deviation was less than 0·1‰ for $\delta^{13}C$ and $\delta^{18}O$. If possible, facies rich in micritic matrix and poor in cements were chosen in order to obtain an average signal with minimal late-diagenetic influence.

For the sequence-stratigraphic interpretation of the facies evolution, the nomenclature of Vail *et al.* (1991) is

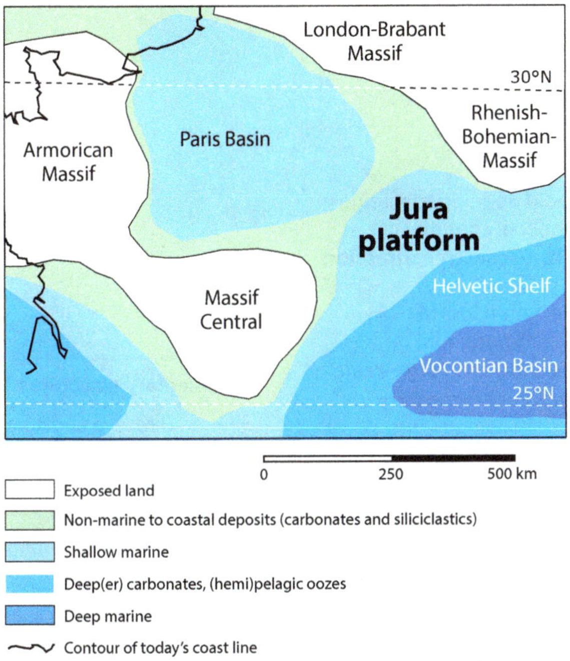

Fig. 3. Simplified palaeogeographic map to situate the Jura platform in Berriasian times (based on Dercourt *et al.*, 2000).

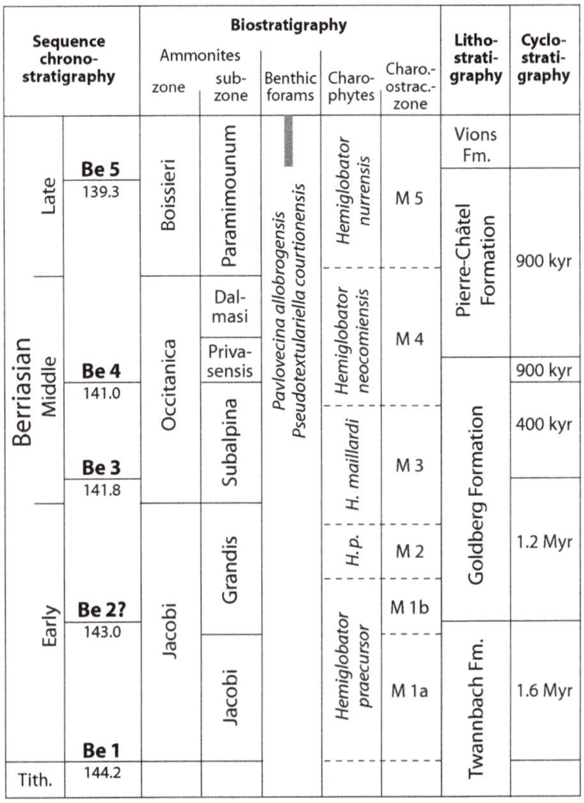

Sequence chrono-stratigraphy
Biostratigraphy
Ammonites
zone
sub-zone
Benthic forams
Charo-phytes
Charo.-ostrac.-zone
Litho-strati-graphy
Cyclo-strati-graphy
Berriasian
Late
Middle
Early
Tith.
Be 5
139.3
Be 4
141.0
Be 3
141.8
Be 2?
143.0
Be 1
144.2
Boissieri
Occitanica
Jacobi
Paramimounum
Dal-masi
Priva-sensis
Subalpina
Grandis
Jacobi
Pavlovecina allobrogensis
Pseudotextulariella courtionensis
Hemiglobator nurrensis
Hemiglobator neocomiensis
H. maillardi
H.p.
Hemiglobator praecursor
M 5
M 4
M 3
M 2
M 1b
M 1a
Vions Fm.
Pierre-Châtel Formation
Goldberg Formation
Twannbach Fm.
900 kyr
900 kyr
400 kyr
1.2 Myr
1.6 Myr

Fig. 4. Stratigraphy of the early, middle and the lower part of the late Berriasian. Sequence chronostratigraphy (ages of sequence boundaries Be 1 to Be 5 in Ma) and ammonite zones and subzones according to Hardenbol *et al.* (1998). The two benthic foraminifera of stratigraphic value are coeval to the Paramimounum subzone (Clavel *et al.*, 1986). Charophyte biostratigraphy and the zones based on charophyte-ostracode assemblages are from Mojon (2002). The Goldberg and Pierre-Châtel formations are tied to the chronostratigraphy via the available fossil content (Clavel *et al.*, 1986; Mojon, 2002). Cyclostratigraphic estimation of time according to Strasser & Hillgärtner (1998) and Strasser *et al.* (2004). Note that the time intervals between sequence boundaries are tied to the lithostratigrahy and not to the bio- and chronostratigraphy (e.g. the Privasensis subzone lasted only 500 kyr; the inferred 900-kyr time interval above sequence boundary Be 4 corresponds to a thin rock interval due to reduced sediment accumulation rate). For more discussion, refer to text. Tith.: Tithonian.

used. Although large-scale systems tracts cannot be defined in the studied outcrops, the sequence-stratigraphic principles can be applied even to the smallest building blocks of the stratigraphic record if their formation was controlled by relative sea-level changes (Mitchum & Van Wagoner, 1991). Vertical facies changes define deepening-shallowing depositional sequences, which are hierarchically stacked (example in Fig. 5). Elementary sequences are defined as the smallest units where facies evolution indicates a cycle of environmental change, including sea-level change (Strasser *et al.*, 1999; the term 'sequence' is used to describe the sedimentary record that formed during a 'cycle' of processes). In some cases, there is no facies evolution discernable within a bed but marls or discontinuity surfaces delimiting the bed suggest an environmental change (Strasser & Hillgärtner, 1998). Commonly, two to seven elementary sequences compose a small-scale sequence, which generally displays a deepening then shallowing trend and exhibits the relatively shallowest facies at its boundaries. For example, birdseye structures, microbial laminites or penecontemporaneous dolomitization suggest tidal-flat environments, lithoclasts imply erosion of previously exposed and consolidated sediment, and calcrete or desiccation cracks indicate subaerial exposure. Four small-scale sequences make up a medium-scale sequence, which again displays a general deepening-shallowing trend of facies evolution and the relatively shallowest facies at its boundaries (Fig. 5). Furthermore, the elementary sequences are typically thinner around the small-scale and medium-scale sequence boundaries, which suggests reduced accommodation space. Thick elementary sequences in the central parts of these sequences imply higher accommodation space. The medium-scale sequences then group into large-scale sequences, which can be compared with the sequences of Hardenbol *et al.* (1998) (Fig. 4). In some cases, no unique sequence boundaries or maximum-flooding surfaces can be identified; there, sequence-boundary zones covering intervals of lowest accommodation and maxi-

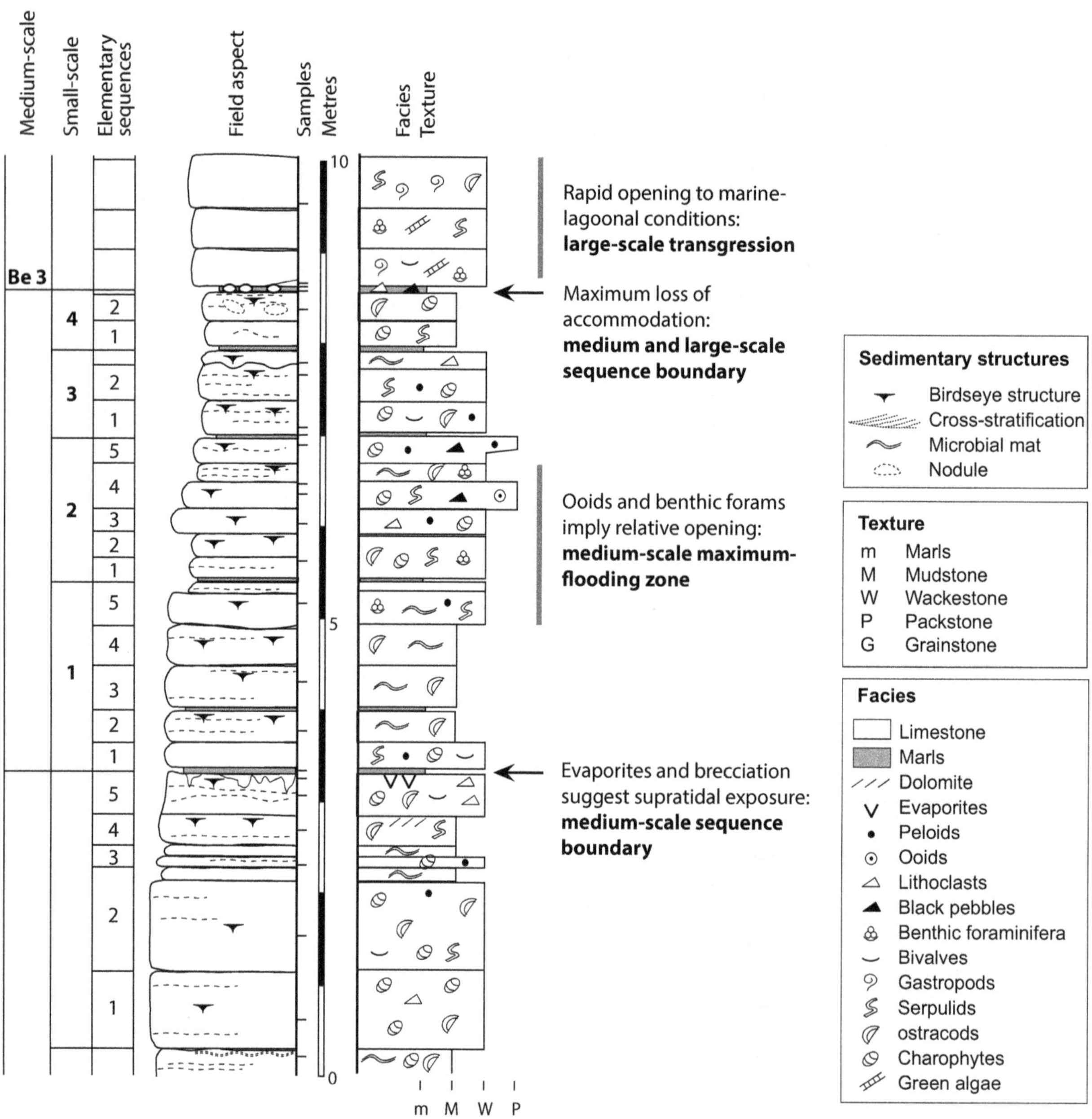

Fig. 5. Part of the Goldberg Formation in the Salève section (see also Fig. 6). Individual beds (separated by thin marly layers) are interpreted as elementary sequences. These are hierarchically stacked into small-scale and medium-scale sequences. The upper boundary of the medium-scale sequence is at the same time large-scale sequence boundary Be 3 (Fig. 4). Note the general thinning-up trend of beds up to Be 3, which indicates a general loss of accommodation space and corresponds to the late highstand of the previous large-scale sequence. Small-scale sequences 3 and 4 contain less than five elementary sequences, and the lithoclasts at sequence boundary Be 3 imply early cementation of sediment and subsequent reworking. Note also that it is useful to define a maximum-flooding zone because there is no unique surface that would have formed during maximum accommodation gain. For more discussion, refer to text (modified from Strasser & Hillgärtner, 1998).

mum-flooding zones with evidence for relatively deep or open-marine water are defined (Fig. 5; Strasser *et al.*, 1999).

As explained above, it is suggested that one small-scale sequence lasted about 100 kyr and thus formed in tune with the short eccentricity cycle of the Earth's orbit. The medium-scale sequences, commonly holding four small-scale sequences, then would represent the long eccentricity cycle of 400 kyr. Where five elementary sequences compose one small-scale sequence, they probably correspond to the precession cycle of about 20 kyr (Berger *et al.*, 1989). However, autocyclical processes could also have led

Fig. 6. Outcrop photograph of part of the Salève section, including sequence boundaries Be 3 and Be 4 (compare with Figs 4 and 5). Small-scale sequences below Be 3 are numbered as in Fig. 5. Note the slight dip of the stratification towards the right at the base of the Pierre-Châtel Formation (top of picture), which implies lateral migration of a large subtidal dune.

to the formation of such sequences, and their unequivocal attribution to an orbital cycle in many cases is not possible (Strasser, 1991; Hill *et al.*, 2012). Even if some question marks remain, the cyclostratigraphic interpretation nevertheless offers a high-resolution time frame within which the sedimentological processes can be discussed.

DEPOSITIONAL ENVIRONMENTS

The predominant depositional environments on the shallow Jura platform have been reconstructed based on detailed microfacies analyses (Strasser, 1988; Pasquier, 1995; Hillgärtner, 1999; Tresch, 2007). These environments include:

- Emergent land characterized by calcrete crusts, root traces and circumgranular cracks. Black pebbles, commonly reworked in inter- or subtidal deposits, imply impregnation by organic matter that partly resulted from forest fires (Strasser & Davaud, 1983).
- Coastal lakes rich in charophytes (stems and oogons) and ostracods.
- Sabkhas as suggested by gypsum and anhydrite pseudomorphs as well as by brecciation that probably resulted from collapse after dissolution of evaporites.
- Tidal flats represented by mudstones bearing birdseye structures, desiccation polygons, and microbial lamination. Some of these stromatolitic laminae are dolomitized.
- Beaches composed of ooids, bioclasts and lithoclasts, and characterized by parallel lamination and keystone vugs. Locally, beachrock blocks are found.
- Shallow lagoons with peloids and oncoids. Abundance and diversity of benthic fauna and flora (foraminifera, gastropods, bivalves, brachiopods, echinoderms, serpulids, ostracods, dasycladalean algae) are variable, suggesting that some lagoons were restricted in terms of oxygenation and water energy, whereas others had open-marine conditions.

- Ooid and bioclastic shoals with unidirectional foresets implying lateral migration, or with herringbone cross-stratification created by tidal currents.
- Coral patch reefs.

Grainstones imply high-energy conditions under the influence of currents or waves. Packstones, wackestones and mudstones formed in quiet water below wave base. Marls are mostly found in millimetre- to centimetre-thick layers between the limestone beds but may also form metre-thick intervals in the Goldberg Formation. Marls also constitute the matrix of conglomerates or breccias that occur locally. Traces of bioturbation are common in all facies and mostly represented by *Thalassinoides*. In mudstones, millimetre-wide circular tubes are attributed to worm burrows.

Estimation of water depth based on facies is difficult (Immenhauser, 2009; Purkis *et al.*, 2015). Indicators of intertidal conditions such as birdseye structures, desiccation cracks, or keystone vugs can be taken as zero depth. However, in the subtidal realm, wave base or tidal currents can be very shallow in protected positions behind a barrier island but deep when facing the open ocean; thus, they can give only an appreciation of relative water depth. Stenohaline organisms such as brachiopods and echinoderms inform only about the salinity of the sea water but not about water depth. Corals on the shallow platform are stenohaline and have to live in the photic zone, the depth of which depends on water transparency. Based on reconstructions of high-frequency sea-level changes in the Berriasian, it is estimated that water depth was no more than a few metres for the facies encountered here (Strasser *et al.*, 2004).

On the Jura platform, these different environments were juxtaposed and shifted through time (Strasser *et al.*, 2004). Outcrop conditions do not allow access to the platform-to-slope transition to evaluate how much of the sediment produced on the platform was shed onto the slope. The limestone-marl alternations in the slope and basin sections sampled by Pasquier (1995) are composed of hemipelagic facies and do not contain typical platform-derived grains. However, Pasquier & Strasser (1997) discussed the possibility that carbonate mud was exported from the platform to the basin by highstand shedding (Schlager, 1991; Milliman *et al.*, 1993).

HIATAL SURFACES AND CONDENSED INTERVALS

In sequence-stratigraphic terminology, a hiatus is an interval of geological time that is not recorded in strata, and the corresponding surface along which this time is missing is an unconformity (Mitchum, 1977; Catuneanu *et al.*, 2009). For this study, however, where also small-scale and short-term interruptions of sedimentation are discussed, the term 'hiatal surface' is preferred and defined as a discontinuity surface along which sedimentation has been interrupted, or which results from mechanical or chemical erosion of previously deposited sediment. Facies below and above the hiatal surface may be the same or different, and the time not recorded at these surfaces can vary from hours to millions of years.

In sequence stratigraphy, condensed sections are defined as thin stratigraphic intervals characterized by very low sediment accumulation rates that are mainly found in deep-water settings (Loutit *et al.*, 1988; Catuneanu *et al.*, 2009). Here, the term 'condensed interval' is used to describe a thin sediment body on the platform, in which much geologic time is represented. It may be composed of a single facies or be a composite of several thin sediment layers separated by closely spaced hiatal surfaces.

In the studied outcrops, beds are commonly separated by marly layers that are a few millimetres to a few centimetres thick. The beds can be followed over hundreds of metres where outcrop conditions allow (for example in the cliffs at Salève; Fig. 6). At the base of the beds the limestones may evolve gradually from the marls or else they are separated from them by sharp surfaces. The tops of the beds may pass into marls or again represent sharp surfaces. However, surfaces occur also within the beds where they are made visible through abrupt facies changes. These latter surfaces are of limited lateral extent. Table 1 summarizes the different types of surface recognized and their interpretation.

Simple surfaces of limited lateral extent

In oolitic and bioclastic grainstones, bed-parallel or oblique lamination can be observed that is created through differences in grain size. Erosion or cracking may follow softer lithologies and create visible surfaces (Fig. 7A). These structures are interpreted as reactivation surfaces having formed on laterally migrating dunes (Gonzalez & Eberli, 1997). Tidal influence is suggested by thin clay seams that occur locally and are interpreted as flaser bedding, and by bidirectionally dipping laminae (herringbone cross-bedding).

In limestone beds with a micritic matrix, irregular surfaces may occur, again rendered visible by facies contrasts. The best example is found in the Salève section (Fig. 7B) where an undulating surface separates bioturbated wackestone from dark-coloured floatstone with irregularly rounded clasts of the underlying facies. This feature is interpreted as the floor of a tidal channel. The relief of the surface was probably created by bioturbation and by

Table 1. Comparison of short-term and long-term processes, associated sedimentary features, potentially produced hiatal surfaces and their lateral extent, and estimations of geological time lost in these surfaces

	Processes	Products	Hiatal surfaces	Lateral extent	Time lost in surfaces
Short-term	Waves and weak currents	Ripples	Reactivation surfaces	Tens of centimetres	Hours
	Tides	Tidal laminites, microbial mats, desiccation polygons	Tidal-flat laminae	Tens of centimetres to tens of metres	Hours to days
	Strong tidal currents	Ooid and bioclastic shoals, grainstones, flaser bedding, herringbone cross-bedding, tidal channels	Reactivation surfaces, scour surfaces	Metres to tens of metres	Hours to tens of years
	Storm waves, storm-induced currents	Tempestites, HCS, SCS, rip-up clasts, flat-pebble conglomerates	Reactivation surfaces, scour surfaces	Metres to tens of metres	Hours to hundreds of years
Long-term	Sediment bypass, sediment starvation	Softgrounds, firmgrounds, hardgrounds, bioturbation, biogenic encrustation, mineralization, reworked clasts	Firmground and hardground surfaces, maximum-flooding surfaces	Metres to tens of kilometres	Years to millions of years
	Transgression	Lag deposits, rip-up clasts, deepening-up facies evolution	Transgressive surfaces, ravinement surfaces	Tens of metres to tens of kilometres	Hundreds to thousands of years
	Subaerial exposure	Karst, pedogenesis, soils, brecciation, calcrete, root traces, coal, freshwater diagenesis	Surfaces of maximum regression, sequence boundaries	Tens of metres to tens of kilometres	Hundreds to millions of years

tidal current scouring, which ripped up the cohesive sediment to form the soft pebbles. When current activity stopped, the bottom waters became stagnant to permit the partial preservation of organic matter, explaining the dark colouring. The channel fill then occurred in oxygenated conditions, and erosion of the channel flanks continued furnishing soft pebbles.

Around sequence boundary Be 1 in the Salève section, some surfaces marked by ripples and micrite-filled ripple troughs spread over a few metres but then disappear into thinly laminated, dolomitized beds (Fig. 7C and D). This configuration is interpreted as a tidal-flat environment with microbial mats and localized shallow depressions where currents and waves formed ripple marks (Bover-Arnal & Strasser, 2013). Thin marly flasers are also observed within these beds, again indicating tidal influence (Fig. 7D). Microbial lamination and desiccation into polygons are clearly visible below sequence boundary Be 2 at Salève (Fig. 7E). Chips break off along the lamina planes and furnish flat pebbles when reworked. The lamination is interpreted as being due to daily cycles of microbial growth on tidal flats (Hardie & Ginsburg, 1977).

Broken and tilted oolitic grainstone slabs occur at sequence boundary Be 2 at Salève (Fig. 7F). They are overlain by a badly sorted conglomerate, the components of which include the grainstone facies from the slabs below but also black pebbles and wacke- to packstones (Bover-Arnal & Strasser, 2013). Keystone vugs and parallel lamination in the oolites point to a beach environment, and the formation of the slabs is explained by early cementation to form beachrock, which was then broken by strong wave action. Locally, the broken oolite is covered by calcrete, implying exposure in subaerial conditions before the deposition of the conglomerate.

In all cases described above, the lateral extent of the surfaces is limited to a few tens of centimetres to a few tens of metres. The processes that created these surfaces were of rather short duration: a few hours or days in the case of reactivation surfaces and laminites created by tidal activity, a few months to a few tens of years in the case of an abandoned tidal channel, and a few hundreds of years to cement carbonate beach sand and then break it into slabs (Halley & Harris, 1979; Strasser & Davaud, 1986). It has to be remembered, however, that the preserved sediments are snap shots of the geological history. For example, a tidal dune can migrate back and forth for hundreds of years before the current regime changes and the bedform is stabilized and recorded.

Correlation of sections

Outcrop conditions in the Jura Mountains do not allow walking out beds and surfaces over large distances. Therefore,

Fig. 7. Outcrop photographs of short-term hiatal surfaces. (A) Reactivation surfaces in the subtidal dune at the base of the Pierre-Châtel Formation, dipping to the right (see also Fig. 6). These surfaces are enhanced by cracks formed along grain-size contrasts. Pen is 14 cm long. (B) Bottom of tidal channel in the Goldberg Formation, Salève section. The lower part of the bed is strongly bioturbated (vertical traces filled with calcite cement). The bottom surface of the channel is irregular and wavy (to the right of pen). The lower part of the channel fill is dark-coloured and contains light-coloured soft pebbles reworked from the underlying sediment. Brownish-coloured soft pebbles occur in the upper part of the channel fill. The top of the bed appears in the upper left corner of the photograph. Visible part of pen is 6 cm long. (C) Lateral pinch-out of beds at sequence boundary Be 1 in the Salève section. Some of the surfaces are laterally continuous, others disappear. Arrow points to ripple trough shown in D. Hammer is 33 cm long. (D) Detail of C, showing a ripple trough filled with brownish micrite. Thin marly seams (flasers) are visible below and above (arrows) but are not continuous laterally. (E) Desiccation polygons below sequence boundary Be 2 in the Salève section. Note chips breaking off along tidal laminae. Visible part of red pen is 3 cm long. (F) Broken beachrock slabs overlain by conglomerate at sequence boundary Be 2, Salève section. Note the dark-grey or black pebbles. Visible part of pen is 10 cm long.

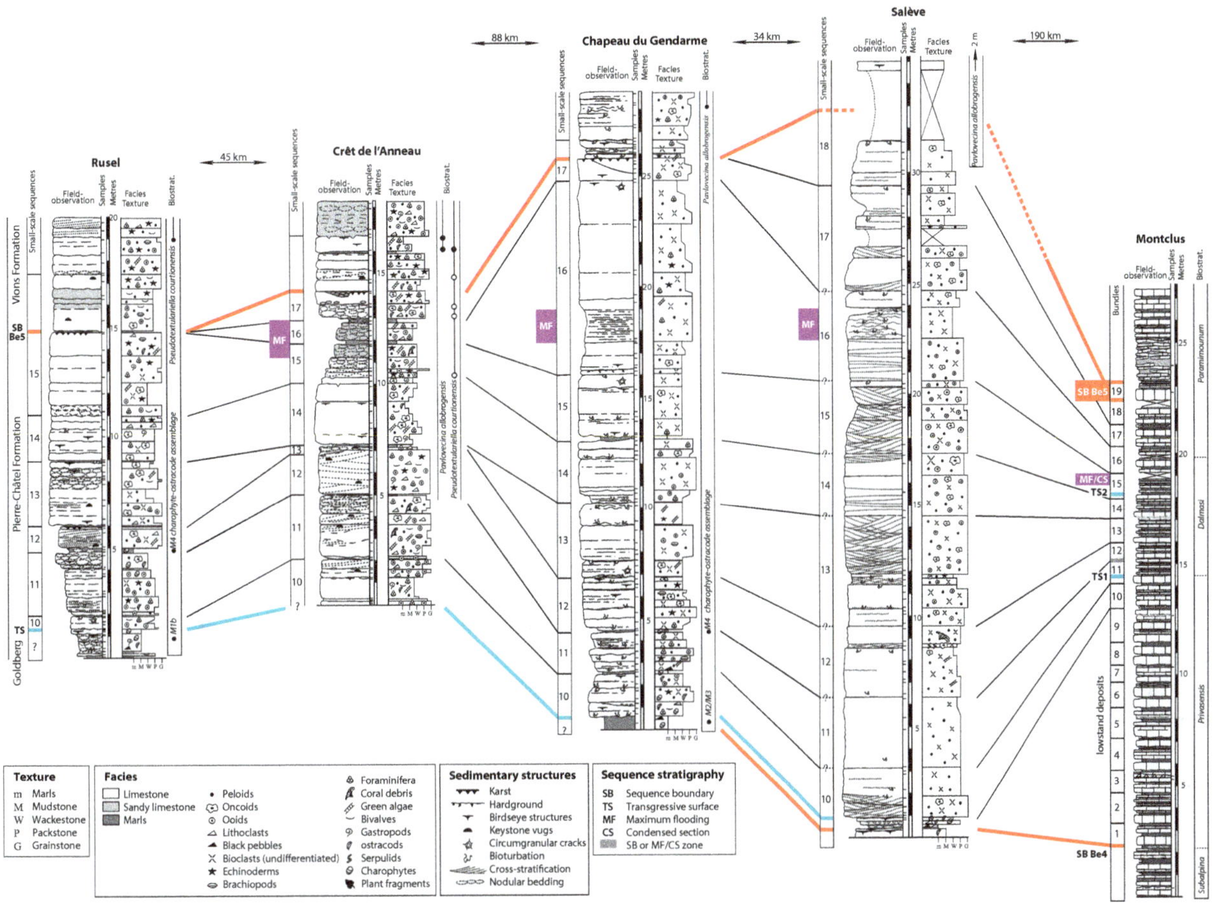

Fig. 8. Sequence- and cyclostratigraphic correlation of four platform sections (indicated in Fig. 2) and the hemipelagic Montclus section in the Vocontian Basin where ammonite biostratigraphy is available. For discussion refer to text.

measured sections have first been interpreted as described in the Methods chapter and then correlated. Figure 8 shows such a correlation between four sections representing the shallow-water platform carbonates of the Pierre-Châtel Formation. A correlation with the hemipelagic facies of the Montclus section in the Vocontian Basin (southern France) is added, where the ammonite biostratigraphy is well established and allows the age of the shallow-water sections to be constrained (Fig. 4; Strasser *et al.*, 2004). The limestone-marl couplets at Montclus are interpreted to have formed in tune with the 20 kyr orbital precession cycle. These couplets commonly group into bundles of five, indicative of the 100 kyr short eccentricity cycle. The bundles correspond to the small-scale sequences identified in the platform sections. Thus, major sequence boundaries identified and dated by Hardenbol *et al.* (1998) in many European basins can be identified and correlated from deep-water to shallow-water sections.

In Fig. 8, sequence boundaries (SB) Be 4 and Be 5 are indicated. The base of the Pierre-Châtel Formation corresponds to the transgressive surface (TS) above SB Be 4. At Montclus, however, transgressive surfaces (TS 1 and TS 2 in Fig. 8) are suggested by rapid changes from more limestone-dominated couplets to more marl-dominated ones, implying that the physical expression of transgression did not occur at the same time in the hemipelagic basin as on the platform (Pasquier & Strasser, 1997). At Monclus, the maximum flooding is interpreted to be represented by the most marly interval of this outcrop, which corresponds to marly and strongly bioturbated facies in the platform sections. Sequence boundary Be 5 at Montclus can be placed either at the base of a thick limestone bed (bundle 19) or else at the base of the slumps (23·3 to 25·2 m); consequently, a sequence-boundary zone is proposed that covers this interval of uncertainty. The nine hemipelagic bundles at Montclus (representing 900 kyr) above SB Be 4 are interpreted to be condensed into half a metre of sediment at Salève. At Montclus, 1·8 to 1·9 Myr are recorded between SB Be 4 and SB Be 5, whereas at Chapeau du Gendarme and Crêt de l'Anneau the record of this large-scale sequence ends with small-scale sequence 17, and at Rusel with number

15. At Salève, a gap in the outcrop does not allow the placement of SB Be 5.

In the following, four sequence-boundary zones and one maximum-flooding zone will be discussed in detail.

Complex intervals of wide lateral extent

The irregular beds with lateral pinch-outs in the lower part of the Salève section (Fig. 7C) can be followed laterally over a few tens of metres. They show a dramatic facies change from high-energy, subtidal, marine grainstones in the massive bed below passing upwards to inter- to supratidal dolomitized microbial laminites including black pebbles and wood fragments, followed by lacustrine wackestones (Bover-Arnal & Strasser, 2013). The charophyte-ostracode assemblage M 1a suggest an early Berriasian age, and the sequence- and cyclostratigraphic interpretation of Strasser & Hillgärtner (1998) suggests that this interval corresponds to sequence boundary Be 1 (Fig. 4). Unfortunately, outcrop conditions in the other studied sections do not allow this boundary to be confidently correlated over large distances. However, even at the decametre scale, the complexity of this sequence-boundary zone is evident, with some surfaces and beds being laterally continuous, others having a limited extent.

The boundary between the Goldberg and the Pierre-Châtel formations is very well-developed and can be followed over the entire study area. In the Salève section, this boundary corresponds to a complex interval comprising peritidal facies with microbial laminites and desiccation polygons followed by broken beds of various marine, peritidal and lacustrine facies (Fig. 9A; Bover-Arnal & Strasser, 2013). Some clasts are blackened, suggesting supratidal conditions (Strasser & Davaud, 1983). In other sections, similar breaking and reworking of pre-existing limestones is found (Fig. 9B). This interval is attributed to sequence boundary Be 4, well dated by ammonites (Fig. 4; Clavel *et al.*, 1986). The transgression leading to the open-marine, high-energy facies at the base of the Pierre-Châtel Formation occurs in two pulses in the Salève section (Fig. 9A), whereas at Rusel, it is more gradual (Fig. 9B).

At Crêt de l'Anneau, the top of small-scale sequence 12 is developed as an undulating, reddish surface bearing dinosaur tracks (Pasquier, 1995). It is covered by a bioturbated wacke- to packstone that pinches out laterally (Figs 8 and 9C, D, E). Assuming that the correlation presented in Fig. 8 is correct, this same surface terminating sequence 12 shows karst features due to subaerial exposure at Rusel (Fig. 9F) and birdseye structures of a tidal-flat environment at Chapeau du Gendarme. At Salève, its precise correlation is uncertain as the whole section is composed of subtidal facies (Fig. 8). The wedge of bioturbated limestone at Crêt de l'Anneau develops laterally into a full-fledged small-scale sequence (number 13 in Fig. 8). In parts of the outcrop at Crêt de l'Anneau, only a thin marly layer separates small-scale sequences 12 and 14; sequence 13 is missing completely (Fig. 9 E). At Rusel, the top of sequence 12 not only contains karst features and yellow-red staining but also borings (Fig. 9F), indicating that the surface was hardened before marine flooding (Tresch, 2007).

The best candidate for the maximum-flooding zone of large-scale sequence Be 4 is the strongly bioturbated interval in small-scale sequence 16 that is commonly expressed by a softer lithology in the field (Figs 8 and 10A, B). Concentration of bioturbation is here interpreted as being a sign of condensation: at low sediment accumulation rates, the same sediment is reworked by many generations of burrowing organisms. The low accumulation rate may be the result of low carbonate production in deeper and/or more turbid water where photosynthesis of the carbonate-producing organisms (e.g. green algae) is reduced, and/or of removal of sediment by currents. The increased marl content reflects water depth below wave base where the clay minerals were not winnowed away (Strasser & Hillgärtner, 1998). At Crêt de l'Anneau, deepening of the environment must have occurred rapidly because the top of small-scale sequence 14 still shows karst features and the overlying bed pinches out laterally, probably a sign of channelling (Fig. 10B). At Chapeau du Gendarme, the top of small-scale sequence 15 (Fig. 10C) is interpreted as a small-scale sequence boundary because the bed below contains circumgranular cracks implying pedogenesis (Fig. 8). The perforated hardground capping the thin bed above formed through sediment starvation, announcing the maximum flooding in the overlying sequence 16. In the hemipelagic section of Montclus, the most marly part is found in sequence 15. This discrepancy can be explained by the fact that not all places on the platform and in the basin reacted in the same way and at the same time to a rapid sea-level rise (Strasser *et al.*, 2000). The position of this maximum-flooding zone in the Be 4 sequence is strongly asymmetric (Fig. 8). The same asymmetry is observed in other European basins (Hardenbol *et al.*, 1998), suggesting that the associated long-term sea-level change also was asymmetric. At Rusel, this maximum-flooding interval is not recorded. Nevertheless, the relatively thick small-scale sequence 15 there (Fig. 8) indicates that accommodation was high, which might have been due to an increased rate of sea-level rise (but see also the discussion on synsedimentary tectonics below).

Fig. 9. Outcrop photographs of sequence boundaries and transgressive surfaces with a wide lateral extent (for numbering of sequences and correlations see Fig. 8). (A) Sequence-boundary zone Be 4 at Salève, at the top of the Goldberg Formation. Note bed limit above hammer to the left, while to the right of photograph the beds are completely dismantled. A laterally continuous bed delimited by two sharp transgressive surfaces at the base of the Pierre-Châtel Formation covers the sequence-boundary zone. Hammer is 33 cm long. (B) Sequence-boundary zone Be 4 at Rusel. The marly Goldberg Formation below and the yellowish, massive Pierre-Châtel Formation above are separated by dismantled limestone beds, and no clear transgressive surface is developed. (C) Sharp reddish surface at the top of small-scale sequence 12 in the Crêt de l'Anneau section. Irregular depressions in the surface are possibly due to dinosaur tracks. Bioturbated wacke- to packstones are onlapping on this surface from the right (arrow at contact). The bioturbated facies constitutes small-scale sequence 13. (D) Detail of bioturbated facies (upper half of photograph) overlying the sharp surface. Pen is 14 cm long. (E) Only a thin marly layer separates the sharp surface (interpreted as sequence boundary) at the top of small-scale sequence 12 and the transgressive surface forming the base of small-scale sequence 14. Sequence 13 is not recorded. (F) Red-stained surface on top of small-scale sequence 12 in the Rusel section. Circular structures are borings (probably *Gastrochaenolites*), suggesting marine flooding after subaerial exposure.

Fig. 10. Outcrop photographs of maxiumum-flooding zone and surface, and sequence boundaries with a wide lateral extent (for numbering of sequences and correlations see Fig. 8). (A) Nodular, marly, strongly bioturbated interval in the Crêt de l'Anneau section, composing the upper half of small-scale sequence 15 and sequence 16. The limit between these two sequences is not well defined but suggested by a softer, more marly passage that is laterally consistent (dashed white line). This interval is interpreted as the maximum-flooding zone of large-scale sequence Be 4. (B) Irregular, reddish surface at top of small-scale sequence 14 in the Crêt de l'Anneau section. Note lateral pinch-out of the overlying bed (arrow). Hammer is 33 cm long. (C) Sharp, bioperforated hardground surface (at the base of hammer) on top of the thin bed at the base of sequence 16 at Chapeau du Gendarme. It is interpreted as a first manifestation of the large-scale maximum flooding. (D) Interval of thin beds at Chapeau du Gendarme, with a hardground on top of the thick bed below (small-scale sequence 17). This hardground surface is interpreted as sequence boundary Be 5. The surface below the hardground pinches out laterally and possibly was a channel floor (arrow). (E) Yellow-reddish, irregular surface (at base of hammer) interpreted as large-scale sequence-boundary Be 5 in the Crêt de l'Anneau section. (F) Same surface as in E, showing the deep penetration of the iron staining (face of bed in lower left corner of picture).

Sequence-boundary zone Be 5 is placed below a slumped interval in the hemipelagic Montclus section and well dated by ammonites (Figs 4 and 8). In the platform sections of Rusel, Crêt de l'Anneau, and Chapeau du Gendarme, SB Be 5 separates the Pierre-Châtel Formation from the Vions Formation and is developed as a karstified surface (unfortunately, it is covered in the Salève section). At Chapeau du Gendarme, this surface caps lagoonal sediments with an oblique stratification suggesting a tidal channel (Figs 8 and 10D). It is overlain by thin, bioturbated beds with marine fauna (echinoderms), implying rapid flooding after the subaerial exposure. At Rusel and Crêt de l'Anneau, the surface displays intense red-yellow iron staining that penetrates a few tens of centimetres into the underlying limestone (Fig. 10E and F). Stable-isotope analyses carried out in these two sections (Pasquier, 1995) show a positive shift in $\delta^{18}O$ just below the surface, while $\delta^{13}C$ has a negative shift (Fig. 11). This is explained by subaerial exposure that increased evaporation (concentrating the heavy oxygen isotopes), while soil gas led to concentration of light carbon in the early-diagenetic fluids. This particular composition was then preserved through mineralogic stabilization and cementation of the sediments (Joachimski, 1994). Such a signature is not present at the surface on top of small-scale sequence 14 (Fig. 11), probably due to a shorter subaerial exposure time there than at the surface attributed to sequence-boundary Be 5.

From the examples presented above it is evident that surfaces of wide lateral extent are rarely simple surfaces but complex intervals that change their aspect from one outcrop to another. In many cases they are amalgamations or stacks of closely spaced simple surfaces, each of which has only local extent (e.g. Figs 7C and 10D). Correlation based on physical expression alone therefore can be difficult, and it is rather the position in the stratigraphic column that demonstrates their relationship. Sequence boundaries of basin-wide scale are commonly composites of several types of simple surface that, however, all express loss of accommodation space on the platform. Transgressive surfaces in the studied sections commonly are sharp, whereas maximum-flooding conditions can be expressed by a single hardground but also by an interval with dense bioturbation and increased clay content.

DISCUSSION

From the living platform to the depositional record

On modern, shallow carbonate platforms it is easy to directly observe the short-term processes that lead to the formation of distinct surfaces: waves and tidal currents erode, transport, and re-deposit carbonate grains and mud, and strong tidal or storm-induced currents scour channels into previously deposited sediment (Table 1). These processes lead to short-term hiatuses in which a few hours (in the case of ripple migration or tidal-flat lamination) to a few hundreds or thousands of years (in

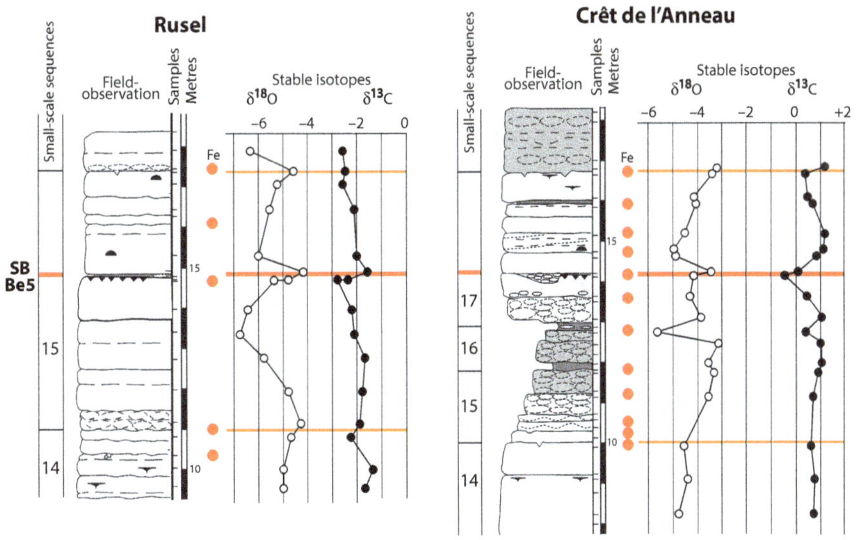

Fig. 11. Stable O and C isotopes measured across large-scale sequence boundary Be 5 in the Rusel and Crêt de l'Anneau sections. Note the opposite shifts of O and C values below the sequence boundary but then the parallel shifts for the samples situated at the boundary itself. Shifts at small-scale sequence boundaries are not or are less well expressed. For discussion refer to text. For symbols, see legend in Fig. 8.

the case of deep scour surfaces) of sedimentation history are lost. On active ooid or bioclastic dunes, the sediment is mobile and the hiatuses are represented by reactivation surfaces (McCabe & Jones, 1977; Gonzalez & Eberli, 1997). Incipient hardgrounds having formed by microbial binding and rapid cementation on the lagoon floor or on inactive dunes can be reworked to produce lithoclasts that are incorporated into the overlying sediment package (Hillgärtner *et al.*, 2002). Erosion of cohesive and commonly microbially colonized carbonate mud on tidal flats and in low-energy lagoons produces soft pebbles that are again incorporated into the overlying deposit. The corresponding hiatal surfaces are cutting irregularly into the underlying soft sediment. On the upper parts of tidal flats, non-deposition occurs at low tide during a few hours. Strong tides or storms ripping up the microbially stabilized tidal laminae produce flat pebbles (Hardie & Ginsburg, 1977).

Bioturbation leads to homogenisation and time-averaging of the uppermost sediment layer (Shinn, 1968; Flessa *et al.*, 1993). Thus, hydrodynamic sedimentary structures and hiatal surfaces may be destroyed and detailed information about a few years to hundreds of years of sedimentary history is lost. Sediment bypass and sediment starvation lead to the formation of softgrounds that turn into firmgrounds and hardgrounds when stabilized by biogenic and/or mineral encrustations and by early diagenesis (Christ *et al.*, 2015). Years to hundreds of years of sedimentary history may thus be condensed into a thin interval or a single surface (Table 1). With continued sediment starvation, hardgrounds may also represent millions of years but in this case represent an amalgamation of multiple processes such as repeated phases of flooding and subaerial emergence (Rameil *et al.*, 2012).

Hiatal surfaces produced by short-term processes and covering hours to a few hundreds of years commonly are of limited lateral extent (Table 1). This is due to the facies mosaics typical of shallow carbonate platforms where different depositional environments are closely juxtaposed (Rankey, 2002; Rankey & Reeder, 2010). For example, a tidal channel produces a scoured surface that is only a few metres wide, and an individual reactivation surface in an ooid shoal covers at the most a few tens of square-metres, corresponding to the size of the inter- or subtidal dunes forming the shoal. The same holds for short-term condensation, for example expressed by intense bioturbation over a few hundreds of square-metres in a quiet and sediment-starved depression behind the shoals.

With the knowledge of short-term processes and products observed on modern carbonate platforms, it is relatively straightforward to interpret the features seen in the sedimentary record. Fossil examples of such hiatal surfaces with limited lateral extent have been presented in a previous section (Fig. 7).

When long-term (i.e. involving thousands to hundreds of thousands to millions of years) rises and falls of relative sea-level flood or expose a carbonate platform, the above-mentioned short-term processes interact and produce complex amalgamations and stacks of hiatal surfaces or condensed intervals (examples in Figs 9 and 10). These surfaces and intervals can be interpreted in terms of sequence stratigraphy (Vail *et al.*, 1991; Catuneanu *et al.*, 2009).

Sequence boundaries on a shallow carbonate platform form when sediment has completely filled the available space (created by eustatic sea-level and subsidence) and reached intertidal to supratidal conditions (at the top of shallowing-upward highstand deposits). If eustatic sea-level falls below the previously accumulated sediment surface or if there is tectonic uplift, erosion of mobile sediment, cementation in a fresh-water lens, karstification and pedogenesis occur. Depending on the amplitude of relative sea-level fall and on the time of subaerial exposure, important amounts of previously deposited sediment may be lost through mechanical erosion and/or chemical dissolution. In the sedimentary record, this can lead to superposition of subaerial exposure surfaces directly on subtidal facies, or to the erosion of entire previously deposited sequences. In the case of a deep lagoon and a low-amplitude sea-level fall, facies may stay subtidal throughout the sequence. Sequence boundaries there may be expressed only indirectly by input of siliciclastics eroded from the hinterland (Osleger, 1991; Strasser & Hillgärtner, 1998).

Transgressive surfaces form when relative sea-level rise leads to flooding of the platform. On tidal flats, accommodation is created but quickly filled by flourishing microbial mats until sea-level rise outpaces the accumulation potential. On beaches, storm surges will create ravinement surfaces and push back the beach ridges until the system is drowned by continued rising sea-level (Donselaar, 1989). On previously subaerially exposed surfaces, the flooding will pick up lithoclasts derived from the underlying cemented sequence, pedogenic elements and plant fragments, and incorporate them at the base of the deepening-up transgressive deposits (Strasser & Davaud, 1983). A certain lag time will pass before the organic carbonate factory starts up again and sediment production keeps up with rising sea-level (Tipper, 1997; Kemp & Sadler, 2014).

Maximum-flooding conditions on a shallow carbonate platform are indicated by the turnaround from a deepening-up to a shallowing-up facies evolution. This change may be gradual and produce an interval of sediment,

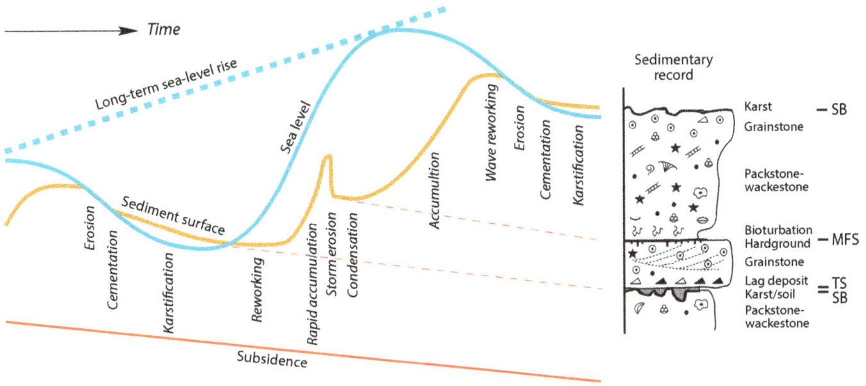

Fig. 12. Hypothetical evolution of one point on a shallow carbonate platform, governed by one sea-level cycle superimposed on a longer term sea-level rise and constant subsidence. The resulting sediment stack corresponds to a metre-scale elementary sequence as seen in the studied sections. For explanation, refer to text. For symbols and abbreviations see legend in Fig. 8.

without the development of a specific surface. However, when the fastest relative sea-level rise is accompanied by sediment starvation, a strongly bioturbated surface may form in lagoons, which likely will develop into a hardground (Reolid *et al.*, 2014). In the studied sections, the condensed intervals are enriched in clay minerals, suggesting that terrigenous input possibly hampered the carbonate-producing organisms and thus contributed to a low sedimentation rate.

Figure 12 schematically illustrates the formation of two sequence boundaries, a transgressive surface, and a maximum-flooding surface on a shallow carbonate platform. It is assumed that one symmetrical sea-level cycle is superimposed on a longer term rising trend that, together with subsidence, creates accommodation space. When sea-level first drops below the pre-existing sediment surface, the soft sediment is eroded. Cementation in the fresh-water lens has in the meantime stabilized the sediment, and karstification occurs. This karstified surface will be recognized as the first sequence boundary. When sea-level rise leads to flooding of the karst surface (placing the transgressive surface directly over the sequence boundary), lithoclasts will be reworked, forming a lag deposit. Carbonate production and sediment accumulation will pick up after a lag time. Water depth at first is ideal to produce ooid dunes, shifted by tidal currents. Further deepening leads to lagoonal facies, which keeps up with rising sea-level. If a storm event happens, the sediment surface may be lowered through erosion of the lagoonal sediment. Sediment starvation may lead to a hardground (that will be interpreted as the maximum-flooding surface) and a strongly bioturbated condensed interval, corresponding to the deepest water. With diminishing water depth, carbonate production picks up again and sediment starts filling the lagoon, until falling sea-level forces erosion and a second phase of karstification (the second sequence boundary). This hypothetical scenario describes the evolution of only one point on the platform and may (at the same time) be entirely different at other points of the same platform, depending on the substrate morphology and the lateral facies distribution. Also, rates of carbonate production and sediment accumulation vary strongly throughout this one cycle.

In a two-dimensional vision of the platform (Fig. 13), it is schematically illustrated how areas of sediment accumulation and areas of concomitant erosion and sediment reworking are juxtaposed. Sea-level fluctuations with two superimposed frequencies control the gain or loss of accommodation space. Depending on the initial platform morphology and the sediment produced in lakes, on beaches, on shoals, in lagoons or on reefs, the distribution of facies and of hiatal surfaces will vary dramatically. Accordingly, the resulting sedimentary records will differ depending on their position on the platform. It is also demonstrated that the sequence boundaries and transgressive surfaces as identified according to facies changes in each record will not necessarily have formed during the same sea-level cycle, or are amalgamations when the record of one or more cycles is missing. For example, record D represents all five sea-level cycles, while record E shows only two. In the position of record C, a beach formed during cycle 1. Its preservation is possible because of a first transgressive pulse above the sequence boundary. A second transgressive surface then leads to the installation of ooid dunes. In record D, the lagoon was filled in during cycle 1. The presence of lithoclasts and plant fragments in the overlying bed (2) implies the presence of a sequence boundary (position b on the sea-level curve), directly overlain by a second transgressive surface. Lowstand deposits are missing because of the low accommodation space. The maximum-flooding interval always appears within cycle 4 when much accommodation is

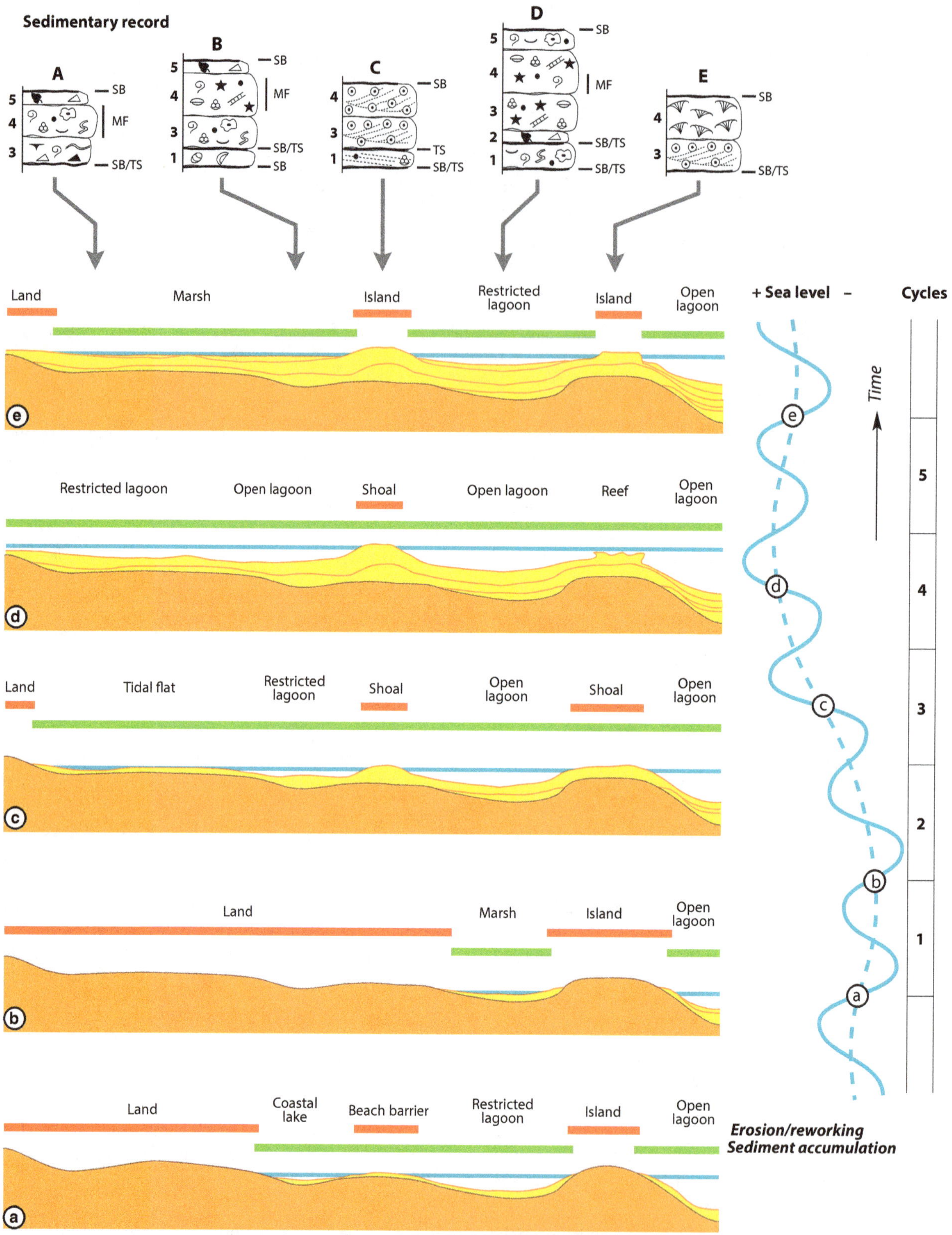

Fig. 13. Hypothetical evolution of a platform transect during five high-frequency sea-level cycles. Note that the completeness of the sedimentary record changes strongly depending on the position on the platform. Cross-sections a to e reflect the platform morphologies and the facies distributions at the time steps a to e indicated on the sea-level curve. For explanation, refer to text.

created by the superposition of a short-term sea-level rise on the high of the lower frequency curve (position d). However, in record C there is no facies change that would allow identification of this maximum flooding. Note that on shoals and on beaches erosion and accumulation may occur at the same time as sediment is reworked during any time of a sea-level cycle. It is clear that the picture is even more complex if a real-world, three-dimensional platform is considered. In addition, synsedimentary tectonics may modify the accommodation space at irregular intervals and differentially from one place on the platform to another.

Not only cyclical sea-level changes but also random processes influence the sedimentation on shallow carbonate platforms. For example, lateral migration of sediment bodies may create shallowing-up facies successions at constant sea-level (Pratt & James, 1986; Satterley, 1996) that look similar to the ones produced by sea-level fluctuations (Strasser, 1991). Storm events may create erosion surfaces at any time, independent of a sea-level cycle. Burgess & Wright (2003) and Burgess (2006) showed by forward modelling that changes in carbonate production rate and changes in sediment transport direction may produce repetitive strata, independent of sea-level changes. Nevertheless, the hierarchical stacking of depositional sequences (Fig. 5) and the relatively good fit with the time frame given by Hardenbol *et al.* (1998) (Fig. 4) suggest that the periodicities of the orbital cycles are reproduced in the studied sections. In Berriasian times, continental ice probably was present (Fairbridge, 1976; Frakes *et al.*, 1992; Eyles, 1993) but ice volumes were small and the glacio-eustatic fluctuations (resulting from orbitally controlled insolation changes) of low amplitude. However, these same insolation changes also caused thermal expansion and retraction of the uppermost layer of ocean water (Gornitz *et al.*, 1982; Wigley & Raper, 1987), thermally induced volume changes in deep-water circulation (Schulz & Schäfer-Neth, 1998), and/or water retention and release in lakes and aquifers (Jacobs & Sahagian, 1993). These processes contributed to high-frequency, low-amplitude sea-level changes (Plint *et al.*, 1992; Conrad, 2013). Consequently, it is assumed that the cyclical insolation changes translated rather directly into sea-level fluctuations that were therefore more or less symmetrical (in contrast to the asymmetric glacio-eustatic fluctuations during ice-house times when there was slow waxing and rapid waning of polar ice caps). Based on the Chapeau du Gendarme section, Strasser *et al.* (2004) reconstructed amplitudes of a few metres for a sea-level cycle related to the orbital short eccentricity (100 kyr) cycle. On a shallow platform, even such low amplitudes will lead to significant facies changes. In addition, orbitally controlled climate changes may influence rainfall in the hinterland and thus modulate siliciclastic and nutrient input onto the platform. Not only orbitally induced processes in the Milankovitch frequency band but also millennial-scale cyclical processes may create laterally consistent bedding planes. For example, Zühlke *et al.* (2003) identified high-frequency cyclicity producing shallowing-upward sequences in the Triassic Latemar platform, and Tucker *et al.* (2009) explained mid-Carboniferous cyclothems by high-frequency arid-humid climate changes that were superimposed on climate and sea-level changes induced by the orbital precession cycle. These high-frequency climate changes were possibly controlled by variations in solar activity (Elrick & Hinnov, 2007).

Consequently, the final sedimentary record results from a combination of allocyclic (externally controlled) and random processes. Whereas allocyclic processes such as orbitally controlled climate and sea-level changes, or millennial-scale climate changes, influence the entire platform, random processes are of limited lateral extent. In the studied sections, it is assumed that the formation of small- and medium-scale sequences that can be correlated across the entire Jura platform was at least partly controlled by orbitally induced eustatic sea-level cycles. In the case of some elementary sequences, however, random processes dominated and blurred the low-amplitude sea-level signal. The lateral variability in the thickness of sequences (Fig. 8) is attributed not only to the variable potential of sediment accumulation in the different environments (Fig. 13) but also to synsedimentary tectonics that structured the Jura platform into high and low areas (Hillgärtner, 1999). This structure changed through time, thus modifying the substrate morphology and the accommodation at irregular intervals.

Duration of hiatuses and condensation

Estimation of time lost in hiatal surfaces or condensed in sequence-boundary or maximum-flooding zones is not an easy task. Time lost means that parts of the sedimentary history are not recorded at all, and time condensed means that many events of a long history are amalgamated and overprint each other in a thin layer of sediment that may be difficult to disentangle.

For simple surfaces, comparisons with present-day processes are helpful. For example, reactivation surfaces in tidally influenced ooid dunes are controlled by switches in current direction and current strength, and time lost at such surfaces is in the order of a few hours (Hine, 1977; Gonzalez & Eberli, 1997). However, reactivation surfaces may also be induced by spring tides (twice-monthly), equinoxial tides (twice-yearly), or episodic storms, which then increases the time span manifold. Estimation of the geological time actually represented by a tidal dune is

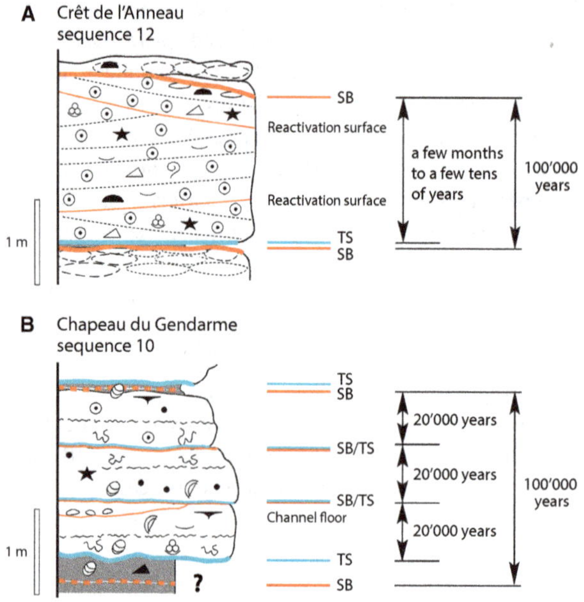

Fig. 14. Examples of unequal time distribution in a sequence formed in a high-energy setting (A: small-scale sequence 12 at Crêt de l'Anneau) and in a sequence reflecting a low-energy context (B: small-scale sequence 10 of the Chapeau du Gendarme section). Symbols and abbreviations as in Fig. 8. For discussion, refer to text.

even more difficult. The fact that it is preserved in the depositional record implies that the sedimentary system must have changed in order to stop the dune's active migration. This can happen when, for example, a barrier forms that shelters the bedform from the influence of tidal currents, or when sea-level rise puts the dune below the level where the tidal currents can move the sediment grains. Examples for this latter case are the drowned, relict bedforms of Lily Bank in the Bahamas (Hine, 1977). Once inactive, the bedforms will be stabilized by benthic organisms and cemented in the marine phreatic zone. In another scenario, a sea-level drop will lead first to intertidal exposure of the dune and possibly to the formation of a beach, then to fresh-water cementation that stabilizes it. Time not recorded within the bedform itself will be the sum of time lost at each hiatal surface but it is unknown how much time is not recorded at the surface over which the dune migrated and on the top of the dune.

Small-scale sequence 12 of the Crêt de l'Anneau section illustrates such a case (Fig. 14A). It is a massive bed of cross-stratified oolitic and bioclastic grainstone. Its base is sharp and interpreted as a transgressive surface, the sequence boundary being situated in the thin marly layer below. The top of sequence 12 exhibits iron staining and dinosaur tracks (Fig. 9C). Lateral correlation suggests that this sequence represents 100 kyr (Fig. 8). Reactivation surfaces within the massive bed imply dune migration. Keystone vugs indicate that the dune occasionally entered the intertidal zone and developed a beach. Lithoclasts composed of the same facies are present, suggesting that incipient hardgrounds had formed and were reworked. Consequently, time lost in such surfaces could have been a few months to a few tens of years. The conclusion is that almost all of the 100 kyr available between the two sequence boundaries is lost or condensed in the marly layer below the bed and in the reddish surface on top of it. Additional, short time intervals are lost along the reactivation surfaces. It is impossible to say at which time within the 100-kyr cycle the dune was deposited, and if (and how much) other sediment was deposited but then eroded during this time interval.

In low-energy settings, simple surfaces form for example when tidal channels are cut into tidal flats. In the Bahamas, tidal channels stay active over tens of years with only little lateral migration (Rankey & Morgan, 2002), and their bottom will thus form a hiatal surface where this time interval is not recorded. Time is lost also on supratidal flats where sediment accumulates only during storm washovers, although algal-microbial growth may contribute to some accretion (Hardie & Ginsburg, 1977). In shallow lagoons, time may be condensed when sediment production and accumulation is low, which will result in strongly bioturbated intervals and hardgrounds.

An example of a low-energy setting is shown in Fig. 14B. Small-scale sequence 10 of the Chapeau du Gendarme section is composed of three beds of wacke- to packstones. Echinoderms and benthic foraminifera indicate a marine environment but charophytes were washed in from nearby coastal lakes, or reworked from now eroded lacustrine facies. Birdseye structures imply periodic inter-to supratidal exposure. The lower boundary of this sequence is not clearly developed but interpreted to sit in the thick marly bed that carries charophytes and black pebbles. The upper boundary is placed in the marly, charophyte-bearing layer separating it from sequence 11. Lateral correlation suggests that the time available to form small-scale sequence 10 was 100 kyr (Fig. 8). The base of the lowermost limestone bed is seen as an important transgressive surface that can be correlated over the entire study area (base of the Pierre-Châtel Formation; Fig. 8). According to the cyclostratigraphic scheme discussed above and illustrated in Fig. 5, it is assumed that each limestone bed represents 20 kyr. The beds are separated by thin marly layers, implying a change in environmental conditions and input of clay minerals (Strasser & Hillgärtner, 1998). At the bottom of the tidal channel (top of the first bed), a few tens to a few hundreds of years were possibly lost. There is no evidence for other hiatal surfaces, but the slightly more marly and bioturbated levels within the beds may express some condensation. The partings between the beds are interpreted as sequence boundaries directly followed by transgressive surfaces of the elementary sequences. At transgressive surfaces, a lag time may be lost, which has been estimated to be some 3000 years for shallow carbonate platforms (Kemp & Sadler, 2014). However, considering the various facies and laterally changing environments in shallow carbonate systems, this lag time may strongly vary from one position on the platform to another. Today, tropical lagoonal carbonate sediments accumulate with rates of a few tenths of a millimetre to a few millimetres per year (Enos, 1991; Strasser & Samankassou, 2003). Each of the three beds is about 60 cm thick, but decompacted it would have corresponded to about 1·2 m of sediment (Goldhammer, 1997). At an assumed average accumulation rate of 1 mm yr^{-1}, it would have taken about 1200 years to make one bed. Consequently, it can be concluded that most of the 20 kyr available for the formation of one bed must be contained in the partings at the bed limits. As only three beds are recorded but the entire small-scale sequence lasted 100 kyr, additional time was lost and/or condensed in the marly intervals below and above the three limestone beds (Fig. 14B).

Complex, composite surfaces that correspond to sequence-boundary zones are in fact an amalgamation of different types of simple hiatal surfaces. As a result of reduced accommodation space on the shallow platform, processes creating hiatal surfaces are common: reactivation of dunes and beaches, ravinement surfaces related to storms, cutting of tidal channels, and subaerial exposure. As long as the sediment is not consolidated, erosion may partly or completely remove previously deposited material. Once the sediment is cemented, erosion and reworking will produce clasts that are incorporated into the overlying deposit. Cementation of beachrock and coastal carbonate dunes can happen within a few hundreds to thousands of years (Halley & Harris, 1979; Strasser & Davaud, 1986). Estimation of time distribution in such complex intervals is difficult. Sequence-boundary zone Be 1 (Fig. 7C) is characterized by well-defined beds that, however, pinch-out laterally. These beds may have formed under the influence of low-amplitude sea-level fluctuations driven by the 20 kyr precession cycle, or else they may be related to autocyclical, random processes of unknown duration (Strasser, 1991). As there is no lateral continuity of these beds, it is difficult to evaluate the time lost in the surfaces. In the case of sequence-boundary zone Be 4, about 900 kyr are represented by just half a metre of sediment (Figs 8 and 9A). Even by analysing all relict beds and the reworked clasts it is not possible to say at which time within these 900 kyr which facies was formed.

Sequence boundary Be 5 on the platform is developed as a sharp, unique surface (Figs 8 and 10D, E) although it certainly contains a complex history of deposition, erosion, and subaerial exposure (the latter evidenced by C and O isotopes; Fig. 11). According to the correlation in Fig. 8, 100 to 300 kyr are missing depending on the palaeogeographical position of the outcrops, which is explained by a tectonic tilt of the platform (Hillgärtner, 1999). A more local tectonic event may have been responsible for the thin or even absent small-scale sequence 13 in the Crêt de l'Anneau section (Figs 8 and 9C, E).

Maximum-flooding zones in the studied shallow-water outcrops commonly display intense bioturbation (Fig. 10A). The only way to estimate the time represented within these zones is lateral correlation and a cyclostratigraphic analysis. In the case of sequence 16 of the Crêt de l'Anneau section, 100 kyr are condensed in about 1 m of sediment. There is no evidence of hardgrounds or other hiatal surfaces, implying that it was a low sedimentation rate that led to this concentration of bioturbation. Also sequence 13 at Crêt de l'Anneau is strongly bioturbated. In addition to low sediment accumulation rate, it was reduced accommodation that induced the very thin deposit and the lateral pinch-out (Figs 8 and 9C).

Sedimentation rates

For the purpose of this paper, three types of 'sedimentation rate' are distinguished:

1 'Carbonate production rate' describes the sediment furnished by the carbonate-producing organisms (measured in kg m^{-2} yr^{-1} for weight or mm per year for vertical accretion in modern environments; Enos, 1991).

2 'Sediment accumulation rate' refers to the sediment that is accumulated in one spot on the platform (measured in mm per year in modern systems or estimated in ancient rocks). Sediment accumulation rate may be equal to carbonate production (accretion) rate if there is no sediment transport away from or into the production area, and if there are no hiatuses. To compare production and accumulation rates, the sedimentary record must be decompacted (Strasser & Samankassou, 2003). Sediment accumulation rates in ancient rocks should be estimated over short time intervals in order to better compare them with modern rates and interpret the sedimentary processes and facies evolution on the platform.

3 'Preservation rate' is calculated by dividing the thickness of the rock record by the time interval it covers (mm yr^{-1} to m Myr^{-1}). Preservation rate is generally

less than accumulation rate because hiatuses occur in the rock record and the originally accumulated sediment has been compacted. Preservation rates are useful to describe the long-term evolution of a platform, including long-term changes in sea-level and subsidence. In many publications, 'sediment accumulation rate' is calculated based on the preserved rock record representing millions of years and would thus correspond to the 'preservation rate' introduced here (Sadler, 1981, 1999; Bosscher & Schlager, 1993; Schlager, 1999).

Considering the abundance of hiatuses and condensations in shallow-water carbonate sediments, it is of course difficult to calculate sediment accumulation rates. Bosscher & Schlager (1993) compiled published accumulation rates and estimated a maximum of 200 m Myr^{-1} for Phanerozoic carbonate platforms, but they did not consider compaction or the presence of hiatuses. Sadler (1981, 1999) and Schlager (1999) discussed the completeness of the stratigraphic record and the scaling of sedimentation rates: apparently, sediment accumulation rates decrease as the time interval over which they are averaged increases. Schlager *et al.* (1998) stated that this is due to 'the fact that sedimentation is an episodic process and that the sediment record is riddled with hiatuses on all scales'. If ancient sediment accumulation rates want to be estimated and compared to the ones in modern systems, it is more appropriate to consider only short time intervals in which hiatal surfaces and condensed intervals are absent or at least minimal. Furthermore, the sedimentary record must be decompacted according to facies (Goldhammer, 1997; Strasser & Samankassou, 2003). In the examples of the Jura Mountains discussed here, where overburden has never been more than about 1 km (Trümpy, 1980) and tectonic compaction can be excluded, one metre of sedimentary record of an ooid shoal would have corresponded to 1·1 to 1·2 m of mobile ooid sand. However, as discussed above, this represents only the final accumulation and not the production potential of an ooid shoal where most of the grains are washed back and forth and often are also exported to deeper basins (Rankey & Reeder, 2010). A bed of lagoonal wackestone measuring 1 m today would have corresponded to some 2 m of soft sediment. If such a bed is interpreted to have formed under the control of an orbital precession cycle of 20 kyr, then a very low sediment accumulation rate is calculated (0·1 mm yr^{-1}), which is ten times less than values seen in modern lagoons (average of 1 mm yr^{-1}; Enos, 1991). Assuming that the ecological conditions for the carbonate-producing organisms in the Berriasian were similar to the ones today (and thus the accumulation rates are comparable), it has to be postulated that much time is lost at the bedding planes. Furthermore, carbonate production and accumulation rates may vary significantly throughout the time interval during which the bed was accumulated, and also laterally across the platform (Figs 12 and 13; see also Strasser *et al.*, 2012, for an Oxfordian case study).

These two examples illustrate that, to calculate a sedimentation rate, it is misleading to just divide the thickness of a section by the number of years it represents. Such a calculation informs at the most about an average preservation rate. It is much more informative to calculate the rates of specific intervals (decompacted according to facies) in the sedimentary record where it can be assumed that sedimentation was continuous. Even so, it will be no more than an average sediment accumulation rate because lateral changes of facies and sediment thickness can strongly vary on a shallow platform. Furthermore, sediment accumulation rate may be much less than carbonate production rate because much of the material produced in one site may be shifted across the platform by currents or exported to a deeper basin (e.g. Schlager, 1991; Milliman *et al.*, 1993). In many cases, on a shallow platform, accumulation is limited not by carbonate production but by accommodation once the available space has been filled by sediment.

To illustrate this issue, the section of Chapeau du Gendarme is analysed in detail (Table 2). Facies are decompacted assuming that marls and mudstones have been more compacted than textures containing grains such as wackestones, packstones, and grainstones (Strasser & Samankassou, 2003). Then, modern sedimentation rates (averaged from Enos, 1991) are used to calculate the time theoretically needed to accumulate the decompacted facies, and this time is compared with the total time span between sequence boundaries Be 4 and Be 5. While Be 4 has not been identified at Chapeau du Gendarme it is assumed to be situated in the marls below the transgressive surface (Fig. 8) where black pebbles and charophytes occur. The lowstand deposits identified in the hemipelagic Montclus section comprise 900 kyr and, at Chapeau du Gendarme, are condensed to a few tens of centimetres (as seen in the Salève section). At sequence boundary Be 5, at least 100 kyr are missing (corresponding to small-scale sequence 18; Fig. 8). The 25·85 m of non-decompacted section would thus represent the history of 1·7 Myr (Fig. 4), which gives an average preservation rate of 0·015 mm yr^{-1}. This number is meaningful only when the long-term history of the Jura platform is discussed. When considering the (non-decompacted) sediment stack preserved and using the time span suggested by the cyclostratigraphical interpretation in Fig. 8, then the same 25·85 m would represent only 800 kyr, implying an average preservation rate of 0·033 mm yr^{-1}. Still, this number is meaningless when discussing the rates of the sedimen-

Table 2. Texture-dependent decompaction of the Chapeau du Gendarme section (Fig. 8) and estimation of time required to accumulate the observed deposits. Decompaction factors according to Strasser & Samankassou (2003), average Holocene sedimentation rates from Enos (1991)

Texture	***metres in section per texture***	***Decompaction factor***	***metres decompacted***	***Holocene sedimentaion rate***	***Time required***
marls (m)	1.27	3.0	3.81	1 mm/yr	43'060 years
mud-wackestone (M/W)	0.57	2.25	1.28		
wackestone (W)	9.81	2.0	19.62		
wacke-packstone (W/P)	3.69	1.75	6.46		
packstone (P)	1.85	1.5	2.78		
pack-grainstone (P/G)	6.75	1.35	9.11		
grainstone (G)	1.91	1.2	2.29	2 mm/yr	1145 years
	Total: 25.85 m		Total: 45.35 m		Total: 44'205 years

tary processes that ruled on the platform in Berriasian times. According to the values presented in Table 2, the time needed to accumulate the preserved sediment was about 44 kyr. As accommodation on the shallow platform was generally low and intertidal to supratidal conditions were frequent, it can be estimated that some 756 (800 minus 44) kyr were spent in non-deposition, condensation, and/or erosion. Hiatuses and condensation occurred mainly during lowstands of relative sea-level when accommodation was low, which corresponds to the boundaries of elementary and small-scale sequences (Fig. 5). However, time may be lost also in maximum-flooding surfaces and intervals, or by episodic erosive events such as storms (Fig. 12). These numbers are conservative estimates because the smallest units considered are the beds formed under the control of the 20 kyr precession cycle, and because it is assumed that most of the time was lost at the bedding planes. Depending on the depositional environment, additional hiatuses and condensations may occur within the bed, making the sediment accumulation rates higher as the time interval over which they are averaged decreases (Schlager *et al.*, 1998; Sadler, 1999).

The approach proposed here to estimate sediment accumulation rates – and the durations of hiatuses and condensations – depends of course on the validity of the cyclostratigraphic interpretation that attributes numerical time spans to elementary and small-scale sequences (20 kyr and 100 kyr, respectively). However, even if there are many uncertainties and assumptions (numerical ages, cyclostratigraphic interpretation, decompaction factors, modern sedimentation rates), such considerations allow a realistic picture of sedimentary history on a shallow carbonate platform to be imagined.

CONCLUSIONS

The detailed sedimentological, sequence-stratigraphic and cyclostratigraphic analysis of Berriasian sections in the Swiss and French Jura reveals a complex but structured record of shallow-water carbonate facies. Their vertical evolution allows deepening-up and shallowing-up trends to be distinguished that define depositional sequences of different orders, the boundaries of which are furthermore enhanced by marly layers. The hierarchical stacking of these sequences and the time-frame given by bio- and sequence-stratigraphic correlation with other, well-dated basins in Europe suggest that high-frequency, low-amplitude eustatic sea-level changes were a controlling factor and that they were related to insolation changes induced by the cyclical perturbations of the Earth's orbit. The precession cycle (20 kyr) controlled the formation of the smallest units (elementary sequences), and the short eccentricity cycle (100 kyr) was responsible for small-scale sequences. These then group into medium-scale sequences related to the long eccentricity cycle (400 kyr). Although random processes such as lateral migration of sediment bodies or storm events were superimposed, the interpreted orbital cyclicity offers a time frame within which the durations of hiatuses and condensation intervals can be discussed, and sedimentation rates can be estimated.

As the encountered facies resemble those seen in modern shallow carbonate environments, average sediment accumulation rates can be assumed to be comparable. For the studied Berriasian example it is thus estimated that 44 kyr would have been sufficient to produce a sedimentary record of about 45 m, which corresponds today to about 26 m of compacted limestone and marls. However, the time interval suggested by lateral correlation with numerically dated sequences is about 800 kyr. This means that the missing 756 kyr are lost in surfaces or condensed in thin sedimentary layers.

Short-term processes such as tidally controlled migration of ripples and dunes, scouring of tidal channels, or storm-induced erosion lead to hiatal surfaces where the record of a few hours to a few hundreds of years is lost. On the longer term, thousands to millions of years are not recorded or represented only by thin layers of sediment at sequence boundaries, at transgressive surfaces, and during maximum flooding. In the studied outcrops, reactivation surfaces have been identified within grainstones, whereas the boundaries of elementary and small-scale sequences in many cases display clasts that suggest reworking of previously cemented and now missing sediment. Iron staining and shifts in C and O isotopes point to subaerial exposure where again geological time has been lost. At the base of the studied sections, about 900 kyr are condensed in a few tens of centimetres of sediment, whereas at the top of the sections 100 to 300 kyr are not recorded. Such long-term gaps and condensations commonly are amalgamations or stacks of narrowly spaced shorter term hiatal surfaces. The absence of entire small-scale sequences and the lateral changes of thickness of the preserved sequences are attributed to synsedimentary tectonics that structured the Jura platform: lows favoured sediment accumulation while highs were prone to emergence followed by non-deposition or erosion.

Despite the many uncertainties related to numerical dating and the sequence- and cyclostratigraphic interpretations, this study allows the durations of hiatuses and condensations to be discussed in a narrow time frame on the order of a few tens of thousands of years. Within this time frame, facies evolution and hiatal surfaces of individual depositional sequences can be interpreted in detail, although additional uncertainties are introduced by the comparison with modern sediment accumulation rates. Nevertheless, it becomes clear that, when estimating sedimentation rates in the fossil record, the shortest possible time spans must be considered to minimize the effects of hiatuses and condensation. Although this is still far from the time resolution available in the Holocene and the Recent, a relatively realistic and dynamic picture of an ancient carbonate-dominated platform can be gained. The depositional record stays fragmentary, but the missing time can be well interpreted by comparison with processes seen today in modern carbonate systems.

ACKNOWLEDGEMENTS

This study is the fruit of many years of research on Jurassic and Cretaceous carbonate platforms in the Swiss and French Jura Mountains, as well as on modern carbonate systems. I thank reviewers Maurice Tucker and Wolfgang Schlager for their constructive comments, and the members of the editorial office of The Depositional Record for their help. This research has been funded mainly by the Swiss National Science Foundation whose support is gratefully acknowledged.

References

Ager, D.V. (1980) *The Nature of the Stratigraphical Record*, 2nd edn. John Wiley, New York, 122 pp.

Allenbach, R.P. (2002) The ups and downs of "tectonic quiescence" – recognizing differential subsidence in the epicontinental sea of the Oxfordian in the Swiss Jura Mountains. *Sed. Geol.*, **150**, 323–342.

Barrell, J. (1917) Rhythm and the measurement of time. *Geol. Soc. Am. Bull.*, **28**, 745–904.

Berger, A., Loutre, M.F. and **Dehant, V.** (1989) Astronomical frequencies for pre-Quaternary palaeoclimate studies. *Terra Nova*, **1**, 474–479.

Bernier, P. (1984) Les formations carbonatées du Kimméridgien et du Portlandien dans le Jura méridional. Stratigraphie, micropaléontologie, sédimentologie. *Doc. Lab. Géol. Lyon*, **92**, 730.

Bosscher, H. and **Schlager, W.** (1993) Accumulation rates of carbonate platforms. *J. Geol.*, **101**, 345–355.

Bover-Arnal, T. and **Strasser, A.** (2013) Relative sea-level change, climate, and sequence boundaries: insights from the Kimmeridgian to Berriasian platform carbonates of Mount Salève (E France). *Int. J. Earth Sci.*, **102**, 493–515.

Brlek, M., Korbar, T., Košir, A., Glumac, B., Grizelj, A. and **Otoničar, B.** (2014) Discontinuity surfaces in Upper Cretaceous to Paleogene carbonates of central Dalmatia (Croatia): *Glossifungites* ichnofacies, biogenic calcretes, and stratigraphic implications. *Facies*, **60**, 467–487.

Burgess, P.M. (2006) The signal and the noise: forward modeling of allocyclic and autocyclic processes influencing peritidal carbonate stacking patterns. *J. Sed. Res.*, **76**, 962–977.

Burgess, P.M. and **Wright, V.P.** (2003) Numerical forward modeling of carbonate platform dynamics: an evaluation of complexity and completeness in carbonate strata. *J. Sed. Res.*, **73**, 637–652.

Catuneanu, O., Abreu, V., Bhattacharya, J.P., Blum, M.D., Dalrymple, R.W., Eriksson, P.G., Fielding, C.R., Fisher, W.L., Galloway, W.E., Gibling, M.R., Giles, K.A., Holbrook, J.M., Jordan, R., Kendall, C.G., St, C., Macurda, B., Martinsen, O.J., Miall, A.D., Neal, J.E., Nummedal, D., Pomar, L., Posamentier, H.W., Pratt, B.R., Sarg, J.F., Shanley, K.W., Steel, R.J., Strasser, A., Tucker, M.E. and **Winker, C.** (2009) Towards the standardization of sequence stratigraphy. *Earth-Sci. Rev.*, **92**, 1–33.

Christ, P.A., Immenhauser, A., Wood, R.A., Darwich, K. and **Niedermayr, A.** (2015) Petrography and environmental controls on the formation of Phanerozoic marine carbonate hardgrounds. *Earth-Sci. Rev.*, **151**, 176–226.

Clari, P.A., Dela Pierre, F. and **Martire, L.** (1995) Discontinuities in carbonate successions: identification, interpretation and classification of some Italian examples. *Sed. Geol.*, **100**, 97–121.

Clavel, B., Charollais, J., Busnardo, R. and **Le Hégarat, G.** (1986) Précisions stratigraphiques sur le Crétacé inférieur basal du Jura méridional. *Eclogae Geol. Helvetiae*, **79**, 319–341.

Conrad, C.P. (2013) The solid Earth's influence on sea level. *GSA Bull.*, **125**, 1027–1052.

Dercourt, J., Gaetani, M., Vrielynck, B., Barrier, E., Biju-Duval, B., Brunet, M.F., Cadet, J.-P., Crasquin, S. and **Sandulescu, M.** (2000) *Atlas Peri-Tethys, Palaeogeographical Maps*. CCGM/CGMW, Paris.

Donselaar, M.E. (1989) The Cliff House Sandstone, San Juan Basin, New Mexico: model for the stacking of 'transgressive' barrier complexes. *J. Sed. Petrol.*, **59**, 13–27.

Dott, R.H.J. (1983) Episodic sedimentation – how normal is average? How rare is rare? Does it matter? *J. Sed. Petrol.*, **53**, 5–23.

Dunham, R.G. (1962) Classification of carbonate rocks according to depositional texture. *AAPG Mem.*, **1**, 108–121.

Elrick, M. and **Hinnov, L.A.** (2007) Millennial-scale paleoclimate cycles recorded in widespread Palaeozoic deeper water rhythmites of North America. *Palaeogeogr. Palaeoclimatol. Palaeoecol.*, **243**, 348–372.

Enos, P. (1991) Sedimentary parameters for computer modeling. In: *Sedimentary Modeling, Computer Simulations and Methods for Improved Parameter Definition* (Eds E.K. Franseen, W.L. Watney, C.G.S.T.C. Kendall and W. Ross), *Kansas Geol. Survey Bull.*, **233**, pp. 63–99.

Eyles, N. (1993) Earth's glacial record and its tectonic setting. *Earth-Sci. Rev.*, **35**, 1–248.

Fairbridge, R.W. (1976) Convergence of evidence on climatic change and ice ages. *Ann. New York Acad. Sci.*, **91**, 542–579.

Fischer, A.G., Hilgen, F.J. and **Garrison, R.E.** (2009) Mediterranean contributions to cyclostratigraphy and astrochronology. *Sedimentology*, **56**, 63–94.

Flessa, K.W., Cutler, A.H. and **Meldahl, K.H.** (1993) Time and taphonomy: quantitative estimates of time-averaging and stratigraphic disorder in a shallow marine habitat. *Paleobiology*, **19**, 266–286.

Frakes, L.A., Francis, J.E. and **Syktus, J.I.** (1992) *Climate Modes of the Phanerozoic*. Cambridge University Press, Cambridge, 274 pp.

Goldhammer, R.K. (1997) Compaction and decompaction algorithms for sedimentary carbonates. *J. Sed. Petrol.*, **67**, 26–35.

Gómez, J.J. and **Fernández-López, S.** (1994) Condensation processes in shallow platforms. *Sed. Geol.*, **92**, 147–159.

Gonzalez, R. and **Eberli, G.P.** (1997) Sediment transport and bedforms in a carbonate tidal inlet; Lee Stocking Island, Exumas, Bahamas. *Sedimentology*, **44**, 1015–1030.

Gornitz, V., Lebedeff, S. and **Hansen, J.** (1982) Global sea-level trend in the past century. *Science*, **215**, 1611–1614.

Gradstein, F.M., Agterberg, F.P., Ogg, J.G., Hardenbol, J., van Veen, P., Thierry, J. and **Huang, Z.** (1995) A Triassic, Jurassic and Cretaceous time scale. In: *Geochronology, Time Scales, and Global Stratigraphic Correlation* (Eds W.A. Berggren, D.V. Kent, M.-P. Aubry and J. Hardenbol), *SEPM Spec. Publ.*, **54**, pp. 95–126.

Häfeli, C. (1966) Die Jura/Kreide-Grenzschichten im Bielerseegebiet (Kt. Bern). *Eclogae Geol. Helvetiae*, **59**, 565–695.

Halley, R.B. and **Harris, P.M.** (1979) Fresh-water cementation of a 1,000-year-old oolite. *J. Sed. Petrol.*, **49**, 969–988.

Hardenbol, J., Thierry, J., Farley, M.B., Jacquin, T., De Graciansky, P.-C. and **Vail, P.R.** (1998) Cretaceous sequence chronostratigraphy. In: *Mesozoic and Cenozoic Sequence Stratigraphy of European Basins* (Eds P.-C. De Graciansky, J. Hardenbol, T. Jacquin and P.R. Vail), *SEPM Spec. Publ.*, **60**, chart.

Hardie, L.A. and **Ginsburg, R.N.** (1977) Layering; the origin and environmental significance of lamination and thin bedding. In: *Sedimentation on the Modern Carbonate Tidal Flats of Northwest Andros Island, Bahamas* (Ed. L.A. Hardie), *John Hopkins University, Stud. Geol.*, **22**, pp. 50–123.

Hill, J., Wood, R., Curtis, A. and **Tetzlaff, D.M.** (2012) Preservation of forcing signals in shallow water carbonate sediments. *Sed. Geol.*, **275–276**, 79–92.

Hillgärtner, H. (1998) Discontinuity surfaces on a shallow-marine carbonate platform (Berriasian – Valanginian, France and Switzerland). *J. Sed. Res.*, **68**, 1093–1108.

Hillgärtner, H. (1999) The evolution of the French Jura Platform during the Late Berriasian to Early Valanginian: controlling factors and timing. *GeoFocus*, **1**, 203.

Hillgärtner, H., Dupraz, C. and **Hug, W.** (2002) Microbially induced cementation of carbonate sands: are micritic meniscus cements good indicators of vadose diagenesis? *Sedimentology*, **48**, 117–131.

Hine, A.C. (1977) Lily Bank, Bahamas: history of an active oolite sand shoal. *J. Sed. Petrol.*, **47**, 1554–1581.

Immenhauser, A. (2009) Estimating palaeo-water depth from the physical rock record. *Earth-Sci. Rev.*, **96**, 107–139.

Jacobs, D.K. and **Sahagian, D.L.** (1993) Climate-induced fluctuations in sea level during non-glacial times. *Nature*, **361**, 710–712.

Joachimski, M.M. (1994) Subaerial exposure and deposition of shallowing-up sequences: evidence from stable isotopes of Purbeckian peritidal carbonates (basal Cretaceous), Swiss and French Jura Mountains. *Sedimentology*, **41**, 805–824.

Kemp, D. (2012) Stochastic and deterministic controls on stratigraphic completeness and fidelity. *Int. J. Earth Sci.*, **101**, 2225–2238.

Kemp, D.B. and **Sadler, P.M.** (2014) Climatic and eustatic signals in a global compilation of shallow marine carbonate accumulation rates. *Sedimentology*, **61**, 1286–1297.

Loutit, T.S., Hardenbol, J., Vail, P.R. and **Baum, G.R.** (1988) Condensed sections: the key to age-dating and correlation of continental margin sequences. In: *Sea Level Changes – An Integrated Approach* (Eds C.K. Wilgus, B.S. Hastings, C.G.St.C. Kendall, H.W. Posamentier, C.A. Ross and J.C. Van Wagoner), *SEPM Spec. Publ.*, **42**, 183–213.

McCabe, P.J. and **Jones, C.M.** (1977) Formation of reactivation surfaces within superimposed deltas and bedforms. *J. Sed. Petrol.*, **47**, 707–715.

Miall, A. (2014) The emptiness of the stratigraphic record: a preliminary evaluation of missing time in the Mesa Verde Group, Book Cliffs, Utah, U.S.A. *J. Sed. Res.*, **84**, 457–469.

Milliman, J.D., Freile, D., Steinen, R.P. and **Wilber, R.J.** (1993) Great Bahama Bank aragonitic muds: mostly inorganically precipitated, mostly exported. *J. Sed. Petrol.*, **63**, 589–595.

Mitchum, R.M., Jr (1977) Glossary of terms used in seismic stratigraphy. In: *Seismic Stratigraphy: Applications to Hydrocarbon Exploration* (Ed. C.E. Payton), AAPG Mem., **26**, 205–212..

Mitchum, R.M., Jr and **Van Wagoner, J.C.** (1991) High-frequency sequences and their stacking patterns: sequence-stratigraphic evidence of high-frequency eustatic cycles. *Sed. Geol.*, **70**, 131–160.

Mojon, P.-O. (2002) *Les formations mésozoïques à charophytes (Jurassique moyen – Crétacé inférieur) de la marge téthysienne nord-occidentale (Sud-Est de la France, Suisse occidentale, Nord-Est de l'Espagne): Sédimentologie, micropaléontologie, biostratigraphie.* Unpublished PhD Dissertation, University of Grenoble, Grenoble, France, 386 pp.

Osleger, D. (1991) Subtidal carbonate cycles: implications for allocyclic vs. autocyclic controls. *Geology*, **19**, 917–920.

Pasquier, J.-B. (1995) *Sédimentologie, stratigraphie séquentielle et cyclostratigraphie de la marge nord-téthysienne au Berriasien en Suisse occidentale (Jura, Helvétique et Ultrahelvétique; comparaison avec les séries de bassin des domaines Vocontien et Subbriançonnais).* Unpublished PhD Dissertation, University of Fribourg, Fribourg, Switzerland, 274 pp.

Pasquier, J.-B. and **Strasser, A.** (1997) Platform-to-basin correlation by high-resolution sequence stratigraphy and cyclostratigraphy (Berriasian, Switzerland and France). *Sedimentology*, **44**, 1071–1092.

Plint, A.G., Eyles, N., Eyles, C.H. and **Walker, R.G.** (1992) Control of sea level change. In: *Facies Models – Response to Sea Level Change* (Eds R.G. Walker and N.P. James), *Geol. Assoc. Canada*, St. John's, 15–25.

Pratt, B.R. and **James, N.P.** (1986) The St George Group (Lower Ordovician) of western Newfoundland: tidal flat island model for carbonate sedimentation in shallow epeiric seas. *Sedimentology*, **33**, 313–343.

Purkis, S.J., Rowlands, G.P. and **Kerr, J.M.** (2015) Unravelling the influence of water depth and wave energy on the facies diversity of shelf carbonates. *Sedimentology*, **62**, 541–565.

Rameil, N., Immenhauser, A., Csoma, A.E. and **Warrlich, G.** (2012) Surfaces with a long history: the Aptian top Shu'aiba Formation unconformity, Sultanate of Oman. *Sedimentology*, **59**, 212–248.

Rankey, E.C. (2002) Spatial patterns of sediment accumulation on a Holocene carbonate tidal flat, northwest Andros Island, Bahamas. *J. Sed. Res.*, **72**, 591–601.

Rankey, E.C. and **Morgan, J.** (2002) Quantified rates of geomorphic change on a modern tidal flat, Bahamas. *Geology*, **30**, 583–586.

Rankey, E.C. and **Reeder, S.L.** (2010) Controls on platform-scale patterns of surface sediments, shallow Holocene platforms, Bahamas. *Sedimentology*, **57**, 1545–1565.

Reineck, H.-E. (1960) Über Zeitlücken in rezenten Flachsee-Sedimenten. *Geol. Rdsch.*, **49**, 149–161.

Reolid, M., Marok, A. and **Lasgaa, I.** (2014) Taphonomy and ichnology: tools for interpreting a maximum flooding interval in the Berriasian of Tlemcen Domain (Western Tellian Atlas, Algeria). *Facies*, **60**, 905–920.

Rodrigo-Gámiz, M., Martínez-Ruiz, F., Rodríguez-Tovar, F.J., Jiménez-Espejo, F.J. and **Pardo-Igúzquiza, E.** (2014) Millennial- to centennial-scale climate periodicities and forcing mechanisms in the westernmost Mediterranean for the past 20,000 yr. *Quat. Res.*, **81**, 78–93.

Sadler, P.M. (1981) Sediment accumulation rates and the completeness of stratigraphic sections. *J. Geol.*, **89**, 569–584.

Sadler, P.M. (1999) The influence of hiatuses on sediment accumulation rates. *GeoResearchForum*, **5**, 15–40.

Satterley, A.K. (1996) The interpretation of cyclic successions of the Middle and Upper Triassic of the Northern and Southern Alps. *Earth-Sci. Rev.*, **40**, 181–207.

Sattler, U., Immenhauser, A., Hillgärtner, H. and **Esteban, M.** (2005) Characterization, lateral variability and lateral extent of discontinuity surfaces on a carbonate platform (Barremian to Lower Aptian, Oman). *Sedimentology*, **52**, 339–361.

Schlager, W. (1991) Depositional bias and environmental change – important factors in sequence stratigraphy. *Sed. Geol.*, **70**, 109–130.

Schlager, W. (1999) Scaling of sedimentation rates and drowning of reefs and carbonate platforms. *Sed. Geol.*, **70**, 109–130.

Schlager, W., Marsal, D., van der Geest, P.A.G. and **Sprenger, A.** (1998) Sedimentation rates, observation span, and the problem of spurious correlation. *Math. Geol.*, **30**, 547–556.

Schulz, M. and **Schäfer-Neth, C.** (1998) Translating Milankovitch climate forcing into eustatic fluctuations via thermal deep water expansion: a conceptual link. *Terra Nova*, **9**, 228–231.

Shinn, E.A. (1968) Burrowing in recent lime sediments of Florida and the Bahamas. *J. Paleont.*, **42**, 879–894.

Steinhauser, N. and **Lombard, A.** (1969) Définition de nouvelles unités lithostratigraphiques dans le Crétacé inférieur du Jura méridional (France). *Comptes Rendus Soc. Physique et D'Hist. Nat. Genève*, **4**, 100–113.

Stockhecke, M., Sturm, M., Brunner, I., Schmincke, H.-U., Sumita, M., Kipfer, R., Cukur, D., Kwiecien, O. and **Anselmetti, F.S.** (2014) Sedimentary evolution and environmental history of Lake Van (Turkey). *Sedimentology*, **61**, 1830–1861.

Strasser, A. (1988) Shallowing-upward sequences in Purbeckian peritidal carbonates (lowermost Cretaceous, Swiss and French Jura Mountains). *Sedimentology*, **35**, 369–383.

Strasser, A. (1991) Lagoonal-peritidal sequences in carbonate environments: autocyclic and allocyclic processes. In: *Cycles and Events in Stratigraphy* (Eds G. Einsele, W. Ricken and A. Seilacher), pp. 709–721. Springer-Verlag, Heidelberg.

Strasser, A. and **Davaud, E.** (1983) Black pebbles of the Purbeckian (Swiss and French Jura): lithology, geochemistry and origin. *Eclogae Geol. Helvetiae*, **76**, 551–580.

Strasser, A. and **Davaud, E.** (1986) Formation of Holocene limestone sequences by progradation, cementation, and erosion: two examples from the Bahamas. *J. Sed. Petrol.*, **56**, 422–428.

Strasser, A. and **Hillgärtner, H.** (1998) High-frequency sea-level fluctuations recorded on a shallow carbonate platform (Berriasian and Lower Valanginian of Mount Salève, French Jura). *Eclogae Geol. Helvetiae*, **91**, 375–390.

Strasser, A. and **Samankassou, E.** (2003) Carbonate sedimentation rates today and in the past: Holocene of Florida Bay, Bahamas, and Bermuda vs. Upper Jurassic and Lower Cretaceous of the Jura Mountains (Switzerland and France). *Geol. Croatica*, **56**, 1–18.

Strasser, A., Pittet, B., Hillgärtner, H. and **Pasquier, J.-B.** (1999) Depositional sequences in shallow carbonate-dominated sedimentary systems: concepts for a high-resolution analysis. *Sed. Geol.*, **128**, 201–221.

Strasser, A., Hillgärtner, H., Hug, W. and **Pittet, B.** (2000) Third-order depositional sequences reflecting Milankovitch cyclicity. *Terra Nova*, **12**, 303–311.

Strasser, A., Hillgärtner, H. and **Pasquier, J.-B.** (2004) Cyclostratigraphic timing of sedimentary processes: an example from the Berriasian of the Swiss and French Jura Mountains. In: *Cyclostratigraphy: Approaches and Case Histories* (Eds B. D'Argenio, A.G. Fischer, I. Premoli Silva, H. Weissert and V. Ferreri), *SEPM Spec. Publ.*, **81**, 135–151.

Strasser, A., Védrine, S. and **Stienne, N.** (2012) Rate and synchronicity of environmental changes on a shallow carbonate platform (Late Oxfordian, Swiss Jura Mountains). *Sedimentology*, **59**, 185–211.

Tipper, J.C. (1997) Modeling carbonate platform sedimentation – lag comes naturally. *Geology*, **25**, 495–498.

Tresch, J. (2007) History of a Middle Berriasian transgression (France, Switzerland, and southern England). *GeoFocus*, **16**, 271.

Trümpy, R. (1980) *Geology of Switzerland, A Guide-Book. Part A: An Outline of the Geology of Switzerland.* Wepf & Co., Basel, 104 pp.

Tucker, M.E., Gallagher, J. and **Leng, M.J.** (2009) Are beds in shelf carbonates millennial-scale cycles? An example from the mid-Carboniferous of northern England. *Sed. Geol.*, **214**, 19–34.

Vail, P.R., Audemard, F., Bowen, S.A., Eisner, P.N. and **Perez-Cruz, C.** (1991) The stratigraphic signatures of tectonics, eustasy and sedimentology – an overview. In: *Cycles and Events in Stratigraphy* (Eds G. Einsele, W. Ricken and A. Seilacher), pp. 617–659. Springer-Verlag, Heidelberg.

Waite, R., Marty, D., Strasser, A. and **Wetzel, A.** (2013) The lost paleosols: masked evidence for emergence and soil formation on the Kimmeridgian Jura platform (NW Switzerland). *Palaeogeogr. Palaeoclimatol. Palaeoecol.*, **376**, 73–90.

Wigley, T.M.L. and **Raper, S.C.B.** (1987) Thermal expansion of sea water associated with global warming. *Nature*, **330**, 127–131.

Zühlke, R., Bechstädt, T. and **Mundil, R.** (2003) Sub-Milankovitch and Milankovitch forcing on a model Mesozoic carbonate platform – the Latemar (Middle Triassic, Italy). *Terra Nova*, **15**, 69–80.

4

Intercontinental correlation of organic carbon and carbonate stable isotope records: evidence of climate and sea-level change during the Turonian (Cretaceous)

IAN JARVIS*, JOÃO TRABUCHO-ALEXANDRE†,‡, DARREN R. GRÖCKE†, DAVID ULIČNÝ§ and JIŘÍ LAURIN§

*Department of Geography and Geology, Kingston University London, Kingston upon Thames, KT1 2EE, UK
†Department of Earth Sciences, Durham University, Durham, DH1 3LE, UK
‡Institute of Earth Sciences, Utrecht University, Budapestlaan, 43584 CD, Utrecht, Netherlands
§Institute of Geophysics, Academy of Sciences of the Czech Republic, 141 31, Prague, Czech Republic

Keywords
Carbon isotopes, chemostratigraphy, climate change, Cretaceous, oxygen isotopes, pCO_2, sea-level change.

ABSTRACT

Carbon ($\delta^{13}C_{org}$, $\delta^{13}C_{carb}$) and oxygen ($\delta^{18}O_{carb}$) isotope records are presented for an expanded Upper Cretaceous (Turonian–Coniacian) hemipelagic succession cored in the central Bohemian Cretaceous Basin, Czech Republic. Geophysical logs, biostratigraphy and stable carbon isotope chemostratigraphy provide a high-resolution stratigraphic framework. The $\delta^{13}C_{carb}$ and $\delta^{13}C_{org}$ profiles are compared, and the time series correlated with published coeval marine and non-marine isotope records from Europe, North America and Japan. All previously named Turonian carbon isotope events are identified and correlated at high-resolution between multiple sections, in different facies, basins and continents. The viability of using both carbonate and organic matter carbon isotope chemostratigraphy for improved stratigraphic resolution, for placing stage boundaries, and for intercontinental correlation is demonstrated, but anchoring the time series using biostratigraphic data is essential. An Early to Middle Turonian thermal maximum followed by a synchronous episode of stepped cooling throughout Europe during the Middle to Late Turonian is evidenced by bulk carbonate and brachiopod shell $\delta^{18}O_{carb}$ data, and regional changes in the distribution and composition of macrofaunal assemblages. The Late Turonian Cool Phase in Europe was coincident with a period of long-term sea-level fall, with significant water-mass reorganization occurring during the mid-Late Turonian maximum lowstand. Falling $\Delta^{13}C$ ($\delta^{13}C_{carb}$ – $\delta^{13}C_{org}$) trends coincident with two major cooling pulses, point to pCO_2 drawdown accompanying cooling, but the use of paired carbon isotopes as a high-resolution pCO_2 proxy is compromised in the low-carbonate sediments of the Bohemian Basin study section by diagenetic overprinting of the $\delta^{13}C_{carb}$ record. Carbon isotope chemostratigraphy is confirmed as a powerful tool for testing and refining intercontinental and marine to terrestrial correlations.

INTRODUCTION

The global carbon cycle constitutes one of the most fundamental biogeochemical systems affecting all surface reservoirs on our planet, with complex biosphere–atmosphere–hydrosphere–lithosphere interactions that modulate and drive climate change on both short and long timescales (Archer, 2010; Ciais *et al.*, 2013; Schlesinger & Berhardt, 2013). Secular variation in stable carbon isotope ratios determined from fossil carbonate and organic matter provides evidence that the sizes of, and fluxes between, global carbon reservoirs have changed significantly throughout the geological record (Veizer *et al.*, 1999). A residence time of *ca* 100 kyr for carbon in the ocean-atmosphere system (Walker, 1986; Kump & Arthur, 1999; Berner, 2006) ensures that the rock record has potential to capture a robust global signal of palaeo-environmental change affecting the carbon cycle.

Stable carbon isotope ($\delta^{13}C$) chemostratigraphy is increasingly being used as a tool for regional to global correlation of Cretaceous successions (Wendler, 2013 and references therein). It offers higher precision than possible using conventional biostratigraphy (Paul & Lamolda, 2009), potentially down to 10 kyr, and as a result it has been adopted as one of the criteria for the definition of Cretaceous Global Boundary Stratotype Section and Points (GSSPs, Kennedy *et al.*, 2014; Lamolda *et al.*, 2014).

Most stratigraphic studies of Cretaceous carbon isotopes have focussed on $\delta^{13}C$ time series obtained from marine bulk pelagic or hemipelagic carbonates (Scholle & Arthur, 1980; Jenkyns *et al.*, 1994; Weissert *et al.*, 1998, 2008; Herrle *et al.*, 2004; Katz *et al.*, 2005; Sprovieri *et al.*, 2006, 2013; Jarvis *et al.*, 2002, 2006; Wendler, 2013). However, a unique feature of carbon isotope chemostratigraphy is the ability to compare records derived from oxidized carbon (carbonate, $\delta^{13}C_{carb}$) and reduced carbon (organic matter, $\delta^{13}C_{org}$) reservoirs (Jarvis *et al.*, 2011), and between marine and non-marine (terrestrial) environments (Gröcke *et al.*, 1999, 2005; Uramoto *et al.*, 2013).

Multiple complementary $\delta^{13}C$ time series may be produced by analysing a wide range of carbon-bearing materials, including bulk sedimentary carbonate or organic matter, carbonate fine fraction (micrite), early diagenetic cements, fossils (skeletal carbonate, leaves, wood, charcoal) and individual organic compounds (biomarkers). However, particularly in Mesozoic and older sediments, facies variation and diagenesis commonly limit the ability to obtain reliable multiple $\delta^{13}C$ records from the same interval within a single section.

The Cenomanian–Turonian boundary (CTB) interval (*ca* 94 Ma) is characterized by a large global positive excursion of $\delta^{13}C$ spanning *ca* 500 kyr that occurs in marine carbonates (values reaching >5‰ $\delta^{13}C_{carb}$), and both marine and terrestrial organic matter (Schlanger *et al.*, 1983, 1987; Arthur *et al.*, 1988; Jarvis *et al.*, 1988a, b, 2006, 2011; Jenkyns *et al.*, 1994; Hasegawa, 1997; Takashima *et al.*, 2011; Uramoto *et al.*, 2013; Joo & Sageman, 2014). This phenomenon is an expression of Oceanic Anoxic Event 2 (OAE2; Schlanger & Jenkyns, 1976), one of the best developed and geographically most extensive of the Mesozoic OAEs (Jenkyns, 2010), which represents an episode of widespread 'black shale' deposition and a major change in the dynamics of the global carbon cycle. It is generally considered that increased burial of organic matter in black shales and other organic reservoirs during OAE2 sequestered ^{12}C, leading to ^{13}C enrichment of all surface carbon reservoirs, and development of the global positive carbon isotope anomaly preserved in multiple archives.

Following OAE2, $\delta^{13}C$ values declined, but carbon isotopes continued to display greater short-term and long-term variation through the Turonian, 93·89 to 89·75 Ma, than any other Late Cretaceous stage (Jarvis *et al.*, 2006; Wendler, 2013). The earliest Turonian represented one of the highest sea-level stands in the Phanerozoic (Hancock & Kauffman, 1979; Haq *et al.*, 1987; Haq, 2014 and references therein), coincident with the highest ocean water temperatures of the last 110 Myr (Friedrich *et al.*, 2012). Major episodes of sea-level and climate change characterized the later Turonian, and hence, the stage provides an excellent opportunity to evaluate interactions between a range of stable isotope and other palaeo-environmental proxies within an interval representing the most extreme Late Cretaceous super-greenhouse.

In this paper, the first continuous high-resolution paired $\delta^{13}C_{carb}$ and $\delta^{13}C_{org}$ records for the Turonian (uppermost Cenomanian–Lower Coniacian) are presented; the similarities and differences between the two time series are critically assessed and correlated with published coeval marine and non-marine records from Europe, North America and Japan (Fig. 1). The viability of using both carbonate and organic matter carbon isotope chemostratigraphy for improved stratigraphic resolution, for placing stage boundaries and for intercontinental correlation is demonstrated, with calibration of the time series using biostratigraphic data. The uses of bulk-sediment carbonate $\delta^{18}O_{carb}$ as a sea-surface temperature (SST) proxy, and of $\Delta^{13}C$ ($\delta^{13}C_{carb}$ − $\delta^{13}C_{org}$) as a pCO_2 proxy, are critically assessed. In addition, evidence is presented for a Europe-wide Late Turonian cool phase that was associated with a drawdown in pCO_2.

CARBON ISOTOPES AS A STRATIGRAPHIC TOOL

The isotopic composition of all surface global carbon reservoirs is considered to be broadly in equilibrium on geological timescales, with rapid exchange of carbon between atmospheric and oceanic carbon dioxide, marine bicarbonate and carbonate tests, and between carbon dioxide and both marine and terrestrial biota (Kump, 1991; Holser, 1997; Kump & Arthur, 1999). Each reservoir displays different $\delta^{13}C$ values due to fractionation effects. The most extreme of these is associated with plant photosynthesis favouring ^{12}C, which today leads to an offset of around −18‰ between dissolved CO_2 and marine phytoplankton, and about −26‰ between dissolved inorganic carbon (DIC; largely bicarbonate HCO_3^-) and phytoplankton (Killops & Killops, 2005). The amount of

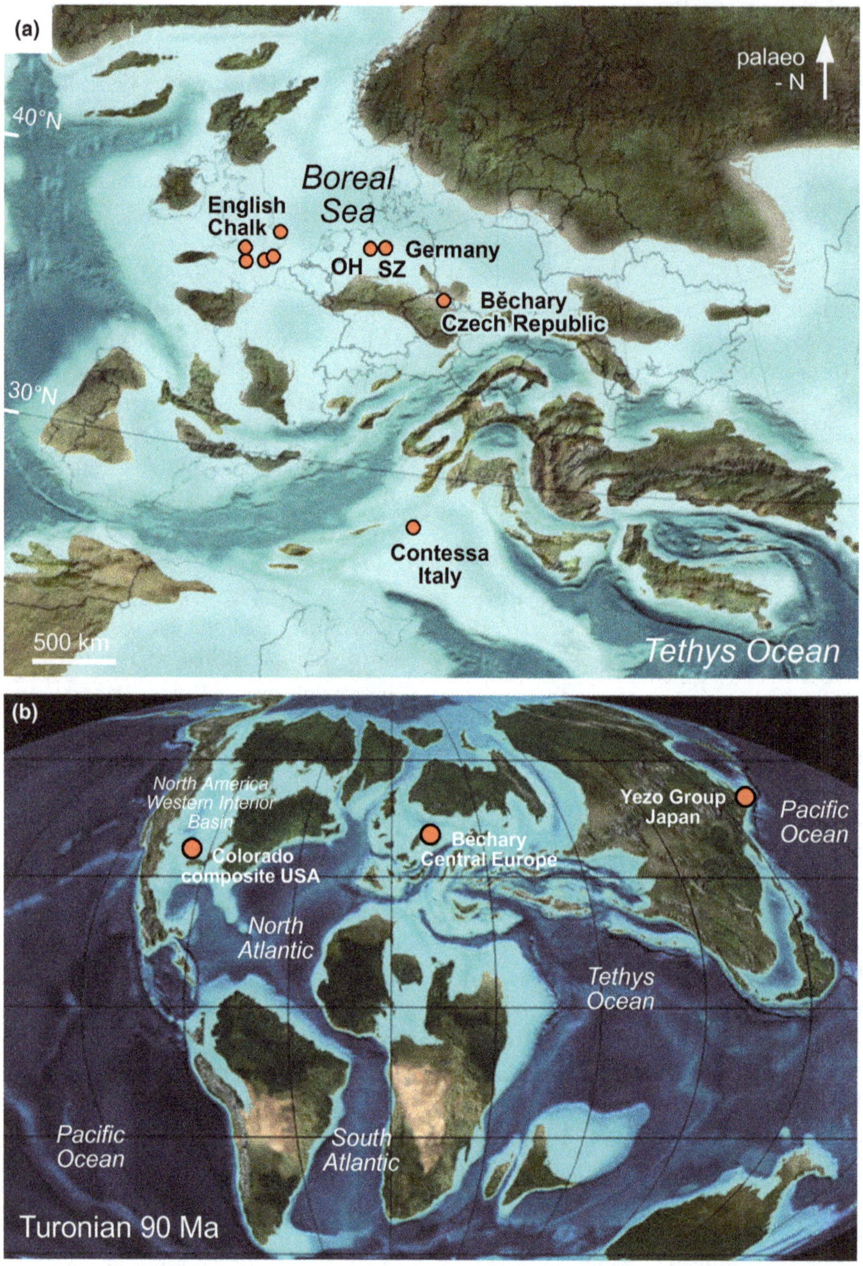

Fig. 1. Turonian palaeogeography and location of sites. (A) Late Cretaceous palaeogeography of Europe showing location of the main regional study sites (filled circles). OH = Oerlinghausen-Halle; SZ = Saltzgitter-Salder. (B) Global palaeogeography at 90 Ma showing location of non-European sections discussed in text. Reconstructions after R.C. Blakey, NAU Geology (http://cpgeosystems.com/75_Cret_EurMap_sm.jpg; http://cpgeosystems.com/90moll.jpg).

isotopic fractionation depends on the photosynthetic 'pathway' (e.g. C3 versus C4 plants; Kump & Arthur, 1999; Gröcke, 2002), on whether photosynthesis takes place in the marine (sea water) or terrestrial (air) environment, on the DIC concentration (or the partial pressure of CO_2), growth rate and on temperature (Rau *et al.*, 1991; Holser, 1997). By contrast, fractionation during precipitation of biogenic or abiogenic calcite from marine bicarbonate (dissolved inorganic carbon) is minor, at around $-1‰$ (Killops & Killops, 2005).

Carbon isotope chemostratigraphy

Stratigraphic variation in $\delta^{13}C$ values preserved in sedimentary archives is controlled principally by the fraction of carbon that is buried on land and in the oceans relative to marine carbonate carbon (Kump & Arthur, 1999). An increase in $\delta^{13}C$, for example, implies the burial of a higher fraction of organic carbon or, alternatively, a decrease in the oxidation of organic matter relative to the weathering of carbonate rocks. Variation in the amount of isotopically light authigenic carbonate precipitated in marine sediments has been proposed as an additional mechanism for driving stratigraphic variation (Schrag *et al.*, 2013), although the significance of this remains unproven. However, in addition to stratigraphic changes, geographical variation occurs in the isotopic composition of individual carbon reservoirs. The $\delta^{13}C$ of inorganically precipitated carbonate in the oceans is close to that of DIC, but $\delta^{13}C$ values of modern marine carbonates differ by 1‰ or more, depending on mineralogy, 'vital effects' and water-mass type and 'age' (Rohling & Cooke, 1999).

Different host materials yield different absolute $\delta^{13}C$ values. Upper Cretaceous marine carbonate has typical $\delta^{13}C_{carb}$ values of 1 to 3‰ (Wendler, 2013). By contrast, carbon isotope fractionation during photosynthesis under high pCO_2 conditions has led to Cretaceous marine organic matter exhibiting low $\delta^{13}C_{org}$ values ranging from −26‰ to −28‰ (Hayes *et al.*, 1999; Meyers, 2014), which are lower than coeval land plant or terrestrial organic carbon, with average $\delta^{13}C_{org}$ values of −23‰ to −25‰ (Dean *et al.*, 1986; Gröcke, 2002; Hasegawa, 2003; Uramoto *et al.*, 2013).

Does bulk carbonate preserve a sea water $\delta^{13}C$ record?

The carbonate component of pelagic and hemipelagic Late Cretaceous and younger sediments are typically dominated by mixed assemblages of coccolithophores and other calcareous nannofossils derived from the photic zone (0 to 200 m depth, and with highest abundance around 50 m; Tappan, 1980), which is reflected in positive $\delta^{13}C_{carb}$ values. This condition is a result of the preferential uptake of ^{12}C from surface waters by phytoplankton and the downward export of ^{12}C-enriched marine organic matter out of the photic zone.

Significant isotopic variation between different coccolith species ('vital effects') and between populations living in different environmental conditions (e.g. pH, temperature, nutrient levels) has been observed in laboratory experiments (Ziveri *et al.*, 2003), which raises concerns about stratigraphic variation in bulk-sediment $\delta^{13}C_{carb}$ values being driven by variations in nannofossil assemblage composition or by local environmental changes. However, compared to modern examples, a very small range of vital effects has been observed in Palaeocene coccoliths (Stoll, 2005), which has been attributed to larger cell diameters and more similar carbon acquisition strategies among different fossil species, perhaps in response to higher atmospheric CO_2 concentrations at that time. This condition suggests that bulk carbonate-carbon isotope data from pelagic sediments probably provide reliable records of surface water $\delta^{13}C$ for other Early and pre-Cenozoic sediments (Bolton *et al.*, 2012).

As with all carbonate systems, diagenesis remains a concern (Swart, 2015 and references therein). Carbon isotopes are much less prone to diagenetic alteration than oxygen isotopes in marine carbonates (Hudson, 1977; Anderson & Arthur, 1983; Banner & Hanson, 1990; Marshall, 1992) because porewater in the sediments generally contains little organic matter, the carbon isotope system is rock-dominated, and carbon isotopes show no significant temperature-controlled fractionation during burial. Notable exceptions occur in association with subaerial exposure surfaces where soil zone CO_2 in meteoric porewaters commonly drives bulk-sediment $\delta^{13}C$ to low values, producing local negative excursions of up to several ‰ (Gross, 1964; Allan & Matthews, 1982; Marshall, 1992; Immenhauser *et al.*, 2008). Furthermore, even though the processes that produce these types of excursion are local, they can be synchronously distributed on a global scale as a result of eustatic sea-level fall. Indeed, in some situations, coupled negative excursions in carbonate and organic $\delta^{13}C$ may result from multiple periods of meteoric alteration of the carbonate $\delta^{13}C$ record combined with an increased contribution of isotopically negative terrestrial organic matter to the sediment (Oehlert & Swart, 2014).

Subaerial exposure cannot be invoked for the hemipelagic and pelagic settings considered here. Excellent agreement between the trends of biostratigraphically well-constrained Upper Cretaceous $\delta^{13}C_{carb}$ profiles from hemipelagic and pelagic successions throughout Europe supports the synchroneity of changes in the isotope record, and illustrates the potential of using a composite $\delta^{13}C_{carb}$ reference curve as a primary criterion for trans-continental correlation (Jarvis *et al.*, 2006; Voigt *et al.*, 2010; Wendler, 2013).

The presence of both pervasive and bed-scale diagenesis is clearly discernable in most Cretaceous pelagic and hemipelagic successions. With the exception of synsedimentary sea floor lithification accompanying nodular chalk and hardground formation (Kennedy & Garrison, 1975; Surlyk *et al.*, 2003; Christ *et al.*, 2015), early diagenesis in most Cretaceous pelagic sections is limited by the organic matter-poor, fine-grained, low permeability, low-Mg calcite-dominated composition of the primary sediment and deposition in sub-storm-wave base environments isolated from exposure to meteoric water.

However, during burial, pressure-solution driven redistribution of carbonate typically affects smaller particles that, in Late Cretaceous pelagic carbonates, consist mainly of coccoliths. These generally have lower $\delta^{13}C$ values than most other co-occurring fossils, so a distinct isotopic pattern that is often observed, with higher $\delta^{13}C$ values in marls than in adjacent chalks (Jarvis *et al.*, 1988a; Frank *et al.*, 1999; Paul *et al.*, 1999; Jeans *et al.*, 2012), may be attributed in part to coccolith depletion in the former lithology. At the same time, pressure-solution driven redistribution of marl-derived carbonate into adjacent chalks as cement (Jarvis *et al.*, 1988a; Mitchell *et al.*, 1997; Jeans *et al.*, 2012) will cause an offset to lower $\delta^{13}C$ values in the chalks, and lead to $\delta^{13}C$ versus $\delta^{18}O$ covariance. However, kinetic fractionation effects during precipitation of the original biogenic calcite may also produce covariance (Wendler *et al.*, 2013), so this is not an unequivocal sign of diagenesis. On the other hand, oxygen isotopes display far greater fractionation than carbon during burial diagenesis, and so offer sensitive indicators of alteration for screening on a sample-by-sample basis (see Anderson & Arthur, 1983; Marshall, 1992).

Can bulk organic matter preserve a primary $\delta^{13}C$ record?

The stable carbon isotope composition of bulk organic matter is affected by changes in the terrestrial contribution to the total organic carbon (TOC) (Kuypers *et al.*, 2004): Cretaceous marine organic matter typically yields $\delta^{13}C$ values that are lower by 3 to 4‰ compared to coeval land plant or terrestrial organic C, so changing ratios of the two components in samples may lead to significant variation in $\delta^{13}C_{org}$ values of bulk organic matter. Varying proportions of different terrestrial constituents such as charcoal (wood), leaf and cuticle may also influence bulk $\delta^{13}C_{org}$ values, although such effects are relatively minor (Heimhofer *et al.*, 2003; Gröcke *et al.*, 2006). Primary chemostratigraphic trends, therefore, will be best obtained from bulk samples where either the contribution of marine or terrestrial organic matter predominates, or the ratio of the two components remains constant.

Regional differences in the degree of carbon isotope fractionation in coeval marine organic matter may be caused by varying rates of productivity. Southern proto-North Atlantic CTB sites, for example, display 2‰ greater amplitude positive $\delta^{13}C_{org}$ excursions than sections elsewhere (Arthur *et al.*, 1988; Sinninghe Damsté *et al.*, 2008). This difference has been interpreted as being a result of higher rates of local CO_2 uptake in surface waters accompanying enhanced primary production, driven by shallowing of the chemocline and increased nutrient input into the photic zone (Laws *et al.*, 1995; Kuypers *et al.*, 2002).

Marine organic matter is also much more reactive than carbonate, a large part of the exported organic material is remineralized in the upper part of the sediment column. Changes in the degree of preservation of components with higher isotope values, such as carbohydrate carbon (Sinninghe Damsté & Köster, 1998; Forster *et al.*, 2008; Zonneveld *et al.*, 2010), may affect the preserved signal. Furthermore, terrestrial organic matter may be preferentially preserved during early diagenesis (de Lange *et al.*, 1994; Hatch & Leventhal, 1997; Prahl *et al.*, 1997), which can produce shifts in $\delta^{13}C_{org}$ records of up to 4 to 5‰, which has implications for the ratio of marine to terrestrial organic matter in bulk deposits, and the subsequent $\delta^{13}C_{org}$ record. Overall, the processes influencing the $\delta^{13}C_{org}$ record are complex and less well understood than those for carbonate (Werne & Hollander, 2004). For stratigraphic purposes, therefore, the form and shape of isotope profiles preserved in geological archives is more reliable than the absolute values, and the potential effects of varying terrestrial versus marine organic matter ratios need to be critically assessed.

The ultimate test for the preservation of a primary signal is whether the same pattern of variation in $\delta^{13}C$ can be recognized in different host materials (e.g. carbonate and organic matter), in different sections, and in different basins, where good stratigraphic constraints are available using biostratigraphy, magnetostratigraphy, astrochronology and/or geochronology. Even then, care is required. For example, Oehlert *et al.* (2012) demonstrated how local carbonate and organic $\delta^{13}C$ covariance may be caused by mixing between pelagic and platform-derived carbonate and organic matter. However, the main focus of the present paper is on European and North American shallow-buried hemipelagic to pelagic carbonate successions that were never subject to subaerial exposure, and lacked adjacent carbonate platforms. These successions should offer optimum conditions for deriving robust primary chemostratigraphic data.

Turonian carbon isotope chemostratigraphy

The most prominent feature of the Late Cretaceous carbon isotope record is the large positive $\delta^{13}C$ excursion representing OAE2, spanning the CTB. The carbon isotope record of the CTB interval has been studied extensively (Schlanger *et al.*, 1987; Jarvis *et al.*, 1988a,b, 2001, 2006, 2011; Arthur *et al.*, 1990; Gale *et al.*, 1993, 2005; Pratt *et al.*, 1993; Jenkyns *et al.*, 1994; Hasegawa, 1997, 2003; Voigt & Hilbrecht, 1997; De Cabrera *et al.*, 1999; Hasegawa & Hatsugai, 2000; Voigt, 2000a; Keller *et al.*, 2001; Wang *et al.*, 2001; Tsikos *et al.*, 2004; Amédro *et al.*, 2005; Bowman & Bralower, 2005; Erbacher *et al.*, 2005; Kolonic *et al.*, 2005; Kuhnt *et al.*, 2005; Li *et al.*, 2006; Parente *et al.*, 2007; Scopelliti *et al.*, 2008; Elrick *et al.*, 2009; Takashima *et al.*, 2011; van Bentum *et al.*, 2012; Hasegawa *et al.*, 2013; Elderbak *et al.*, 2014; Eldrett *et al.*, 2014; Joo & Sageman, 2014; Nagm *et al.*, 2014; Wohlwend *et al.*, 2015) since the pioneering work of Scholle and Arthur (1980).

By marked contrast, a review of carbon isotope stratigraphy from the Archaean to present-day by Saltzman and Thomas (2012) showed an absence of data from almost the entire Turonian. This major gap is due to a paucity of Turonian data from deep-sea isotope records (Katz *et al.*, 2005). Nonetheless, a carbon isotope stratigraphy for the Turonian, based on bulk carbonate records from the English Chalk, was erected by Jarvis *et al.* (2006), who reviewed earlier work on CTB sections in England (Jarvis *et al.*, 1988a,b, 2001; Jeans *et al.*, 1991; Gale *et al.*, 1993, 2005; Lamolda *et al.*, 1994; Paul *et al.*, 1999; Keller *et al.*, 2001; Tsikos *et al.*, 2004), and broader isotopic studies of the Turonian in England (Jenkyns *et al.*, 1994; Pearce *et al.*, 2003), Germany (Voigt & Hilbrecht, 1997; Wiese, 1999; Wiese & Kaplan, 2001), northern Spain (Wiese, 1999) and Italy (Corfield *et al.*, 1991; Jenkyns *et al.*, 1994; Stoll & Schrag, 2000).

Subsequent work has included new carbonate $\delta^{13}C$ data for the Turonian of Germany (Voigt *et al.*, 2007, 2008; Richardt & Wilmsen, 2012) and Italy (Sprovieri *et al.*, 2013; Gambacorta *et al.*, 2015). Turonian $\delta^{13}C_{carb}$ profiles have also been presented from southern Tethyan sections in Tibet (Li *et al.*, 2006; Wendler *et al.*, 2009, 2011), although these display generally lower values and more erratic patterns in the Lower and Middle Turonian than European successions (Wendler, 2013). Most recently, complete Turonian organic carbon $\delta^{13}C$ records have been published from the Czech Republic (Uličný *et al.*, 2014; Olde *et al.*, 2015a) and the US Western Interior Basin (Joo & Sageman, 2014).

MATERIAL AND METHODS

The isotope data presented in this paper were obtained from a research core drilled during 2010 through a thick (405 m) uppermost Cenomanian to Lower Coniacian hemipelagic succession in the Bohemian Cretaceous Basin (Fig. 1; Uličný *et al.*, 2014). This NW–SE oriented 280 km long elongate basin extends between Saxony, Bohemia and Moravia (Czech Republic).

The Bch-1 core site (50·31506°N 15·29497°E), located in the village of Běchary, is situated in the central basin between two depocentres, one adjacent to the Most-Teplice High and Western Sudetic Island in the north-west, the other bordering the Bohemian Massif to the south-east (Uličný *et al.*, 2014, fig. 1). These terrestrial source areas contributed varying amounts of sediment through the Turonian, with the Western Sudetic Island being by far the most prominent. Fine-grained siliciclastic sediment transported via basin-margin deltas and shorefaces became mixed with autochthonous pelagic carbonate to generate calcareous hemipelagic successions in the central basin. No carbonate platform facies are developed within the region, which was dominated by siliciclastic sand facies in near-shore settings.

Bch-1 study core

The main lithofacies at Běchary (Fig. 2) consists of very dark grey marlstones and calcareous mudstones with a varying proportion of quartz silt (coarsest intervals occurring between 360 to 380 m and 140 to 220 m). The mean percentage of $CaCO_3$ through the core is *ca* 35% (range: 4 to 71%), and carbonate is generally represented by a micritic component, some mm-scale bioclasts and microspar in horizons with concretionary cement. These are mostly prominently developed in the low to mid-Upper Turonian (Fig. 2). TOC contents average 0·42% (range: 0·17 to 0·80%) in the bulk sediments (TOC_{WR}) and 0·68% (0·18 to 1·28%) in the acid insoluble residues (TOC_{IR}). Turonian lithofacies show abundant bioturbation throughout the core, dominated by a distal *Cruziana* ichnofacies (cf. MacEachern *et al.*, 2010).

The core was described by Uličný *et al.* (2014), who used geophysical logs combined with lithological and biostratigraphic data to correlate the succession to neighbouring cores and outcrops. The section has been placed in a regional famework via a basin-scale correlation grid developed using well-log correlation (gamma-ray, resistivity, neutron porosity logs) and core data from >700 boreholes, where possible, calibrated by outcrop sedimentology and gamma-ray logging (Uličný *et al.*, 2009, 2014).

The Turonian–Coniacian of the Bohemian Cretaceous Basin has been subdivided into a number of genetic sequences, termed TUR1–TUR7, CON1 and CON2, which were detailed by Uličný *et al.* (2009). The sequences record long-term cycles of regression and subsequent transgression, within which there are multiple smaller scale events; the positions of these sequences in the Bch-1 core are shown in Fig. 2, revised from Uličný *et al.* (2014), following Olde *et al.* (2015a,b). Basin-wide sediment geometries and transgressive-regressive (shore proximity) curves have additionally been used to construct an inferred eustatic sea-level curve, which has been correlated with the Bch-1 well (Uličný *et al.*, 2014).

A precise chronostratigraphic framework for Bch-1 has been developed using macrofossil, calcareous nannofossil and dinoflagellate cyst records from the core (Fig. 2; Uličný *et al.*, 2014; Olde *et al.*, 2015a), combined with

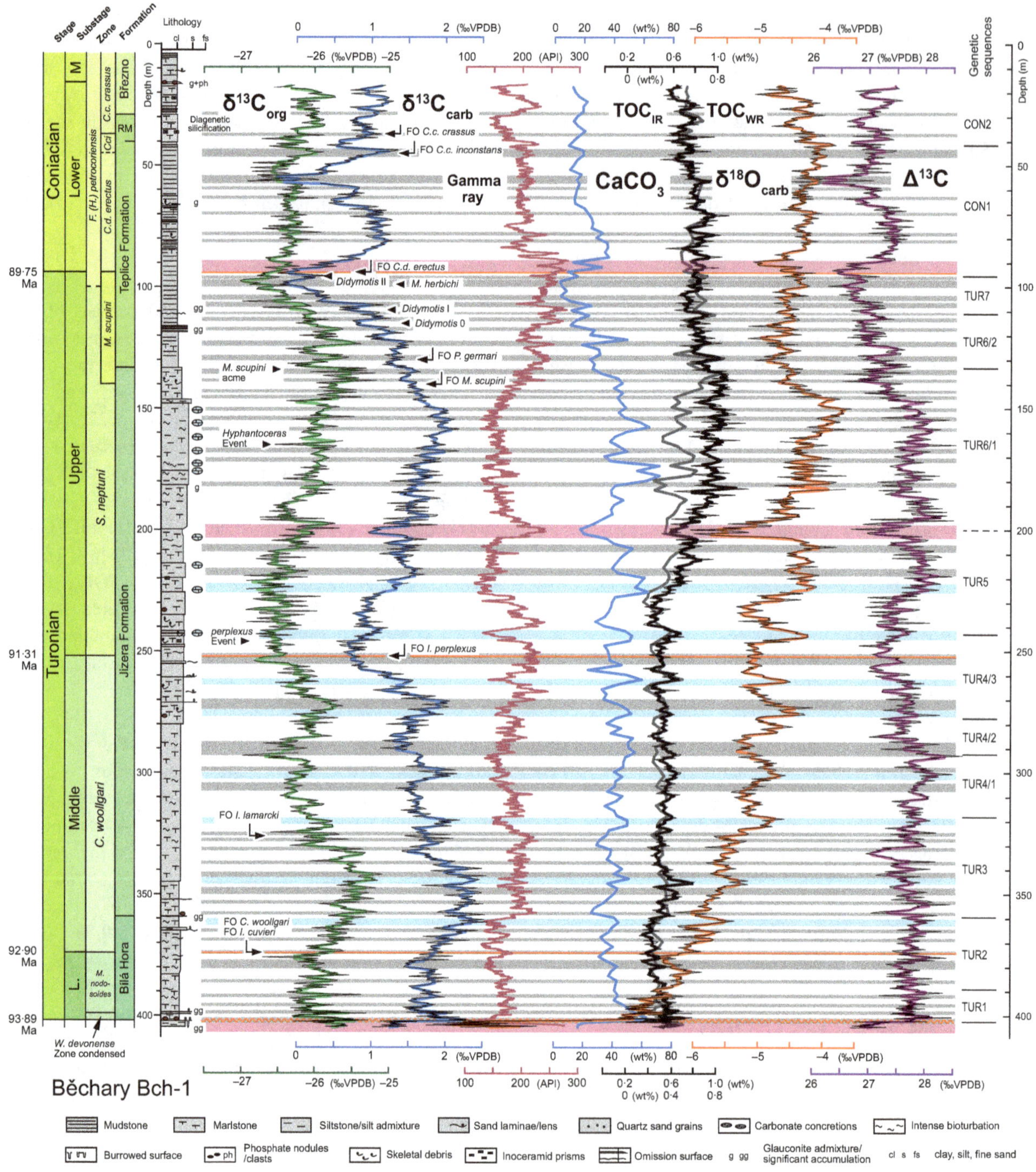

Fig. 2. Lithology, stratigraphy and geochemistry of the Bch-1 well. Carbon isotopes of bulk organic matter ($\delta^{13}C_{org}$) and bulk carbonate ($\delta^{13}C_{carb}$), gamma ray, $CaCO_3$, insoluble residue total organic carbon (TOC_{IR}; black high-resolution profile and black numerals) and whole-rock TOC (TOC_{WR}; grey low-resolution profile and grey numerals), bulk carbonate oxygen isotopes ($\delta^{18}O_{carb}$), and the offset between $\delta^{13}C_{org}$ and $\delta^{13}C_{carb}$ ($\Delta^{13}C$) are shown. Thin black lines represent all data; associated smoothed coloured curves are three-point moving averages. Gamma ray, $CaCO_3$ and TOC_{WR} curves are unsmoothed. Ages of stage and substage boundaries derived from Ogg *et al.* (2012), Laurin *et al.* (2014) and Sageman *et al.* (2014). Biostratigraphy and fossils datum levels after Olde *et al.* (2015b); lithostratigraphic terminology after Čech *et al.* (1980). Basin-scale genetic sequences are modified from Uličný *et al.* (2014), following Olde *et al.* (2015b). Grey bands highlight coincident peaks and troughs in the paired $\delta^{13}C$ profiles. Blue bars are carbonate-rich intervals displaying positive $\delta^{18}O_{carb}$ values. Purple bands highlight levels with a significant diagenetic overprint (see text for details). FO = first occurrence.

geophysical log correlation of key macrofossil biostratigraphic datum levels from adjacent cores and outcrops. Biostratigraphic tie points and age controls are summarized in Appendix S1. The CTB near the base of the core (402 m) is marked by an omission surface. A major hiatus at this level (Uličný *et al.*, 1993, 2014) is confirmed by the absence of calcareous nannofossil zones UC 5a-b, which correlates to the upper part of the *Metoicoceras geslinianum* and *Neocardioceras juddii* ammonite zones (Burnett *et al.*, 1998). This hiatus has been attributed to a major flooding episode (Valečka & Skoček, 1991).

The first occurrence (FO) of the ammonite *Collignoceras woollgari* (Mantell), which marks the base of the Middle Turonian, occurs in the middle of Sequence TUR2, and is correlated with 374 m in Bch-1 (Uličný *et al.*, 2014). This level corresponds to a major regional sea-level lowstand, which was followed by a marked early Middle Turonian transgression. The FO of *Inoceramus perplexus* Whitfield, the Upper Turonian index taxon, is correlated with 252 m (Uličný *et al.* 2014). The end-Middle Turonian marks a long-term regressive maximum, following a general Middle Turonian sea-level fall.

A transgressive event at the base of TUR5 (243 m in Bch-1) in the lowest Upper Turonian is prominent basin wide, and marks a shift to intervals with higher carbonate contents and more widespread cementation in all facies (Fig. 2; Uličný *et al.*, 2014). This event correlates to an acme of *I. perplexus* (= *perplexus* Event). A coarsening-upward trend within TUR6/1, above, provides evidence of shallowing, with high-energy and probably very shallow-water (close to fair-weather wave base) conditions. Subsequent drowning during the latest Turonian, is indicated by a fining-upwards trend accompanying sharply falling $CaCO_3$ contents towards the top of Sequence TUR6/2, but starting from around from 135 m (Fig. 2). Uppermost Turonian Sequence TUR7 is marked by low-carbonate contents (Fig. 2) that fall to a minimum at the top of the sequence, immediately below the stage boundary. The FO of *Cremnoceramus deformis erectus* (Meek), the base Coniacian marker (Kauffman *et al.* 1996), is found towards the bottom of Sequence CON1, correlated with 94 m in Bch-1, with specimens recorded from the core a short distance above (Uličný *et al.*, 2014).

Analytical methods

Samples of *ca* 20 g were taken every 50 cm through the 406 m Bch-1 core for bulk carbonate ($\delta^{13}C_{carb}$, $\delta^{18}O_{carb}$) and bulk organic matter ($\delta^{13}C_{org}$) stable isotope analysis, and TOC determination. New $\delta^{13}C_{carb}$, $\delta^{18}O_{carb}$ data are reported here (803 samples); $\delta^{13}C_{org}$ and TOC results for the same samples were presented previously by Uličný *et al.* (2014). Analytical methodologies are described in Appendix S2. Isotopic ($\delta^{13}C_{carb}$, $\delta^{13}C_{org}$, $\delta^{18}O_{carb}$) and TOC data are provided in Appendix S3. Based on an average compacted sedimentation rate for the Middle and Upper Turonian of 9 cm kyr^{-1} (Uličný *et al.*, 2014), sampling resolution is on the order of 5·6 kyr.

The elemental geochemistry and palynology of the section were studied by Uličný *et al.* (2014) and Olde *et al.* (2015a,b) using a larger size (50 g) lower resolution sample set taken at 2 m intervals (22 kyr) through the core. The isotope data presented here include results from splits of these larger samples, they do not constitute a separate sample set.

GEOCHEMICAL VARIATION IN BCH-1

Stable isotope ($\delta^{13}C_{org}$, $\delta^{13}C_{carb}$, $\delta^{18}O_{carb}$) and TOC data are plotted as chemostratigraphic profiles in Fig. 2, together with the gamma-ray log and a carbonate curve from the lower resolution data set. It is evident that the organic-carbon and carbonate-carbon isotope profiles show very similar long-term trends, offset by 26 to 28‰ ($\Delta^{13}C = \delta^{13}C_{carb} - \delta^{13}C_{org}$), with low-Middle Turonian and mid-Upper Turonian maxima, and lowest Upper Turonian and Turonian–Coniacian boundary minima. Neither carbon isotope curve bears any similarity to the long-term trend of $\delta^{18}O_{carb}$, which displays generally rising values from the base upwards, peaking in the mid-Upper Turonian, followed by a basal *M. scupini* Zone minimum, and then rises again thereafter.

Predictably, $CaCO_3$ shows an inverse correlation with downhole gamma-ray (sourced from K, Th, U radionuclides located principally in the aluminosilicate fraction) values, the lower resolution profile of the former capturing the main stratigraphic patterns seen in the high-resolution (5 cm; 500 years) gamma-ray data (Fig. 2; Uličný *et al.*, 2014). The TOC_{WR} and TOC_{IR} contents are both low throughout the Lower and Middle Turonian. They rise through the Upper Turonian to coincident long-term maxima at the top of the Jizera Formation (around the level of a $\delta^{13}C_{org}$ maximum), then remain at higher levels throughout the remainder of the section; TOC_{WR} contents are generally low, rarely exceeding those of average shale (0·8%; Mason & Moore, 1982). The $\Delta^{13}C$ profile displays a relatively flat long-term trend with values of *ca* 27·7‰ through most of the Turonian, but with an interval of lower values spanning the Middle–Upper Turonian boundary interval. The $\Delta^{13}C$ values fall to *ca* 27·1‰ towards the top of the Upper Turonian, at the summit of the Jizera Formation.

In addition to the long-term (>400 kyr) interrelationships described above, several noteworthy medium-term correlations are evident; these are considered below.

Carbon and oxygen isotopes

There is good correspondence between many short-term (10 to 50 kyr) peaks and troughs developed in the $\delta^{13}C_{carb}$ and $\delta^{13}C_{org}$ curves; these are highlighted by the grey shaded bands in Fig. 2. These coincident peaks and troughs probably reflect an original palaeo-environmental signal, and offer the greatest potential to provide robust datum levels in carbon isotope chemostratigraphy.

In a few cases, the correspondence between the $\delta^{13}C_{carb}$ and $\delta^{13}C_{org}$ curves is poor, or the curves are anti-correlated. Three levels (purple shaded bands in Fig. 2) in particular show major discrepancies: (1) the Cenomanian section at the base of the core below 402 m; (2) a low-carbonate interval in the Upper Turonian around 200 m; (3) the basal Coniacian section below 90 m. All three intervals are characterized by low-carbonate contents with coincident gamma-ray peaks, and exhibit negative excursions in $\delta^{13}C_{carb}$, $\delta^{18}O_{carb}$ and $\Delta^{13}C$. The low absolute values of $\delta^{13}C_{carb}$, $\delta^{18}O_{carb}$ and coincident depletion in both isotopes, point to significant local diagenetic overprinting of carbonate at these levels. This condition was probably caused by carbonate dissolution and the addition of burial microspar precipitated under elevated pore-fluid temperatures (cf. Choquette & James, 1987) or late stage interaction with freshwater aquifer fluids (see below for discussion).

A second type of divergence between the carbon isotope profiles occurs in the Middle Turonian and lower Upper Turonian section. Here, a series of *ca* 0·5‰ positive excursion in $\delta^{18}O_{carb}$ correspond to carbonate-rich levels with small increases in $\delta^{13}C_{carb}$ that are not matched exactly by corresponding peaks in $\delta^{13}C_{org}$ (blue shaded bands in Fig. 2). In many cases this manifests as a small peak offset (e.g. peaks around 320 m), while in others, the disparity is large (e.g. peaks around 300 m). Some of these horizons correlate to concretionary horizons identified in the core (Fig. 2). It is likely, therefore, that the locally elevated values in $\delta^{18}O_{carb}$ and $\delta^{13}C_{carb}$ reflect the addition of early diagenetic calcite cements that have protected the host sediment from later diagenetic alteration. The possibility of significant interaction with organic-derived bicarbonate, which characterizes many concretion carbonates (Hudson, 1977), is excluded by the modest changes observed in $\delta^{13}C_{carb}$.

Geochemical interrelationships

Stratigraphic and diagenetic interrelationships in the geochemical data may be visualized using bivariate plots (Figs 3 and 4). The Al_2O_3 versus $CaCO_3$ plot (Fig. 3A to D) illustrates an inverse relationship between carbonate and clay mineral contents that control the bulk-sediment composition. A linear 'dilution' trend is well displayed by the Upper Turonian interval (Fig. 3C) that contains the widest range of carbonate contents. These data produce a well-defined mixing line between a non-calcareous mudrock with 15% Al_2O_3 and a marly limestone with 80% carbonate (green line in Fig. 3A to D). The line parallels the mixing line between 'average shale' (Wedepohl, 1971) and a pure carbonate end-member (dashed line in Fig. 3A to D). The same regression lines plotted on the other stratigraphic intervals reveals similar tends but with greater scatter below the line, particularly in the Lower Coniacian (Fig. 3D). This phenomenon is attributed to the presence of varying amounts of biogenic silica and/or detrital quartz in the samples; both calcitized silicisponge spicules and detrital quartz grains are visible in thin sections from several intervals.

The TOC_{WR} is positively correlated with Al_2O_3 but with considerable scatter around the 'average shale' mixing line (Fig. 3E to H). This correlation is characteristic of modern shelf sediments and is attributed to the high surface area of the clay mineral fraction (represented here by the Al proxy) favouring the adsorption and preservation of organic matter (Keil *et al.*, 1994; Mayer, 1994; Hedges & Keil, 1995; Hedges *et al.*, 1997). Iron minerals also enhance the preservation of organic matter in marine sediments (Berner, 1970; Lalonde *et al.*, 2012); this may be an additional factor, since Fe is positively correlated with Al in Bch-1 (R^2 = 0·88, data presented in Olde *et al.*, 2015b). The top Cenomanian–Lower Turonian and Middle Turonian intervals are distinguished by having lower whole-rock TOC contents and being TOC-depleted relative to their Al contents (Figs 2 and 3E to F), when compared to the stratigraphically higher parts of the section. This condition is consistent with lower sedimentation rates (Uličný *et al.*, 2014; see footnote in Table S2), greater sea floor oxidation, and reduced burial efficiency driving a lower organic matter burial flux (Hedges & Keil, 1995; Hedges *et al.*, 1999).

With the exception of the Cenomanian–Turonian boundary interval, $\delta^{13}C_{org}$ is generally positively correlated with $\delta^{13}C_{carb}$ throughout the succession (Fig. 4) but with considerable scatter. A reference regression line calculated for the Middle Turonian samples (Fig. 4B) and transposed onto the other three age groups (Fig. 4A and C to D), illustrates a shift to lower $\delta^{13}C_{carb}$ values (<0·5‰) for low-carbonate samples in the uppermost Upper Turonian and Lower Coniacian intervals (Fig. 4C and D). This phenomenon is attributed to the addition of isotopically lighter cement associated with the microbial decomposition of organic matter in this part of the section (cf. Hudson, 1977). Samples in this category display significant shifts to lower $\delta^{13}C_{carb}$ compared to their corresponding $\delta^{13}C_{org}$ and $\delta^{18}O_{carb}$ values, which are compa-

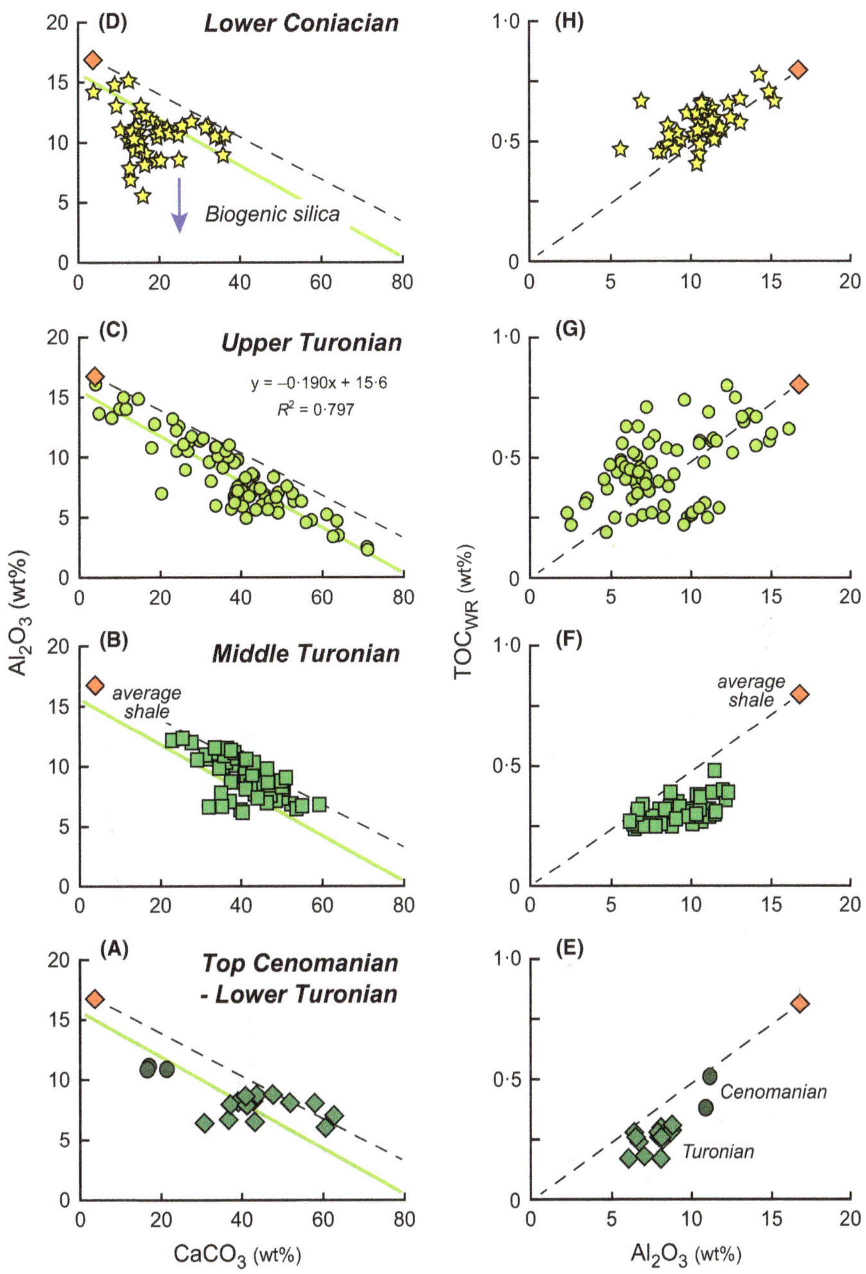

Fig. 3. Elemental cross-plots showing inter-relationships and stratigraphic trends in Bch-1 samples. (A to D) Al_2O_3 versus $CaCO_3$ for the: (A) top Cenomanian–Lower Turonian; (B) Middle Turonian; (C) Upper Turonian; (D) Lower Coniacian intervals. (E to F) Whole-rock total organic carbon (TOC_{WR}) versus Al_2O_3 for the: (E) top Cenomanian–Lower Turonian; (F) Middle Turonian; (G) Upper Turonian; (H) Lower Coniacian intervals. Green filled circles in (A) and (E) are top Cenomanian; green filled diamonds are Lower Turonian samples. Red-filled diamonds show the composition of average shale (Wedepohl, 1971). Green lines in (A to D) are the least squares regression line derived from the Upper Turonian sample set (C). Black dashed lines are mixing lines between average shale and a pure carbonate end-member. Blue arrow in (D) indicates trend offset towards lower Al_2O_3 values caused by the addition of biogenic silica.

rable to values of less altered samples within the same stratigraphic interval (Fig. 4C to D and G to H). The lack of a corresponding shift to lower $\delta^{18}O_{carb}$ values coincident with $\delta^{13}C_{carb}$ depletion (e.g. Fig. 4H) points to an early burial origin for these organic matter-derived cements.

The carbonate carbon versus oxygen isotope plots (Fig. 4) emphasize the stratigraphic trends noted from

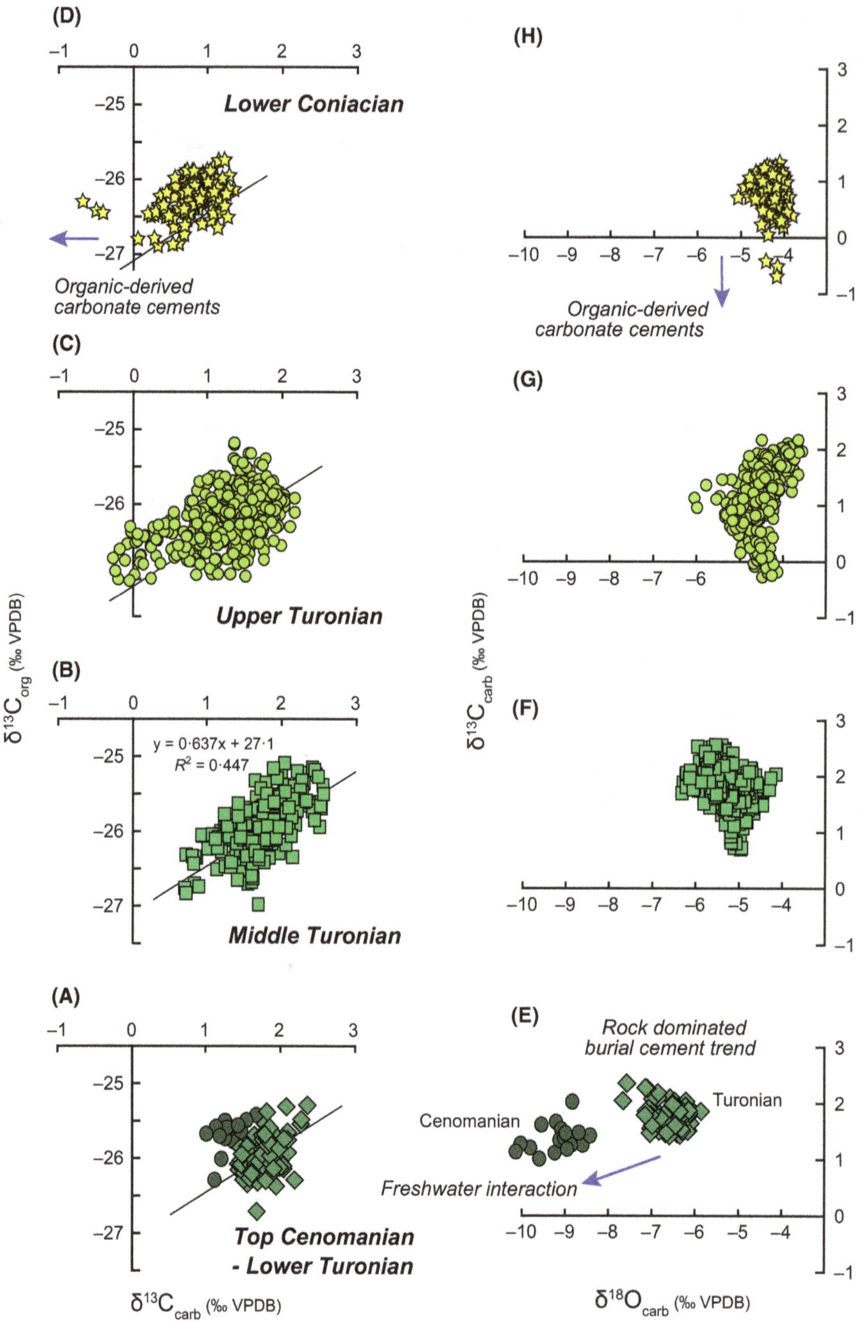

Fig. 4. Isotope cross-plots showing inter-relationships and stratigraphic trends in Bch-1 samples. (A to D) Organic-carbon versus carbonate-carbon stable isotopes ($\delta^{13}C_{org}$ versus $\delta^{13}C_{carb}$) for the: (A) top Cenomanian–Lower Turonian; (B) Middle Turonian; (C) Upper Turonian; (D) Lower Coniacian intervals. (E to F) Carbonate carbon versus oxygen isotopes ($\delta^{13}C_{carb}$ versus $\delta^{18}O_{carb}$) for the: (E) top Cenomanian–Lower Turonian; (F) Middle Turonian; (G) Upper Turonian; (H) Lower Coniacian intervals. Green filled circles in (A) and (E) are top Cenomanian; green filled diamonds are Lower Turonian samples. Black least squares regression lines in (A to D) are derived from the Middle Turonian sample set (B). Major diagenetic trends (blue arrows) are indicated; see text for discussion.

the isotope profiles (Fig. 2), with increasing $\delta^{18}O_{carb}$ values stratigraphically upwards (Fig. 4E to H), and extremely low values (−6 to −10‰ $\delta^{18}O_{carb}$ and *ca* 1‰ $\delta^{13}C_{carb}$) at the base of the section in the uppermost Cenomanian and Lower Turonian (Fig. 4E). No covariance is apparent between $\delta^{18}O_{carb}$ and $\delta^{13}C_{carb}$ in any of the four stratigraphic intervals. Interpretation of these signatures is discussed further below.

CARBON ISOTOPE STRATIGRAPHY AND EUROPEAN CORRELATION

Uličný *et al.* (2014) presented a $\delta^{13}C_{org}$ curve for Bch-1 that was used to place the positions of key named Turonian–Coniacian carbon isotope events (CIEs) in the section. Biostratigraphic data, predominantly macrofossil and nannofossil records from the core and correlated FO datum levels of macrofossils, were employed to pin the isotope curve to a time framework, principally using the bases of the Lower, Middle and Upper Turonian and the base Lower Coniacian (Fig. 2). The placement of named CIEs was based on the recognition of key positive and negative excursions, and inflection and turning points on the $\delta^{13}C_{org}$ curve, and the correlation of these to corresponding features on the English Chalk $\delta^{13}C_{carb}$ reference curve of Jarvis *et al.* (2006), taking into account available biostratigraphic data. Good correlations were demonstrated between the Běchary $\delta^{13}C_{org}$ curve and $\delta^{13}C_{carb}$ curves and faunal records from England, Liencres northern Spain and Salzgitter-Salder NW Germany (Uličný *et al.*, 2014, fig. 3), except for a divergence of trends in the upper part of the Upper Turonian, which was attributed to the more expanded nature of the Běchary section.

The placement of CIEs at Běchary may be re-evaluated using the new $\delta^{13}C_{carb}$ results obtained during the present study (Fig. 5). The $\delta^{13}C_{carb}$ and $\delta^{13}C_{org}$ profiles show generally good agreement and both compare well to the English Chalk reference curve (Jarvis *et al.*, 2006). Confidence in the assignment of the CIEs is further demonstrated by detailed correlation with $\delta^{13}C_{carb}$ curves from corresponding sections at Oerlinghausen-Halle and Saltzgitter-Salder in Germany (Wiese, 1999; Voigt *et al.*, 2007) and Contessa Quarry in Italy (Stoll & Schrag, 2000), where additional biostratigraphic data are available (Figs 1 and 5). The four $\delta^{13}C_{carb}$ curves display parallel trends but varying offsets in absolute values, with Běchary being the lowest by *ca* 1‰. Good correlation has also been achieved to stratigraphically less complete or biostratigraphically less well-constrained $\delta^{13}C_{carb}$ curves from: Anröchte (Richardt & Wilmsen, 2012), Söhlde (Voigt & Hilbrecht, 1997) and Wunstorf (Voigt *et al.*, 2008), Germany; Bottaccione (Corfield *et al.*, 1991; Jenkyns *et al.*, 1994; Sprovieri *et al.*, 2013) and Gubbio (Tsikos *et al.*, 2004) Italy; Santa Ines (Stoll & Schrag, 2000), Spain; and even beyond Europe to Gongzha (Li *et al.*, 2006) and Guru (Wendler *et al.*, 2011), Tibet (data not shown; see Wendler, 2013 for a review).

Precise placement of the CIEs at Běchary is hampered by greater sample-to-sample $\delta^{13}C_{carb}$ variation than exhibited by the smoother $\delta^{13}C_{carb}$ curves obtained from more carbonate-rich and diagenetically more uniform successions elsewhere (Fig. 5). As is generally the case, the complementary $\delta^{13}C_{org}$ curve is even noisier due the greater number of factors influencing bulk organic matter isotopic composition. Using the data from both curves, however, and giving the greatest confidence to features that are coincident in both, it is possible to confirm the placements of Uličný *et al.* (2014) for the Holywell, Pewsey, Navigation, Beeding and Light Point CIEs, and many unnamed correlatable excursions. Taking into account the new $\delta^{13}C_{carb}$ data, which in some intervals show much less variance and closer agreement with the correlative carbonate curves than $\delta^{13}C_{org}$ (Fig. 5), the Round Down and Low-woollgari CIEs are placed one carbon peak lower (equivalent to *ca* 125 kyr), and small shifts upwards or downwards (equivalent to *ca* 50 kyr) are applied to the placement of the Lulworth, Glynde, Caburn, Bridgewick and Hitch Wood CIEs (as adopted previously by Olde *et al.*, 2015a,b; compare Fig. 4 to Uličný *et al.* 2014, fig. 3).

The greatest divergence between the Běchary $\delta^{13}C_{carb}$ and $\delta^{13}C_{org}$ profiles is seen in the Upper Turonian above the Hitch Wood CIE. This CIE, which corresponds to a small positive excursion and turning point at peak $\delta^{13}C_{carb}$ values in the Upper Turonian (Jarvis *et al.* 2006), is immediately preceded by the *Hyphantoceras* faunal Event at Běchary, Saltzgitter-Salder and in southern England (Fig. 5), and elsewhere in Europe. A turning point at an equivalent stratigraphic level (based on biostratigraphy) is not present in the $\delta^{13}C_{org}$ curve, which continues to rise to a stratigraphically higher maximum at the HW2 CIE of Uličný *et al.* (2014). Those authors recognized the lack of correspondence between the Běchary $\delta^{13}C_{org}$ profile and published $\delta^{13}C_{carb}$ curves from elsewhere and, taking into account the biostratigraphic data, placed the Hitch Wood CIE at a minor peak in the organic carbon curve which was designated HW1. This placement falls a short distance below the Hitch Wood Event defined here using the $\delta^{13}C_{carb}$ profile, and coincides with the *Hyphantoceras* Event.

The amplitude of $\delta^{13}C_{carb}$ variation through the Upper Turonian–Lower Coniacian boundary interval is greater at Běchary than in the other European sections illustrated here (Fig. 5), which is attributed to greater diagenetic overprinting of the carbonate-poor sediments in Bch-1. Nonetheless, the general shape of the curve agrees closely with those from elsewhere, and corresponds well to the $\delta^{13}C_{org}$ curve. In this interval, the Běchary $\delta^{13}C_{org}$ curve shows lower amplitude medium-term variation than $\delta^{13}C_{carb}$, and the former is closer to the scale of variation seen in $\delta^{13}C_{carb}$ curves in other sections. Minor Upper Turonian $\delta^{13}C$ excursions (including HW2 and HW3 of Uličný *et al.*, 2014) that are well-constrained by biostratigraphy show good correspondence to peaks identified at other localities (Fig. 5).

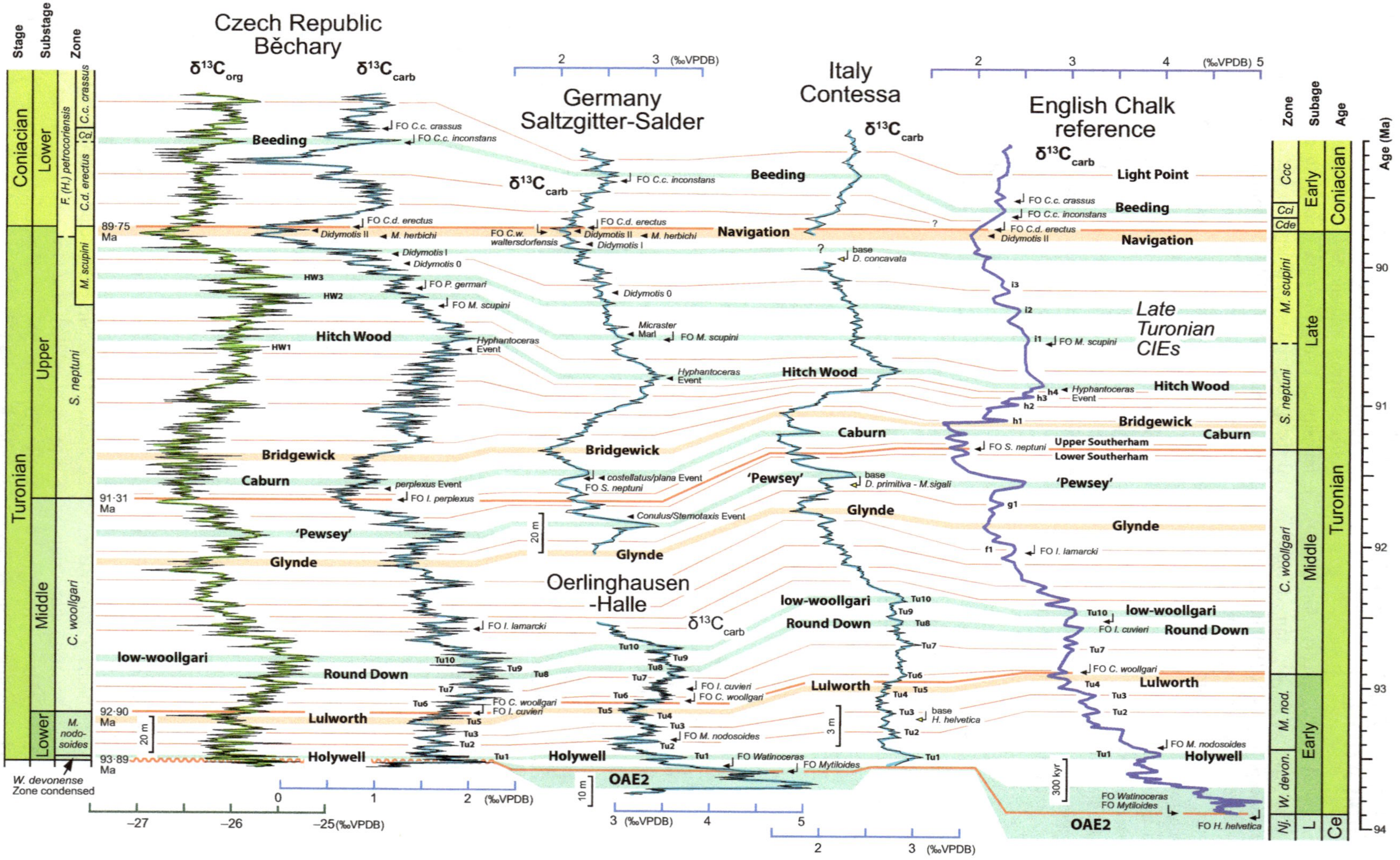

Fig. 5. Chemostratigraphic correlation of European Turonian sections using bulk-sediment carbonate-carbon stable isotopes. Location of the sites is shown in Fig. 1. High-resolution paired $\delta^{13}C_{carb}$ and $\delta^{13}C_{org}$ data from the Běchary core illustrate generally good agreement between the two profiles. Thin black lines represent all data; associated smoothed coloured curves are three-point moving averages. The English Chalk reference curve displays smoothed data only. Isotope data sources: Běchary $\delta^{13}C_{carb}$, this study; $\delta^{13}C_{org}$, Uličný *et al.* (2014); Saltzgitter-Salder (Voigt and Hilbrecht, 1997); Oerlinghausen-Halle, Voigt *et al.* (2007); Contessa, Stoll and Schrag (2000); English Chalk, Jarvis *et al.* (2006), recalibrated to GTS2012 after Laurin *et al.* (2014). Ages of stage and substage boundaries derived from Ogg *et al.* (2012), Laurin *et al.* (2014) and Sageman *et al.* (2014). Placement of Contessa biostratigraphic datum levels based on Sprovieri *et al.* (2013). Break in the Contessa profile represents an unsampled section correlated with immediately below the Navigation CIE (Sprovieri *et al.*, 2013). Correlation of positive (green) and negative (cream) carbon isotope excursions defining named carbon isotope events (Jarvis *et al.*, 2006) is shown by horizontal coloured bands. Thin red horizontal lines indicate other isotope correlations; thick red horizontal lines are stage and substage boundaries.

Jeans *et al.* (2012) have argued that the CIEs defined by Jarvis *et al.* (2006) from the English Chalk might be attributed largely to diagenesis, driven by the presence of varying proportions of fine-grained (<2 μm) cements having $\delta^{13}C_{carb}$ values ranging from 3·5 to −8‰. A lack of correlation between carbonate and organic matter $\delta^{13}C$ trends in a short section of English Lower Campanian Chalk was offered as evidence that the carbonate is diagenetically altered, despite the well-preserved nature of the sediments and their enclosed fossils. As a result, most previously defined Upper Cretaceous CIEs (Jarvis *et al.*, 2006) were considered to be untenable by Jeans *et al.* (2012). However, the new high-resolution carbon isotope data presented here fully support our previous work and assumptions.

Our correlations confirm the utility of carbon isotope stratigraphy, even in sections that have been overprinted by significant diagenesis, but they demonstrate that it is essential to have complementary biostratigraphic data to anchor the chemostratigraphic framework. Correlation precision is limited by sample-to-sample variability in the isotope data, but robust correlations may still be achieved when both $\delta^{13}C_{carb}$ and $\delta^{13}C_{org}$ curves are available. Despite the inherent differences expected between $\delta^{13}C_{carb}$ and $\delta^{13}C_{org}$ profiles obtained from successions showing a significant diagenetic overprint, correlation using a $\delta^{13}C_{org}$ curve from such sections, compared to $\delta^{13}C_{carb}$ curves from elsewhere, is able to achieve a precision for the correlation of major CIEs of better than 40 kyr, in most cases.

OXYGEN ISOTOPES AND THE LATE TURONIAN COOL PHASE

Cretaceous pelagic and hemipelagic carbonate bulk sediment and carbonate fine-fraction oxygen isotopes commonly preserve consistent stratigraphic trends that have been used to interpret variation in past SST (Ditchfield & Marshall, 1989; Jenkyns *et al.*, 1994; Schrag *et al.*, 1995; Clarke & Jenkyns, 1999; Stoll & Schrag, 2000). The suitability of $\delta^{18}O$ trends derived from bulk-sediment analyses as a palaeo-environmental proxy, is supported by compatible $\delta^{18}O$ data obtained from pristine brachiopod shells (Voigt, 2000b; Voigt *et al.*, 2004, 2006) and planktonic foraminifera (Voigt *et al.*, 2010) enclosed in the same sediments, and from coincident changes in macrofossil assemblages (Jarvis *et al.*, 2011).

Bulk pelagic carbonate $\delta^{18}O$ data for the Cretaceous generally exhibit the lowest values around the CTB extending into the Lower Turonian, pointing to a SST maximum and associated climate optimum, followed by general long-term cooling through the remainder of the Late Cretaceous (Scholle & Arthur, 1980; Jenkyns *et al.*, 1994; Clarke & Jenkyns, 1999). This interpretation is supported by isotopic analysis of pristine benthic and planktonic foraminifera in deep-sea sediments, and by TEX_{86} biomarker studies (Huber *et al.*, 2002; Friedrich *et al.*, 2012; MacLeod *et al.*, 2013; Linnert *et al.*, 2014).

Běchary oxygen isotope record

The long-term oxygen isotope profile at Bch-1 displays background rising values from −7 to −5‰ $\delta^{18}O$ through the Lower to mid-Middle Turonian, then a relatively flat background but with high-amplitude *ca* 1‰ medium-term peaks and troughs through the upper Middle to low Upper Turonian (Fig. 2). A shift to higher values of −4·5‰ $\delta^{18}O$ occurs in the Upper Turonian, reaching a maximum of −3·6‰ immediately above the Hitch Wood CIE and *Hyphantoceras* Event (Fig. 6). Falling values characterize the uppermost Upper Turonian, with a minimum of around −5‰ $\delta^{18}O$ in the mid-*M. scupini* Zone, then values rise again through the uppermost Turonian–Lower Coniacian. They reach a maximum of −3·8‰ $\delta^{18}O$ above the Beeding CIE, followed by a decline in the uppermost Lower Coniacian. The amplitude of medium-term $\delta^{18}O$ variation is generally lower (*ca* 0·5‰) in the upper half of the section (Fig. 2).

The CTB sediments at the base of the cored section in the Bch-1 well (402 to 405 m) display anomalously low $\delta^{18}O$ values of −8 to −10‰ (Figs 2 and 4E; Appendix S3). Similar low values have been reported for CTB sediments from other sections in the Bohemian Cretaceous Basin (Uličný *et al.*, 1993). These indicate a substantial diagenetic overprint when compared to values of −3 to −4‰ $\delta^{18}O$ that characterize most shallow-buried CTB pelagic carbonates elsewhere (Jenkyns *et al.*, 1994; Jarvis *et al.*, 2006, 2011). Carbonate in this interval also has $\delta^{13}C$ values that are up to 1‰ lower than the overlying Lower Turonian sediments (Figs 2 and 4E; Appendix S3), despite lying at a stratigraphic level characterized by a global positive $\delta^{13}C$ excursion of *ca* 2‰ amplitude (Jarvis *et al.*, 2006; Figs 4 and 5), generating $\delta^{13}C_{carb}$ values of up to 5‰ elsewhere. These isotope trends (Figs 2 and 4E) and values of coincident $\delta^{13}C_{carb}$ and $\delta^{18}O_{carb}$ depletion are consistent with substantial carbonate recrystallization and cementation occurring in a rock-dominated semi-closed porewater system (Hudson, 1977; Choquette & James, 1987; Marshall, 1992). Thin-section and SEM studies of equivalent levels at other localities in the area (Uličný *et al.*, 1993) provide evidence of significant dissolution of biogenic calcite and precipitation of microspar cement.

The cored interval in Bch-1 is underlain by an Upper Cenomanian calcareous sandstone aquifer with high fluid pressures that provides a major groundwater resource in

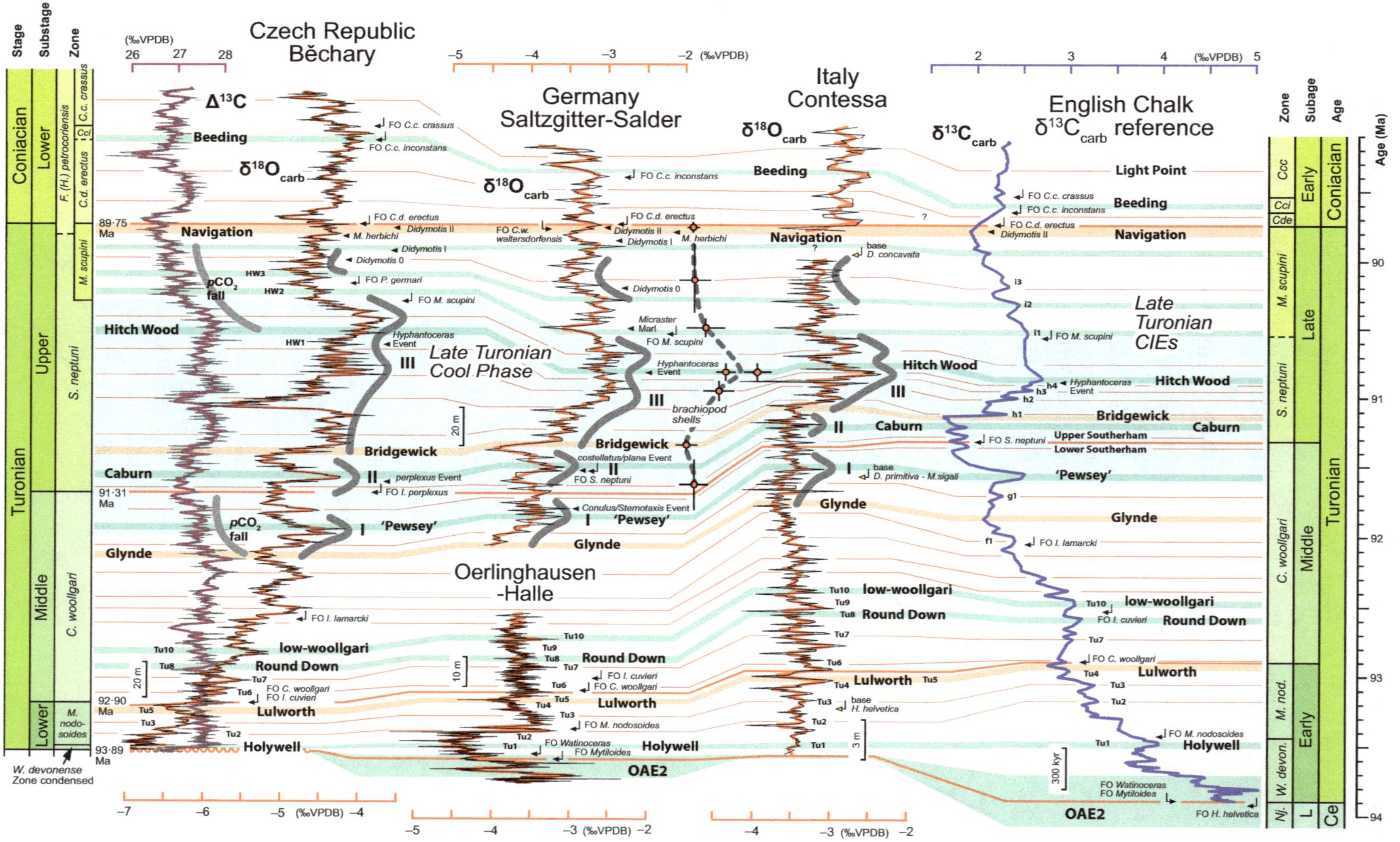

Fig. 6. Oxygen isotope profiles of European Turonian sections derived from bulk-sediment carbonate and brachiopod shells. Correlation of sections follows the carbon isotope stratigraphy illustrated in Fig. 5 (English Chalk $\delta^{13}C_{carb}$ reference curve shown on far right of diagram). Note coherent cooling tends (thick grey curves) towards higher $\delta^{18}O_{carb}$ values in the Mid-Turonian to Upper Turonian of the Czech Republic, Germany and Italy. Crosses (mean values with error bars) and red-filled symbols are brachiopod shell data from southern England (circles) and NW Germany (diamond), after Voigt (2000b). Brachiopod shells are typically offset to *ca* 1‰ higher values than their enclosing sediment but display identical trends (Voigt, 2000b, fig. 8). The three cooling intervals (I to III) of Voigt and Wiese (2000) are shown. Sediment isotope data sources: Běchary, this study; Saltzgitter-Salder (Voigt & Hilbrecht, 1997); Oerlinghausen-Halle, Voigt *et al.* (2007); Contessa, Stoll and Schrag (2000); English Chalk, Jarvis *et al.* (2006). Break in the Contessa profile represents an unsampled section correlated with immediately below the Navigation CIE (Sprovieri *et al.*, 2013). See Fig. 5 for explanation of other symbols.

the Bohemian Cretaceous Basin (Paces *et al.*, 2008). Fluid mixing of groundwater with porewaters in the overlying sediments would be limited by the low permeability of the carbonate-poor (17 to 21% $CaCO_3$) fine-grained calcareous mudstones (an aquiclude) at the base of the section. However, diffusional exchange with the aquifer waters and enhanced pressure solution (cf. Bloomfield, 1997) might be invoked to explain extensive carbonate recrystallization and the resetting of isotope values in the CTB section. A similar pattern of extreme $\delta^{18}O_{carb}$ depletion, with values of −10 to −14‰ and $\delta^{13}C_{carb}$ values *ca* 2‰, has been previously reported from Upper Cenomanian calcareous sandstones and overlying calcareous siltstones by Voigt and Hilbrecht (1997) in the Dresden-Blasewitz borehole of Saxony, northern Germany.

Present-day groundwater in Bohemian Cenomanian aquifers has $\delta^{18}O_{water}$ VSMOW values of around −10‰ (Jiráková *et al.*, 2010), close to the mean annual value of regional precipitation (−9·4‰ IAEA, 2009). Groundwater temperatures in the aquifer are of the order of 30°C which, using the equation of Anderson and Arthur (1983), generates a calcite equilibrium value of −13‰ ($\delta^{18}O_{carb}$ VPDB), or lower if higher temperatures developed in the geothermally influenced aquifer system (cf. Jiráková *et al.*, 2011). The isotopic composition of total dissolved inorganic carbon in the aquifer waters is also significantly lower than the host sediment, but is much more variable than for oxygen (−4·5 to −16·3‰ $\delta^{13}C_{carb}$, Jiráková *et al.*, 2010). For carbon, the potential impact of these very low values on sediment $\delta^{13}C_{carb}$ during recrystallization is limited by high rock: fluid inorganic carbon ratios.

Salinity in the Bohemian Cretaceous Basin

The Turonian $\delta^{18}O$ profile in Bch-1 shows a steeply rising trend from −8 to −5‰ VPDB through the Lower to mid-Middle Turonian (Figs 2 and 6). Sharp rises in $\delta^{18}O$ of *ca* 1·5‰ following a lowest Turonian minimum (Early Turonian thermal maximum) occur at the level of the Lower Turonian Tu2 CIE at Halle North Germany (Fig. 5; Voigt *et al.*, 2007) and at Dover England (Jenkyns *et al.*, 1994, fig. 4), but in those sections background values above Tu2 are relatively constant through the remainder of the Lower–Middle Turonian (Fig. 6). The background $\delta^{18}O$ trend at Contessa Italy remains flat throughout the Lower–Middle Turonian interval (Fig. 6), falling slightly in the upper Middle Turonian below the Glynde CIE.

The anomalous sharply rising $\delta^{18}O$ profile through the Lower–Middle Turonian at Bch-1 (Fig. 6), which is also evident in the record from the Dresden-Blasewitz (Voigt & Hilbrecht, 1997, fig. 4), might be attributed to carbonate recrystallization and cementation accompanying diffusion exchange with the underling aquifer waters and/or enhanced pressure-solution effects in the deeper buried section. However, this would require isotopic exchange over an 80 m section of low permeability claystones, which is unlikely. It is notable that spar-filled moulds after silicisponge spicules are common in the Lower Turonian section, decreasing in abundance upwards. A falling proportion of coarse burial-cement filled mouldic pores through the Lower to mid-Middle Turonian may partly explain the observed upward increase in $\delta^{18}O$.

An alternative explanation is that the increasing $\delta^{18}O$ trend reflects rising salinity of the Bohemian Cretaceous Basin and waters bordering the Bohemian Massif through the early to mid-Middle Turonian. Oxygen isotope values in the Upper Turonian–Coniacian section at Běchary are lower by *ca* 1‰ than equivalent sections in North Germany (Lower Saxony Basin) and Italy (Fig. 6). This depletion may be attributed to greater effect of diagenesis on bulk carbonate $\delta^{18}O$ in the relative carbonate-lean sediments. The Lower Turonian section, by contrast, is offset by −2·5‰ compared to the other sections. Taking into account a −1‰ $\delta^{18}O$ diagenetic offset, leaves −1·5‰ unaccounted for; this might potentially be attributed to reduced salinity of Bohemian Cretaceous Basin surface water compared to NW Germany Boreal and central Italy Tethyan waters.

The Bohemian Cretaceous Basin was surrounded by landmasses that during the Turonian sourced substantial volumes of siliciclastic sediments via prograding delta systems to the NW and SE of the study section (Uličný *et al.*, 2009, 2014). Although strongly tidally influenced with vigorous circulation (Mitchell *et al.*, 2010), basin waters received substantial freshwater input, and the occurrence of shallow-water dinoflagellate cyst species in the central basin points to transport by hypopycnal flows carrying low-salinity surface water across the basin (Olde *et al.*, 2015b). Water depths of <50 m are estimated for most of the basin area.

Results from a parallel ocean climate model for the Middle Cretaceous suggests salinities of *ca* 31·5 g kg^{-1} for the European area, a value that is 3 g kg^{-1} below the assumed global average of 34·7 g kg^{-1} (Poulsen *et al.*, 1998). Shackleton and Kennett (1975) estimated a mean $\delta^{18}O$ value of −1·0‰ VSMOW for sea water of a Cretaceous ice-free world. In the northern hemisphere where freshwater runoff was high, open-ocean sea water isotope values may have been lowered to less than −4‰ in the Arctic, while modelled Tethyan surface water has been estimated at +0·3 to +0·5‰ (Zhou *et al.*, 2008). A general value of −1·5‰ $\delta^{18}O$ has been proposed by Turonian epicontinental sea water in Western European basins (Voigt, 2000a).

At the assumed Turonian palaeolatitude of Central Europe *ca* 35°N (Fig. 1), mean annual zonal average precipitation is estimated to have been around −6‰ (Zhou *et al.*, 2008). The amount of altitude-effect $\delta^{18}O$ depletion in runoff is probably to have been modest due to the restricted topography on the adjacent landmasses. A simple mass balance calculation shows that addition of *ca* 22% freshwater of −6‰ $\delta^{18}O$ is required to generate the −1‰ offset observed in the earliest Turonian $\delta^{18}O$, with surface waters becoming progressive more saline thereafter. Using this estimate and assuming mixing freshwater of 0·5 g kg^{-1} with sea water of 31·5 g kg^{-1} generates a mildly brackish salinity of 24·7 g kg^{-1} for the earliest Turonian Bohemian Cretaceous Basin surface water. A lower volume of freshwater is required and the salinity would have been correspondingly higher if local runoff was more enriched in the light isotope.

A problem with invoking increasing salinity to explain rising $\delta^{18}O$ values through the Early–mid-Middle Turonian is that this period corresponds to a period of medium-term to long-term sea-level fall rather than sea-level rise (Uličný *et al.*, 2014). Lowest sea-levels are projected for the mid-Late Turonian. Increased marine water influence would be expected to accompany sea-level rise not fall.

The cause of rising $\delta^{18}O$ values through the Lower–Middle Turonian in Bch-1 remains uncertain. Oxygen isotope data from other sections in the Bohemian Cretaceous Basin are required to assess the potential influence of diagenetic versus salinity effects on the $\delta^{18}O$ records.

The Late Turonian Cool Phase

By contrast to the lower beds, oxygen isotope records from the upper Middle Turonian–Lower Coniacian at Běchary (Fig. 6) show very similar medium-term to long-term trends to published $\delta^{18}O_{carb}$ records from Salzgitter-Salder (Voigt & Hilbrecht, 1997) and Söhlde NW Germany (Voigt & Hilbrecht, 1997), Dover England (Jenkyns *et al.*, 1994, fig. 4), Liencres northern Spain (Wiese, 1999), and Contessa Italy (Stoll & Schrag, 2000). The oxygen stable isotope profile for the Upper Turonian at Běchary shows a strong trend of increasing values upwards to a maximum immediately above the *Hyphantoceras* Event and Hitch Wood CIE in the Upper Turonian mid-*S. neptuni* Zone (Figs 2 and 6). A $\delta^{18}O$ maximum at the same level is displayed by a less stratigraphically extensive low-resolution $\delta^{18}O$ curve for the mid-Upper Turonian at Úpohlavy (Wiese *et al.*, 2004), located on the NW margin of the Bohemian Cretaceous Basin, 90 km NW of Běchary.

Bulk-sediment oxygen isotope profiles for the Upper Turonian of the Bohemian Cretaceous Basin and elsewhere in Europe exhibit punctuated multi-stage increases in $\delta^{18}O$ interpreted to represent a marine cooling trend (Wiese, 1999; Voigt, 2000b; Wiese & Voigt, 2002), starting with a $\delta^{18}O$ rise in the upper Middle Turonian around the 'Pewsey' CIE (Fig. 6). Values increase sharply by *ca* 1‰ above the Bridgewick CIE, and peak above and below the Hitch Wood CIE, representing maximum Late Turonian cooling, before decreasing again to a minimum, indicative of temporarily warmer conditions in the latest Turonian *M. scupini* Zone. Oxygen isotope values increase above this, demonstrating continued cooling into the Early Coniacian.

Bulk $\delta^{18}O$ curves are inherently noisy due to the susceptibility of oxygen isotopes to diagenetic overprinting (Marshall, 1992; Schrag *et al.*, 1995; Frank *et al.*, 1999) and commonly display a strong lithological control. Unsurprising, therefore, unlike carbon isotopes, short-term peaks and troughs in $\delta^{18}O$ show relatively poor agreement between different sections (Fig. 6). In addition, $\delta^{18}O$ values in the Bohemian Cretaceous Basin hemipelagic successions are lower by *ca* 1‰ than those at equivalent levels in the pelagic German and Italian sections, and show high-amplitude (up to 1‰) short-term variation, evidencing more extensive and more variable lithology-controlled diagenesis in the relatively carbonate-poor sediments (Fig. 2; 4 to 71% $CaCO_3$, average 35%). Lower $\delta^{18}O$ values and higher amplitude isotopic variation are also observed in other low-carbonate Upper Turonian hemipelagic sections (e.g. at Liencres, Voigt & Wiese, 2000).

An anomalous large negative excursion of *ca* 2‰ $\delta^{18}O$ occurs in the lower *S. neptuni* Zone at the facies change marking the boundary between Sequences TUR 5 and TUR6/1 in Bch-1. This excursion is confined to a low-carbonate interval (18% $CaCO_3$), and corresponds to a $\delta^{13}C_{carb}$ negative excursion that coincides with a $\delta^{13}C_{org}$ peak (Fig. 2). The level marks a significant lithological and geochemical facies change upwards to coarser grained silty and more calcareous marlstones (Fig. 2) with high Si/Al, Ti/Al and Zr/Al ratios (Uličný *et al.*, 2014; Olde *et al.*, 2015b). No comparable negative $\delta^{18}O$ isotope excursion is observed in equivalent oxygen isotope profiles from elsewhere in Europe (Fig. 6). Coupled oxygen and carbon isotope depletion in bulk carbonate points to a diagenetic origin for this feature, probably caused by enhanced pressure solution and potentially fluid flow focussed along the facies boundary. A similar but much lower amplitude (*ca* 0·5‰ $\delta^{18}O$) negative excursion occurs in low-carbonate claystones (9% $CaCO_3$), immediately below the transition to marlstones (35% $CaCO_3$) near the base of the Lower Coniacian, above the Navigation CIE (Figs 2 and 6).

Identical medium-term to long-term Middle Turonian–Lower Coniacian $\delta^{18}O$ trends, constrained by a combination of high-quality biostratigraphy, tephrostratigraphy and carbon isotope stratigraphy, are seen not only at Běchary, Saltzgitter-Salder and Contessa (Fig. 6), but also in curves from Dover southern England (Jenkyns *et al.*, 1994), Söhlde NW Germany (Voigt & Hilbrecht, 1997), Dubivtsi western Ukraine (Dubicka & Peryt, 2012), Liencres northern Spain (Wiese, 1999), and Santa Ines southern Spain (Stoll & Schrag, 2000), with variable values and amplitudes reflecting varying lithological compositions, and different burial and diagenetic histories. The remarkable consistency of the stratigraphic trends mediates against them being purely a diagenetic artefact. Essentially synchronous diagenetic effects in multiple sections might be invoked if diagenesis were linked to eustatic sea-level change via an influence on the amount of authigenic cements (cf. the carbon isotope argument of Schrag *et al.*, 2013), or via sediment coarsening and associated higher permeability (and thus susceptibility to diagenesis) of sediments deposited during periods of lower eustatic sea-level. However, it is difficult to envisage how so similar $\delta^{18}O$ trends and values might be generated in such a wide range of depositional settings, from tidally influenced shallow-water epicontinental basins (e.g. Běchary) to pelagic slope environments at 1 to 1·5 km water depth (e.g. Contessa; Premoli Silva & Sliter, 1995).

Three stages of Mid-Turonian to Late Turonian medium-term (*ca* 250 kyr) stepped cooling were postulated by Voigt and Wiese (2000) using $\delta^{18}O$ data from England, Germany and Spain, which they designated as 'Phases' I to III (termed 'Intervals' I to III here, Fig. 6). These intervals peak around the 'Pewsey', Caburn, and Hitch Wood positive CIEs, with significant $\delta^{18}O$ increases starting above the Glynde (Interval I) and Bridgewick (Interval III) negative CIEs. Significantly, oxygen isotope analysis of well-preserved brachiopod shells from southern England and NW Germany follow bulk-rock $\delta^{18}O$ trends in their enclosing sediments (Fig. 5; Voigt, 2000b), with *ca* 1‰ lower values in the latter attributable to the addition of a pervasive isotopically lighter cement, and/or kinetic 'vital' effects influencing brachiopod shell values. Brachiopod data indicate *ca* 2°C of bottom-water cooling from 18·2° to 16·0°C during Interval III, with lower minimum temperatures attained in NW Germany (14·2°C) than in southern England (Fig. 5; Voigt, 2000b). Voigt (2000b) attributed this to a greater influence of cool northern Boreal North Sea waters in Germany compared to the proto-Atlantic influenced Anglo-Paris Basin.

The palaeo-environmental significance of the medium-term and long-term $\delta^{18}O$ isotope trends is demonstrated by temporary influxes of Boreal and temperate taxa into the 'Northern Transitional Subprovince' during peak cooling of Intervals I and III (Voigt & Wiese, 2000; Wiese & Voigt, 2002; Wiese *et al.*, 2004). This area extended from northern Spain, through southern France, SE Germany, and the Czech Republic to Austria, and is characterized by faunal assemblages that are a hybrid of northern and southern affinity ammonites, bivalves and echinoids (Wiese & Voigt, 2002).

Interval I terminated with the short-term immigration of northern affinity echinoids and ammonites into the Spanish North Cantabrian Basin (Voigt & Wiese, 2000). Contemporaneous with the southward spread of collignoniceratid ammonites into northern Spain, taxa more indicative of southern areas (*Romaniceras*, *Coilopoceras*) migrated northwards, suggesting weakening of former provincialism. Interval III is characterized by the progressive establishment of typical northern echinoid and inoceramid assemblages in northern and central Europe; in northern Spain these assemblages become dominant within the interval of highest $\delta^{18}O$ values above the Hitch Wood CIE. Collignoniceratid ammonites occur in southern France at the same level (Devalque *et al.*, 1982). The combination of isotopic and faunal evidence for Late Turonian cooling throughout Europe, points to a temporary shift of northern waters southwards during Intervals I and III, together with other water-mass reorganization (Voigt & Wiese, 2000; Wiese & Voigt, 2002; Wiese *et al.*, 2004). Coincident faunal changes during Interval III include the migration of the North American ammonite *Prionocyclus* into Europe and North Africa.

The Bohemian Cretaceous Basin was a key area sensitive to climate change during the Turonian, as it represented a gateway between northern Tethys and the southern Boreal Sea (Fig. 1A). Faunally, the area lay at the northern limit of the Northern Transitional Subprovince. The ammonite assemblage or '*reussianum* fauna' characterizing the *Hyphantoceras* Event [Fig. 6; named after the distinctive heteromorph ammonite *Hyphantoceras reussianum* (d'Orbigny)] and peak Interval III cooling, is widely distributed in the Czech Republic, England, parts of France, Germany, Poland and, to some extent, Kazakhstan (Wiese *et al.*, 2004). However, the relative rarity of allocrioceratids and collignoniceratids in the Bohemian faunas point to a southern affinity for the ammonite assemblage there (Wiese *et al.*, 2004).

In Bohemia, warm-water Nerineacean gastropod assemblages of the Cenomanian–Middle Turonian are replaced by a Boreal Pleurotomariacean fauna at the level of the *Hyphantoceras* Event (Kollmann *et al.*, 1998), along with an influx of rare Boreal belemnites from North America via Greenland and Scandinavia

(Košťák *et al.*, 2004; Wiese *et al.*, 2004; Košťák & Wiese, 2011). The nautiloid *Deltocymatoceras rugatum* (Fritsch & Schönbach) is recorded solely from the *Hyphantoceras* Event, and is restricted to shallow-water facies on the northern margins of the Bohemian Massif, and in the vicinity of the adjacent Lausitz and Sudetic blocks (Frank *et al.*, 2013). Bohemian Cretaceous Basin records, therefore, provide strong evidence of regional faunal changes accompanying the Late Turonian Cool Phase.

LINKED CLIMATE AND SEA-LEVEL CHANGE

A tentative eustatic sea-level curve for the Turonian based on the analysis of sediment geometries and the delineation of transgressive/regressive maxima across the Bohemian Cretaceous Basin was presented by Uličný *et al.* (2014). The medium-term to long-term transgressive–regressive framework and inferred sea-level model (Fig. 7) is supported by complementary chemostratigraphic and palynological studies (Olde *et al.*, 2015b). For example, well-defined maxima in bulk-sediment manganese content are associated with maximum flooding zones, and troughs with intervals of lowstand; falling Mn contents accompany regression and rising values transgression (Jarvis *et al.*, 2001, 2008; Olde *et al.*, 2015b, fig. 11). The most prominent feature of the Mn profile is a major long-term symmetrical trough centred on the *Hyphantoceras* Event and Hitch Wood CIE, interpreted to represent a peak regional lowstand on a long timescale. The Si/Al, Ti/Al and Zr/Al ratios, tracers for input of a more proximal siliciclastic fine fraction, display the opposite trend to Mn, with rising values accompany long-term sea-level fall, and declining values following sea-level rise, as exemplified by the Ti/Al profile in Fig. 7.

Dinoflagellate cyst species richness provides another excellent sea-level proxy in the Bohemian Cretaceous Basin (Olde *et al.*, 2015b). Intervals yielding low-diversity dinocyst assemblages in the Běchary succession correlate to sea-level minima. Sharp increases in species richness accompany transgression, with maxima coincident with periods of maximum flooding. The majority of well-defined short-term transgressive–regressive sea-level cycles in the Middle–Upper Turonian at Běchary (Fig. 7) are clearly expressed by peaks and troughs in the dinocyst species richness record (Olde *et al.*, 2015b, fig. 10).

The Late Turonian Cool Phase coincided, therefore, with evidence of a major third-order fall of sea-level that started in the early-Middle Turonian and terminated in the mid-Late Turonian (Fig. 7). The hardground complexes of the Chalk Rock were deposited at that time in England (Bromley & Gale, 1982; Hancock, 1989; Gale, 1996), massively bedded, in-part nodular, limestones formed in northern Germany (Wood *et al.*, 1984), and nodular glauconitic limestones and turbidite successions were deposited in northern Spain (Wiese, 1997).

Interval I, around the level of the 'Pewsey' CIE, represents an initial period of medium-term shallowing (e.g. Fig. 7). Coarser grained, silty marlstones with common skeletal debris and sand laminae characterize this level at Běchary (270 to 276 m, Fig. 2). The onset of massive stacked hardground development on basin margins occurred in southern England, a thinning upwards sequence of nodular and bedded limestones terminating in the *Conulus*/*Sternotaxis* Event was deposited in northern Germany (Fig. 6), and glauconitic turbidite sequences, hardgrounds and regional hiatuses characterize the interval in northern Spain (Voigt & Wiese, 2000; Wiese & Voigt, 2002).

Increased sedimentation in all areas characterizes Interval II, reflecting a medium-term transgression. In Bohemia, a transgression at the base of Upper Turonian Sequence TUR 5 is prominent basin wide, and marks the onset of higher carbonate contents and more widespread cementation in all facies that culminate around the level of the *Hyphantoceras* Event (Fig. 2). An acme of *Inoceramus perplexus* Whitfield in the Bohemian sections (*perplexus* Event) correlates to the *costellatus/plana* Event in NW Germany (cf. Richardt & Wilmsen, 2012), at the base of the Caburn CIE (Fig. 6).

Interval III is associated with evidence of progressive shallowing in many European basins: a regressive maximum is seen throughout the Bohemian Cretaceous Basin, corresponding to the summit of the coarsening-upward succession at Běchary (Fig. 2). The uppermost hardgrounds of the English Chalk Rock developed at this time, terminating in the Hitch Wood Hardground (Bromley & Gale, 1982; Gale, 1996). There is an increased abundance of benthic carbonate producers, and the development of nodular limestones with regional hiatuses in northern Germany, and in northern Spain, a calciturbidite sequence occurs, terminated by calcarenitic channel-fill deposits (Voigt & Wiese, 2000; Wiese & Voigt, 2002). In Bohemia and western Ukraine, a transition to more calcareous dinoflagellate cyst-rich (principally pithonellids) sediments and the temporary disappearance of keeled planktonic foraminifera further indicate shallower water and more oligotrophic conditions (Wiese *et al.*, 2004; Dubicka & Peryt, 2012).

Renewed transgression characterized the latest Turonian basal *M. scupini* Zone throughout Europe, and the biota of the Northern Transitional Subprovince retained a more Boreal affinity thereafter. Sedimentological evidence suggests, therefore, a relation between regional cooling of bottom and surface waters, the southward spread of Bor-

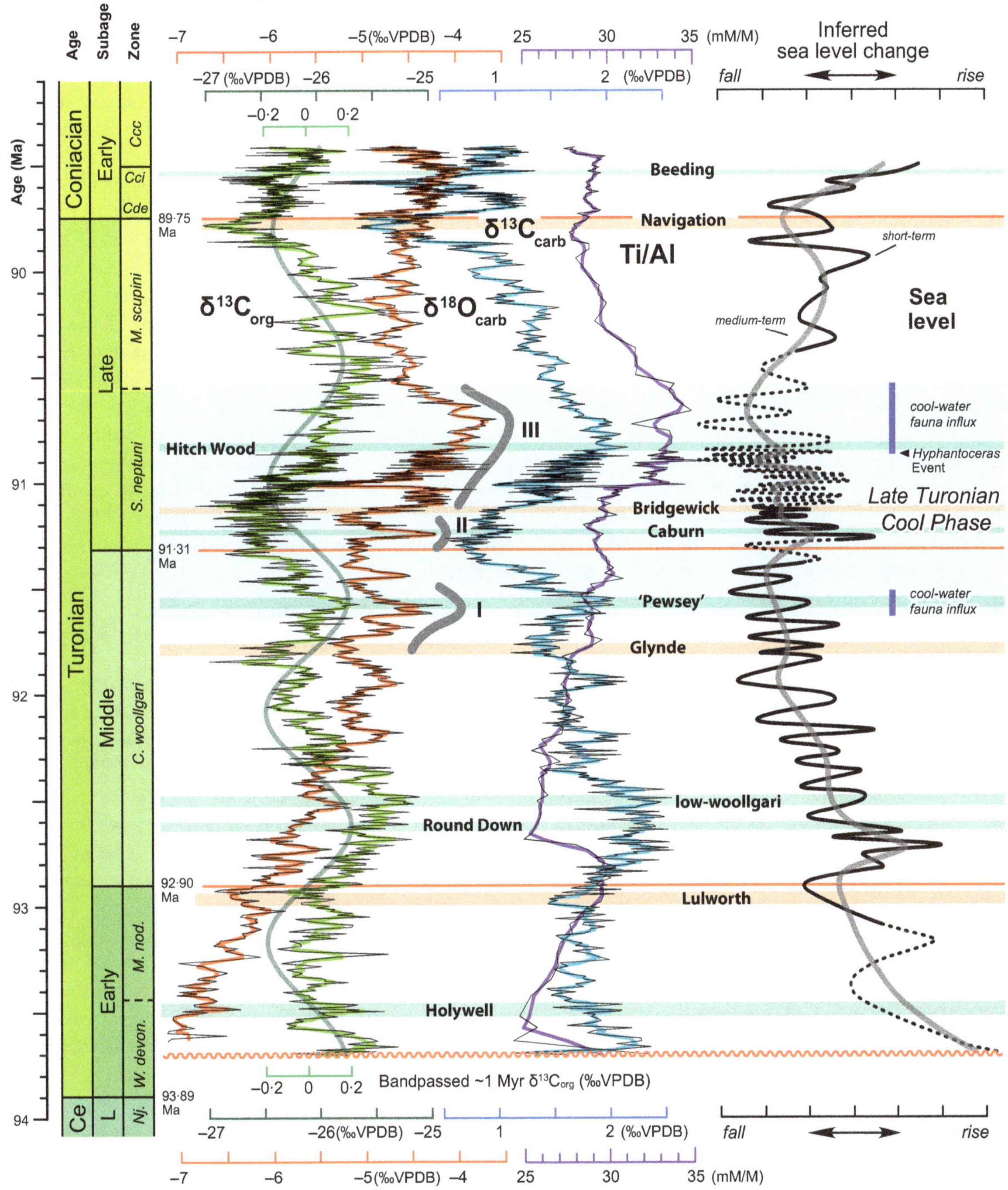

Fig. 7. Stable isotope profiles for the Turonian–Coniacian at Běchary compared to an inferred sea-level curve. Sea-level trends derived from transgressive–regressive maxima and basin-scale sediment geometries in the Bohemian Cretaceous Basin (Uličný *et al.*, 2014). Late Turonian cooling intervals after Voigt and Wiese (2000). A bandpassed *ca* 1 Myr signal in $\delta^{13}C_{org}$ interpreted as a signature of axial-obliquity modulation (Laurin *et al.*, 2015) is indicated by a thick green curve. Data were calibrated in the time domain using a modified age model from Laurin *et al.* (2015): the stratigraphic position of the Hitch Wood Event was reinterpreted based on new $\delta^{13}C_{carb}$ data (this study); the FO *P. germari* was not used as an age control point due to uncertainties in the local distribution of this taxon (cf. Laurin *et al.* 2014); for the Lower Turonian, a 200 kyr hiatus was applied at the base of the Turonian, above this hiatus, sedimentation rates were linearly increased from 0 cm kyr^{-1} to 7·5 cm kyr^{-1} (mean sedimentation rate for the Middle Turonian) at the top of the substage. Ages of stage and substage boundaries derived from Ogg *et al.* (2012), Laurin *et al.* (2014) and Sageman *et al.* (2014); age controls are summarized in Appendix S1. Ti/Al ratio curve, a potential sea-level proxy, from Olde *et al.* (2015b). Basin-scale genetic sequences are modified from Uličný *et al.* (2014), following Olde *et al.* (2015b). Note that Late Turonian cooling Intervals I and III observed in the oxygen isotope profiles throughout Europe (Fig. 5), correspond to packages displaying influxes of cool-water fauna in the Bohemian Cretaceous Basin, and more widely across the Northern Transitional faunal Subprovince (Voigt & Wiese, 2000; Wiese & Voigt, 2002). Cool-water faunal influxes occurred during episodes of short-term transgression, approaching levels of maximum regression and inferred sea-level lowstand.

eal taxa and low-order transgressions following sea-level falls.

The wider impact of the Late Turonian Cool Phase in Europe in relation to global climate and palaeoceanography remains to be adequately tested. However, northern hemisphere-wide water-mass reorganization is indicated by the a temporary influx during Interval III of *Prionocyclus* ammonites from the North American Western Interior Seaway (WIS), which are widely distributed in northern Germany and the Czech Republic, through southern Spain to Tunisia (Robaszynski *et al.*, 2000). At the same time, several belemnite taxa migrated eastwards from Greenland (Košťák & Wiese, 2011), and a connection to Japan is suggested by the occurrence of *Mytiloides incertus* Jimbo in both Europe and Japan (Voigt, 1995; Takahashi, 2005; Hayakawa & Hirano, 2013).

Pronounced Late Turonian cooling is indicated by $\delta^{18}O$ trends in the fine-fraction curves from the Southern Hemisphere Exmouth Plateau (Clarke & Jenkyns, 1999). However, the Late Turonian cooling discussed above appears to be significantly younger than a 'middle' Turonian glacial episode and sea-level lowstand postulated by Miller *et al.* (2004) in New Jersey, by Bornemann *et al.* (2008) from Demerara Rise western Equatorial Atlantic, and by Galeotti *et al.* (2009) on the Apulian margin Italy. Unfortunately, in all of these cases, age control is poor compared to the high-resolution multi-stratigraphic correlations achievable in European pelagic and hemipelagic sections.

Biostratigraphic correlation from northern European to Atlantic and Italian sections remains ambiguous: the correspondences between Turonian CC and UC calcareous nannofossil zones (Sissingh, 1977; Burnett *et al.*, 1998), planktonic foraminiferal zones, regional macrofossil zones, and CIEs are all relatively poorly constrained (Lees, 2008; Švábenická, 2012). The correlation by Bornemann *et al.* (2008) of a $\delta^{18}O$ peak and inferred sea-level lowstand at western equatorial Atlantic ODP Site 1259 with the 'Pewsey' CIE is probably erroneous (see Uličný *et al.*, 2014 for discussion). The possibility of a Middle Turonian glacial episode was rejected by MacLeod *et al.* (2013), based on the uniformity of $\delta^{18}O$ values obtained from multiple benthic and planktonic foraminifera species collected through a Lower to Middle Turonian section in Tanzania. However, their data do not extend into the Upper Turonian, so the possibility of a glacial influence on Late Turonian global cooling and sea-level fall remains untested.

Possible inter-relationships between climate and sea-level change remain controversial for a greenhouse climate system. In the absence of evidence for significant polar ice in the Turonian, Wendler and Wendler (2016) suggested that aquifer-eustatic (cf. Hay & Leslie, 1990) rather than glacio-eustatic forcing of sea-level might occur. This condition challenges the general assumption, based on changes in polar ice volume, that transgression will necessarily accompany warming (with falling sea water $\delta^{18}O$), and regression will accompany cooling (rising sea water $\delta^{18}O$). In an aquifer-eustatic system, an enhanced hydrological cycle during periods of climate warming may lead to increased aquifer storage volumes, sea-level fall and rising sea water $\delta^{18}O$ values. Our Bohemian Cretaceous Basin records, however, show a clear association between medium-term to long-term sea-level fall, inferred from detailed analysis of sediment sequence geometries, and a Europe-wide southward spread of cooler water masses with elevated $\delta^{18}O$ values. In the short term, however, major influxes of cool-water faunas appear to accompany transgression. Clearly, comparable high-resolution data sets from other basins with independently derived sea-level curves are needed to address these issues further.

CARBON ISOTOPES AS A PCO_2 PROXY

Photosynthetic carbon stable isotope fractionation (ε_p) by marine phytoplankton increases with ocean conditions that promote high CO_2 availability in surface waters, such as elevated atmospheric CO_2 concentrations (Dean *et al.*, 1986). This phenomenon explains why Cretaceous marine organic matter typically has $\delta^{13}C$ values that are up to 5 to 7‰ lower than its modern equivalent (Arthur *et al.*, 1985). It has been proposed that the larger amplitude of the CTB $\delta^{13}C_{org}$ excursion (as much as 4 to 6‰) compared to the $\delta^{13}C_{carb}$ excursion (typically *ca* 2‰) in sections worldwide may be attributed to reduced isotopic fractionation between dissolved inorganic carbon and marine organic matter as a consequence of lower atmospheric carbon dioxide (CO_2 drawdown) and increased marine productivity during OAE2 (Arthur *et al.*, 1988; Freeman & Hayes, 1992; Kuypers *et al.*, 1999, 2002; Tsikos *et al.*, 2004; Sinninghe Damsté *et al.*, 2008; Jarvis *et al.*, 2011).

Given the relationship between ε_p and pCO_2, stratigraphic variation in the offset between covarying $\delta^{13}C_{carb}$ and $\delta^{13}C_{org}$ curves, expressed by $\Delta^{13}C$, offers a potential tool for tracing palaeo-pCO_2 change (cf. Kump & Arthur, 1999; Jarvis *et al.*, 2011), assuming: (1) no significant diagenetic alteration of the carbonate or organic carbon $\delta^{13}C$ values, or a uniform systematic overprinting of these; (2) an overwhelmingly marine or terrestrial organic matter fraction, or a constant proportion of these; and (3) limited temporally restricted productivity effects.

Paired carbonate and organic matter $\delta^{13}C$ records have been reported from several CTB sections (Freeman & Hayes, 1992; Tsikos *et al.*, 2004; Sageman *et al.*, 2006; Voigt *et al.*, 2006, 2007; Scopelliti *et al.*, 2008; Jarvis *et al.*,

2011). However, in many cases, the reliability of one of the data sets is questionable due to: (1) an absence of carbonate in the most organic-rich layers and/or insufficient organic matter in some limestones (e.g. the Bonarelli Level at Bottaccione); (2) the occurrence of erratic anomalously low values in $\delta^{13}C_{carb}$ profiles, indicative of locally precipitated organic matter-derived carbonate cements (e.g. Tarfaya) and (3) uniform low $\delta^{13}C_{carb}$ values, suggesting pervasive overprinting by recrystallization or the addition of extensive homogenous calcite cement (e.g. Bonarelli equivalent, Novara di Sicilia). However, stratigraphic variation in paired records presented by Jarvis *et al.* (2011) from a CTB section in the Vocontian Basin of SE France, compare favourably to the data across the same interval in England (Paul *et al.*, 1999; Gale *et al.*, 2005) and Germany (Voigt *et al.*, 2006), and offsets between the carbonate and organic curves ($\Delta^{13}C = \delta^{13}C_{carb} - \delta^{13}C_{org}$) were used to interpret a pCO_2 record for the interval.

Turonian $\Delta^{13}C$ values at Běchary are relatively constant at 27·7 ± 0·3‰ up to the level of the Hitch Wood CIE (Fig. 6) and then fall to a minimum of *ca* 26·4‰ at the Navigation CIE and Turonian–Coniacian boundary. This fall is initiated at the main influx of cool-water fauna around the *Hyphantoceras* Event. The ensuing medium-term $\Delta^{13}C$ falling trend spans an interval of falling then rising $\delta^{18}O$ values at Běchary (temporary warming), although a comparable $\delta^{18}O$ minimum is less clearly expressed at Saltzgitter-Salder and Contessa (Fig. 6). In these cases, the main feature of the long-term $\delta^{18}O$ trends is a step change to cooler temperatures by the later Late Turonian. This phenomenon implies that the main pCO_2 fall followed rather than preceded cooling. Significantly, a less marked $\Delta^{13}C$ fall occurs in the upper Middle Turonian at the level of the Glynde CIE, immediately preceding the first influx of cool-water biota around the 'Pewsey' CIE and the first step upwards in $\delta^{18}O$ (Fig. 6).

Any interpretation of $\Delta^{13}C$ trends at Běchary as a pCO_2 proxy must be treated with caution, given the observed diagenetic modification of primary $\delta^{18}O$ values, and the coincidence of the main $\Delta^{13}C$ shift with a facies change to finer-grained less-calcareous sediments (Fig. 2). The decreasing offset between $\delta^{13}C_{carb}$ and $\delta^{13}C_{org}$ records might be related to an increased proportion of authigenic carbonate, either locally at Běchary, or more generally in the oceans (cf. Schrag *et al.*, 2013). An increase in the proportion of organically influenced isotopically lighter carbonate cement will lower $\Delta^{13}C$. This fall would be amplified if marine organic matter was preferentially oxidized, leaving a higher proportion of isotopically heavier terrestrial organic matter. However, this is unlikely to be the case in Bch-1, where the terrestrial/marine palynomorph ratio falls significantly through the uppermost Turonian–lowest Coniacian (Olde *et al.*, 2015b); this would be expected to increase rather than decrease $\Delta^{13}C$ values. Nonetheless, the coincidence between the faunal and geochemical proxies of climate change is intriguing and warrants further investigation in other sections.

CARBON ISOTOPE RECORDS AND SEA-LEVEL CHANGE

A number of short-term, basin-wide regressions in the Bohemian Cretaceous Basin, most probably reflecting eustatic falls, have been documented with a recurrence interval of 100 kyr or less (Uličný *et al.*, 2009, 2014; Mitchell *et al.*, 2010). The estimated magnitude of these sea-level falls is typically 10 to 20 m and generally <40 m. The correspondence between $\delta^{13}C_{org}$ at Běchary and an inferred sea-level curve for the Basin was examined by Uličný *et al.* (2014). They noted that a long-term 'background' cycle of $\delta^{13}C_{org}$ (Fig. 7), shows a duration close to the 2·4 Myr long-eccentricity cycle, and shorter-term (1 Myr scale) highs and lows in $\delta^{13}C_{org}$ appear to broadly correspond to intervals characterized by more pronounced short-term sea-level highs and lows, respectively. However, despite a number of individual matches, neither a systematic in-phase nor out-of-phase correlation with interpreted sea-level cycles could be demonstrated at the level of either short-term (≤100 kyr) or intermediate-term (100 to 500 kyr) $\delta^{13}C_{org}$ fluctuations.

Comparison of the new $\delta^{13}C_{carb}$ and published $\delta^{13}C_{org}$ profiles to the sea-level model of Uličný *et al.* (2014) indicates a better correlation for the former than for the latter (Fig. 7), with approximately two-thirds of the inflection points on the short-term sea-level curve (transgressive surfaces) corresponding to the bases of $\delta^{13}C_{carb}$ peaks, and over half of the regressive maxima corresponding to $\delta^{13}C_{carb}$ minima. However, only half of the positive correlations of $\delta^{13}C_{carb}$ to sea-level show coincident shifts in $\delta^{13}C_{org}$, so a clear relationship with the global carbon cycle remains unproven.

A number of factors may influence the differing relationships between $\delta^{13}C_{carb}$ and $\delta^{13}C_{org}$ at Běchary and the sea-level record. First, both isotope records are relatively noisy, prejudicing the exact placement of maxima and minima. Second, uncertainty remains in the correlation of transgressive/regressive maxima between the NW and SE basin fills and the Bch-1 core; the sea-level model requires further refinement. Third, elemental chemostratigraphy (Olde *et al.*, 2015b) demonstrates short-term changes in sediment composition that accompany transgressive pulses, exemplified by intervals with increased Ti/Al (Fig. 7), Si/Al and Zr/Al ratios in the core. The combination of physical and mineralogical changes at these levels would probably affect carbonate diagenesis, which impacts $\delta^{13}C_{carb}$. At the same time, subtle changes in organic matter provenance, composition or preservation might also cause a divergence between the $\delta^{13}C_{org}$ versus $\delta^{13}C_{carb}$ records. Additional work is required to rigorously test relationships between carbon isotope records and sea-level change.

The main interval of divergence between the $\delta^{13}C_{carb}$ and $\delta^{13}C_{org}$ profiles lies in the higher part of the Upper Turonian where falling $\delta^{13}C_{carb}$ accompanies a rising $\delta^{13}C_{org}$ trend. Declining $\delta^{13}C_{carb}$ values might be interpreted to indicate increasing carbonate diagenesis accompanying falling carbonate values (Fig. 2). However, an identical trend is seen in carbonate isotope curves throughout Europe (Fig. 5), so it is unlikely to be a diagenetic artefact. Declining isotopic fractionation in marine organic matter due to falling pCO_2 (Fig. 5) offers a possible explanation for the divergence between $\delta^{13}C_{org}$ and $\delta^{13}C_{carb}$ trends (cf. Jarvis *et al.*, 2011).

CARBONATE AND ORGANIC CARBON FLUXES

Simplistically, the long-term Turonian carbon isotope record at Bch-1 (Figs 5 and 7) implies a moderately high organic matter versus carbonate burial flux in the Early Turonian, a period of enhanced burial of organic matter in the early-Middle Turonian, then a falling burial flux through the remainder of the Middle Turonian to a minimum during the earliest Late Turonian. The Late Turonian shows distinct rising then falling organic matter burial, peaking in the middle of the subzone and with a period of minimum burial spanning the Turonian–Coniacian boundary then a modest recovery thereafter.

Laurin *et al.* (2015) employed spectral analysis of $\delta^{13}C_{org}$ data from the Bch-1 core, together with $\delta^{13}C_{carb}$ data from other European Cretaceous sections, to propose that transfers between surface carbon reservoirs may be controlled by external forcing, principally *ca* 1 Myr changes in the amplitude of axial obliquity. The authors argued that the astronomical control causes transient storage of organic matter or methane in quasi-stable reservoirs such as terrestrial peat, soils and lakes, marginal zones of marine euxinic strata and, potentially, permafrost. These reservoirs responded nonlinearly to obliquity-driven changes in high-latitude insolation and/or the meridional insolation gradient, resulting in the *ca* 1 Myr cyclic $\delta^{13}C$ pattern observed in the Turonian–Coniacian (e.g. Figs 5 and 7), and potentially driving the multi-Myr-scale cyclicity observed in the Cenozoic (Boulila *et al.*, 2012).

The balance between the carbonate-carbon and organic-carbon burial fluxes is not controlled solely by the efficiency of organic matter preservation; both fluxes must balance the terrestrial carbon input flux on a 10^6 timescale and longer due to the small size of the ocean-atmosphere reservoir in comparison with the observed flux rates. At steady state, with a constant terrestrial carbon input by chemical weathering, an increase in inorganic carbon burial will lead to decreased organic carbon burial. This condition will release CO_2 to the ocean–atmosphere system and decrease $\delta^{13}C$ values. The marked expansion of chalk sedimentation in the Early Turonian (Voigt, 2000a, fig. 5) might explain the long-term $\delta^{13}C$ decline evident in Lower and Middle Turonian records (Fig. 5).

GLOBAL CORRELATION OF TURONIAN CARBON ISOTOPE CURVES

US Western Interior Basin: a $\delta^{13}C_{org}$ correlation

The Cretaceous Western Interior Basin of North America (Fig. 1B) has a well-established inoceramid bivalve and ammonite biostratigraphy (Kauffman *et al.*, 1993; Cobban *et al.*, 2006), has excellent geochonological control from radiometrically dated volcanic ash bands (Obradovich, 1993; Meyers *et al.*, 2012; Sageman *et al.*, 2014), includes key intervals with astrochronological time scales (Sageman *et al.*, 2006; Meyers *et al.*, 2012), and hosts the GSSP for the base Turonian Stage (Kennedy *et al.*, 2005). The occurrence of common bentonites throughout the Western Interior succession offers unique potential for geochronological calibration of the Cretaceous timescale using paired $^{40}Ar/^{39}Ar$ sanidine and U–Pb zircon ages.

Unfortunately, intercontinental correlation of Western Interior successions has been hampered by: taxonomic issues with inoceramid assemblages; a dominance of endemic ammonite faunas in the post-Cenomanian section; the general absence of echinoderms and other stenohaline taxa; and basin restriction and the presence of siliciclastic sediments in many intervals limiting the use of planktonic foraminifera and calcareous nannofossil biostratigraphy. However, carbon isotope chemostratigraphy has been successfully applied for high-resolution correlation of the base Turonian GSSP to Europe (Gale *et al.*, 1993; Kennedy *et al.*, 2005), and offers great potential to develop more refined Turonian correlations between North America and other successions globally.

A composite $\delta^{13}C_{org}$ reference curve for Cenomanian–Campanian of the United States Western Interior Basin was published by Joo and Sageman (2014), based on analysis of three cored boreholes and correlation of macrofossil biostratigraphic datum levels to these from outcrop. in addition, a temporal framework was developed by correlation with current geochronological and astrochronological timescales (Meyers *et al.*, 2012). However, biostratigraphic control is limited above the CTB interval, and inter-core correlation and the placement of zonal boundaries was based largely on lithostratigraphy. The Turonian section of the North American composite curve is compared to the age-calibrated English Chalk $\delta^{13}C_{carb}$ reference curve and age-calibrated $\delta^{13}C_{carb}$ and $\delta^{13}C_{org}$ curves from Běchary in Fig. 8.

Correlation of the CTB interval between the Western Interior and Europe is well-established (Kennedy *et al.*, 2005). The boundary section at Běchary is thin and incomplete (Fig. 2), and cannot be compared to the expanded North American succession in detail. The Lower Turonian in the Western Interior can be anchored by the FO *Watinoceras devonense* and *Mytiloides puebloensis* at the bottom and FO *C. woollgari* at the top of the substage, with the base of the *Mammites nodosoides* Zone coincident with the top of the Holywell CIE (Fig. 7; = excursion T1 of Joo & Sageman, 2014).

The Western Interior Basin Middle Turonian composite curve shows a characteristic rising trend to the Low-woollgari CIE maximum, and declines above. A step increase of 1‰ at the base of the *Collignoniceras praecox-Prionocyclus hyatti/Inoceramus howellii* Zones (lower profile break in Fig. 8) is almost certainly an artefact caused by stacking data from the CL-1 core on top of the Portland core. The profile across the interval in the Portland core alone shows progressively declining values upwards with no offset (Joo & Sageman, 2014, fig. 2).

Placement of the base Upper Turonian in the Western Interior is uncertain compared to its equivalent level in Europe (i.e. first appearance of *I. perplexus*). It is conventionally placed at the base of the *Scaphites whitfieldi* Zone (Joo & Sageman, 2014) in the Western Interior, but the carbon isotope correlation with the CL-1 core indicates that positioning the substage boundary lower, at the base of the *Inoceramus dimidius* Zone offers better consistency (Fig. 8). This positioning conforms to the base Upper Turonian as defined by Kauffman *et al.* (1993), although those authors did not indicate their reason for placing it at that level. The recorded FO of *I. dimidius* in the CL-1 core (Ball *et al.*, 2010), used to construct this part of the composite curve, is consistent with the position of the Upper Turonian substage boundary based on our carbon isotope correlation (Fig. 8). Records of *I.* aff. *perplexus* in the mid-*I. dimidius* Zone of CL-1 (Ball *et al.*, 2010) also support this interpretation, although these are above the $\delta^{13}C_{org}$ minimum marking the Bridgewick CIE. The shape of the Late Turonian Western Interior curve corresponds very closely to the Běchary record (Fig. 8), but the best-fit isotope correlation shows an apparent age offset of around 300 kyr at the level of the HW3 CIE. This condition is a consequence of the age model used by Joo and Sageman (2014), who pinned the base Coniacian to the bottom of a $\delta^{13}C_{org}$ minimum (which they interpreted to represent the Navigation CIE) below their Co1 negative excursion. A different interpretation of the CIE stratigraphy is proposed here (see below).

The Western Interior $\delta^{13}C_{org}$ curve lacks resolution across the Turonian–Coniacian boundary interval (Joo & Sageman, 2014). The composite curve is based on splicing the uppermost Turonian of the CL-1 core to the basal Coniacian in the Aristocrat Angus core. Data from the former section does not extend to the base Coniacian, which corresponds to a minor disconformity surface at the base of the Niobrara Formation (Fort Hayes Limestone Member; Ball *et al.*, 2010). The Fort Hays Limestone in the Aristocrat Angus core rests on a major disconformity surface, and represents a thin, highly condensed and probably incomplete, Upper Turonian to basal Coniacian succession, as documented more generally for the US Western Interior (Walaszczyk *et al.*, 2014). The exact placement of the base of the Lower Coniacian *Scaphites preventricosus* ammonite zone with respect to the isotope curve is therefore uncertain.

The FO of *S. preventricosus* coincides with the FO of *C. deformis erectus* in the Western Interior (Walaszczyk & Cobban, 2000). This inoceramid is first recorded at the top of the Navigation CIE in Europe (Fig. 5), placing the Turonian–Coniacian boundary at the top of the negative isotope excursion, not at its base, as indicated by Joo and Sageman (2014, fig. 4). It is probably that either the base of the *S. preventricosus* Zone has been misplaced on their isotope curve, or the identification of the Navigation CIE immediately below their isotope peak Co1 is incorrect. Overall, the isotope profile indicates the presence of a significant hiatus spanning the Turonian–Coniacian boundary interval in the Western Interior composite curve.

Within the constraints of the available data, there is excellent agreement between both long-term and short-term variation in the Turonian $\delta^{13}C_{org}$ curves from Central Europe and North America (Fig. 8). It is particularly noteworthy that the long-term reversal of rising to falling $\delta^{13}C_{org}$ values lies within the mid-*M. scupini* Zone in both sections, rather than the turning point occurring lower in the mid-*S. neptuni* Zone, as seen universally in $\delta^{13}C_{carb}$ records (Fig. 5), which supports our previous argument that the divergence in paired carbon isotope trends with falling $\Delta^{13}C$ in the Late Turonian, may be attributed to pCO_2 drawdown rather than regional differences or diagenetic factors.

Correlation with the terrestrial record: Yezo Group, Japan

The isotopic linkage between marine and terrestrial carbon reservoirs means that carbon isotope stratigraphy is

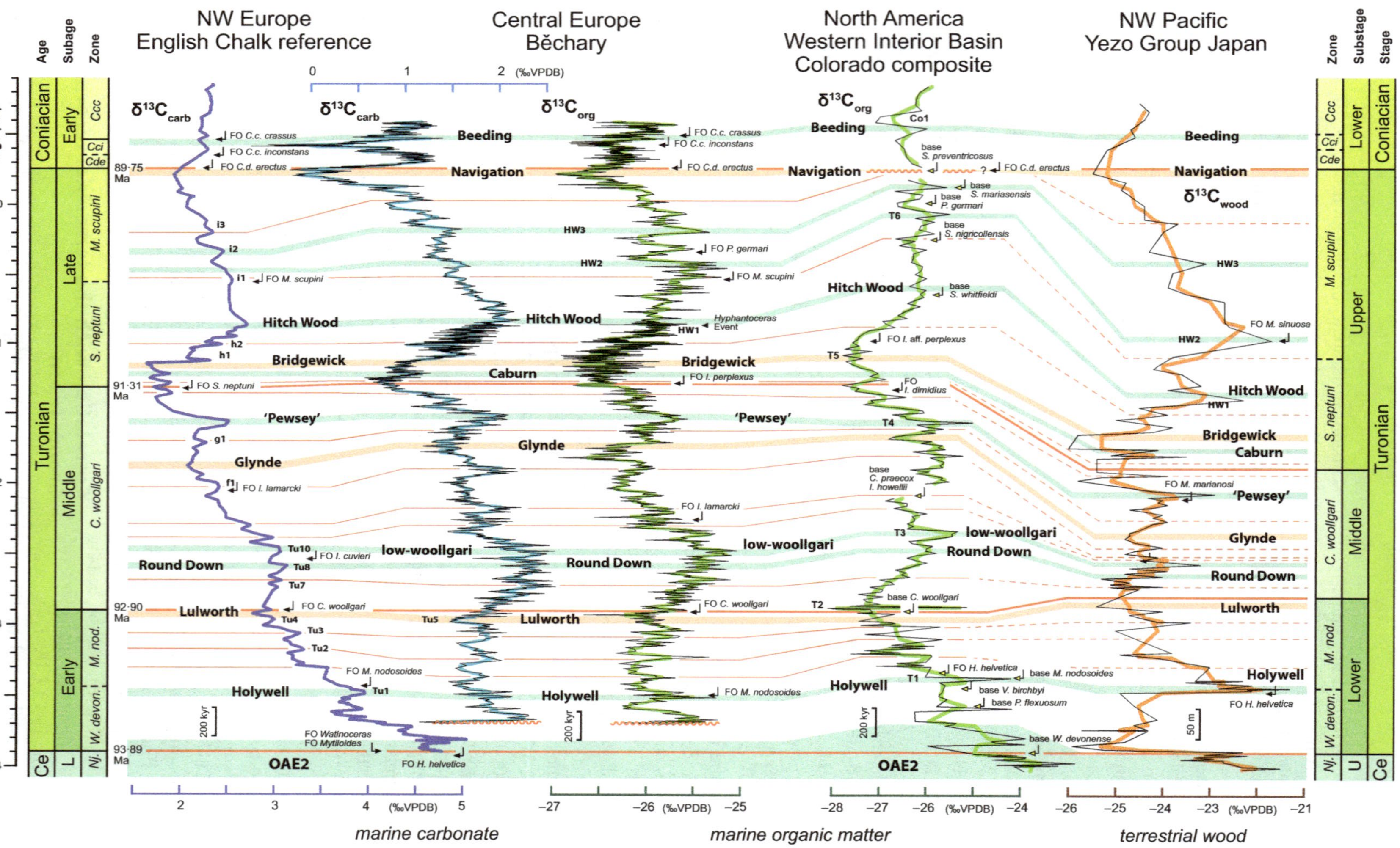

Fig. 8. Correlation of age-calibrated Turonian $\delta^{13}C_{carb}$ profiles for Europe (English Chalk reference curve and Běchary), with bulk marine $\delta^{13}C_{org}$ curves from Europe (Běchary) and North America (US Western Interior Basin composite), and the terrestrial wood record ($\delta^{13}C_{wood}$) from the NW Pacific (Yezo Group, Japan). Locations of the sites are shown in Fig. 1. Breaks in the North America profile in the mid-Middle Turonian and at the Turonian–Coniacian boundary indicate positions of core changes in the stacked composite profile (see text for discussion). Data sources: Běchary, Uličný *et al.* (2014), age calibration as in Fig. 7; US WIS composite, isotope data and zonal bases (yellow arrows) from Joo and Sageman (2014), faunal records (black arrows) from Ball *et al.* (2010); Japan, Takashima *et al.* (2010); English Chalk, Jarvis *et al.* (2006), age recalibrated after Laurin *et al.* (2014). Ages of stage and substage boundaries derived from Ogg *et al.* (2012), Laurin *et al.* (2014) and Sageman *et al.* (2014). For the Lower Turonian at Běchary, a 200 kyr hiatus was applied at the base of the Turonian, above this hiatus, sedimentation rates were linearly increased from 0 cm kyr^{-1} to 7·5 cm kyr^{-1} (mean sedimentation rate for the Middle Turonian) at the top of the substage.

potentially a powerful tool for high-resolution correlation between marine and non-marine successions. Extensive work has been undertaken on characterizing $\delta^{13}C$ trends in bulk terrestrial organic matter in the Turonian Yezo Group of Japan (Hasegawa & Saito, 1993; Hasegawa, 1997, 2003; Hasegawa & Hatsugai, 2000; Hasegawa *et al.*, 2003; Tsuchiya *et al.*, 2003; Uramoto *et al.*, 2009, 2013, 2015; Hayakawa & Hirano, 2013; Takashima *et al.*, 2010, 2011), enabling comparison to marine $\delta^{13}C$ records.

The Upper Cenomanian–Lower Coniacian Yezo Group crops out widely along the NW Pacific margin, constituting the sedimentary fill of a forearc basin (Fig. 1B; Uramoto *et al.*, 2013, 2015). Elevated sedimentation rates (20 to 40 cm kyr^{-1}) offer high stratigraphic resolution through the Turonian but, with the exception of the CTB interval, biostratigraphic control is poor. The most detailed Turonian terrestrial $\delta^{13}C$ record available to date has been derived from the analysis of wood fragments ($\delta^{13}C_{wood}$), sampled at 5 to 20 m intervals throughout *ca* 1 km thickness of a section at Kotanbetsu, northern Hokkaido, Japan (Takashima *et al.*, 2010). Wood was chosen in an attempt to overcome possible limitations of previous work using bulk organic matter ($\delta^{13}C_{TOM}$), in which the inclusion of variable amounts of marine organic material, the presence of varying floral components (e.g. leaf, stem, root) or material originating from different environments with different isotopic signatures, might have overprinted global stratigraphic trends (see Gröcke *et al.*, 2005 for discussion).

Tethyan planktonic foraminifera marker species are rare or absent from the Yezo Group, necessitating the application of a regional foraminifera biostratigraphy (Nishi *et al.*, 2003), calibrated using records of occasional international markers. The regional macrofossil biostratigraphy principally employs endemic inoceramid bivalve assemblages, calibrated to the international timescale using occasional co-occurring Tethyan ammonites and inoceramid species (Toshimitsu *et al.*, 1995; Hayakawa & Hirano, 2013). The latter have only been recorded sporadically from any of the carbon isotope study sections. In addition, Turonian stable isotope values of bulk wood fluctuate widely between −26 and −21‰ (Takashima *et al.*, 2010, 2011), with an average value of −24‰ $\delta^{13}C_{wood}$, producing a very noisy isotope profile (Fig. 8). Similar values have been obtained by lower resolution $\delta^{13}C_{TOM}$ studies of the Turonian aged Yezo Group (Uramoto *et al.*, 2013, 2015).

The base of the Turonian at Kotanbetsu is relatively well-constrained by a + 4‰ positive $\delta^{13}C_{wood}$ isotope excursion located a short distance above the LO *Rotalipora cushmani* (Morrow) (Fig. 7; Takashima *et al.*, 2010). The CTB interval is similarly characterized by a + 3‰ excursion in complementary $\delta^{13}C_{TOM}$ profiles (Uramoto *et al.*, 2013, 2015). The exact placement of the stage boundary within the Yezo Group has been confirmed recently by osmium isotope stratigraphy and U–Pb geochronology (Du Vivier *et al.*, 2015).

Takashima *et al.* (2010, fig. 4) highlighted a number of other planktonic foraminifera ranges as being of stratigraphic value in the Yezo Group, which were used to pin the $\delta^{13}C_{wood}$ isotope curve. The FO *Helvetoglobotruncana helvetica* (Bolli) occurs at a second large positive $\delta^{13}C_{wood}$ excursion (Fig. 8) below the base of the Saku Formation. The FO of this species has been widely used as a Lower Turonian biostratigraphic marker, but this has proven problematic due to an apparently diachronous FO (Desmares *et al.*, 2007; Robaszynski *et al.*, 2010). This diachroneity is attributable to a number of factors, including inconsistency in the taxonomic concept of the nominate species, its rarity in the lower part of its stratigraphic range, its rarity in or absence from higher latitude and nearshore settings, and its high degree of morphological variability (Huber & Petrizzo, 2014). At the GSSP site for the base of the Turonian at Pueblo, following the concept of Caron *et al.* (2006), the FO of *H. helvetica* occurs just above the base of the *M. nodosoides* Zone (Sageman *et al.*, 2006), at the level of the Holywell CIE (Fig. 8).

A short total range of *Marginotruncana marianosi* Douglas within to the mid-*H. helvetica* Zone has been documented at Kotanbetsu (Takashima *et al.*, 2010). These records are of little stratigraphic value. At the Exmouth Plateau, offshore NW Australia, the species appears in the Coniacian and extends into the lower Santonian (Wonders, 1992; Petrizzo, 2000, 2002). In the tropical western Tethyan Realm it extends through the lower Turonian to a LO at the top of the Middle Turonian (Robaszynski & Caron, 1979a,b; Caron, 1985). In Tibet, *M. marianosi* is also present in the lower and middle Turonian (Wendler *et al.*, 2011) but extends well into the Coniacian. The species has been recorded from the Upper Turonian in the Western Interior (Sikora *et al.*, 2004; note modified age assignment of Wagon Mound section by Walaszczyk *et al.*, 2012).

Takashima *et al.* (2010) documented the FO *Marginotruncana sinuosa* Porthault in the Kotanbetsu section, which they placed in the uppermost Turonian. This taxon was proposed as a base Coniacian marker by Walaszczyk and Peryt (1998), but the species has subsequently been recorded from the higher Upper Turonian *M. scupini* Zone at Saltzgitter–Salder, and at similar levels in the base Coniacian proposed GSSP of Słupia Nadbrzeżna (Walaszczyk *et al.*, 2010), and in the Western Interior (Walaszczyk *et al.*, 2012). It appears to be a consistent higher Upper Turonian marker, based on the current inoceramid definition for the base Coniacian.

Takashima *et al.* (2010) recorded the LO *H. helvetica* above the FO *Marginotruncana sinuosa* at Kotanbetsu, which is potentially problematic because of recent suggestions (Huber & Petrizzo, 2014) that the LO *H. helvetica* is a reliable Middle Turonian marker, which is supported by the $\delta^{13}C_{carb}$ records and planktonic foraminifera ranges reported by Wendler (2013, fig. 7). The data presented by Takashima *et al.* (2010) points to a diachronous LO between Tethys and the NW Pacific. Similarly in the WIS, *H. helvetica* has been observed towards the top of the Iona-1 core, well into the Lower Coniacian (Eldrett *et al.*, 2015). The FO *Dicarinella concavata* Brotzen was recorded at Kotanbetsu in the Santonian (Takashima *et al.*, 2010, fig. 4), considerably above the stratigraphic interval being discussed here. The index species of the *D. concavata* Zone first appears in the Upper Turonian, but it is usually more abundant and consistently present in the Coniacian, making it a relatively unreliable stratigraphic marker (Robaszynski & Caron, 1995; Wendler, 2013).

Biostratigraphic constraints on the Yezo Group $\delta^{13}C_{wood}$ Turonian curve are therefore ambiguous. This situation is not improved by comparison with low-resolution regional $\delta^{13}C_{TOM}$ profiles (Uramoto *et al.*, 2013, 2015). These are similarly poorly constrained, with calibration attempted using published macrofossil records from the study sections. Ages derived from these when placed within the lithostratigraphic framework agree poorly with the planktonic foraminifera data. Interpretation is complicated further by evidence of sediment condensation and local slumping of the Middle Turonian interval (Uramoto *et al.*, 2015). Further work is required to integrate microfossil data and macrofossil data with the $\delta^{13}C_{wood}$ and $\delta^{13}C_{TOM}$ curves.

Despite the stratigraphic limitations discussed above, a coherent correlation between the terrestrial and marine $\delta^{13}C$ records may be achieved using the limited age constraints provided by planktonic foraminiferal ranges (Takashima *et al.*, 2010), combined with matching medium-term and long-term trends in the $\delta^{13}C$ profiles. The resulting correlation (Fig. 8) differs from previous work (Takashima *et al.*, 2010; Hayakawa & Hirano, 2013; Uramoto *et al.*, 2013, 2015) in recognizing a higher Upper Turonian turning point in isotope profiles from organic matter compared to $\delta^{13}C_{carb}$ curves, and correlating the turning point at Kotanbetsu to the Běchary HW2 CIE, rather than the older Hitch Wood CIE. This reinterpretation is supported by the FO *M. sinuosa*, a high Upper Turonian marker, coincident with HW2.

Synchronous isotope excursions in marine and terrestrial organic matter that diverge from marine carbonate trends would be predicted by the fundamental observation that carbon isotope fractionation increases in both marine plankton (Dean *et al.*, 1986; Kump & Arthur, 1999) and terrestrial C_3 land plants (Schubert & Jahren, 2013) in response to increasing pCO_2 levels. Other terrestrial photosynthetic pathways are not believed to be important here, because classical C_4 photosynthesis is a recent evolutionary innovation, becoming significant only in the Miocene, since 13 Ma (Edwards *et al.*, 2001), and CAM photosynthesis is largely limited to aqueous and desert environments. It is notable that the post-Hitch Wood pCO_2 fall interpreted from our Central European high-resolution $\Delta^{13}C$ curve (Fig. 5) is consistent with low-resolution Turonian terrestrial organic matter-based $\Delta^{13}C$ trends derived from the NW Pacific (Uramoto *et al.*, 2013), and general models indicating a long-term Late Cretaceous $_{p}CO_2$ fall (Tajika, 1999; Berner, 2006).

The recognition of key Turonian carbon CIEs in the terrestrial carbon record offers an opportunity to further refine the calibration of the biostratigraphy of the endemic mollusc faunas in the NW Pacific to the international timescale (cf. Hayakawa & Hirano, 2013). In addition, there is potential for dating fully non-marine Turonian successions barren of fossils, although here a long and detailed time series will be required, and an absence of major hiatuses is essential to correctly identify major isotope shifts that can be assigned to specific CIEs.

CONCLUSIONS

Trends in Turonian carbon isotope curves derived from hemipelagic sediments sampled at 5·6 kyr average resolution enable the recognition of 10 major Turonian CIEs and more than 20 secondary correlation levels in the Bohemian Cretaceous Basin. Mismatches between the $\delta^{13}C_{carb}$ versus $\delta^{13}C_{org}$ profiles are attributed principally to diagenesis and compositional variation in the two fractions, except for a divergence of the medium-term trends in the Upper Turonian interval, where atmospheric CO_2 drawdown may have been responsible.

Carbonate-carbon isotope curves allow the precise correlation of Turonian successions throughout Europe, despite diagenetic overprinting in some low-carbonate sections. Calibration using biostratigraphic datum levels is essential to ensure the robust unambiguous correlation of $\delta^{13}C$ profiles; following calibration, a correlation resolution of around 40 kyr is achievable where sampling density is sufficient.

Bulk-sediment oxygen isotopes provide evidence of Turonian climate change. Sea-surface temperatures were highest in the Early to Middle Turonian, coincident with high eustatic sea-levels. Medium-term to long-term trends in $\delta^{18}O_{carb}$ profiles indicate a Europe-wide trend of stepped cooling that accompanied long-term sea-level fall, beginning in the late-Middle Turonian and culminating in the mid-Late Turonian – the *Late Turonian Cool Phase.*

Brachiopod $\delta^{18}O_{carb}$ shell data indicate up to 4°C cooling of bottom waters. Coincident faunal changes include the southward spread of Boreal taxa in Europe, and evidence of major water-mass reorganization accompanying a eustatic lowstand, prior to renewed sea-level rise in the latest Turonian.

Turonian marine carbonate $\delta^{13}C$ records have been successfully correlated with marine organic matter and terrestrial wood carbon isotope records from Europe ($\delta^{13}C_{org}$), the North American Western Interior Basin ($\delta^{13}C_{org}$) and the NW Pacific ($\delta^{13}C_{wood}$), which offers opportunities for the improved intercalibration of regional biostratigraphic schemes, necessitated by the presence of endemic faunas, and for the correlation of marine to fully terrestrial records.

Correlation of $\delta^{13}C_{org}$ profiles from Central Europe to the North American Western Interior Basin demonstrates consistent trends, with the identification of key CIEs in both records, but with a hiatus spanning the Turonian–Coniacian boundary in the North American composite section.

Consistent marine carbonate and organic carbon isotope records on two continents, and comparable trends in terrestrial wood, evidence strong coupling of isotope signatures in the ocean–atmosphere–biosphere carbon reservoirs during the Late Cretaceous. Convergence in $\delta^{13}C_{carb}$ and $\delta^{13}C_{org}$ values (falling $\Delta^{13}C_{org}$) during the latest Turonian may represent a period of pCO_2 decline, beginning during the final stages of the Late Turonian Cool Phase.

ACKNOWLEDGEMENTS

This paper benefited from detailed reviews by Amanda Oehlert and Ines Wendler, and comments from Editor Peter Swart. IJ and DRG acknowledge funding by UK Natural Environment Research Council (NERC) grants NE/H020756/1 and NE/H021868/1, respectively. This research was supported by the Czech Science Foundation (GACR) grant P210/10/1991 and research programme AV0Z30120515 of the Academy of Sciences of the Czech Republic. JL acknowledges support from KONTAKT II Programme, grant LH12041.

References

Allan, J.R. and **Matthews, R.K.** (1982) Isotope signatures associated with early meteoric diagenesis. *Sedimentology*, **29**, 797–817.

Amédro, F., Accarie, H. and **Robaszynski, F.** (2005) Position de la limite Cénomanien – Turonien dans la Formation Bahloul de Tunisie centrale: apports intégrés des ammonites et des isotopes du carbone ($\delta^{13}C$). *Eclogae Geol. Helv.*, **98**, 151–167.

Anderson, T.F., and **Arthur M.A.** (1983) Stable isotopes of oxygen and carbon and their application to sedimentologic and environmental problems. In: *Stable Isotopes in Sedimentary Geology* (Eds M.A. Arthur, T.F. Anderson, I.R. Kaplan, J. Veizer and L.S. Land), *Short Course, Soc. Econ. Paleontol. Mineral.*, **10**, 1–151.

Archer, D. (2010) *The Global Carbon Cycle*. Princeton University Press, Princeton, 205 pp.

Arthur, M.A., Dean, W.E. and **Schlanger, S.O.** (1985) Variations in the global carbon cycle during the Cretaceous related to climate, volcanism and changes in atmospheric CO_2. In: *The Carbon Cycle and Atmospheric CO2: Natural Variations Archean to Present* (Eds E.T. Sundquist and W.S. Broecker), *Am. Geophys. Union Monogr.*, **32**, 504–529. Washington, DC.

Arthur, M.A., Dean, W.E. and **Pratt, L.M.** (1988) Geochemical and climatic effects of increased marine organic carbon burial at the Cenomanian/Turonian boundary. *Nature*, **335**, 714–717.

Arthur, M.A., Jenkyns, H.C., Brumsack, H.J. and **Schlanger, S.O.** (1990) Stratigraphy, geochemistry, and paleoceanography of organic carbon-rich Cretaceous sequences. In: *Cretaceous Resources Events and Rhythms: Background and Plans for Research* (Eds R.N. Ginsburg and B. Beaudoin), *NATO Sci. Ser. C Math. Phys. Sci.*, **304**, 75–119. Kluwer Academic Publishers, Dordrecht, The Netherlands.

Ball, B.A., Cobban, W.A., Merewether, E.A., Grauch, R.I., McKinney, K.C. and **Livo, K.E.** (2010) *Fossils, Lithologies, and Geophysical Logs of the Mancos Shale from Core Hole USGS CL-1 in Montrose County, Colorado*. U.S. Geol. Surv., Reston, Virginia, 38 pp.

Banner, J.L. and **Hanson, G.N.** (1990) Calculation of simultaneous isotopic and trace-element variations during water-rock interaction with applications to carbonate diagenesis. *Geochim. Cosmochim. Acta*, **54**, 3123–3137.

van Bentum, E.C., Reichart, G.J., Forster, A. and **Sinninghe Damsté, J.S.** (2012) Latitudinal differences in the amplitude of the OAE-2 carbon isotopic excursion: pCO_2 and paleo productivity. *Biogeoscience*, **9**, 717–731.

Berner, R.A. (1970) Sedimentary pyrite formation. *Am. J. Sci.*, **268**, 1–23.

Berner, R.A. (2006) GEOCARBSULF: a combined model for Phanerozoic atmospheric O_2 and CO_2. *Geochim. Cosmochim. Acta*, **70**, 5653–5664.

Bloomfield, J.P. (1997) The role of diagenesis in the hydrogeological stratification of carbonate aquifers: an example from the Chalk at Faircross, Berkshire, UK. *Hydrol. Earth Syst. Sci.*, **1**, 19–33.

Bolton, C.T., Stoll, H.M. and **Mendez-Vicente, A.** (2012) Vital effects in coccolith calcite: cenozoic climate-pCO_2 drove the diversity of carbon acquisition strategies in coccolithophores? *Paleoceanography*, **27**, PA4204, doi:10.1029/2012PA002339.

Bornemann, A., **Norris, R.D.**, **Friedrich, O.**, **Beckmann, B.**, **Schouten, S.**, **Damste, J.S.S.**, **Vogel, J.**, **Hofmann, P.** and **Wagner, T.** (2008) Isotopic evidence for glaciation during the Cretaceous supergreenhouse. *Science*, **319**, 189–192.

Boulila, S., **Galbrun, B.**, **Laskar, J.** and **Paelike, H.** (2012) A ~9 myr cycle in Cenozoic $\delta^{13}C$ record and long-term orbital eccentricity modulation: is there a link? *Earth Planet. Sci. Lett.*, **317**, 273–281.

Bowman, A.R. and **Bralower, T.J.** (2005) Paleoceanographic significance of high-resolution carbon isotope records across the Cenomanian-Turonian boundary in the Western Interior and New Jersey coastal plain, USA. *Mar. Geol.*, **217**, 305–321.

Bromley, R.G. and **Gale, A.S.** (1982) The lithostratigraphy of the English Chalk Rock. *Cret. Res.*, **3**, 273–306.

Burnett, J.A., **Gallagher, L.T.** and **Hampton, M.J.** (1998) Upper Cretaceous. In: *Calcareous Nannofossil Biostratigraphy* (Ed. P.R. Bown), *Br. Micropalaeontol. Soc. Publ. Ser.*, 132–199. Kluwer, Dordrecht.

Caron, M. (1985) Cretaceous planktonic foraminifera. In: *Plankton Stratigraphy* (Eds H.M. Bolli, M. Saunders and K. Perch-Nielsen), pp. 17–86. Cambridge University Press, Cambridge.

Caron, M., **Dall'Agnolo, S.**, **Accarie, H.**, **Barrera, E.**, **Kauffman, E.G.**, **Amédro, F.** and **Robaszynski, F.** (2006) High-resolution stratigraphy of the Cenomanian-Turonian boundary interval at Pueblo (USA) and wadi Bahloul (Tunisia) stable isotope and bio-events correlation. *Geobios*, **39**, 171–200.

Čech, S., **Klein, V.**, **Kříž, J.** and **Valečka, J.** (1980) Revision of the Upper Cretaceous stratigraphy of the Bohemian Cretaceous Basin. *Věst. Ústřed. Úst. Geol. (Bull. Geol. Surv. Prague)*, **55**, 277–296.

Choquette, P.W. and **James, N.P.** (1987) Diagenesis #12. Diagenesis in limestones 3. The deep burial environment. *Geosci. Canada*, **14**, 3–35.

Christ, N., **Immenhauser, A.**, **Wood, R.A.**, **Darwich, K.** and **Niedermayr, A.** (2015) Petrography and environmental controls on the formation of Phanerozoic marine carbonate hardgrounds. *Earth-Sci. Rev.*, **151**, 176–226.

Ciais, P., **Sabine, C.**, **Bala, G.**, **Bopp, L.**, **Brovkin, V.**, **Canadell, J.**, **Chhabra, A.**, **DeFries, R.**, **Galloway, J.**, **Heimann, M.**, **Jones, C.**, **Le Quéré, C.**, **Myneni, R.B.**, **Piao, S.** and **Thornton, P.** (2013) Carbon and other biogeochemical cycles. In: *Climate Change 2013: The Physical Science Basis. Contribution of Working Group I to the Fifth Assessment Report of the Intergovernmental Panel on Climate Change* (Eds T.F. Stocker, D. Qin, G.-K. Plattner, M. Tignor, S.K. Allen, J. Boschung, A. Nauels, Y. Xia, V. Bex and P.M. Midgley), pp. 465–570. Cambridge University Press, Cambridge.

Clarke, L.J. and **Jenkyns, H.C.** (1999) New oxygen isotope evidence for long-term Cretaceous climatic change in the Southern Hemisphere. *Geology*, **27**, 699–702.

Cobban, W.A., **Walaszczyk, I.**, **Obradovich, J.D.** and **McKinney, K.C.** (2006) *A USGS Zonal Table for the Upper Cretaceous Middle Cenomanian–Maastrichtian of the Western Interior of the United States Based on Ammonites, Inoceramids, and Radiometric Ages.* U.S. Geological Survey, Reston, Virginia, 45 pp.

Corfield, R.M., **Cartlidge, J.E.**, **Premoli Silva, I.P.** and **Housley, R.A.** (1991) Oxygen and carbon isotope stratigraphy of the Palaeogene and Cretaceous limestones in the Bottaccione Gorge and the Contessa Highway sections, Umbria, Italy. *Terra Nova*, **3**, 414–422.

De Cabrera, S.C., **Sliter, W.V.** and **Jarvis, I.** (1999) Integrated foraminiferal biostratigraphy and chemostratigraphy of the Querecual Formation (Cretaceous), Eastern Venezuela. *J. Foram. Res.*, **29**, 487–499.

Dean, W.E., **Arthur, M.A.** and **Claypool, G.E.** (1986) Depletion of ^{13}C in Cretaceous marine organic matter: source, diagenetic or environmental signal? *Mar. Geol.*, **70**, 119–157.

Desmares, D., **Grosheny, D.**, **Beaudoin, B.**, **Gardin, S.** and **Gauthier-Lafaye, F.** (2007) High resolution stratigraphic record constrained by volcanic ash beds at the Cenomanian-Turonian boundary in the Western Interior Basin, USA. *Cret. Res.*, **28**, 561–582.

Devalque, C., **Amédro, F.**, **Philip, J.** and **Robaszynski, F.** (1982) État des corrélation litho et biostratigraphiques dans le Turonien supérieur des massifs d'Uchaux et de la Céze. Les zones d'ammonites et de rudistes. *Mém. Mus. Nation. Hist. Nat. C Sci. Terre*, **49**, 57–69.

Ditchfield, P. and **Marshall, J.D.** (1989) Isotopic variation in rhythmically bedded chalks: paleotemperature variation in the Upper Cretaceous. *Geology*, **17**, 842–845.

Du Vivier, A.D.C., **Selby, D.**, **Condon, D.J.**, **Takashima, R.** and **Nishi, H.** (2015) Pacific $^{187}Os/^{188}Os$ isotope chemistry and U-Pb geochronology: synchroneity of global Os isotope change across OAE 2. *Earth Planet. Sci. Lett.*, **428**, 204–216.

Dubicka, Z. and **Peryt, D.** (2012) Foraminifers and stable isotope record of the Dubivtsi chalk (upper Turonian, Western Ukraine): palaeoenvironmental implications. *Geol. Q.*, **56**, 199–214.

Edwards, G.E., **Furbank, R.T.**, **Hatch, M.D.** and **Osmond, C.B.** (2001) What does it take to be C4? Lessons from the evolution of C4 photosynthesis. *Plant Physiol.*, **125**, 46–49.

Elderbak, K., **Leckie, R.M.** and **Tibert, N.E.** (2014) Paleoenvironmental and paleoceanographic changes across the Cenomanian-Turonian Boundary Event (Oceanic Anoxic Event 2) as indicated by foraminiferal assemblages from the eastern margin of the Cretaceous Western Interior Sea. *Palaeogeogr. Palaeoclimatol. Palaeoecol.*, **413**, 29–48.

Eldrett, J.S., **Minisini, D.** and **Bergman, S.C.** (2014) Decoupling of the carbon cycle during Ocean Anoxic Event 2. *Geology*, **42**, 567–570.

Eldrett, J.S., **Ma, C.**, **Bergman, S.C.**, **Lutz, B.**, **Gregory, F.J.**, **Dodsworth, P.**, **Phipps, M.**, **Hardas, P.**, **Minisini, D.**,

Ozkan, A., Ramezani, J., Bowring, S.A., Kamo, S.L., Ferguson, K., Macaulay, C. and **Kelly, A.E.** (2015) An astronomically calibrated stratigraphy of the Cenomanian, Turonian and earliest Coniacian from the Cretaceous Western Interior Seaway, USA: implications for global chronostratigraphy. *Cret. Res.*, **56**, 316–344.

Elrick, M., Molina-Garza, R., Duncan, R. and **Snow, L.** (2009) C-isotope stratigraphy and paleoenvironmental changes across OAE2 (mid-Cretaceous) from shallow-water platform carbonates of southern Mexico. *Earth Planet. Sci. Lett.*, **277**, 295–306.

Erbacher, J., Friedrich, O., Wilson, P.A., Birch, H. and **Mutterlose, J.** (2005) Stable organic carbon isotope stratigraphy across Oceanic Anoxic Event 2 of Demerara Rise, western tropical Atlantic. *Geochem. Geophys. Geosyst.*, **6**, Q06010. doi:10.1029/2004GC000850.

Forster, A., Kuypers, M.M.M., Turgeon, S.C., Brumsack, H.J., Petrizzo, M.R. and **Sinninghe Damsté, J.S.** (2008) The Cenomanian/Turonian oceanic anoxic event in the South Atlantic: new insights from a geochemical study of DSDP Site 530A. *Palaeogeogr. Palaeoclimat. Palaeoecol.*, **267**, 256–283.

Frank, T.D., Arthur, M.A. and **Dean, W.E.** (1999) Diagenesis of Lower Cretaceous pelagic carbonates, North Atlantic: paleoceanographic signals obscured. *J. Foram. Res.*, **29**, 340–351.

Frank, J., Wilmsen, M. and **Košťák, M.** (2013) The endemic and morphologically remarkable nautilid genus *Deltocymatoceras* Kummel, 1956 from the Late Cretaceous of Central Europe. *Bull. Geosci.*, **88**, 793–812.

Freeman, K.H. and **Hayes, J.M.** (1992) Fractionation of carbon isotopes by phytoplankton and estimates of ancient CO_2 levels. *Biogeochem. Cycles*, **6**, 185–198.

Friedrich, O., Norris, R.D. and **Erbacher, J.** (2012) Evolution of middle to Late Cretaceous oceans – a 55 m.y. record of Earth's temperature and carbon cycle. *Geology*, **40**, 107–110.

Gale, A.S. (1996) Turonian correlation and sequence stratigraphy of the Chalk in southern England. In: *Sequence Stratigraphy in British Geology* (Eds S.P. Hesselbo and D.N. Parkinson), *Geol. Soc. London. Spec. Publ.*, **103**, 177–195. Geological Society of London, Bath.

Gale, A.S., Jenkyns, H.C., Kennedy, W.J. and **Corfield, R.M.** (1993) Chemostratigraphy versus biostratigraphy: data from around the Cenomanian-Turonian boundary. *J. Geol. Soc. London*, **150**, 29–32.

Gale, A.S., Kennedy, W.J., Voigt, S. and **Walaszczyk, I.** (2005) Stratigraphy of the Upper Cenomanian – Lower Turonian Chalk succession at Eastbourne, Sussex, UK: ammonites, inoceramid bivalves and stable carbon isotopes. *Cret. Res.*, **26**, 460–487.

Galeotti, S., Rusciadelli, G., Sprovieri, M., Lanci, L., Gaudio, A. and **Pekar, S.** (2009) Sea-level control on facies architecture in the Cenomanian-Coniacian Apulian margin (Western Tethys): a record of glacio-eustatic fluctuations during the Cretaceous greenhouse? *Palaeogeogr. Palaeoclimatol. Palaeoecol.*, **276**, 196–205.

Gambacorta, G., Jenkyns, H.C., Russo, F., Tsikos, H., Wilson, P.A., Faucher, G. and **Erba, E.** (2015) Carbon- and oxygen-isotope records of mid-Cretaceous Tethyan pelagic sequences from the Umbria–Marche and Belluno Basins (Italy). *Newsl. Stratigr.*, **48**, 299–323.

Gröcke, D.R. (2002) The carbon isotope composition of ancient CO_2 based on higher-plant organic matter. *Phil. Trans. R. Soc. Lond. Ser. A Math. Phys. Eng. Sci.*, **360**, 633–658.

Gröcke, D.R., Hesselbo, S.P. and **Jenkyns, H.C.** (1999) Carbon-isotope composition of Lower Cretaceous fossil wood: ocean-atmosphere chemistry and relation to sea-level change. *Geology*, **27**, 155–158.

Gröcke, D.R., Price, G.D., Robinson, S.A., Baraboshkin, E.Y., Mutterlose, J. and **Ruffell, A.H.** (2005) The Upper Valanginian (Early Cretaceous) positive carbon-isotope event recorded in terrestrial plants. *Earth Planet. Sci. Lett.*, **240**, 495–509.

Gröcke, D.R., Ludvigson, G.A., Witzke, B.L., Robinson, S.A., Joeckel, R.M., Ufnar, D.F. and **Ravn, R.L.** (2006) Recognizing the Albian-Cenomanian (OAE1d) sequence boundary using plant carbon isotopes: Dakota Formation, Western Interior Basin, USA. *Geology*, **34**, 193–196.

Gross, M.G. (1964) Variations in the $^{18}O/^{16}O$ and $^{13}C/^{12}C$ ratios of diagenetically altered limestones in the Bermuda islands. *J. Geol.*, **72**, 172–193.

Hancock, J.M. (1989) Sea-level changes in the British region during the Late Cretaceous. *Proc. Geol. Assoc.*, **100**, 565–594.

Hancock, J.M. and **Kauffman, E.G.** (1979) The great transgressions of the Late Cretaceous. *J. Geol. Soc. London*, **136**, 175–186.

Haq, B.U. (2014) Cretaceous eustasy revisited. *Glob. Planet. Change*, **113**, 44–58.

Haq, B.U., Hardenbol, J. and **Vail, P.R.** (1987) Chronology of fluctuating sea levels since the Triassic. *Science*, **235**, 1156–1167.

Hasegawa, T. (1997) Cenomanian-Turonian carbon isotope events recorded in terrestrial organic matter from northern Japan. *Palaeogeogr. Palaeoclimatol. Palaeoecol.*, **130**, 251–273.

Hasegawa, T. (2003) A global carbon-isotope event in the Middle Turonian (Cretaceous) sequences in Japan and Russian Far East. *Proc. Jpn. Acad. Ser. B Phys. Biol. Sci.*, **79**, 141–144.

Hasegawa, T. and **Hatsugai, T.** (2000) Carbon-isotope stratigraphy and its chronostratigraphic significance for the Cretaceous Yezo Group, Kotanbetsu area, Hokkaido, Japan. *Paleontol. Res.*, **4**, 95–106.

Hasegawa, T. and **Saito, T.** (1993) Global synchroneity of a positive carbon isotope excursion at the Cenomanian/Turonian boundary: validation by calcareous microfossil biostratigraphy of the Yezo Group, Hokkaido, Japan. *Island Arc.*, **3**, 181–191.

Hasegawa, T., **Pratt, L.M.**, **Maeda, H.**, **Shigeta, Y.**, **Okamoto, T.**, **Kase, T.** and **Uemura, K.** (2003) Upper Cretaceous stable carbon isotope stratigraphy of terrestrial organic matter from Sakhalin, Russian Far East: a proxy for the isotopic composition of paleoatmospheric CO_2. *Palaeogeogr. Palaeoclimatol. Palaeoecol.*, **189**, 97–115.

Hasegawa, T., **Crampton, J.S.**, **Schioler, P.**, **Field, B.**, **Fukushi, K.** and **Kakizaki, Y.** (2013) Carbon isotope stratigraphy and depositional oxia through Cenomanian/ Turonian boundary sequences (Upper Cretaceous) in New Zealand. *Cret. Res.*, **40**, 61–80.

Hatch, J.R. and **Leventhal, J.S.** (1997) Early diagenetic partial oxidation of organic matter and sulfides in the Middle Pennsylvanian (Desmoinesian) Excello Shale Member of the Fort Scott Limestone and equivalents, northern Midcontinent region, USA. *Chem. Geol.*, **134**, 215–235.

Hay, W.W. and **Leslie, M.A.** (1990) Could possible changes in global groundwater reservoir cause eustatic sea level fluctuations? In: *Sea Level Change: Studies in Geophysics* (Ed. R. Ravelle), pp. 161–170. National Academy Press, Washington, DC.

Hayakawa, T. and **Hirano, H.** (2013) A revised inoceramid biozonation for the Upper Cretaceous based on high-resolution carbon isotope stratigraphy in northwestern Hokkaido, Japan. *Acta Geol. Pol.*, **63**, 239–263.

Hayes, J.M., **Strauss, H.** and **Kaufman, A.J.** (1999) The abundance of ^{13}C in marine organic matter and isotopic fractionation in the global biogeochemical cycle of carbon during the past 800 Ma. *Chem. Geol.*, **161**, 103–125.

Hedges, J.I. and **Keil, R.G.** (1995) Sedimentary organic matter preservation – an assessment and speculative synthesis. *Mar. Chem.*, **49**, 81–115.

Hedges, J.I., **Keil, R.G.** and **Benner, R.** (1997) What happens to terrestrial organic matter in the ocean? *Org. Geochem.*, **27**, 195–212.

Hedges, J.I., **Hu, F.S.**, **Devol, A.H.**, **Hartnett, H.E.**, **Tsamakis, E.** and **Keil, R.G.** (1999) Sedimentary organic matter preservation: a test for selective degradation under oxic conditions. *Am. J. Sci.*, **299**, 529–555.

Heimhofer, U., **Hochuli, P.A.**, **Burla, S.**, **Andersen, N.** and **Weissert, H.** (2003) Terrestrial carbon-isotope records from coastal deposits (Algarve, Portugal): a tool for chemostratigraphic correlation on an intrabasinal and global scale. *Terra Nova*, **15**, 8–13.

Herrle, J.O., **Kossler, P.**, **Friedrich, O.**, **Erlenkeuser, H.** and **Hemleben, C.** (2004) High-resolution carbon isotope records of the Aptian to Lower Albian from SE France and the Mazagan Plateau (DSDP Site 545): a stratigraphic tool for paleoceanographic and paleobiologic reconstruction. *Earth Planet. Sci. Lett.*, **218**, 149–161.

Holser, W.T. (1997) Geochemical events documented in inorganic carbon isotopes. *Palaeogeogr. Palaeoclimatol. Palaeoecol.*, **132**, 173–182.

Huber, B.T. and **Petrizzo, M.R.** (2014) Evolution and taxonomic study of the Cretaceous planktic foraminiferal genus *Helvetoglobotruncana* Reiss, 1957. *J. Foram. Res.*, **44**, 40–57.

Huber, B.T., **Norris, R.D.** and **MacLeod, K.G.** (2002) Deep-sea paleotemperature record of extreme warmth during the Cretaceous. *Geology*, **30**, 123–126.

Hudson, J.D. (1977) Stable isotopes and limestone lithification. *J. Geol. Soc. London*, **133**, 637–660.

IAEA (2009) *Czech Republic.* Global Network of Isotopes in Precipitation (GNIP), International Atomic Energy Agency, Vienna. Available at: http://www.univie.ac.at/cartography/project/wiser/.

Immenhauser, A., **Holmden, C.** and **Patterson, W.P.** (2008) Interpreting the carbon-isotope record of ancient shallow epeiric seas: lessons from the recent. In: *Dynamics of Epeiric Seas* (Eds B.R. Pratt and C. Holmden), *Geol. Assoc. Canada Spec. Pap.*, **48**, 137–174. Geological Association of Canada.

Jarvis, I., **Carson, G.A.**, **Cooper, M.K.E.**, **Hart, M.B.**, **Leary, P.N.**, **Tocher, B.A.**, **Horne, D.** and **Rosenfeld, A.** (1988a) Microfossil assemblages and the Cenomanian – Turonian (late Cretaceous) oceanic anoxic event. *Cret. Res.*, **9**, 3–103.

Jarvis, I., **Carson, G.A.**, **Hart, M.B.**, **Leary, P.N.** and **Tocher, B.A.** (1988b) The Cenomanian – Turonian (late Cretaceous) anoxic event in SW England: evidence from Hooken Cliffs near Beer, SE Devon. *Newsl. Stratigr.*, **18**, 147–164.

Jarvis, I., **Murphy, A.M.** and **Gale, A.S.** (2001) Geochemistry of pelagic and hemipelagic carbonates: criteria for identifying systems tracts and sea-level change. *J. Geol. Soc. London*, **158**, 685–696.

Jarvis, I., **Mabrouk, A.**, **Moody, R.T.J.** and **De Cabrera, S.C.** (2002) Late Cretaceous (Campanian) carbon isotope events, sea-level change and correlation of the Tethyan and Boreal realms. *Palaeogeogr. Palaeoclimatol. Palaeoecol.*, **188**, 215–248.

Jarvis, I., **Gale, A.S.**, **Jenkyns, H.C.** and **Pearce, M.A.** (2006) Secular variation in Late Cretaceous carbon isotopes and sea-level change: evidence from a new $\delta^{13}C$ carbonate reference curve for the Cenomanian – Campanian (99.6 – 70.6 Ma). *Geol. Mag.*, **143**, 561–608.

Jarvis, I., **Mabrouk, A.**, **Moody, R.T.J.**, **Murphy, A.M.** and **Sandman, R.I.** (2008) Applications of carbon isotope and elemental (Sr/Ca, Mn) chemostratigraphy to sequence analysis: sea-level change and the global correlation of pelagic carbonates. In: *Geology of East Libya* (Eds M.J. Salem and A.S. El-Hawat), *Earth Sci. Soc. Libya Tripoli*, **I**, 369–396.

Jarvis, I., **Lignum, J.S.**, **Groecke, D.R.**, **Jenkyns, H.C.** and **Pearce, M.A.** (2011) Black shale deposition, atmospheric CO_2 drawdown, and cooling during the Cenomanian-Turonian Oceanic Anoxic Event. *Paleoceanography*, **26**, Pa3201. doi:10.1029/2010pa002081.

Jeans, C.V., **Long, D.**, **Hall, M.A.**, **Bland, D.J.** and **Cornford, C.** (1991) The geochemistry of the Plenus Marls at Dover, England: evidence of fluctuating oceanographic conditions and of glacial control during the development of the Cenomanian-Turonian $\delta^{13}C$ anomaly. *Geol. Mag.*, **128**, 603–632.

Jeans, C.V., Hu, X. and **Mortimore, R.** (2012) Calcite cements and the stratigraphical significance of the marine delta $\delta^{13}C$ carbonate reference curve for the Upper Cretaceous Chalk of England. *Acta Geol. Pol.*, **62**, 173–196.

Jenkyns, H.C. (2010) Geochemistry of oceanic anoxic events. *Geochem. Geophys. Geosyst.*, **11**, Q03004. doi:10.1029/2009GC002788.

Jenkyns, H.C., Gale, A.S. and **Corfield, R.M.** (1994) Carbon- and oxygen-isotope stratigraphy of the English Chalk and Italian Scaglia and its palaeoclimatic significance. *Geol. Mag.*, **131**, 1–34.

Jiráková, H., Huneau, F., Hrkal, Z., Celle-Jeanton, H. and **Le Coustumer, P.** (2010) Carbon isotopes to constrain the origin and circulation pattern of groundwater in the north-western part of the Bohemian Cretaceous Basin (Czech Republic). *Appl. Geochem.*, **25**, 1265–1279.

Jiráková, H., Procházka, M., Dědeček, P., Kobr, M., Hrkal, Z., Huneau, F. and **Le Coustumer, P.** (2011) Geothermal assessment of the deep aquifers of the northwestern part of the Bohemian Cretaceous basin, Czech Republic. *Geothermics*, **40**, 112–124.

Joo, Y.J. and **Sageman, B.B.** (2014) Cenomanian to Campanian carbon isotope chemostratigraphy from the Western Interior Basin, U.S.A. *J. Sed. Res.*, **84**, 529–542.

Katz, M.E., Wright, J.D., Miller, K.G., Cramer, B.S., Fennel, K. and **Falkowski, P.G.** (2005) Biological overprint of the geological carbon cycle. *Mar. Geol.*, **217**, 323–338.

Kauffman, E.G., Sageman, B.B., Elder, W.P., Kirkland, J.I. and **Villamil, T.** (1993) Cretaceous molluscan biostratigraphy and biogeography, Western Interior Basin, North America. In: *Evolution of the Western Interior Basin* (Eds W.G.E. Caldwell and E.G. Kauffman), *Spec. Pap. Geol. Assoc. Canada*, **39**, 397–434.

Kauffman, E.G., Kennedy, W.J. and **Wood, C.J.** (1996) The Coniacian stage and substage boundaries. In: *Proceedings "Second International Symposium on Cretaceous Stage Boundaries" Brussels 8-16 September 1995* (Eds P.F. Rawson, A.V. Dhondt, J.M. Hancock and W.J. Kennedy), *Bulletin de l'Institut Royal des Sciences Naturelles de Belgique Sciences de la Terre*, **66** (Suppl), 81–94.

Keil, R.G., Hu, F.S., Tsamakis, E.C. and **Hedges, J.I.** (1994) Pollen in marine sediments as an indicator of oxidation of organic matter. *Nature*, **369**, 639–641.

Keller, G., Han, Q., Adatte, T. and **Burns, S.J.** (2001) Palaeoenvironment of the Cenomanian-Turonian transition at Eastbourne, England. *Cret. Res.*, **22**, 391–422.

Kennedy, W.J. and **Garrison, R.E.** (1975) Morphology and genesis of nodular chalks and hardgrounds in the Upper Cretaceous of southern England. *Sedimentology*, **22**, 311–386.

Kennedy, W.J., Walaszczyk, I. and **Cobban, W.A.** (2005) The Global Boundary Stratotype Section and Point for the base of the Turonian Stage of the Cretaceous: Pueblo, Colorado, U.S.A. *Episodes*, **28**, 93–104.

Kennedy, W.J., Gale, A.S., Huber, B.T., Petrizzo, M.R., Bown, P., Barchetta, A. and **Jenkyns, H.C.** (2014) Integrated stratigraphy across the Aptian/Albian boundary at Col de Pre-Guittard (southeast France): a candidate Global Boundary Stratotype Section. *Cret. Res.*, **51**, 248–259.

Killops, S. and **Killops, V.** (2005) *Introduction to Organic Geochemistry*. Blackwell, Oxford, 393 pp.

Kollmann, H.A., Peza, L.H. and **Čech, S.** (1998) Upper Cretaceous Nerineacea of the Bohemian Basin (Czech Republic) and the Saxonian Basin (Germany) and their significance for Tethyan environments. *Abh. Staat. Mus. Mineral. Geol. Dresden*, **43/44**, 151–172.

Kolonic, S., Wagner, T., Forster, A., Sinninghe Damsté, J.S.S., Walsworth-Bell, B., Erba, E., Turgeon, S., Brumsack, H.-J., Chellai, E.H., Tsikos, H., Kuhnt, W. and **Kuypers, M.M.M.** (2005) Black shale deposition on the northwest African Shelf during the Cenomanian/Turonian oceanic anoxic event: climate coupling and global organic carbon burial. *Paleoceanography*, **20**, PA1006. doi:10.1029/2003PA000950.

Košťák, M. and **Wiese, F.** (2011) Extremely rare Turonian belemnites from the Bohemian Cretaceous Basin and their palaeogeographical importance. *Acta Geol. Pol.*, **56**, 433–437.

Košťák, M., Čech, S., Ekrt, B., Mazuch, M., Wiese, F., Voigt, S. and **Wood, C.J.** (2004) Belemnites of the Bohemian Cretaceous Basin in a global context. *Acta Geol. Pol.*, **54**, 511–533.

Kuhnt, W., Luderer, F., Nederbragt, S., Thurow, J. and **Wagner, T.** (2005) Orbital-scale record of the late Cenomanian-Turonian oceanic anoxic event (OAE-2) in the Tarfaya Basin (Morocco). *Int. J. Earth Sci. (Geol. Rundsch.)*, **94**, 147–159.

Kump, L.R. (1991) Interpreting carbon-isotope excursions – Stangelove oceans. *Geology*, **19**, 299–302.

Kump, L.R. and **Arthur, M.A.** (1999) Interpreting carbon-isotope excursions: carbonates and organic matter. *Chem. Geol.*, **161**, 181–198.

Kuypers, M.M.M., Pancost, R.D. and **Sinninghe Damsté, J.S.** (1999) A large and abrupt fall in atmospheric CO_2 concentration during Cretaceous times. *Nature*, **399**, 342–345.

Kuypers, M.M.M., Pancost, R.D., Nijenhuis, I.A. and **Sinninghe Damsté, J.S.** (2002) Enhanced productivity led to increased organic carbon burial in the euxinic North Atlantic basin during the late Cenomanian oceanic anoxic event. *Paleoceanography*, **17**, 1051. doi:10.1029/2000PA000569.

Kuypers, M.M.M., Lourens, L.J., Rijpstra, W.R.C., Pancost, R.D., Nijenhuis, I.A. and **Sinninghe Damsté, J.S.** (2004) Orbital forcing of organic carbon burial in the proto-North Atlantic during oceanic anoxic event 2. *Earth Planet. Sci. Lett.*, **228**, 465–482.

Lalonde, K., **Mucci, A.**, **Ouellet, A.** and **Gelinas, Y.** (2012) Preservation of organic matter in sediments promoted by iron. *Nature*, **483**, 198–200.

Lamolda, M.A., **Gorostidi, A.** and **Paul, C.R.C.** (1994) Quantitative estimates of calcareous nannofossil changes across the Plenus Marls (latest Cenomanian), Dover, England – implications for the generation of the Cenomanian-Turonian boundary event. *Cret. Res.*, **15**, 143–164.

Lamolda, M.A., **Paul, C.R.C.**, **Peryt, D.** and **Pons, J.M.** (2014) The Global Boundary Stratotype and Section Point (GSSP) for the base of the Santonian Stage, "Cantera de Margas", Olazagutia, northern Spain. *Episodes*, **37**, 2–13.

de Lange, G.J., **van Os, B.**, **Pruysers, P.A.**, **Middelburg, J.J.**, **Castradori, D.**, **van Santvoort, P.**, **Müller, P.J.**, **Eggenkamp, H.** and **Prahl, F.G.** (1994) Possible early diagenetic alteration of palaeo proxies. In: *Carbon Cycling in the Glacial Ocean: Constraints on the Ocean's Role in Global Change* (Eds R. Zahn, T.F. Pederson, M.A. Kaminski and L. Labeyrie), *NATO ASI Ser.*, **117**, 225–258. Springer-Verlag, Berlin, Heidelberg.

Laurin, J., **Čech, S.**, **Uličný, D.**, **Štaffen, Z.** and **Svobodová, M.** (2014) Astrochronology of the Late Turonian: implications for the behavior of the carbon cycle at the demise of peak greenhouse. *Earth Planet. Sci. Lett.*, **394**, 254–269.

Laurin, J., **Meyers, S.R.**, **Uličný, D.**, **Jarvis, I.** and **Sageman, B.B.** (2015) Axial obliquity control on the greenhouse carbon budget through middle- to high-latitude reservoirs. *Paleoceanography*, **30**, 133–149. doi:10.1002/2014PA002736.

Laws, E.A., **Popp, B.N.**, **Bidigare, R.R.**, **Kennicutt, M.C.** and **Macko, S.A.** (1995) Dependence of phytoplankton carbon isotope composition on growth rate and $(CO_2)_{aq}$: theoretical considerations and experimental results. *Geochim. Cosmochim. Acta*, **59**, 1131–1138.

Lees, J.A. (2008) The calcareous nannofossil record across the Late Cretaceous Turonian/Coniacian boundary, including new data from Germany, Poland, the Czech Republic and England. *Cret. Res.*, **29**, 40–64.

Li, X., **Jenkyns, H.C.**, **Wang, C.**, **Hu, X.**, **Chen, X.**, **Wei, Y.**, **Huang, Y.** and **Cui, J.** (2006) Upper Cretaceous carbon- and oxygen-isotope stratigraphy of hemipelagic carbonate facies from southern Tibet, China. *J. Geol. Soc. London*, **163**, 375–382.

Linnert, C., **Robinson, S.A.**, **Lees, J.A.**, **Bown, P.R.**, **Perez-Rodriguez, I.**, **Petrizzo, M.R.**, **Falzoni, F.**, **Littler, K.**, **Antonio Arz, J.** and **Russell, E.E.** (2014) Evidence for global cooling in the Late Cretaceous. *Nat. Comm.*, **5**, 1–7.

MacEachern, J.A., **Pemberton, S.G.**, **Gingras, M.K.** and **Bann, K.L.** (2010) Ichnology and facies models. In: *Facies Models 4* (Eds N.P. James and R.W. Dalrymple), pp. 19–58. Geol. Assoc. Canada, St John's, Newfoundland.

MacLeod, K.G., **Huber, B.T.**, **Jimenez Berrocoso, A.** and **Wendler, I.** (2013) A stable and hot Turonian without glacial $\delta^{18}O$ excursions is indicated by exquisitely preserved Tanzanian foraminifera. *Geology*, **41**, 1083–1086.

Marshall, J.D. (1992) Climatic and oceanographic isotopic signals from the carbonate rock record and their preservation. *Geol. Mag.*, **129**, 143–160.

Mason, B. and **Moore, C.B.** (1982) *Principles of Geochemistry*. Wiley, New York, 344 pp.

Mayer, L.M. (1994) Surface-area control of organic-carbon accumulation in continental-shelf sediments. *Geochim. Cosmochim. Acta*, **58**, 1271–1284.

Meyers, P.A. (2014) Why are the $\delta^{13}C_{org}$ values in Phanerozoic black shales more negative than in modern marine organic matter? *Geochem. Geophys. Geosyst.*, **15**, 3085–3106.

Meyers, S.R., **Siewert, S.E.**, **Singer, B.S.**, **Sageman, B.B.**, **Condon, D.J.**, **Obradovich, J.D.**, **Jicha, B.R.** and **Sawyer, D.A.** (2012) Intercalibration of radioisotopic and astrochronologic time scales for the Cenomanian-Turonian boundary interval, Western Interior Basin, USA. *Geology*, **40**, 7–10.

Miller, K.G., **Sugarman, P.J.**, **Browning, J.V.**, **Kominz, M.A.**, **Olsson, R.K.**, **Feigenson, M.D.** and **Hernandez, J.C.** (2004) Upper Cretaceous sequences and sea-level history, New Jersey Coastal Plain. *GSA Bull.*, **116**, 368–393.

Mitchell, S.F., **Ball, J.D.**, **Crowley, S.F.**, **Marshall, J.D.**, **Paul, C.R.C.**, **Veltkamp, C.J.** and **Samir, A.** (1997) Isotope data from Cretaceous chalks and foraminifera: environmental or diagenetic signals? *Geology*, **25**, 691–694.

Mitchell, A.J., **Uličný, D.**, **Hampson, G.J.**, **Allison, P.A.**, **Gorman, G.J.**, **Piggott, M.D.**, **Wells, M.R.** and **Pain, C.C.** (2010) Modelling tidal current-induced bed shear stress and palaeocirculation in an epicontinental seaway: the Bohemian Cretaceous Basin, Central Europe. *Sedimentology*, **57**, 359–388.

Nagm, E., **El-Qot, G.** and **Wilmsen, M.** (2014) Stable-isotope stratigraphy of the Cenomanian-Turonian (Upper Cretaceous) boundary event (CTBE) in Wadi Qena, Eastern Desert, Egypt. *J. Afr. Earth Sci.*, **100**, 524–531.

Nishi, H., **Takashima, R.**, **Hatsugai, T.**, **Saito, T.**, **Moriya, K.**, **Ennyu, A.** and **Sakai, T.** (2003) Planktonic foraminiferal zonation in the Cretaceous Yezo Group, Central Hokkaido, Japan. *J. Asian Earth Sci.*, **21**, 867–886.

Obradovich, J. (1993) A Cretaceous time scale. In: *Evolution of the Western Interior Basin* (Eds W.G.E. Caldwell and E.G. Kauffman), *Geol. Assoc. Canada Spec. Pap.*, **39**, 379–396.

Oehlert, A.M. and **Swart, P.K.** (2014) Interpreting carbonate and organic carbon isotope covariance in the sedimentary record. *Nat. Comm.*, **5**, 4672. doi:10.1038/ncomms5672.

Oehlert, A.M., **Lamb-Wozniak, K.A.**, **Devlin, Q.B.**, **Mackenzie, G.J.**, **Reijmer, J.J.G.** and **Swart, P.K.** (2012) The stable carbon isotopic composition of organic material in platform derived sediments: implications for reconstructing the global carbon cycle. *Sedimentology*, **59**, 319–335.

Ogg, J.G., Hinnov, L.A. and **Huang, C.** (2012) Cretaceous. In: *The Geological Time Scale 2012* (Eds F.M. Gradstein, J.G. Ogg, M.D. Schmitz and G.M. Ogg), **2**, 793–853. Elsevier, Amsterdam.

Olde, K., Jarvis, I., Pearce, M.A., Uličný, D., Tocher, B.A., Trabucho-Alexandre, J. and **Gröcke, D.** (2015a) A revised northern European Turonian (Upper Cretaceous) dinoflagellate cyst biostratigraphy: integrating palynology and carbon isotope events. *Rev. Palaeobot. Palynol.*, **213**, 1–16.

Olde, K., Jarvis, I., Uličný, D., Pearce, M.A., Trabucho-Alexandre, J., Čech, S., Gröcke, D.R., Laurin, J., Švábenická, L. and **Tocher, B.A.** (2015b) Geochemical and palynological sea-level proxies in hemipelagic sediments: a critical assessment from the Upper Cretaceous of the Czech Republic. *Palaeogeogr. Palaeoclimatol. Palaeoecol.*, **435**, 222–243.

Paces, T., Corcho Alvarado, J.A., Herrman, Z., Kodes, V., Muzak, J., Novak, J., Purtschert, R., Remenarova, D. and **Valecka, J.** (2008) The Cenomanian and Turonian aquifers of the Bohemian Cretaceous Basin, Czech Republic. In: *Natural Groundwater Quality* (Eds W.E. Edmunds and P. Shand), pp. 372–390. Blackwell, Oxford.

Parente, M., Frijia, G. and **Di Lucia, M.** (2007) Carbon-isotope stratigraphy of Cenomanian-Turonian platform carbonates from the southern Apennines (Italy): a chemostratigraphic approach to the problem of correlation between shallow-water and deep-water successions. *J. Geol. Soc. London*, **164**, 609–620.

Paul, C.R.C. and **Lamolda, M.A.** (2009) Testing the precision of bioevents. *Geol. Mag.*, **146**, 625–637.

Paul, C.R.C., Lamolda, M.A., Mitchell, S.F., Vaziri, M.R., Gorostidi, A. and **Marshall, J.D.** (1999) The Cenomanian-Turonian boundary at Eastbourne (Sussex, UK): a proposed European reference section. *Palaeogeogr. Palaeoclimatol. Palaeoecol.*, **150**, 83–121.

Pearce, M.A., Jarvis, I., Swan, A.R.H., Murphy, A.M., Tocher, B.A. and **Edmunds, W.M.** (2003) Integrating palynological and geochemical data in a new approach to palaeoecological studies: Upper Cretaceous of the Banterwick Barn Chalk borehole, Berkshire, UK. *Mar. Micropaleontol.*, **47**, 271–306.

Petrizzo, M.R. (2000) Upper Turonian–lower Campanian planktonic foraminifera from southern mid-high latitudes (Exmouth Plateau, NW Australia): biostratigraphy and taxonomic notes. *Cret. Res.*, **21**, 479–505.

Petrizzo, M.R. (2002) Palaeoceanographic and palaeoclimatic inferences from Late Cretaceous planktonic foraminiferal assemblages from the Exmouth Plateau (ODP Sites 762 and 763, eastern Indian Ocean). *Mar. Micropaleontol.*, **45**, 117–150.

Poulsen, C.J., Seidov, D., Barron, E.J. and **Peterson, W.H.** (1998) The impact of paleogeographic evolution on the surface oceanic circulation and the marine environment within the mid-Cretaceous Tethys. *Paleoceanography*, **13**, 546–559.

Prahl, F.G., De Lange, G.J., Scholten, S. and **Cowie, G.L.** (1997) A case of post-depositional aerobic degradation of terrestrial organic matter in turbidite deposits from the Madeira Abyssal Plain. *Org. Geochem.*, **27**, 141–152.

Pratt, L.M., Arthur, M.A., Dean, W.E. and **Scholle, P.A.** (1993) Paleoceanographic cycles and events during the Late Cretaceous in the Western Interior Seaway of North America. In: *Cretaceous Evolution of the Western Interior Basin of North America* (Eds W.G.E. Caldwell and E.G. Kauffman), *Geol. Assoc. Canada Spec. Pap.*, **39**, 333–353.

Premoli Silva, I. and **Sliter, W.V.** (1995) Cretaceous planktonic foraminiferal biostratigraphy and evolutionary trends from the Bottaccione section, Gubbio, Italy. *Palaeontogr. Ital.*, **82**, 1–89.

Rau, G.H., Froelich, P.N., Takahashi, T. and **Des Marais, D.J.** (1991) Does sedimentary organic $\delta^{13}C$ record variations in Quaternary ocean [$CO_2(aq)$]? *Paleoceanography*, **6**, 335–347.

Richardt, N. and **Wilmsen, M.** (2012) Lower Upper Cretaceous standard section of the southern Münsterland (NW Germany): carbon stable-isotopes and sequence stratigraphy. *Newsl. Stratigr.*, **45**, 1–24.

Robaszynski, F. and **Caron, M.** (1979a) Atlas de Foraminifères planctoniques de Crétacé moyen (Mer Boréal et Téthys). Première partie. *Cah. Micropaléontol.*, **1**, 1–185.

Robaszynski, F. and **Caron, M.** (1979b) Atlas de Foraminifères planctoniques de Crétacé moyen (Mer Boréal et Téthys). Deuxième partie. *Cah. Micropaléontol.*, **2**, 1–181.

Robaszynski, F. and **Caron, M.** (1995) Cretaceous planktonic foraminifera: comments on the Europe-Mediterranean zonation. *Bull. Soc. Géol. France*, **166**, 681–692.

Robaszynski, F., Donoso, J.M.G., Linares, D., Amedro, F., Caron, M., Dupuis, C., Dhondt, A.V. and **Gartner, S.** (2000) The Upper Cretaceous of the Kalaat Senan region, Central Tunisia. Integrated litho-biostratigraphy based on ammonites, planktonic foraminifera and nannofossils zones from Upper Turonian to Maastrichtian. Bull. Centre Rech. Explor.-Prod. *Elf Aquitaine*, **22**, 359–490.

Robaszynski, F., Zagrarni, M.F., Caron, M. and **Amédro, F.** (2010) The global bio–events at the Cenomanian-Turonian transition in the reduced Bahloul Formation of Bou Ghanem (central Tunisia). *Cret. Res.*, **31**, 1–15.

Rohling, E.J. and **Cooke, S.** (1999) Stable oxygen and carbon isotopes in foraminiferal carbonate shells. In: *Modern Foraminifera* (Ed. B.K. Sen Gupta), pp. 239–258. Kluwer, Amsterdam.

Sageman, B.B., Meyers, S.R. and **Arthur, M.A.** (2006) Orbital time scale and new C-isotope record for Cenomanian-Turonian boundary stratotype. *Geology*, **34**, 125–128.

Sageman, B.B., Singer, B.S., Meyers, S.R., Siewert, S.E., Walaszczyk, I., Condon, D.J., Jicha, B.R., Obradovich, J.D. and **Sawyer, D.A.** (2014) Integrating $^{40}Ar/^{39}Ar$, U-Pb, and astronomical clocks in the Cretaceous Niobrara Formation, Western Interior Basin, USA. *GSA Bull.*, **126**, 956–973.

Saltzman, M.R. and **Thomas, E.** (2012) Carbon isotope stratigraphy. In: *The Geological Time Scale 2012* (Eds F.M. Gradstein, J.G. Ogg, M.D. Schmitz and G.M. Ogg), **1**, 207–232. Elsevier, Amsterdam.

Schlanger, S.O. and **Jenkyns, H.C.** (1976) Cretaceous oceanic anoxic events: causes and consequences. *Geol. Mijnb.*, **55**, 179–184.

Schlanger, S.O., Arthur, M.A., Jenkyns, H.C. and **Scholle, P.A.** (1983) Stratigraphic and paleo-oceanographic setting of organic carbon-rich strata deposited during the Cenomanian-Turonian oceanic anoxic event. *AAPG Bull.*, **67**, 545.

Schlanger, S.O., Arthur, M.A., Jenkyns, H.C. and **Scholle, P.A.** (1987) The Cenomanian–Turonian Oceanic Anoxic event, I. Stratigraphy and distribution of organic carbon-rich beds and the marine $d^{13}C$ excursion. In: *Marine Petroleum Source Rocks* (Eds J. Brooks and A.J. Fleet), *Geol. Soc. London. Spec. Publ.*, **26**, 371–399. Blackwell, Oxford.

Schlesinger, W.H. and **Berhardt, E.S.** (2013) *Biogeochemistry. An Analysis of Global Change.* Academic Press, Amsterdam, 672 pp.

Scholle, P.A. and **Arthur, M.A.** (1980) Carbon isotope fluctuation in Cretaceous pelagic limestones: potential stratigraphic and petroleum exploration tool. *AAPG Bull.*, **64**, 67–87.

Schrag, D.P., Depaolo, D.J. and **Richter, F.M.** (1995) Reconstrucing past sea-surface temperatures – correcting for diagenesis of bulk marine carbonate. *Geochim. Cosmochim. Acta*, **59**, 2265–2278.

Schrag, D.P., Higgins, J.A., Macdonald, F.A. and **Johnston, D.T.** (2013) Authigenic carbonate and the history of the global carbon cycle. *Science*, **339**, 540–543.

Schubert, B.A. and **Jahren, A.H.** (2013) Reconciliation of marine and terrestrial carbon isotope excursions based on changing atmospheric CO_2 levels. *Nat. Commun.*, **4**, 1653, doi:10.1038/ncomms2659.

Scopelliti, G., Bellanca, A., Erba, E., Jenkyns, H.C., Neri, R., Tamagnini, P., Luciani, V. and **Masetti, D.** (2008) Cenomanian-Turonian carbonate and organic-carbon isotope records, biostratigraphy and provenance of a key section in NE Sicily, Italy: palaeoceanographic and palaeogeographic implications. *Palaeogeogr. Palaeoclimatol. Palaeoecol.*, **265**, 59–77.

Shackleton, N.J. and **Kennett, J.P.** (1975) Paleotemperature history of the Cenozoic and the initiation of Antarctic glaciation: oxygen and carbon isotope analyses in DSDP sites 277, 279, and 281. (Eds J.P. Kennett and R.E. Houtz *et al.*), *Init. Rept. DSDP*, **29**, 743–755. US Government Printing Office, Washington, DC.

Sikora, P.J., Howe, R.W., Gale, A.S. and **Stein, J.A.** (2004) Chronostratigraphy of proposed Turonian–Coniacian (Upper Cretaceous) stage boundary stratotypes: Salzgitter-Salder, Germany, and Wagon Mound, New Mexico, USA. In: *Palynology and Micropalaeontology of Boundaries* (Eds A.B. Beaudoin and M.J.H. Head), *Geol. Soc. London. Spec. Publ.*, **230**, 207–242.

Sinninghe Damsté, J.S. and **Köster, J.** (1998) A euxinic southern North Atlantic Ocean during the Cenomanian/ Turonian oceanic anoxic event. *Earth Planet. Sci. Lett.*, **158**, 165–173.

Sinninghe Damsté, J.S., Kuypers, M.M.M., Pancost, R.D. and **Schouten, S.** (2008) The carbon isotopic response of algae, (cyano)bacteria, archaea and higher plants to the late Cenomanian perturbation of the global carbon cycle: insights from biomarkers in black shales from the Cape Verde Basin (DSDP Site 367). *Org. Geochem.*, **39**, 1703–1718.

Sissingh, W. (1977) Biostratigraphy of Cretaceous nannoplankton. *Geol. Mijnb.*, **56**, 37–65.

Sprovieri, M., Coccioni, R., Lirer, F., Pelosi, N. and **Lozar, F.** (2006) Orbital tuning of a lower Cretaceous composite record (Maiolica Formation, central Italy). *Paleoceanography*, **21**, PA4212, doi:10.1029/2005PA001224.

Sprovieri, M., Sabatino, N., Pelosi, N., Batenburg, S.J., Coccioni, R., Iavarone, M. and **Mazzola, S.** (2013) Late Cretaceous orbitally-paced carbon isotope stratigraphy from the Bottaccione Gorge (Italy). *Palaeogeogr. Palaeoclimatol. Palaeoecol.*, **379**, 81–94.

Stoll, H.M. (2005) Limited range of interspecific vital effects in coccolith stable isotopic records during the Paleocene Eocene thermal maximum. *Paleoceanography*, **20**, PA1007, doi:10.1029/2004PA001046.

Stoll, H.M. and **Schrag, D.P.** (2000) High-resolution stable isotope records from the Upper Cretaceous rocks of Italy and Spain: glacial episodes in a greenhouse planet? *GSA Bull.*, **112**, 308–319.

Surlyk, F., Dons, T., Clausen, C.K. and **Higham, J.** (2003) Upper Cretaceous. In: *The Millennium Atlas: Petroleum Geology of the Central and Northern North Sea* (Eds D. Evans, C. Graham, A. Armour and P. Bathurst), pp. 213–233. The Geological Society of London, London.

Švábenická, L. (2012) Nannofossil record across the Cenomanian-Coniacian interval in the Bohemian Cretaceous Basin and Tethyan foreland basins (Outer Western Carpathians), Czech Republic. *Geol. Carpath.*, **63**, 201–217.

Swart, P.K. (2015) The geochemistry of carbonate diagenesis: the past, present and future. *Sedimentology*, **62**, 1233–1304.

Tajika, E. (1999) Carbon cycle and climate change during the Cretaceous inferred from a biogeochemical carbon cycle model. *Island Arc.*, **8**, 293–303.

Takahashi, A. (2005) Diversity changes in Cretaceous inoceramid bivalves of Japan. *Paleontol. Res.*, **9**, 217–232.

Takashima, R., Nishi, H., Yamanaka, T., Hayashi, K., Waseda, A., Obuse, A., Tomosugi, T., Deguchi, N. and **Mochizuki, S.** (2010) High-resolution terrestrial carbon isotope and planktic foraminiferal records of the Upper Cenomanian to the Lower Campanian in the Northwest Pacific. *Earth Planet. Sci. Lett.*, **289**, 570–582.

Takashima, R., **Nishi, H.**, **Yamanaka, T.**, **Tomosugi, T.**, **Fernando, A.G.**, **Tanabe, K.**, **Moriya, K.**, **Kawabe, F.** and **Hayashi, K.** (2011) Prevailing oxic environments in the Pacific Ocean during the mid-Cretaceous Oceanic Anoxic Event 2. *Nat. Commun.*, **2**, 234, doi:10.1038/ncomms1233.

Tappan, H. (1980) Haptophyta, coccolithophores and other calcareous nannoplankton. *The Paleobiology of Plant Protista* (Ed. H. Tappan), pp. 678–803. Freeman, San Francisco.

Toshimitsu, S., **Matsumoto, T.**, **Noda, M.**, **Nishida, T.** and **Maiya, S.** (1995) Towards an integrated mega-, micro- and magneto-stratigraphy of the Upper Cretaceous in Japan. *J. Geol. Soc. Japan*, **101**, 19–29 (in Japanese with English abstract).

Tsikos, H., **Jenkyns, H.C.**, **Walsworth-Bell, B.**, **Petrizzo, M.R.**, **Forster, A.**, **Kolonic, S.**, **Erba, E.**, **Premoli-Silva, I.P.**, **Baas, M.**, **Wagner, T.** and **Sinninghe Damsté, J.S.** (2004) Carbon-isotope stratigraphy recorded by the Cenomanian – Turonian Oceanic Anoxic Event: correlation and implications based on three key localities. *J. Geol. Soc. London*, **161**, 711–719.

Tsuchiya, K., **Hasegawa, H.** and **Pratt, L.M.** (2003) Stratigraphic relationship between diagnostic carbon isotope profiles and inoceramid biozones from the Yezo Group, Hokkaido, Japan. *J. Geol. Soc. Japan*, **109**, 30–40 (in Japanese with English abstract).

Uličný, D., **Hladíková, J.** and **Hradecká, L.** (1993) Record of sea-level changes, oxygen depletion and the $\delta^{13}C$ anomaly across the Cenomanian-Turonian boundary, Bohemian Cretaceous Basin. *Cret. Res.*, **14**, 211–234.

Uličný, D., **Laurin, J.** and **Čech, S.** (2009) Controls on clastic sequence geometries in a shallow-marine, transtensional basin: the Bohemian Cretaceous Basin, Czech Republic. *Sedimentology*, **56**, 1077–1141.

Uličný, D., **Jarvis, I.**, **Gröcke, D.R.**, **Čech, S.**, **Laurin, J.**, **Olde, K.**, **Trabucho-Alexandre, J.**, **Švábenická, L.** and **Pedenychouk, N.** (2014) A high-resolution carbon-isotope record of the Turonian stage correlated to a siliciclastic basin fill: implications for mid-Cretaceous sea-level change. *Palaeogeogr. Palaeoclimatol. Palaeoecol.*, **405**, 42–58.

Uramoto, G.-I., **Abe, Y.** and **Hirano, H.** (2009) Carbon isotope fluctuations of terrestrial organic matter for the Upper Cretaceous (Cenomanian–Santonian) in the Obira area of Hokkaido, Japan. *Geol. Mag.*, **146**, 761–774.

Uramoto, G.-I., **Tahara, R.**, **Sekiya, T.** and **Hirano, H.** (2013) Carbon isotope stratigraphy of terrestrial organic matter for the Turonian (Upper Cretaceous) in northern Japan: implications for ocean–atmosphere $\delta^{13}C$ trends during the mid-Cretaceous climatic optimum. *Geosphere*, **9**, 355–366.

Uramoto, G.I., **Tahara, R.** and **Hiranio, H.** (2015) Cretaceous carbon isotope stratigraphy and constraints on the sedimentary patterns of the Turonian forearc successions in Hokkaido, nothern Japan. In: *Chemostratigraphy: Concepts, Techniques, and Applications* (Ed. M. Ramkumar), pp. 173–183. Elsevier, Amsterdam.

Valečka, J. and **Skoček, V.** (1991) Late Cretaceous lithoevents in the Bohemian Cretaceous Basin, Czechoslovakia. *Cret. Res.*, **12**, 561–577.

Veizer, J., **Ala, D.**, **Azmy, K.**, **Bruckschen, P.**, **Buhl, D.**, **Bruhn, F.**, **Carden, G.A.F.**, **Diener, A.**, **Ebneth, S.**, **Godderis, Y.**, **Jasper, T.**, **Korte, G.**, **Pawellek, F.**, **Podlaha, O.G.** and **Strauss, H.** (1999) $^{87}Sr/^{86}Sr$, $\delta^{13}C$ and $\delta^{18}O$ evolution of Phanerozoic seawater. *Chem. Geol.*, **161**, 59–88.

Voigt, S. (1995) Palaeobiogeography of early Late Cretaceous inoceramids in the context of a new global palaeogeography. *Cret. Res.*, **16**, 343–356.

Voigt, S. (2000a) Cenomanian-Turonian composite $\delta^{13}C$ curve for Western and Central Europe: the role of organic and inorganic carbon fluxes. *Palaeogeogr. Palaeoclimatol. Palaeoecol.*, **160**, 91–104.

Voigt, S. (2000b) Stable oxygen and carbon isotopes from brachiopods of southern England and northwestern Germany: estimation of Upper Turonian palaeotemperatures. *Geol. Mag.*, **137**, 687–703.

Voigt, S. and **Hilbrecht, H.** (1997) Late Cretaceous carbon isotope stratigraphy in Europe: correlation and relations with sea level and sediment stability. *Palaeogeogr. Palaeoclimatol. Palaeoecol.*, **134**, 39–59.

Voigt, S. and **Wiese, F.** (2000) Evidence for Late Cretaceous (Late Turonian) climate cooling from oxygen-isotope variations and palaeobiogeographic changes in Western and Central Europe. *J. Geol. Soc. London*, **157**, 737–743.

Voigt, S., **Gale, A.S.** and **Flogel, S.** (2004) Midlatitude shelf seas in the Cenomanian-Turonian greenhouse world: temperature evolution and North Atlantic circulation. *Paleoceanography*, **19**, PA4020. doi:10.1029/2004PA001015.

Voigt, S., **Gale, A.S.** and **Voigt, T.** (2006) Sea-level change, carbon cycling and palaeoclimate during the Late Cenomanian of northwest Europe; an integrated palaeoenvironmental analysis. *Cret. Res.*, **27**, 836–858.

Voigt, S., **Aurag, A.**, **Leis, F.** and **Kaplan, U.** (2007) Late Cenomanian to Middle Turonian high-resolution carbon isotope stratigraphy: new data from the Münsterland Cretaceous Basin, Germany. *Earth Planet. Sci. Lett.*, **253**, 196.

Voigt, S., **Erbacher, J.**, **Mutterlose, J.**, **Weiss, W.**, **Westerhold, T.**, **Wiese, F.**, **Wilmsen, M.** and **Wonik, T.** (2008) The Cenomanian-Turonian of the Wunstorf section (North Germany): global stratigraphic reference section and new orbital time scale for Oceanic Anoxic Event 2. *Newsl. Stratigr.*, **43**, 65–89.

Voigt, S., **Friedrich, O.**, **Norris, R.D.** and **Schoenfeld, J.** (2010) Campanian–Maastrichtian carbon isotope stratigraphy: shelf-ocean correlation between the European shelf sea and the tropical Pacific Ocean. *Newsl. Stratigr.*, **44**, 57–72.

Walaszczyk, I. and **Cobban, W.A.** (2000) Inoceramid faunas and biostratigraphy of the Upper Turonian–Lower Coniacian of the Western Interior of the United States. *Spec. Pap. Palaeontol.*, **64**, 1–118.

Walaszczyk, I. and **Peryt, D.** (1998) Inoceramid-foraminiferal biostratigraphy of the Turonian through Santonian deposits of the Middle Vistula Section, Central Poland. *Zent. Geol. Paläontol.*, **I, 11/12**, 1501–1513.

Walaszczyk, I., Wood, C.J., Lees, J.A., Peryt, D., Voigt, S. and **Wiese, F.** (2010) The Salzgitter-Salder Quarry (Lower Saxony, Germany) and Słupia Nadbrzeżna river cliff section (central Poland): a proposed candidate composite Global Boundary Stratotype Section and Point for the Coniacian Stage (Upper Cretaceous). *Acta Geol. Pol.*, **60**, 445–477.

Walaszczyk, I., Lees, J.A., Peryt, D., Cobban, W.A. and **Wood, C.J.** (2012) Testing the congruence of the macrofossil versus microfossil record in the Turonian-Coniacian boundary succession of the Wagon Mound-Springer composite section (NE New Mexico, USA). *Acta Geol. Pol.*, **62**, 581–594.

Walaszczyk, I., Shank, J.A., Plint, A.G. and **Cobban, W.A.** (2014) Interregional correlation of disconformities in Upper Cretaceous strata, Western Interior Seaway: biostratigraphic and sequence-stratigraphic evidence for eustatic change. *GSA Bull.*, **126**, 307–316.

Walker, J.C.G. (1986) Global geochemical cycles of carbon, sulfur and oxygen. *Mar. Geol.*, **70**, 159–174.

Wang, C.S., Hu, X.M., Jansa, L., Wan, X.Q. and **Tao, R.** (2001) The Cenomanian – Turonian anoxic event in southern Tibet. *Cret. Res.*, **22**, 481–490.

Wedepohl, K.H. (1971) Environmental influences on the chemical composition of shales and clays. In: *Physics and Chemistry of the Earth* (Eds L.H. Ahrens, F. Press, S.K. Runcorn and H.C. Urey), Vol. 8, pp. 307–333. Pergamon, Oxford.

Weissert, H., Lini, A., Föllmi, K.B. and **Kuhn, O.** (1998) Correlation of Early Cretaceous carbon isotope stratigraphy and platform drowning events: a possible link? *Palaeogeogr. Palaeoclimatol. Palaeoecol.*, **137**, 189–203.

Weissert, H., Joachimski, M. and **Sarnthein, M.** (2008) Chemostratigraphy. *Newsl. Stratigr.*, **42**, 145–179.

Wendler, I. (2013) A critical evaluation of carbon isotope stratigraphy and biostratigraphic implications for Late Cretaceous global correlation. *Earth-Sci. Rev.*, **126**, 116–146.

Wendler, J.E. and **Wendler, I.** (2016) What drove sea-level fluctuations during the mid-Cretaceous greenhouse climate? *Palaeogeogr. Palaeoclimatol. Palaeoecol.*, **441**, 412–419.

Wendler, I., Wendler, J., Graefe, K.U., Lehmann, J. and **Willems, H.** (2009) Turonian to Santonian carbon isotope data from the Tethys Himalaya, southern Tibet. *Cret. Res.*, **30**, 961–979.

Wendler, I., Willems, H., Graefe, K.-U., Ding, L. and **Luo, H.** (2011) Upper Cretaceous inter-hemispheric correlation between the Southern Tethys and the Boreal: chemo- and biostratigraphy and paleoclimatic reconstructions from a new section in the Tethys Himalaya, S-Tibet. *Newsl. Stratigr.*, **44**, 137–171.

Wendler, I., Huber, B.T., MacLeod, K.G. and **Wendler, J.E.** (2013) Stable oxygen and carbon isotope systematics of exquisitely preserved Turonian foraminifera from Tanzania – understanding isotopic signatures in fossils. *Mar. Micropaleontol.*, **102**, 1–33.

Werne, J.P. and **Hollander, D.J.** (2004) Balancing supply and demand: controls on carbon isotope fractionation in the Cariaco Basin (Venezuela) Younger Dryas to present. *Mar. Chem.*, **92**, 275–293.

Wiese, F. (1997) Das Turon und Unter-Coniac im nordkantabrischen becken (Provinz Kantabrien, Nordspanien): faziesentwicklung, bio-, event- und sequenzstratigraphie. *Berlin Geowiss. Abh. Reihe E*, **24**, 1–131.

Wiese, F. (1999) Stable isotope data ($\delta^{13}C$, $\delta^{18}O$) from the Middle and Upper Turonian (Upper Cretaceous) of Liencres (Cantabria, northern Spain) with a comparison to northern Germany (Söhlde & Salzgitter-Salder). *Newsl. Stratigr.*, **37**, 37–62.

Wiese, F. and **Kaplan, U.** (2001) The potential of the Lengerich section (Münster Basin, northern Germany) as a possible candidate Global boundary Stratotype Section and Point (GSSP) for the Middle/Upper Turonian boundary. *Cret. Res.*, **22**, 549–563.

Wiese, F. and **Voigt, S.** (2002) Late Turonian (Cretaceous) climate cooling in Europe: faunal response and possible causes. *Geobios*, **35**, 65–77.

Wiese, F., Čech, S., Ekrt, B., Košťák, M., Mazuch, M. and **Voigt, S.** (2004) The Upper Turonian of the Bohemian Cretaceous Basin (Czech Republic) exemplified by the Úpohlavy working quarry: integrated stratigraphy and palaeoceanography of a gateway to the Tethys. *Cret. Res.*, **25**, 329–352.

Wohlwend, S., Hart, M. and **Weissert, H.** (2015) Ocean current intensification during the Cretaceous oceanic anoxic event 2 – evidence from the northern Tethys. *Terra Nova*, **27**, 147–155.

Wonders, A.A.H. (1992) Cretaceous planktonic foraminiferal biostratigraphy, Leg 122, Exmouth Plateau, Australia. In: (Eds von Rad U. and B.U. Haq), *Proc. ODP Sci. Res.*, **122**, 587–599. Ocean Drilling Program, College Station, TX.

Wood, C.J., Ernst, G. and **Rasemann, G.** (1984) The Turonian–Coniacian stage boundary in Lower Saxony (Germany) and adjacent areas: the Salzgitter–Salder Quarry as a proposed international standard section. *Bull. Geol. Soc. Denmark*, **33**, 225–238.

Zhou, J., Poulsen, C.J., Pollard, D. and **White, T.S.** (2008) Simulation of modern and middle Cretaceous marine $\delta^{18}O$ with an ocean-atmosphere general circulation model. *Paleoceanography*, **23**, PA3223, doi:10.1029/2008PA001596.

Ziveri, P., Stoll, H., Probert, I., Klass, C., Geisen, M., Ganssen, G. and **Young, J.** (2003) Stable isotope 'vital effects' in coccolith calcite. *Earth Planet. Sci. Lett.*, **210**, 137–149.

Zonneveld, K.A.F., Versteegh, G.J.M., Kasten, S., Eglinton, T.I., Emeis, K.C., Huguet, C., Koch, B.P., de Lange, G.J., de Leeuw, J.W., Middelburg, J.J., Mollenhauer, G., Prahl, F.G., Rethemeyer, J. and **Wakeham, S.G.** (2010) Selective preservation of organic matter in marine environments; processes and impact on the sedimentary record. *Biogeosciences*, **7**, 483–511.

Table S1. Macrofossil biostratigraphic markers in the Bch-1 borehole, Běchary, Czech Republic, according to S. Čech in Olde *et al.* (2015b).

Table S2. Age control points for the Bch-1 borehole.

Appendix S1. Biostratigraphy and age control.

Appendix S2. Analytical methods.

Appendix S3. Stable isotope and total organic carbon compositions of Bch-1 sediment samples.

5

Organomineralization processes in freshwater stromatolites: a living example from eastern Patagonia

MURIEL PACTON*,[1], GABRIEL HUNGER†, VINCENT MARTINUZZI†,[2], GABRIELA CUSMINSKY‡, BEATRICE BURDIN§, KURT BARMETTLER¶, CRISOGONO VASCONCELOS* and DANIEL ARIZTEGUI†

*Geological Institute, ETH-Zürich, Zurich, Switzerland
†Department of Earth Sciences, University of Geneva, Geneva, Switzerland
‡Departamento de Ecología CRUB UNC-INIBIOMA CONICET, Quintral 1250, 8400, Bariloche, Argentina
§Centre technologique des microstructures, Université Lyon 1, Lyon, France
¶Institute of Biogeochemistry and Pollutant Dynamics, ETH-Zürich, Zurich, Switzerland

Keywords
Bacterial fossils, biomineralization, extracellular polymeric substances, freshwater microbialite, nanoglobules, stromatolites.

[1]Present address: Laboratoire de Géologie de Lyon, Université Lyon 1, France

[2]Present address: Geneva Petroleum, Geneva, Switzerland

ABSTRACT

Living stromatolites have been mostly described within shallow marine and (hyper)saline lacustrine environments. Southernmost South America lacks detailed investigations of these (organo)sedimentary buildups, particularly in regions experiencing extreme and variable environmental conditions. Here, we report and describe living freshwater stromatolites in the Maquinchao region, north-western Patagonia, Argentina. Fossil stromatolites characterized by globular and cauliflower shapes are also present in a continuous palaeoshoreline of a former lake at an altitude of 830 m, whereas their living counterparts only occur in the calm waters of sheltered or meandering sections of the Maquinchao River. The living stromatolites and their host waters have been sampled and studied using various chemical and microscopic techniques to better constrain the environmental versus biological factors controlling their development. Our results indicate that today stromatolites only proliferate in freshwater when Ca^{2+} levels are high. A microscopic inspection of the living stromatolite mat indicates stronger photosynthetic activity in the upper green layer associated with crypto/microcrystalline calcite (nanoglobules) compared to the lower beige-white biofilm. This biofilm contains more low-Mg calcite (rhombohedra) precipitates, which can form millimetre-sized aggregates in the underlying anoxic layer. Although sulphate-reducing bacteria are living in the entire mat, they appear more abundant and widely distributed in the lower beige-white layer and are always associated with Mg calcite. Low salinity and low-turbidity water along with microbial (photosynthetic and heterotrophic) activity are the most important factors promoting low-Mg calcite precipitation in the Maquinchao Basin. These conditions are very different from those proposed for recently described lacustrine stromatolites at high altitude in the subtropical and tropical Andes as well as in Chilean Patagonia. Hence, all these observations in modern freshwater stromatolites show the importance of geomicrobiological studies in identifying proxies of the hydrological conditions prevailing during their formation.

INTRODUCTION

Stromatolites constitute some of the oldest evidence for life on Earth (Hofmann *et al.*, 1999). Commonly defined as laminated benthic organosedimentary structures built by the trapping and binding and/or precipitation of minerals via microbial processes (See Riding, 2011), it is no longer clear that all stromatolites represent biogenic structures (Grotzinger & Rothman, 1996; McLoughlin *et al.*, 2008). They can display a variety of morphologies, including columnar, club, spheroidal, domal, nodular or irregular shape (Logan & Cebulski, 1970). The formation of these organosedimentary structures is a consequence of both microbially induced and microbially influenced mineralization (Burne & Moore, 1987; Trichet & Défarge, 1995; Dupraz *et al.*, 2009) in which cyanobacteria are of primary importance.

Due to the variety of surface environments where they occur, their assignment to particular environmental conditions of formation is not always straightforward. Studies of Bahamian stromatolites, for instance, find that lithified interiors beneath surface mat communities are modified by endolithic microbes (Reid *et al.*, 2000). Based on the morphology, marine living stromatolites have been interpreted as analogues of fossil stromatolites. Today, living stromatolites occur in a few shallow marine and mostly (hyper)saline lacustrine environments. In southernmost South America there is a particular scarcity of detailed studies of lacustrine stromatolites in regions subject to extreme and variable environmental conditions. Lately, thrombolites and stromatolites have been identified in Chilean southern Patagonian lakes and used as indicators of palaeoclimate (Solari *et al.*, 2010). To the best of our knowledge, however, no geomicrobiological investigations of the processes governing carbonate precipitation have been conducted. Previous studies on African freshwater stromatolites have shown that they are often distributed along lake shorelines. Thus, they can be used as excellent indicators of hydrological changes in palaeolakes improving the reconstruction of continental palaeoclimates (Casanova & Hillaire-Marcel, 1993). While it is clear that some stromatolites are built in response to microbial forcing (Reid *et al.*, 2000; Awramik & Grey, 2005), results of very recent studies have shown that the frequency of lamina formation is more closely related to regional climate forcing (Andersen *et al.*, 2011; Petryshyn *et al.*, 2012;). Yet, both aspects can be related and even complementary.

Recent freshwater stromatolites from the Maquinchao Basin (north-western Patagonia, Argentina) were studied in order to determine the processes involved in their formation and accretion, while relating these to environmental parameters during their development. Understanding the actual processes governing the formation of these modern microbialites aids in the interpretation of their fossil counterparts formed in the same basin during the last deglaciation and the early Holocene.

STUDY AREA

Laguna Cari-Laufquen is a closed basin (41°8′47″ S, 69°27′37″ W, *ca* 786 m asl; Fig. 1) located in a tectonic depression surrounded by basalt plateaus of Mesozoic to Tertiary age in northern Patagonia, Argentina (Coira, 1979). Presently, the lake system consists of two separate water bodies locally known as Laguna Cari-Laufquen Chica (LCLC; 2·5 × 1·5 km, averaging 3 m water depth in January, 2011), which is a permanent lake body with pH 8·7 (Schwalb *et al.*, 2002) and a sodium bicarbonate concentration of 230 ppm (Galloway *et al.*, 1988); and Laguna Cari-Laufquen Grande (LCLG; 4 × 2 km, average 2 m depth in January, 2011), an ephemeral lake of brackish water with pH 8·8 (Schwalb *et al.*, 2002) and solute concentration of 4000 ppm (Galloway *et al.*, 1988). LCLC sits at 825 m above sea-level, and periodically flows towards the larger basin through the Maquinchao River (Fig. 1). A 20 years series of meteorological data show a mean annual air temperature of 9·0°C ranging from 16·1°C to 1·7°C in Austral

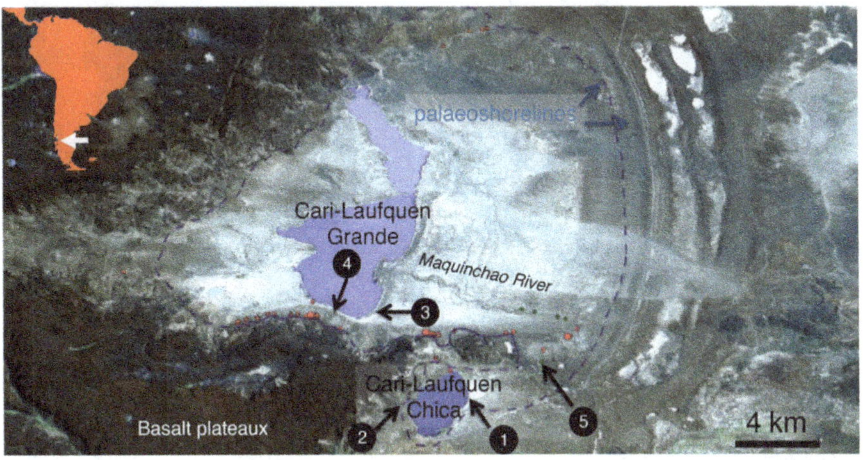

Fig. 1. Satellite image of the Maquinchao Basin (courtesy of GoogleEarth©) showing the location of modern Lagunas Cari-Laufquen Chica and Grande (small and large, respectively). Red and green dots indicate fossil and living stromatolite occurrence, respectively. Numbered black dots indicate sampling sites (see text), whereas blue arrows show palaeoshorelines.

summer and winter seasons, respectively, whereas annual precipitation averages 31·6 cm over the same time interval (source: Argentinean Meteorological Survey). Previous investigations have shown evidence of former lake-level high stands during which both lakes LCLG and LCLC were merged to form a large palaeolake in the late Pleistocene (Del Valle *et al.*, 1993, 1996). This lake encompassed more than 1500 km^2 and was at least 70 m deep during the pluvial phase maximum (González Bonorino & Rabassa, 1973; Volkheimer, 1973; Coira, 1979; Galloway *et al.*, 1988; Tatur *et al.*, 2000; Ariztegui *et al.*, 2008). Several palaeoshorelines indicate higher lake levels than today (Fig. 1). They have been previously dated at *ca* 19 ka (Galloway *et al.*, 1988) and between 14 ka and 10-8 ka BP (Bradbury *et al.*, 2001). Furthermore, lacustrine sediments outcropping within the Maquinchao Basin document a well-defined Late Pleistocene pluvial episode (Whatley & Cuminsky, 1999; Cusminsky *et al.*, 2011) followed by a prograding sedimentary sequence. Stromatolites, also called 'tufa' by Cartwright *et al.* (2011), are widely distributed in the Laguna Cari-Laufquen often tracing the palaeoshorelines. They display a variable structure ranging from cauliflower to more globular with a nucleus of various compositions depending on their location. Fossil stromatolites showing a globular aspect can form plurimetric complexes and are present in a continuous palaeoshoreline at an altitude of 830 m (Figs 1, 2A and 3), dated at *ca* 22·0 ka BP (Cartwright *et al.*, 2011). Globular carbonate laminated structures most commonly surround a basaltic nucleus and do not exhibit any growth preferential axis or other evidence of hydrodynamic conditions.

MATERIALS AND METHODS

Samples

Growing microbial bodies showing a globular structure similar to fossil stromatolites have been found in the Maquinchao River and are the subject of this study (Figs 2B and 4). Five living stromatolites attached to their substrate (basaltic pebbles) were sampled in the field in January, 2011 (Austral summer) when they were subaerially exposed, preserved in sterile 50-ml plastic tubes and stored at 4°C. The biofilms were subsampled by manual separation into a *ca* 1 mm thick surface sample (green biofilm) and an underlying sample taken from *ca* 2 to 3 mm below the surface (beige-white biofilm). Each layer was observed by optical microscopy under natural light (with prior ultrasonic for 30 sec) and air-dehydrated for scanning electron microscopy (SEM), while the entire stromatolite mat was observed using confocal laser scanning microscopy (CLSM).

Surface water samples were taken in five distinct areas as follows (Fig. 1): (1) The LCLC outflow towards the Maquinchao River; (2) LCLC shore area; (3) LCLG shore area; (4) groundwater stream close to a vegetated area on the LCLG shore; and (5) sheltered section of the Maquinchao River containing living stromatolites. Today, the maximum 3 m depth of both lakes results in a water column which is continuously mixed making sampling at different depths unnecessary. Physicochemical data, that is, temperature, pH, salinity and conductivity were measured at all sites before sampling (Fig. 3A and B).

Fig. 2. Photograph of fossil globular stromatolite dated around 22 Ka BP (A); and partially submerged living globular stromatolites growing in a sheltered area beneath calm waters in the Maquinchao River (B).

Fig. 3. Outcrops showing fossil globular stromatolites along a palaeoshoreline.

Methods

Twenty millilitre of each water sample was filtered (0·2 µm nylon filters) and acidified (1% v/v concentrated HCl) for subsequent elemental analysis by inductively coupled plasma – optical emission spectrometry (ICP-OES, Varian Vista MPX, ETH Zürich). The extraction was performed on two to four independently incubated samples.

Water samples were analysed for their stable isotope composition (δD and $\delta^{18}O$, referenced to Vienna Standard Mean Ocean Water (VSMOW) at the Alfred Wegener Institute for Marine and Polar Research (AWI) Potsdam, Germany. A Finnigan MAT Delta-S mass spectrometer equipped with two equilibration units was used for the online determination of hydrogen and oxygen isotopic compositions. The external errors for standard measurements of hydrogen and oxygen are better than 0·8‰ and 0·1‰, respectively (Meyer *et al.*, 2000).

Various microscopic techniques were used to analyse the stromatolite structure down to the nanoscale: optical microscopy (Axioscope, ETH Zürich) under transmitted light, CLSM (Centre Technologique des Microstructures, Université Lyon 1, France), SEM (Jeol Zeiss Supra 50 VP, University of Zurich) on platinum coated samples prior to fixation with glutaraldehyde, and transmission electron microscopy (TEM, Phillips CM100, Centre Technologique des Microstructures, Université Lyon 1, France) after staining sectioned samples (Amann *et al.*, 1995). Investigation of stained samples and control materials was carried out using a Zeiss LSM 510 META. Staining procedures for fluorescence *in situ* hybridization (FISH) were modified from Decho & Kawaguchi (1999). Samples were stained with the 16S rRNA probe SRB385 (CGGCGTCGCTGCGTCAGG) labelled with cyanine 3 to image sulphate-reducing bacteria (SRB). This oligoprobe targets SRB of the **y**-proteobacteria (Amann *et al.*, 1990). Photosynthetic organisms and carbonate minerals were visualized by fluorescence. Unstained green biofilms were excited by the 633 nm laser line with the fluorescence detected above 650 nm, while the unstained beige/white biofilms were excited by the 543 nm laser line with fluorescence detected using a 560 nm long pass filter. Both biofilms were excited by a different wavelength in order to distinguish the autofluorescence of ostracods from that of cyanobacteria in the green biofilm. Autofluorescence from filamentous cyanobacteria in the unstained biofilms has been spectrally separated from the red-shifted autofluorescent calcite and indicated by the development of an artificial green colour. The Cy3-labelled probe SRB385 was excited by the 488 nm laser line with the fluorescence detected between 500 and 555 nm. CLSM Image analyses were performed using Image J and Imaris software version 5·6.

Samples from fossil and living stromatolites were ground to an ultrafine powder in an agate mortar before being placed in a silicon wafer and a plastic sample holder for X-Ray diffraction analyses (XRD). They were

Fig. 4. Location of the living stromatolites in the Maquinchao River: (A) meander-like area; (B) details of (A); (C and D) abundant green slime between stromatolites.

analysed in a Bruker, AXS D8 Advance device, equipped with a Lynx-eye super-speed detector using Cu-K radiation and an anti-scattering slit of 20 mm, while rotating the sample. The sample patterns were recorded from 5° to 85° 2θ in steps of 0·004°, 1 sec counting time per step.

RESULTS

Water chemistry

The water chemistry of LCLG and LCLC, as well as that of the Maquinchao River varies by location. January, 2011 (Austral summer) was particularly hot and dry and the Maquinchao River was partially cut from LCLC being almost entirely fed by groundwater while flowing towards LCLG. This situation implies differential evaporation between the two lakes and the river (e.g. LCLC was almost dry) and consequently supports their different chemistry. Both shore and outflow water of LCLC display temperatures above 33°C while they are around 27°C at the other sites (3, 4 and 5; Fig. 5A). The pH in the water column ranged from 8·1 (where stromatolites are living and in the Maquinchao River) to 9–9·4 in both lakes. Salinity was fairly constant in most of the sampled sites indicating brackish conditions (1·3–12·5‰). However, the riverine waters in which stromatolites are presently growing display more freshwater conditions (0·7‰).

Major and minor ions show some differences between the lakes and the other sites. Waters are dominated by Na with concentrations ranging from 150 up to 4750 ppm, while Ca and Mg concentrations average 19 and 68 ppm, respectively. For cations, the major elemental abundances in both lakes were $Na^+ \gg Mg^{2+} > K^+ > Ca^{2+} > Si^{4+}$, whereas the relative abundances were $Na^+ \gg Ca^{2+} > Mg^{2+} > K^+ > Si^{4+}$ at the other sites (Table 1; Fig. 5B). Sulphur displays similar concentrations to Na and is likely present as SO_4^{2-} (Schwalb *et al.*, 2002). The Mg concentration was as high as 100 ppm and 200 ppm in LCLG and LCLC, respectively, whereas it was less than 25 ppm at the other sites. The highest Ca concentration (42 ppm) was measured where stromatolites are presently living, while minimum values were found in both lakes (around 7 ppm). Silicon concentrations of 3 or 4 ppm in both lakes contrasted with values reaching 20 ppm in the river and groundwater. The Mg:Ca ratio is typically greater than 15 and 25 in both lakes, whereas the other sites display ratios lower than 1.

Figure 6 shows that LCLG and LCLC samples are characterized by ^{18}O-enriched waters (9·53 and 5·82‰), whereas the groundwater and Maquinchao River sites display depleted values (−10‰ and −87‰ for $\delta^{18}O$ and δD, respectively). Lake waters are enriched isotopically compared to those reported by Schwalb *et al.* (2002) and Cartwright *et al.* (2011) which is consistent with the comparatively lower water levels of our sampling year. How-

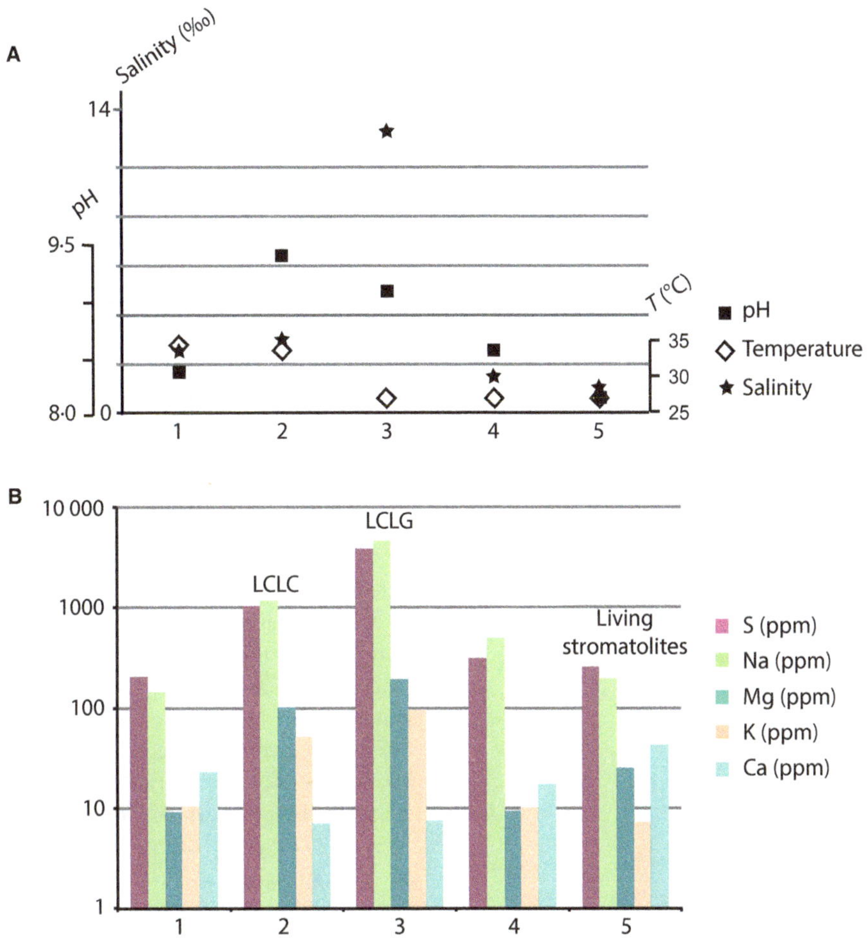

Fig. 5. (A): Temperature, pH and salinity of the sites sampled in January, 2011 (Austral summer). Refer to Figure 1 for sample location. (B) Stacked column plot showing the relative variation in main ions (ppm) for the individual sampled sites (log scale).

ever, the lake waters as a whole form a well-defined δD-δ^{18}O mixing line with river-supplied inflow (Craig *et al.*, 1974). This line is geometrically coincident with and statistically identical to a regression line computed from the lake data alone ($r = 0{\cdot}97$), indicating that lake waters evolved from the same primary water sources. Groundwater and Maquinchao River waters plot along the global meteoric water line, implying very low to non-evaporation. Despite the lack of *in situ* physicochemical measurements throughout the year, the existing meteorological data for the area suggest contrasting seasonal changes. It appears, however, that the biota is adapting to these conditions through time.

Living stromatolites sample description

Five stromatolite samples were collected from the Maquinchao River during mid-January 2011. Living stromatolites are found in several protected areas of the river such as the quiet water of meanders (Fig. 4A). Waters are less turbid here than in either of the lakes and contain abundant green slime between the stromatolites (Fig. 4B–D). These carbonate buildups display a globular morphology (Fig. 7) with a thin (*ca* 1 mm thick) outermost biofilm composed of a green layer overlying an undulating beige-white layer, varying in thickness between 4 and 9 mm (Fig. 4B–D).

A series of XRD analyses in both the green and beige-white layers showed that low-Mg calcite is the main mineral phase (70%) with a few detrital quartz grains and halite, which is an artefact of the sample processing. No crystalline Mg-Si minerals have been found. Similarly, low-Mg calcite is the dominant mineral phase in the fossil stromatolites displaying a Mg/Ca ratio of about 0·03.

Upper green layer of the stromatolite mat

The green layer is composed of mainly filamentous and coccoid bacteria (Fig. 8A) with little granular amorphous organic matter (OM) (white arrow in Fig. 8A). These

Table 1. Chemical characteristics of water samples. Refer to Fig. 1 for sample location

Samples	Ca (ppm)	K (ppm)	Mg (ppm)	Na (ppm)	S (ppm)	Si (ppm)	Al (ppm)	Ni (ppm)	As (ppm)	Ba (ppm)	Cu (ppm)	Fe (ppm)	P (ppm)	Zn (ppm)
1	23·2459	10·6006	9·3463	144·83	207·555	0·9659	0·012	0·002	0·0892	0·05335	0·00045	0·0348	0·3391	0·00215
2	7·0766	51·7111	103·115	1200·81	1029·72	4·8983	0	0·0068	0·1671	0·04355	0·0096	0·00605	1·1572	0·00355
3	7·4915	96·195	194·3125	4750·85	3836·9	3·35175	0	0·0037	0·3354	0·04865	0·00445	0·00335	0·1043	0·0012
4	17·3791	10·2204	9·58155	508·265	315·04	20·05	0·0011	0·0014	0·1639	0·02355	0·0017	0·005	0·753	0·01525
5	42·4272	7·3061	25·5225	198·19	259·775	12·91845	0·0061	0·0066	0·07755	0·01835	0·0014	0·0011	0·0308	0·00425

bacteria are mostly photosynthetic as shown by their green colour under natural light and red autofluorescent pigments due to chlorophyll and phycobiliproteins (Fig. 8A and C), in addition to the presence of *Spirulina* (data not shown). Their abundance is supported by the strong autofluorescence in the first millimetre of the mat compared to the deeper layers (Fig. 8C). Microcrystalline low-Mg calcite is widely observed associated with cyanobacteria as characterized by the mixed, that is, yellow colour. The blue-fluorescing SRB are co-localized with the red-fluorescing cyanobacteria indicating that SRB are living in oxic conditions (Fig. 8E). Microbial communities are trapped with diatoms within an extracellular polymeric substances (EPS) matrix (Fig. 9A). Diatoms are partially to entirely dissolved (Fig. 9B) when completely trapped by EPS. Organic filaments (Fig. 9C) and cryptocrystalline low-Mg calcite are embedded within EPS. Mineralized EPS are Ca-enriched (Fig. 9D) compared to low-Mg calcite (Fig. 9E). The latter is always associated with low Si content (from likely dissolved diatoms), while no Al was detected (Fig. 9E). Micro/cryptocrystalline low-Mg calcites are either characterized by nanoglobules (Fig. 9F) or arranged as platelets, closely associated with EPS suggesting an *in situ* growth. This low-Mg calcite crystal arrangement creates a nanoporosity (Fig. 9F). Examination by TEM confirms that this biofilm is mainly composed of cyanobacteria as shown by multiple rows of tightly stacked thylakoids, that is, membranes where photosynthesis occurs (Fig. 10A), indicating healthy, active photoautotrophic organisms (Stolz, 1991). Extracellular polymeric substances were distributed as nanofilaments creating a mesh-like network that is not restricted to the surface of the cells (Fig. 10B). Other microorganisms such as rod-shaped Gram-negative bacteria have also been identified (Fig. 10C), probably SRB.

Lower beige-white layer of the stromatolite mat

The beige-white biofilm displays more granular amorphous organic matter (AOM) compared to the green biofilm (Fig. 8B) and an association of coccoid cyanobacteria and colourless dead filaments (Fig. 8B). Some filaments were also photosynthetic as shown by their red autofluorescence (Fig. 8D) but most were colourless in contrast to the green colour they displayed in the overlying biofilm. Green, autofluorescent, low-Mg calcite crystals occur locally in conjunction with cyanobacteria (Fig. 8D) but are more abundant when associated with SRB (Fig. 8F) in this layer than they are in the overlying green one. Extracellular polymeric substances and microcrystalline low-Mg calcite form a complex matrix (Fig. 11A and B). These crystals occur as stacks of platelets which combine further to form rhom-

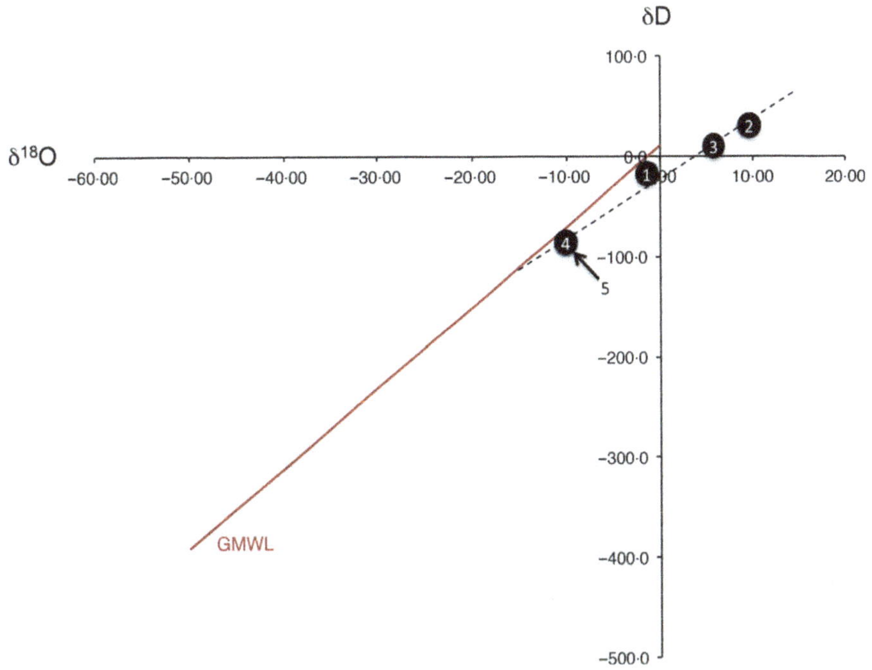

Fig. 6. Variations in δD versus $\delta^{18}O$ of water samples (refer to Fig. 1 and text for location). Global meteoric water line (GMWL) from Craig & Gordon (1965).

bohedra (Fig. 11F). The surface of the platelets is characterized by a nanoglobular structure (Fig. 11C) connected by nanofilaments of EPS (Fig. 11C). These crystals, consisting of low-Mg calcite, were associated with a small amount of silicon possibly resulting from diatom dissolution (Fig. 11D). The ultrastructure of this biofilm consists primarily of dying cyanobacteria surrounded by substantial fibrillar EPS (Fig. 10D) together with the remains of partially dissolved diatoms. Large quantities of fibrillar EPS can form aggregates containing several cells (Fig. 10D). The abundance of empty cyanobacterial sheathes along with the increased density of the collapsed sheath material are indicative of dying cyanobacteria (Fig. 10E) and indicate early to advanced stages of degradation (Stolz *et al.*, 2001). Ample virus-like particles have been found, sometimes icosahedral in shape (Pacton *et al.*, 2014; Fig. 10F).

DISCUSSION

Environmental conditions constraining stromatolite occurrence in the Maquinchao Basin: implications for palaeohydrological reconstructions

Clear differences in water composition have been observed between the presently growing stromatolites and the other sampled sites. These differences allow the environmental parameters controlling stromatolite growth to be identified and, thus, permit the conditions of carbonate precipitation in their fossil counterparts to be estimated. Firstly, the alkaline pH throughout the different sampling sites, that is, 8·1–9·4, indicates that this parameter is not a determining factor for the presence of stromatolites. Secondly, the living stromatolites appear to prefer freshwater rather than the more turbid and brackish lake waters. Thirdly, calcium is the most significant element in the freshwater sites, contrasting with the brackish waters of both present-day lakes where Mg and K dominate (Fig. 5B). Similarly to the high-altitude volcanic lakes of Patagonia, the water of both lakes contains essentially all inorganic nutrients, and at much higher concentrations (except for Ca), than the Maquinchao River, making it unlikely that the input of these chemicals is critical for stromatolite development (Farias *et al.*, 2013). Although nutrient availability has not been investigated in detail, turbidity is one of the most important factors constraining stromatolite development in the Maquinchao Basin. It is widely accepted that the presence of dissolved Mg^{2+} inhibits the precipitation of calcite favouring the precipitation of aragonite, especially with a Mg^{2+}:Ca^{2+} ratio >12 (Bischoff & Fyfe, 1968; Müller *et al.*, 1972; Berner, 1975). Therefore, the palaeolake in which the fossil stromatolites were living most probably had a lower Mg^{2+}:Ca^{2+} ratio than today and was thus closer in composition to the Maquinchao River.

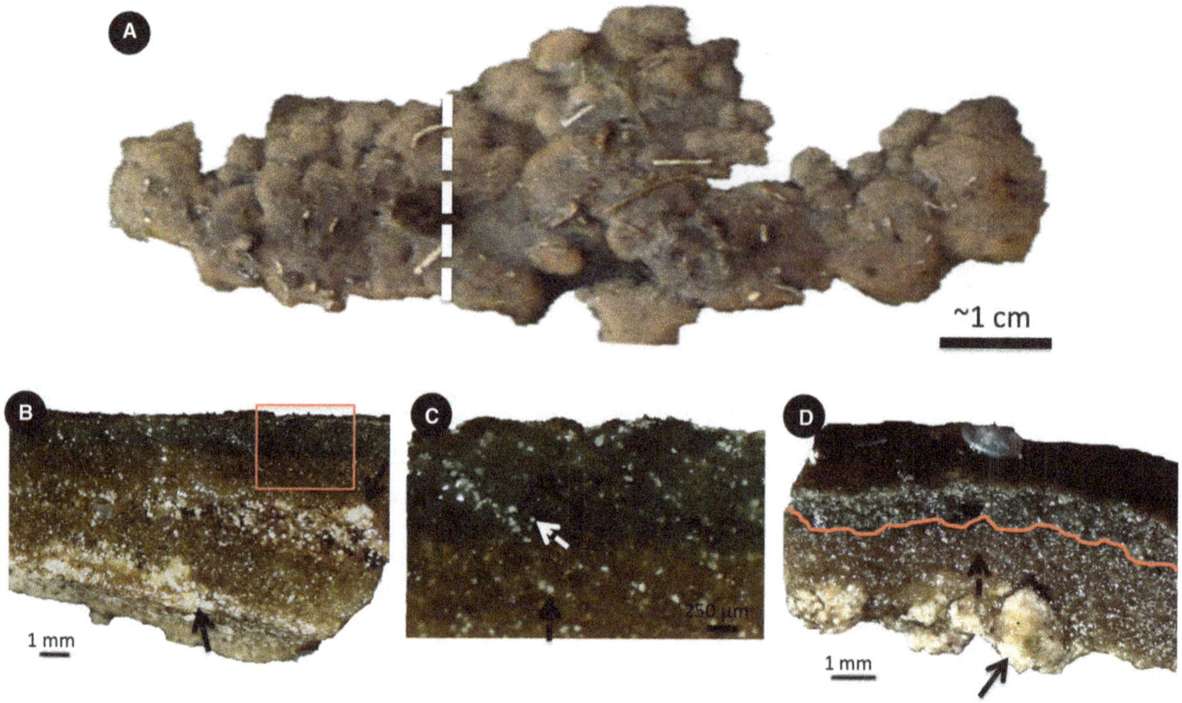

Fig. 7. Top view (A) and cross-section of a recent globular stromatolite mat (dashed line). Note the carbonate concretions in the beige/white layers (arrows).

At present, the Maquinchao region is one of the driest in the world but both lake basins show geomorphological evidence (palaeo-shorelines) of major fluctuations in the water balance since the last deglaciation. According to Bradbury *et al.* (2001) the large Cari-Laufquen palaeolake experienced several high stands during the Late Pleistocene and Early Holocene. These lacustrine basins had low productivity and were less saline and turbid than today. Lakes fed by rivers draining volcanic terrains of dominantly basic composition, coupled with groundwater inflow could create a set of conditions involving high CO_2 input and carbonate alkalinity, high dissolved silica, and high levels of Mg, Ca and Na (Wright, 2012). According to Valero-Garcés *et al.* (2001) Quaternary lacustrine travertines and stromatolites from high-altitude Andean lakes in north-western Argentina were probably formed during low lake levels, when conditions were more favourable for cyanobacterial growth than for plant development. These authors suggested that grazing of microbial mats could be another factor preventing stromatolite formation as microbial mats form the basis of benthic food webs that include metazoan consumers such as gastropods, microcrustaceans, insect larvae and vertebrates (Elser *et al.*, 2005). The geochemical and microscopic studies of the Cari-Laufquen stromatolites presented here clearly indicate that, in the Maquinchao Basin, stromatolites only proliferate under dominantly freshwater conditions corresponding to meander-like fluviatile environments. Although the effect of seasonal variations on stromatolite development has yet to be studied, cold periods might reduce microbial metabolic activity and subsequent carbonate precipitation (Plée *et al.*, 2008). Thus, living stromatolite mats provide a proxy to reconstruct the prevailing environmental conditions influencing growth of their fossil counterparts in the same basin. These results agree with the presence of fossil stromatolites along the late Pleistocene palaeoshorelines.

Stromatolites: the role of microbial communities in carbonate precipitation

There are several significant differences in macroscopic characteristics and species composition between the two biofilm layers constituting the stromatolites. Firstly, both the green and beige/white biofilms contain filamentous and coccoid cyanobacteria, diatoms and other heterotrophs, but in different relative abundances. We observed an increase in dead photosynthetic microbes in the underlying beige/white biofilm as shown by their colourless aspect (Grilli Caiola *et al.*, 1993) along with an increase in SRB and other heterotrophs. Examination by TEM of the beige/white biofilm confirms the increased

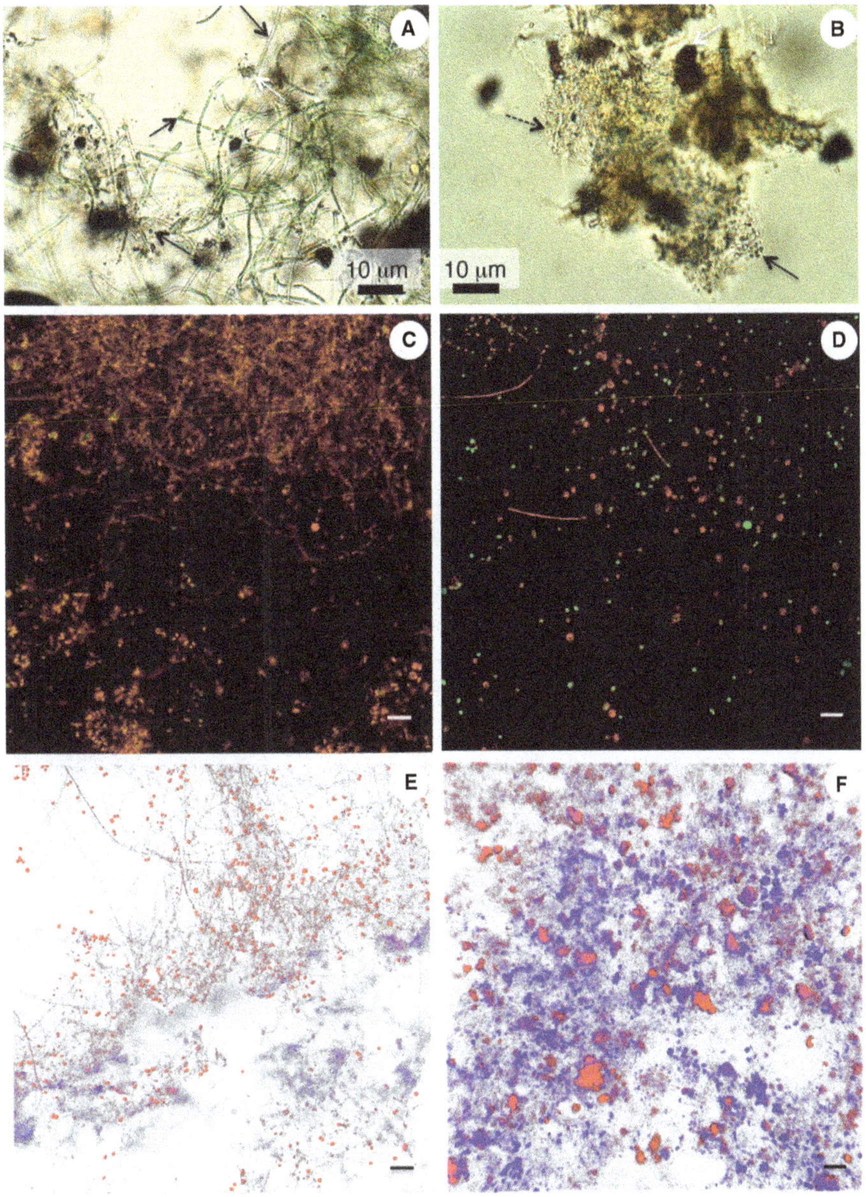

Fig. 8. Recent globular stromatolite mat (A) Light micrographs of the green biofilm showing photosynthetic cyanobacteria (black arrows) and translucent filamentous bacteria (dashed black arrow) associated with amorphous organic matter (AOM, white arrow); Light micrographs of the beige/white biofilm showing coccoid cyanobacteria (black arrow) and translucent dead filamentous bacteria (dashed black arrow); granular AOM (white arrow); (C and D) False colour confocal scanning laser microscopy (CSLM) cross-sectional image of the stromatolite mat showing abundant cyanobacteria (red) in the upper green biofilm and more abundant calcite crystals (green) in the beige/white biofilm (D); (E) some clusters of sulphate-reducing bacteria (SRB, blue) associated with filamentous cyanobacteria and calcite crystals (red autofluorescence) in the green biofilm; (F) Abundant SRB associated with calcite crystals (red autofluorescence) in the beige/white biofilm (F). SRB are detected using a fluorescence *in situ* hybridization (FISH) oligoprobe (SRB 385).

degradation of cyanobacterial cells as well as degradation of EPS into fine fibres. The latter is associated with an increase in granular AOM suggesting OM degradation of microbial origin (Pacton *et al.*, 2011). This associattion might be due to the aerobic respiration and/or sulphate reduction of microorganisms. Such SRB are known not only to survive oxic conditions but also to thrive in the presence of cyanobacteria (Canfield & Des Marais, 1991; Dupraz & Visscher, 2005; Baumgartner *et al.*, 2006) resulting in the partial degradation of EPS and low molecular weight organic carbon (LMW-OC) (Decho, 2000; Glunk *et al.*, 2011). Virus-like particles have also

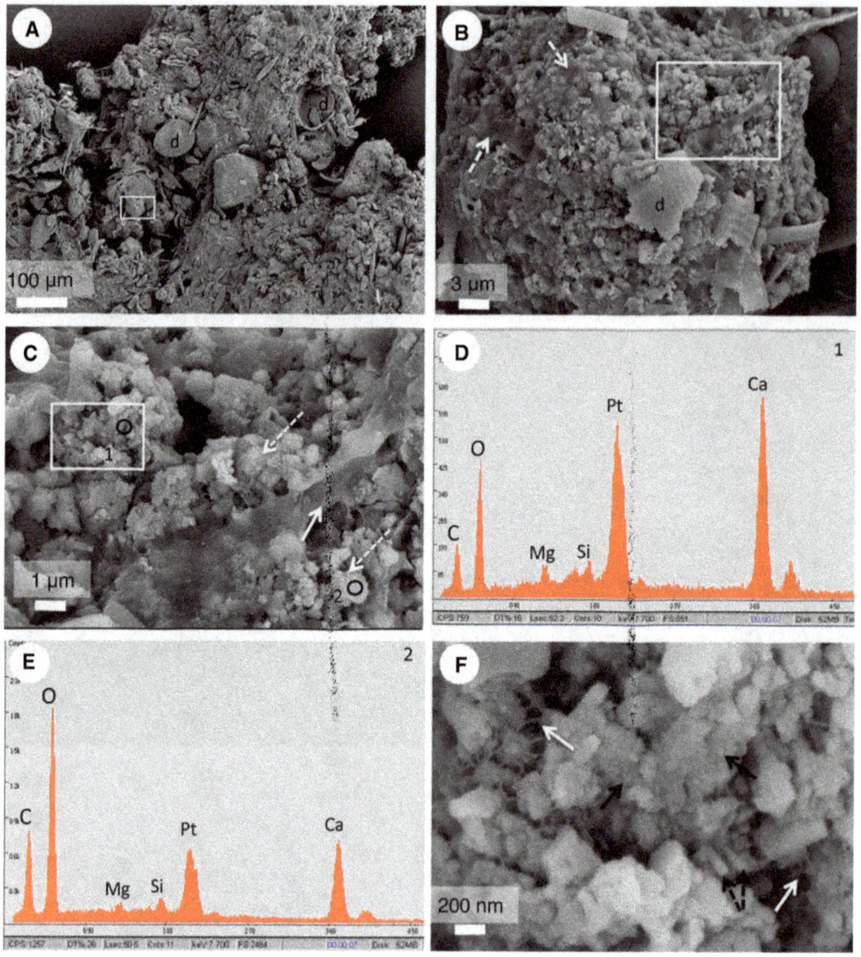

Fig. 9. Scanning electron microscope images of the green biofilm: (A) the biofilm contains different diatom morphospecies (d) which are embedded in an organo-mineral matrix (see boxed area enlarged in B); (B) this organo-mineral matrix is composed of EPS (dashed arrows) and partially dissolved diatoms (d); (boxed area enlarged in (C); (C) EPS (white arrow) are closely associated with single low-Mg calcite crystals (white dashed arrows) and fluffy minerals and enlarged in E (black circles represent spot for EDS analysis in D and E); (D) Elemental analysis showing low-Mg calcite and EPS associated with Si; (E) Elemental analysis of LMC showing a small contribution of Si; (F) Mineralized nanoglobules (black arrows) and platelets (dashed black arrows) closely associated with EPS (white arrows). Nanoporosity characterized by an alveolar network, that is, EPS.

been identified in both biofilms, free within EPS, but appeared more abundant in the beige/white biofilm. It is known that viruses are the most abundant biological entities throughout marine and terrestrial ecosystems (Suttle, 2005), and more specifically in modern microbial mats (Desnues *et al.*, 2008; Pacton *et al.*, 2014). As they have been found in the biofilm associated with degraded cells, we suggest that they may have been involved in the infection of these cells. It is known that viruses represent the largest fraction of nanosized particles that can undergo degradation or, under specific conditions, mineralization processes (Pacton *et al.*, 2014). Viruses have been found as important mineralized entities in microbial mats (Pacton *et al.*, 2014; De Wit *et al.*, 2015) although an alternative origin such as membrane vesicles (MVs) cannot be ruled out. The latter share morphological and chemical similarities and could outnumber viral particles in sea water samples, especially when they are quantified using epifluorescence microscopy (Forterre *et al.*, 2013; Biller *et al.*, 2014). However, according to Soler *et al.* (2015), the abundance of MVs could also be overestimated as they could contain viral genomes. Further studies are required to better discriminate MVs from true viruses in environmental samples.

Although there is no mineralogical difference between the carbonates of both biofilm layers (low-Mg calcite), crystal morphologies are quite distinctive. Low-Mg calcite crystals in the outermost green biofilm are either characterized by nanoglobules (probably mineralized viruses) or arranged as platelets associated with Si and EPS. The

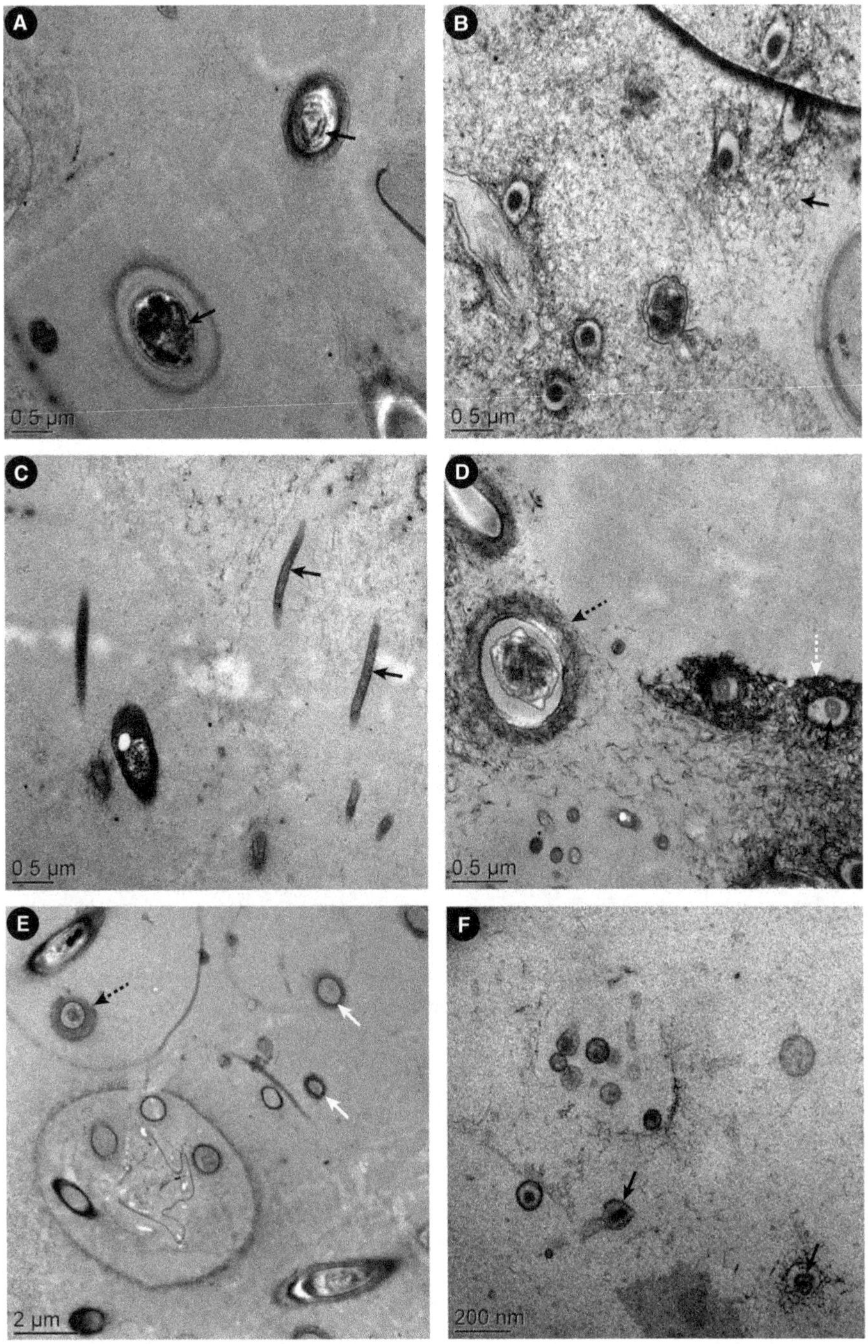

Fig. 10. Transmission electron microscope images of the green (A–C) and beige/white (D–F) biofilms: (A) filaments of cyanobacteria enclosed in their sheath in cross-section with thylakoid membranes (arrows); (B) EPS appear as nanofilaments creating a mesh-like network through the layer and are not restricted to the surface of the cells (arrow); (C) several rod-shape Gram-negative bacteria (arrows); (D) Dying cyanobacteria (black arrow) are surrounded by substantial fibrillar EPS (dashed black arrow) and comprise several cells together (dashed white arrow); (E) thick sheath surrounding cyanobacterial cells in the early stages of degradation (dash arrow) and abundant empty cyanobacterial sheaths in the advanced stages of degradation (white arrows); (F) abundant virus-like particles showing sometimes icosahedral shape (black arrows).

platelets within the beige/white biofilm tend to form euhedral rhombohedra that are often coated or partially filling the pore spaces of the biofilm. The EPS show an enrichment in Si-Ca-Mg elements suggesting that they are mineralized as amorphous Mg-Si ($_{am}$Mg-Si) phases associated with low-Mg calcite (Fig. 11D).

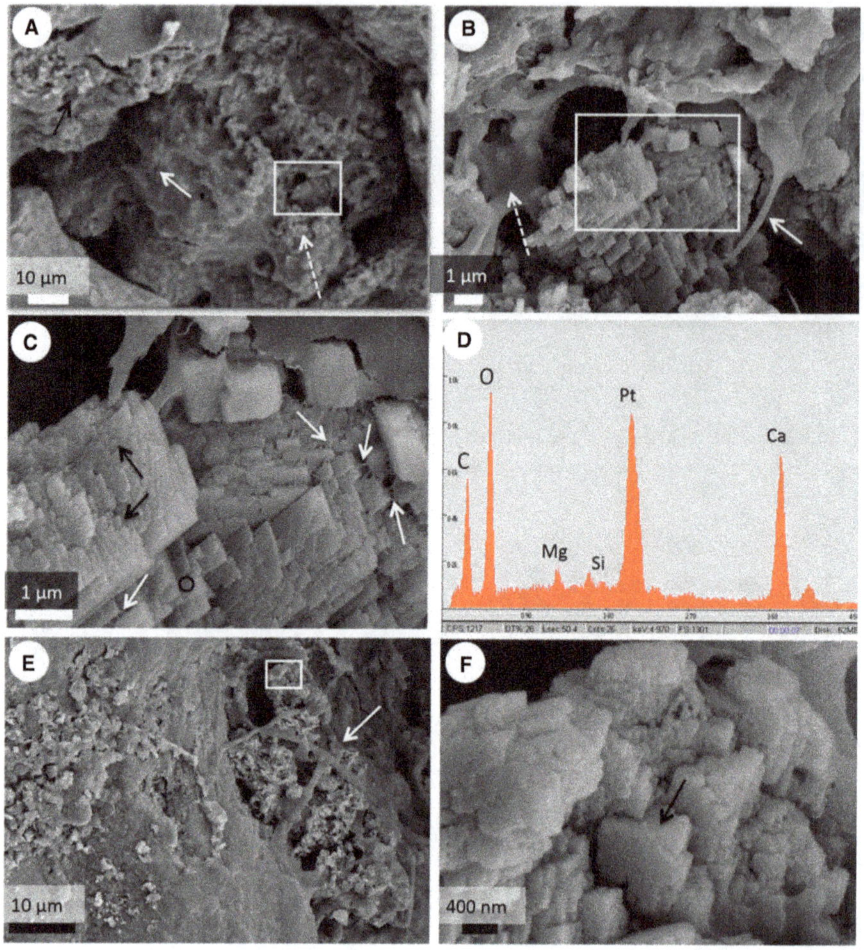

Fig. 11. Scanning electron microscope images of the beige/white biofilm: (A) the biofilm contains thick EPS layers (white arrow) or occurs as an alveolar network (boxed area enlarged in B), which are closely associated with mineral precipitates (dashed arrow); (B): EPS can be filamentous (arrow) or as a thin veil (dashed arrow) associated with low-Mg calcite crystals as oriented platelets (boxed area enlarged in (C); (C) EPS surround platelets (white arrows), which show nanoglobules at surface (black arrows) and the typical 2D orientation suggests a rhombohedra precursor; (D) Elemental analysis of platelets indicates low-Mg calcite. The small amount of Si can be related to surrounding EPS; (E) Single crystals embedded in EPS and enlarged in F; (F) euhedral rhombohedra (black arrow).

Amorphous Mg-Si precipitates have previously been reported in numerous environments, always associated with microbial activity (Arp *et al.*, 2003; Souza-Egipsy *et al.*, 2005; Benzerara *et al.*, 2010; Pacton *et al.*, 2012). It has also been demonstrated that $_{am}$Mg-Si enhances fossilization of microbes, and is commonly found in the rock record. The authigenesis of $_{am}$Mg-Si phases is, however, poorly understood, with previous work suggesting that $_{am}$Mg-Si layers in microbialites replace a primary mineral phase (Arp *et al.*, 2003). In contrast, this data favour a primary origin for these $_{am}$Mg-Si phases. The close association between carbonate crystals, EPS and $_{am}$Mg-Si suggests that the precipitation is microbially mediated. The microbially mediated precipitation pathways are called organomineralization processes and are defined as (i) biologically induced mineralization resulting from the interaction between biological activity and the environment; and (ii) biologically influenced mineralization, which is a passive mineralization of organic matter (biogenic or abiogenic in origin), influencing crystal morphology and composition (Dupraz *et al.*, 2009).

In the green biofilm, most of the mineral phases appear as a complex of Mg-Si and Ca characterized by nanoglobules closely related to EPS. Initially the negatively charged functional groups in fresh EPS bind Ca^{2+} and Mg^{2+}, inhibiting carbonate precipitation (Dupraz *et al.*, 2009). Nanoglobules of EPS could be the first step towards mineralization as EPS have been reported to be permineralized as amorphous Mg-Si phases ($_{am}$Mg-Si) before being calcified (Souza-Egipsy *et al.*, 2005; Pacton *et al.*, 2012).

In the underlying beige/white layer, sulphate reduction and possible respiration would increase availability of Ca^{2+}, promoting carbonate precipitation as rhombohedra (Dupraz & Visscher, 2005). The EPS act as a template for carbonate precipitation arranging platelets into rhombohedra suggesting an EPS-influenced organomineralization. As SRB have been shown to be highly active in lithified zones of microbial mats (Visscher *et al.*, 2000; Dupraz *et al.*, 2004), they might have been involved in lithification of stromatolites. However, SRB and other heterotrophs could have reduced EPS by breaking down exopolymers and inhibiting precipitation in the upper green layer (Arp *et al.*, 2012). This phenomenon is illustrated by the degradation of nanofilaments of EPS into fine fibres. Further investigation of the fossil stromatolites is required in order to decipher the biotic from the abiotic contribution to their formation.

Although cyanobacteria primarily contribute to stromatolite morphogenesis, the metabolic activity of heterotrophic bacteria such as SRB is also significant in microbial mats (Canfield & Des Marais, 1991; Visscher *et al.*, 2000; Des Marais, 2003; Dupraz *et al.*, 2004). The globular morphology of these stromatolites likely reflects specific microbial responses to their environment, which may result from differential growth, migration of microbes with respect to environmental conditions such as light, O_2, etc. or a combination of both (Andersen *et al.*, 2011).

CONCLUSIONS

Modern stromatolites from the Maquinchao Basin (Argentina) are presently formed by microbial communities including filamentous and coccoid cyanobacteria, non-photosynthetic microorganisms capable of respiration and/or sulphate reduction, and diatoms. Combining water chemistry and microscopy data enabled the optimum conditions for stromatolite growth to be constrained.

Unlike recently described stromatolites at high altitude in the subtropical and tropical Andes, a combination of low salinity and turbidity, relatively high Ca and microbial (photosynthetic and heterotrophic) activity appear to be some of the most important factors promoting low-Mg calcite formation in the Maquinchao Basin. Our results further indicate that carbonate formation is mediated by organomineralization processes in which EPS play a fundamental role. Freshwater stromatolites appear to be a useful proxy for changing hydrological conditions. They also highlight the significant role of modern geomicrobiological studies in understanding the mechanisms and environmental conditions underlying their formation.

ACKNOWLEDGMENTS

Fieldwork was accomplished with the generous funding of the *Fondation Augustin Lombard*, Geneva (Switzerland) to G. Hunger and V. Martinuzzi, the logistic support of the Argentinean Research Council (CONICET PIP 00819) National Agency for the Promotion of Science and Technology (PICT 2010-0082), and the University of Comahue (Grant B166) Bariloche, Argentina. We thank the University of Geneva, ETH-Zürich, the Swiss National Science Foundation (grant 2000.112320), and Pethros (Petrobras) for financial support, and the University of Zurich for access to SEM facilities. We also acknowledge Hanno Meyer for isotopic water measurements and Genort Arp for helpful comments on an earlier version of the manuscript.

References

Amann, R.I., **Binder, B.J.**, **Olsen, R.J.**, **Chisholm, S.W.**, **Devereux, R.** and **Stahl, D.A.** (1990) Combination of 16S rRNA-targeted oligonucleotide probes with flow cytometry for analyzing mixed microbial populations. *Appl. Environ. Microbiol.*, **56**, 1919–1925.

Amann, R.I., **Ludwig, W.** and **Schleifer, K.** (1995) Phylogenetic identification and in situ detection of individual microbial cells without cultivation. *Microbiol. Rev.*, **59**, 143–169.

Andersen, D.T., **Sumner, D.Y.**, **Hawes, I.**, **Webster-Brown, J.** and **McKay, P.C.** (2011) Discovery of large conical stromatolites in Lake Untersee, Antarctica. *Geobiology*, **9**, 280–293.

Ariztegui, D., **Anselmetti, F.S.**, **Gilli, A.** and **Waldmann, N.** (2008) Late Pleistocene environmental changes in Patagonia and Tierra del Fuego – a limnogeological approach. In: *Developments in Quaternary Sciences Series 11* (Ed. J. Rabassa), The Late Cenozoic of Patagonia and Tierra del Fuego. ISBN 978-0-444-52954-1. Elsevier Science, pp. 430.

Arp, G., **Reiner, A.** and **Reitner, J.** (2003) Microbialite formation in seawater of increased alkalinity, satonda crater lake, Indonesia. *J. Sediment. Res.*, **73**, 105–127.

Arp, G., **Helms, G.**, **Karlinska, K.**, **Schumann, G.**, **Reimer, A.**, **Reitner, J.** and **Trichet, J.** (2012) Photosynthesis versus exopolymer degradation in the formation of microbialites on the atoll of Kiritimati, Republic of Kiribati, Central Pacific. *Geomicrobiol J.*, **29**, 29–65.

Awramik, S.M. and **Grey, K.** (2005) Stromatolites: biogenicity, biosignatures, and bioconfusion. *Proc. SPIE*, **5906**, 1–9.

Baumgartner, L.K., **Reid, R.P.**, **Dupraz, C.**, **Decho, A.W.**, **Buckley, D.H.**, **Spear, J.R.**, **Przekop, K.M.** and **Visscher, P.T.** (2006) Sulfate reducing bacteria in microbial mats: changing paradigms, new discoveries. *Sed. Geol.*, **185**, 131–145.

Benzerara, K., **Meibom, A.**, **Gautier, Q.**, **Kamierczak, J.**, **Stolarski, J.**, **Menguy, N.** and **Brown, G.E.** (2010) Nanotextures of aragonite in stromatolites from the quasi-marine Satonda crater lake, Indonesia. *Geochem. Soc. London Spec. Publ.*, **336**, 211–224.

Berner, R.A. (1975) The role of magnesium in the crystal growth of calcite and aragonite from sea water. *Geochim. Cosmochim. Acta*, **39**, 489–504.

Biller, S.J., **Schubotz, F.**, **Roggensack, S.E.**, **Thompson, A.W.**, **Summons, R.E.** and **Chisholm, S.W.** (2014) Bacterial vesicles in marine ecosystems. *Science*, **343**, 183–186.

Bischoff, J.L. and **Fyfe, W.S.** (1968) Catalysis, inhibition, and the aragonite-calcite problem I. The aragonite-calcite transformation. *Am. J. Sci.*, **266**, 65–79.

Bradbury, J.P., **Grosjean, M.**, **Stine, S.** and **Sylvestre, F.** (2001) Full and late glacial records along the PEP1 transect: their role in developing interhemispheric paleoclimate interactions. In: *Interhemispheric Climate Linkages* (Ed. V. Markgraf), pp. 265–292. Academic Press, San Diego, CA.

Burne, R.V. and **Moore, L.S.** (1987) Microbialites: organosedimentary deposits of benthic microbial communities. *Palaios*, **2**, 241–254.

Canfield, D.E. and **Des Marais, D.J.** (1991) Aerobic sulfate reduction in microbial mats. *Science*, **251**, 1471–1473.

Cartwright, A., **Quade, J.**, **Stine, S.**, **Adams, K.D.**, **Broecker, W.** and **Cheng, H.** (2011) Chronostratigraphy and lake-level changes of Laguna Cari-Laufquén, Rio Negro, Argentina. *Quatern. Res.*, **76**, 430–440.

Casanova, J. and **Hillaire-Marcel, C.** (1993) Carbon and oxygen isotopes in African lacustrine stromatolites: palaeohydrological interpretation. In: *Climate Change in Continental Isotopic Record* (Ed. P.K. Swart), *Geophys. Monogr.*, **78**, 123–133.

Coira, B.L. (1979) Descripción geológica de la hoja 40 d Ingeniero Jacobacci. *Bol. Serv. Geol. Nac.*, **168**, 101.

Craig, H. and **Gordon, L.I.** (1965) Deuterium and oxygen 18 variations in the ocean and marine atmosphere. In: *Proc. Stable Isotopes in Oceanographic Studies and Paleotemperatures* (Ed. E. Tongiogi), pp. 9–130. V. Lishi e F., Pisa, Spoleto, Italy.

Craig, H., **Dixon, F.**, **Craig, V.**, **Edmond, J.** and **Coulter, G.** (1974) Lake Tanganyika geochemical and hydrographic study: 1973 expedition. *Scripps Inst. Oceanogr. Pub.*, **75**, 1–83.

Cusminsky, G., **Schwalb, A.**, **Pérez, A.P.**, **Pineda, D.**, **Viehberg, F.**, **Whatley, R.**, **Markgraf, V.**, **Gilli, A.**, **Ariztegui, D.** and **Anselmetti, F.S.** (2011) Late Quaternary environmental changes in Patagonia as inferred from lacustrine fossil and extant ostracodes. *Biol. J. Linn. Soc.*, **103**, 397–408.

De Wit, R., **Gautret, P.**, **Bettarel, Y.**, **Roques, C.**, **Marlière, C.**, **Ramonda, M.**, **Nguyen Thanh, T.**, **Tran Quang, H.** and **Bouvier, T.** (2015) Viruses occur incorporated in biogenic high-Mg calcite from hypersaline microbial mats. *PLoS ONE*, **10**, e0130552.

Decho, A.W. (2000) Exopolymer microdomains as a structuring agent for heterogeneity within microbial biofilms. In: *Microbial Sediments* (Eds R.E. Riding and S.M. Awramik), pp. 1–9. Springer-Verlag, Berlin.

Decho, A.W. and **Kawaguchi, T.** (1999) Confocal imaging of natural in situ microbial communities and their extracellular polymeric secretions (EPS) using Nanoplast resin. *Biotechniques*, **27**, 1246–1251.

Del Valle, R.A., **Tatur, A.**, **Amos, A.**, **Ariztegui, D.**, **Bianchi, M.M.**, **Cusminsky, G.C.**, **Hsu, K.**, **Lirio, J.M.**, **Martínez Macchiavello, J.C.**, **Masaferro, J.I.**, **Núñez, H.J.**, **Rinaldi, C.A.**, **Valverde, R.**, **Vigna, S.**, **Vobis, G.** and **Whatley, R.C.** (1993) Laguna Cari Laufquen Grande: registro de una fase climática húmeda del Pleistoceno tardío en la Patagonia septentrional. *Proyecto Pangea Glopals*, comunicaciones, San Juan, 16–19.

Del Valle, R.A., **Lirio, J.M.**, **Nuñez, H.J.**, **Tatur, A.**, **Rinaldi, C.A.**, **Lusky, J.C.** and **Amos, A.J.** (1996) Reconstrucción paleoambiental Pleistoceno-Holoceno en las latitudes medias al este de los Andes. *XIII Congr. Geol. Argent. III Congr. Expl. Hidrocarb Actas*, **IV**, 85–102.

Des Marais, D.J. (2003) Biogeochemistry of hypersaline microbial mats illustrates the dynamics of modern microbial ecosystems and the early evolution of the biosphere. *Biol. Bull.*, **204**, 160–167.

Desnues, C., **Rodriguez-Brito, B.**, **Rayhawk, S.**, **Kelley, S.**, **Tran, T.**, **Haynes, M.**, **Liu, H.**, **Furlan, M.**, **Wegley, L.**, **Chau, B.**, **Ruan, Y.**, **Hall, D.**, **Angly, F.E.**, **Edwards, R.A.**, **Li, L.**, **Vega Thurber, R.**, **Reid, P.R.**, **Siefet, J.**, **Souza, V.**, **Valentine, D.L.**, **Swan, B.K.**, **Breitbart, M.** and **Rohwer, F.** (2008) Biodiversity and biogeography of phages in modern stromatolites and thrombolites. *Nature*, **452**, 340–343.

Dupraz, C. and **Visscher, P.T.** (2005) Microbial lithification in marine stromatolites and hypersaline mats. *Trends Microbiol.*, **13**, 429–438.

Dupraz, C., **Visscher, P.T.**, **Baumgartner, L.K.** and **Reid, R.P.** (2004) Microbe–mineral interactions: early carbonate precipitation in a hypersaline lake (Eleuthera Island, Bahamas). *Sedimentology*, **51**, 745–765.

Dupraz, C., **Reid, R.P.**, **Braissant, O.**, **Decho, A.W.**, **Norman, R.S.** and **Visscher, P.T.** (2009) Process of carbonate precipitation in modern microbial mats. *Earth-Sci. Rev.*, **96**, 141–162.

Elser, J.J., **Schampel, J.H.**, **Garcia-Pichel, F.**, **Wade, B.D.**, **Souza, V.**, **Eguiarte, L.**, **Escalante, A.** and **Farmer, J.D.** (2005) Effects of phosphorus enrichment and grazing snails on modern stromatolitic microbial communities. *Freshw. Biol.*, **50**, 1808–1825.

Farias, M.E., **Rascovan, N.**, **Toneatti, D.M.**, **Albarracin, V.H.**, **Flores, M.R.**, **Poire, D.G.**, **Collavino, M.M.**, **Aguilar, M.O.**, **Vazquez, M.P.** and **Polerecky, L.** (2013) The Discovery of Stromatolites Developing at 3570 m above Sea Level in a High-Altitude Volcanic Lake Socompa, Argentinean Andes. *PLoS ONE*, **8**, e53497.

Forterre, P., **Soler, N.**, **Krupovic, M.**, **Marguet, E.** and **Ackermann, H.W.** (2013) Fake virus particles generated by fluorescence microscopy. *Trends Microbiol.*, **21**, 1–5.

Galloway, R.M., **Markgraf, V.** and **Bradbury, J.P.** (1988) Dating shorelines of lakes in Patagonia Argentina. *J. S. Am. Earth Sci.*, **1**, 195–198.

Glunk, C., **Dupraz, C.**, **Braissant, O.**, **Gallagher, K.L.**, **Verrecchia, E.P.** and **Visscher, P.T.** (2011) Microbially mediated carbonate precipitation in a hypersaline lake, Big Pond (Eleuthera, Bahamas). *Sedimentology*, **58**, 720–738.

González Bonorino, F. and **Rabassa, J.** (1973) La laguna Carilaufquen Grande y el origen de los bajos patagónicos. *Rev. Asoc. Geol. Argent. Rev.*, **28**, 314–318.

Grilli Caiola, M., **Ocampo-Friedmann, R.** and **Friedmann, E.I.** (1993) Cytology of long-term desiccation in the desert cyanobacterium *Chroococcidiopsis* (Chroococcales). *Phycologia*, **32**, 315–322.

Grotzinger, J.P. and **Rothman, D.H.** (1996) An abiotic model for stromatolite morphogenesis. *Nature*, **383**, 423–425.

Hofmann, H.J., **Grey, K.**, **Hickman, A.H.** and **Thorpe, R.I.** (1999) Origin of 3.45 Ga coniform stromatolites inWarrawoona Group, Western Australia. *Geol. Soc. Amer. Bull.*, **111**, 1256–1262.

Logan, B.W. and **Cebulski, D.E.** (1970) Sedimentary environments of Shark Bay, Western Australia. In: *Carbonate Sedimentation and Environments, Shark Bay, Western Australia*, Vol. 13 (Eds B.W. Logan, G.R. Davies, J.F. Read and D.E. Cebulski), pp. 1–37. American Association of Petroleum Geologists, Tulsa, OK.

McLoughlin, N., **Wilson, A.** and **Brasier, M.D.** (2008) Growth of synthetic stromatolites and wrinkle structures in the absence of microbes – implications for the early fossil record. *Geobiology*, **6**, 95–105.

Meyer, H., **Schönicke, L.**, **Wand, U.**, **Hubberten, H.-W.** and **Friedrichsen, H.** (2000) Isotope studies of hydrogen and oxygen in ground ice – experiences with the equilibration technique. *Isot. Environ. Health Stud.*, **36**, 133–149.

Müller, G., **Irion, G.** and **Forstner, U.** (1972) Formation and diagenesis of inorganic Ca-Mg carbonates in the lacustrine environment. *Naturwissenschaften*, **59**, 158–164.

Pacton, M., **Gorin, G.E.** and **Vasconcelos, C.** (2011) Amorphous organic matter – experimental data on formation and the role of microbes. *Rev. Palaeobot. Palynol.*, **166**, 253–267.

Pacton, M., **Ariztegui, D.**, **Wacey, D.**, **Kilburn, M.R.**, **Rollion-Bard, C.**, **Farah, R.** and **Vasconcelos, C.** (2012) Going nano: a new step toward understanding the processes governing freshwater ooid formation. *Geology*, **40**, 547–550.

Pacton, M., **Wacey, D.**, **Corinaldesi, C.**, **Tangherlini, M.**, **Kilburn, M.R.**, **Gorin, G.**, **Danovaro, R.** and **Vasconcelos, C.** (2014) Viruses as a new agent of organomineralisation in the geological record. *Nat. Commun.*, **5**, 4298.

Petryshyn, V.A., **Corsetti, F.A.**, **Berelson, W.M.**, **Beaumont, W.** and **Lund, S.P.** (2012) Stromatolite lamination frequency, Walter Lake, Nevada: implications for stromatolites as biosignatures. *Geology*, **40**, 499–502.

Plée, K., **Ariztegui, D.**, **Martini, R.** and **Davaud, E.** (2008) Unravelling the microbial role in ooids formation – results of an in situ experiment in modern freshwater Lake Geneva in Switzerland. *Geobiology*, **6**, 341–360.

Reid, R.P., **Visscher, P.T.**, **Decho, A.W.**, **Stolz, J.F.**, **Bebout, B.M.**, **Dupraz, C.**, **Macintyre, L.G.**, **Paerl, H.W.**, **Pinckney, J.L.**, **Prufert-Bebout, L.**, **Steppe, T.F.** and **Des Marais, D.J.** (2000) The role of microbes in accretion, lamination and early lithification of modern marine stromatolites. *Nature*, **406**, 989–992.

Riding, R.E. (2011) Microbialites, stromatolites and thrombolites. In: *Encyclopedia of Geobiology* (Eds J. Reitner and V. Thiel), pp. 635–654. Encyclopedia of Earth Science Series. Springer, Heidelberg.

Schwalb, A., **Burns, S.**, **Cusminsky, G.**, **Kelts, K.** and **Markgraf, V.** (2002) Assemblage diversity and isotopic signals of modern ostracodes and host waters from Patagonia, Argentina. *Palaeogeogr. Palaeoclimatol. Palaeoecol.*, **187**, 323–339.

Solari, M.A., **Hervé, F.**, **Le Roux, J.P.**, **Airo, A.** and **Sial, A.N.** (2010) Paleoclimatic significance of lacustrine microbialites: a stable isotope case study of two lakes at Torres del Paine, southern Chile. *Palaeogeogr. Palaeoclimatol. Palaeoecol.*, **297**, 70–82.

Soler, N., **Krupovic, M.**, **Marguet, E.** and **Forterre, P.** (2015) Membrane vesicles in natural environments: a major challenge in viral ecology. *ISME J.*, **9**, 793–796.

Souza-Egipsy, V., **Wierzchos, J.**, **Ascaso, C.** and **Nealson, K.H.** (2005) Mg-silica precipitation in fossilization mechanisms of sand tufa endolithic microbial community, Mono Lake (California). *Chem. Geol.*, **217**, 77–87.

Stolz, J.F. (1991) *Structure of Phototrophic Prokaryotes*. CRC Press, Boca Raton, 131 pp.

Stolz, J.F., **Feinstein, T.N.**, **Salsi, J.**, **Visccher, P.T.** and **Reid, R.P.** (2001) TEM analysis of microbial mediated sedimentation and lithification in a modern marine stromatolite. *Am. Mineral.*, **86**, 826–833.

Suttle, C.A. (2005) Viruses in the sea. *Nature*, **437**, 356–361.

Tatur, A., **del Valle, R.**, **Bianchi, M.M.**, **Outes, V.** and **Villarosa, G.** (2000) Late Pleistocene Pluvial Phase in Patagonia. *Geolines*, **11**, 47–51.

Trichet, J. and **Défarge, C.** (1995) Non-biologically supported organo mineralization. In: *Bulletin de l'Institut Océanographique de Monaco* (Eds D. Allemand and J.P. Cuif), *Proc. 7th Int. Symp. Biomineral.*, **2** (14>), 203–236.

Valero-Garcés, B.L., **Arenas, C.** and **Delgado-Huertas, A.** (2001) Depositional environments of Quaternary lacustrine travertines and stromatolites from high-altitude Andean lakes, northwestern Argentina. *Can. J. Earth Sci.*, **38**, 1263–1283.

Visscher, P.T., **Reid, R.P.** and **Bebout, B.M.** (2000) Microscale observations of sulfate reduction: correlation of microbial activity with lithified micritic laminae in modern marine stromatolites. *Geology*, **28**, 919–922.

Volkheimer, W. (1973) Observaciones geológicas en el área de Ingeniero Jacobacci y adyascencias (provincia de Río Negro). *Rev. Asoc. Geol. Arg.*, **28**, 13–36.

Whatley, R. and **Cuminsky, G.** (1999) Lacustrine Ostracoda and late Quaternary palaeoenvironments from the Lake Cari-Laufquen region, Rio Negro province, Argentina. *Palaeogeogr. Palaeoclimatol. Palaeoecol.*, **151**, 229–239.

Wright, P.V. (2012) Lacustrine carbonates in rift settings: the interaction of volcanic and microbial processes on carbonate deposition. In: *Advances in Carbonate Exploration and Reservoir Analysis* (Eds J. Garland, J.E. Neilson, S.E. Laubach and K.J. Whidden), *Geol. Soc. London. Spec. Publ.*, **370**. doi: 10.1144/SP370.2.

6

Hydrological and sedimentological processes of flood layer formation in Lake Mondsee

LUCAS KÄMPF*,†, PHILIP MUELLER‡, HANNES HÖLLERER§, BIRGIT PLESSEN*, RUDOLF NAUMANN¶, HEIKO THOSS‡, ANDREAS GÜNTNER‡, BRUNO MERZ‡ and ACHIM BRAUER*

*Section 5.2 Climate Dynamics and Landscape Evolution, GFZ German Research Centre for Geosciences, Telegrafenberg, Potsdam, 14473, Germany
†TU Dresden, Faculty of Environmental Sciences, Institute for Soil Science and Site Ecology, Pienner Strasse 19, Tharandt, 01737, Germany
‡Section 5.4 Hydrology, GFZ German Research Centre for Geosciences, Telegrafenberg, Potsdam, 14473, Germany
§Research Institute for Limnology Mondsee, University of Innsbruck, Mondseestrasse 9, Mondsee, 5310, Austria
¶Section 4.2 Inorganic and Isotope Geochemistry, GFZ German Research Centre for Geosciences, Telegrafenberg, Potsdam, 14473, Germany

Keywords
Detrital layers, flood reconstruction, flood-related sediment flux, Lake sediments, monitoring, process understanding.

ABSTRACT

Detrital layers in lake sediments are recorders of extreme flood events. However, their use for establishing time series of past floods is limited by lack in understanding processes of detrital layer formation. Therefore, we monitored hydro-sedimentary dynamics in Lake Mondsee (Upper Austria) and its main tributary, Griesler Ache, over a 3-year period from January 2011 to December 2013. Precipitation, discharge and turbidity were recorded continuously at the river outlet to the lake and compared to sediment fluxes trapped with 3 to 12 days resolution at two locations in the lake basin, in a distance of 0·9 (proximal) and 2·8 km (distal) to the Griesler Ache inflow. Within the 3-year observation period, 26 river floods of different magnitude (10 to 110 $m^3\ s^{-1}$) have been recorded resulting in variable sediment fluxes to the lake (4 to 760 $g\ m^{-2}\ d^{-1}$) including the 'century-scale' flood event in June 2013. The comparison of hydrological and sedimentological data revealed (i) a rapid sedimentation within 3 days after the peak runoff in the proximal and within 6 to 10 days in the distal lake basin; (ii) empirical flood thresholds for triggering sediment flux at the lake floor increasing from the proximal (20 $m^3\ s^{-1}$) to the distal lake basin (30 $m^3\ s^{-1}$) and (iii) various factors that control the detrital sediment transport in the lake. The amount of sediment transported to the lake is controlled by runoff and catchment sediment availability. The distribution of detrital sediment within the lake basin is mainly driven by mesopycnal interflows and closely linked to flood duration and the season in which a flood occurred. The combined hydro-sedimentary monitoring revealed detailed insights into processes of flood layer formation in a meso-scale peri-Alpine lake and, thereby, improves the interpretation of the depositional record of flood layers.

INTRODUCTION

Lakes form ideal sediment traps in the landscape continuously recording land surface processes in the catchment including extreme events (Hsü & Kelts, 1985). Discrete flood-triggered sediment fluxes of detrital catchment material into lakes result in the formation of discrete detrital layers at the lake floor (Sturm & Matter, 1978; Siegenthaler & Sturm, 1991). Therefore, detrital layers in lake sediments are increasingly used to establish long flood chronologies especially in the Alpine (Støren *et al.*, 2010; Glur *et al.*, 2013; Wilhelm *et al.*, 2013; Wirth *et al.*, 2013b), peri-Alpine (Arnaud *et al.*, 2005; Czymzik *et al.*, 2013; Swierczynski *et al.*, 2013) and Arctic realms (Francus *et al.*, 2002; Lamoureux *et al.*, 2006; Lapointe *et al.*, 2012). The recurrence intervals of detrital layers provide information about palaeoflood frequencies (Czymzik *et al.*, 2010; Swierczynski *et al.*, 2012; Schlolaut *et al.*, 2014), whereas flood intensities have been inferred from the thickness (Schiefer *et al.*, 2011; Wilhelm *et al.*, 2013) of individual deposits.

Varved sediment records provide, in addition, the unique opportunity to date detrital layers with seasonal precision (Mangili *et al.*, 2005) and, thereby, (i) determine palaeoflood variability even at seasonal scale (Swierczynski *et al.*, 2012; Wirth *et al.*, 2013a) and (ii) calibrate the sub-recent detrital layer record with instrumental flood data (Francus *et al.*, 2002; Chutko & Lamoureux, 2008; Czymzik *et al.*, 2010).

Commonly, flood reconstruction from lake sediments assume the completeness of the depositional record in the sense that each flood resulted in a well-preserved detrital layer. However, a test of the hypothesis of completeness of the depositional record is still lacking. First detailed comparisons of detrital layer records with instrumental data even questioned the assumption of completeness by providing evidence for both, floods that did not result in detrital layer deposition and detrital layers, which were not triggered by strong floods (Czymzik *et al.*, 2010; Kämpf *et al.*, 2012b). A possible reason might be that the amount and spatial distribution of detrital sediment within a lake basin triggered by flood events might vary (Lamoureux, 1999; Jenny *et al.*, 2014) probably even depending on the season in which a flood occurred (Kämpf *et al.*, 2014). A better knowledge of the hydrological and sedimentary processes of detrital layer formation is required to reduce the bias in interpretation and, thereby, improve the use of depositional records as palaeoflood archives.

To gain a more sophisticated process understanding of detrital layer formation different attempts have been initiated comprising detailed analyses of single flood deposits (Gilbert *et al.*, 2006; Kämpf *et al.*, 2012a) and *in situ* monitoring of flood triggered sediment fluxes (Best *et al.*, 2005; Crookshanks & Gilbert, 2008; Dugan *et al.*, 2009). Most observational studies of detrital sediment fluxes in lakes have been performed in arctic and high mountain lakes with predominantly clastic sedimentation. Despite of the growing number of flood reconstructions from Alpine and peri-Alpine lakes (e.g. Wirth *et al.*, 2013b), in-depth monitoring studies in such lakes with mainly autochthonous sediments (biochemically precipitated calcite and organic components) are still lacking.

Here, we present results of a 3-year integrated lake and catchment monitoring of flood and sediment dynamics in the peri-Alpine Lake Mondsee. The study period comprises the 'century-scale' flood event in June 2013 (Blöschl *et al.*, 2013). We chose Lake Mondsee for this in-depth monitoring since it provides a varved sediment record and a flood layer record covering the last 7100 years (Swierczynski *et al.*, 2013) as well as a good data base of meteorological and hydrological data. In addition, a first calibration of sub-recent detrital layers with instrumental flood data is available (Kämpf *et al.*, 2014). Ultimately, we will contribute new knowledge about flood layer formation in a meso-scale peri-Alpine lake, which is expected to generally improve flood reconstructions from depositional records of lakes.

STUDY SITE

Lake Mondsee is located at the northern fringe of the European Alps in Upper Austria (47°48′N, 13°23′E) at an altitude of 481 m above sea-level (a.s.l.) (Fig. 1). With a surface area of 14 km^2 and a maximum water depth of 68 m Lake Mondsee is a meso-scale peri-Alpine lake characterized by a specific morphometry displaying a significant kink in the generally elongated and NW-SE directed shape (Fig. 1). Lake Mondsee is a meromictic hardwater lake (Dokulil & Skolaut, 1986) with one mixing period in autumn/winter and thermal stratification between May and September (Fig. 2). Episodically, the lake is completely ice-covered which happened once during the observation period in February to March 2012.

The catchment (247 km^2) is subdivided into two major geological units by a main Alpine thrust fault (Van Husen, 1989) following the southern shoreline of the lake (Fig. 1). The northern catchment (ca. 75% of the total catchment) is formed by peri-Alpine hills of up to 1100 m a.s.l. which are built up by Cretaceous Flysch sediments (Sandstone, Argillite). The valleys are covered by moraines formed by latest Pleistocene glacier activity (Van Husen, 1989). Three tributaries drain the northern catchment: the Griesler Ache in the West, the Zeller Ache in the North and the Wangauer Ache in the East. Regarding its size and runoff variability (A_0 = 110 km^2; MQ = 4 m^3 s^{-1}; HHQ = 137 m^3 s^{-1}, ref. BMLFUW (2011)), the Griesler Ache is the largest tributary to Lake Mondsee and the main source for suspended sediments during floods (Kämpf *et al.*, 2014). The southern sub-catchment (ca. 25%) reaches a maximum elevation of 1700 m a.s.l. and is part of the Northern Calcareous Alps. The base rock is composed of Jurassic and Triassic units of limestone and dolomite forming steep slopes at the southern lake shoreline which are drained by small torrents like, for example, the Kienbach creek with a catchment of 2·1 km^2 (Fig. 1). Lake Mondsee drains via the river Seeache into Lake Attersee at the south-eastern end of the lake (Fig. 1).

METHODS

A monitoring network was installed in Lake Mondsee and its catchment between 2011 and 2012, comprising

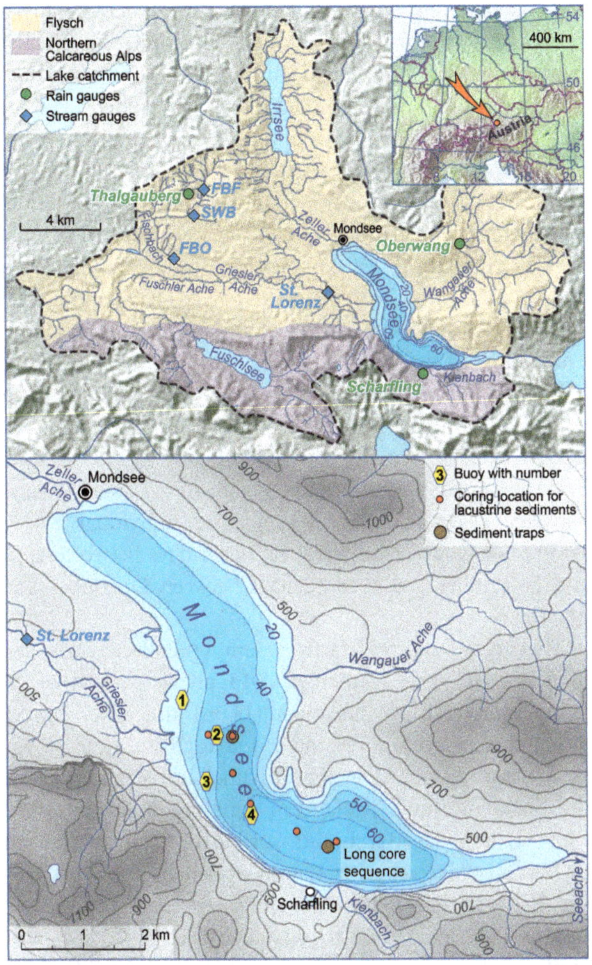

Fig. 1. Lake Mondsee catchment and monitoring set-up including (i) four stream gauges equipped with devices for measuring precipitation, water level, air and water temperature, electrical conductivity, turbidity and automated water sampling, (ii) four monitoring buoys within the lake for multilevel water current and turbidity monitoring (Mueller *et al.*, 2013) and (iii) two sediment trap chains equipped with three integral traps in different water depths, one sequential trap at the lake floor and thermistors in different water depths. Additional rain gauges in the catchment conducted by the hydrographic services of Upper Austria and Salzburg are also indicated. Six surface sediment cores were recovered along a transect from the Griesler Ache inflow towards a long sediment core in the distal lake basin (Swierczynski *et al.*, 2013).

four river gauges along the main tributary to Lake Mondsee, the Griesler Ache, as well as four monitoring buoys and two sediment trap chains within the lake (Fig. 1). The combination of river gauges and monitoring buoys was designed to track hydro-sedimentary dynamics, i.e. runoff generation and sediment transport, continuously from the head catchments to the lake (Mueller *et al.*, 2013).

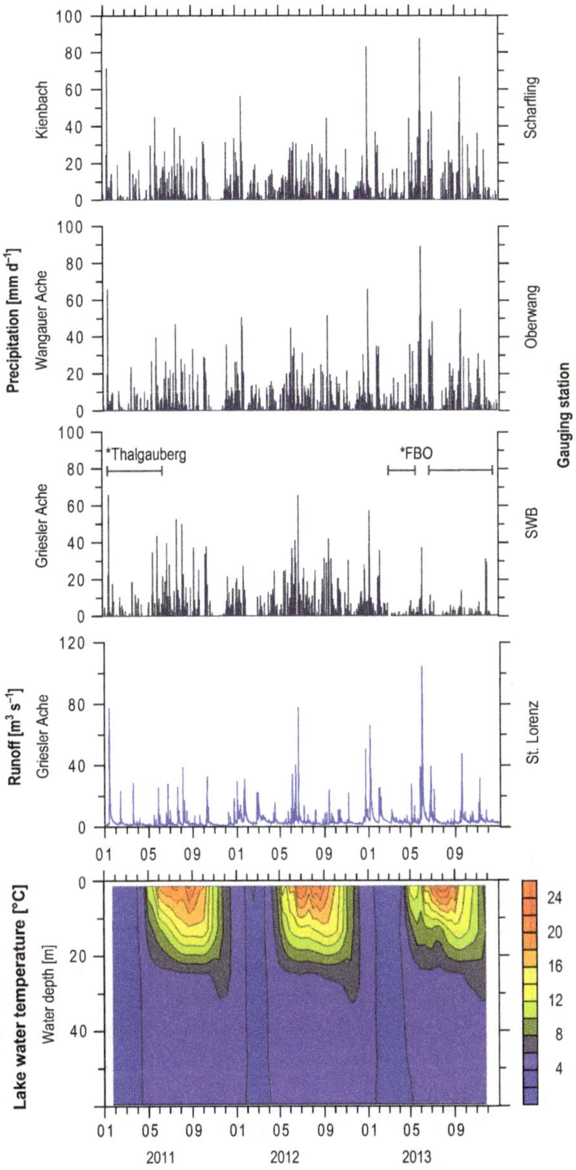

Fig. 2. Hydroclimatic data 2011 to 2013: daily precipitation sums from different catchments: Kienbach (Scharfling rain gauge), Wangauer Ache (Oberwang), and Griesler Ache [Streuwiesenbach (SWB), gaps in the time series were filled with data from Thalgauberg and Fischbach Outlet (FBO) rain gauges]; hourly runoff of the Griesler Ache River at St. Lorenz gauging station and water column temperature at the proximal trap site. Locations of the gauging stations can be found in Fig. 1.

Catchment monitoring

Precipitation in the Griesler Ache catchment is recorded since June 2011 using OTT Pluvio rain gauges (OTT Hydromet, http://www.ott.com) at the gauging stations St. Lorenz located close to Lake Mondsee (482 m a.s.l., 4 km distance to inflow), Fischbach Outlet (FBO, 552 m,

10 km) located at the confluence of the Griesler Ache and the Fischbach, which is the largest tributary to the Griesler Ache, and at two stations in the Fischbach head-catchment, Streuwiesenbach (SWB, 777 m, 11 km) and Fischbach Forest (FBF, 786 m, 12 km). In addition we used data from three precipitation gauges at Thalgauberg in the Griesler Ache catchment (730 m a.s.l., 11 km distance to inflow) (operated by the Hydrographic Survey of Salzburg), Scharfling in the Kienbach catchment (482 m, 0·7 km) and Oberwang in the Wangauer Ache catchment (595 m, 5 km) (both operated by the Hydrographic Survey of Upper Austria). Runoff data were obtained from the stream gauge of St. Lorenz (Hydrographic Survey of Upper Austria), located 4 km upstream of the Griesler Ache inflow to Lake Mondsee (Fig. 1). For monitoring sediment transport in the river, the station was additionally equipped with a FTS DTS-12 turbidity sensor (Forest Technology Systems Inc., http://www.ftsenvironmental.com) and an ISCO 3700 automatic pumping sampler (Teledyne ISCO, http://www.isco.com). Depending on the actual values of discharge and turbidity, 3 to 24 river water samples (1 l) were taken automatically for each of 21 flood events since June 2011 following the turbidity-threshold-sampling (Lewis, 1996).

Lake monitoring

Two moorings were installed in Lake Mondsee, each equipped with one sequencing sediment trap (S-trap) and three integrating sediment traps (I-traps). The locations follow a transect from the inflow of the Griesler Ache river to the location of a long sediment record used for establishing a flood layer chronology over the last 7100 years (Swierczynski *et al.*, 2013), with one mooring in a distance of 900 m to the river mouth (proximal trap: 47°49·21′N, 13°22·78′E, water depth: 56 m) and the other in a distance of 2800 m (distal trap: 47°48·32′N, 13°23·92′E, water depth: 63 m). We compared the sediment trap data with detrital layers investigated in surface sediment cores that were previously retrieved close to the trap locations in the proximal (sediment core MO/10/4: 47°49·16′N, 13°22·79′E) and distal lake basin (MO/05/P3: 47°48·41′N, 13°24·09′E, ref. Kämpf *et al.* (2014)).

The I-traps (UWITEC, http://www.uwitec.at) have two collecting cylinders with an active area of 127 cm^2 in total. The two sequencing traps are equipped with a computer programmable sample bottle carrousel and differ in size: with 500 cm^2 we chose a trap with a smaller active for the proximal location (PPS 4/3, 12 sample bottles, Technicap, http://www.technicap.com) and a larger one for the distal location (1250 cm^2, PPS 3/3, 12 sample bottles) due to expected higher sediment accumulation ratios closer to the inflow of the Griesler Ache (Swierczynski *et al.*, 2009). The S- and one of the I-traps were deployed approximately 3 m above the lake bed surface in a water depth of 53 m (prox.) and 60 m (dist.). The other two I-traps were moored in the upper water column (14 m) and between the upper and lower traps (prox.: 33 m, dist.: 30 m). 12 temperature loggers (Hobo U22 Water Temp Pro, Hobo, http://www.onsetcomp.com) were attached to the proximal mooring at water depths of 1, 3, 5, 8, 11, 14, 18, 22, 33, 43, 53 and 55 m and 4 loggers to the distal mooring at water depths of 1, 14, 30 and 60 m.

The moorings were first deployed at 13 January 2011. The S-trap at the distal mooring was added on 04 April 2012. The traps were recovered in a monthly rhythm between April and November until 03 April 2014. Thus, I-traps collect material on a basis of 21 to 41 days between April and November and of 41 to 123 days between December and March. The individual S-trap samples cover a time interval of 3 to 4 days between April and November and 4 to 12 days between December and March.

This study reports data from samples collected between January 2011 and December 2013 giving a total of 28 I-trap samples from the upper (14 m) and lower (prox.: 53 m, dist.: 60 m) water column at each location as well as 269 S-trap samples at the proximal and 158 samples at the distal location. The S-trap time series exhibit three gaps: (i) between March and April 2012 (37 days) caused by persistent ice cover, (ii) between September and October 2012 (21 days) due to technical reasons and (iii) in June 2013 (6 days) due to very high lake water level that inhibited trap recovery. The third gap was bridged by deploying two additional I-traps close to the mooring locations.

Sediment analyses

The sample bottles of the river water samplers and sediment traps were stored at 4°C after recovery for at least 48 h to ensure that all suspended particles had settled. The samples were freeze dried and the total dry weight was determined. For trap samples, the daily sediment flux (in g m^{-2} d^{-1}) was calculated for each sample. For river water samples, the measured suspended sediment concentration (SSC in g L^{-1}) was used for setting up SSC rating curves of the turbidity sensor at the gauge of St. Lorenz (Figure S2). The rating curves were established individually for the three strongest recorded floods by applying polynomial regression (Lewis & Eads, 2009).

Total carbon (TC), nitrogen (TN) and organic carbon (TOC) were determined for each S-trap sample using an elemental analyser Euro Vector EA (EuroEA 3000, www.eurovector.it). For TC and TN, around 5 mg of powdered sample was loaded in tin capsules and combusted in the elemental analyser. TOC was determined on *in situ* decalcified samples. Around 3 mg of sample was weighted into Ag-capsules, treated with 20% HCl, heated for 3 h at 75°C, and finally wrapped and measured as described above. Replicate determinations showed a standard deviation better than 0·2%. Organic matter (OM) was calculated as OM = 2 × TOC (Meyers & Teranes, 2001) and total inorganic carbon (TIC) as TIC = TC-TOC. The $CaCO_3$ content was calculated stoichiometrically by multiplying the TIC by 8·33, assuming that all inorganic carbon is bound as calcium carbonate. We are aware that minor dolomite contributions from catchment rocks (Figure S1) in the detrital carbonate fraction might cause little inaccuracies.

Grain size was measured for 13 selected trap samples with high sediment flux rates using a laser particle sizer (Fritsch Analysette, Fritsch, http://www.fritsch.de). The samples were sieved at 1 mm and dispersed in an ultrasonic bath before measuring. Image data were automatically transferred to particle distribution by the software Fritsch MaScontrol applying the Fraunhofer model.

The mineralogical composition was determined for those trap samples that were also analysed for their grain size distribution and for four samples of river bed material as well as two detrital layers in sediment cores using a PANanlytical Empyrean X-ray diffractometer equipped with a Cu tube (http://www.panalytical.com). The samples were powdered before measuring.

RESULTS

Hydro-climatic conditions at Lake Mondsee 2011 to 2013

Mean annual precipitation in the Mondsee catchment during the observation period from January 2011 to December 2013 was 1600 mm and thus, in the range of the long-term mean (1981 to 2011, ref. BMLFUW (2011)). Large rainfall amounts (>40 mm d^{-1}) were recorded during 20 days (2011; 4, 2012: 7, 2013: 9) and primarily took place between May and September (15 days) and, secondarily, in winter (5 days >40 mm). With 90 mm d^{-1} the maximum precipitation was recorded at the Oberwang rain gauge on 01 June 2013 (Fig. 2).

The mean runoff between January 2011 and December 2013 was 4·1 m^3 s^{-1} that is close to the long-term mean of 3·9 m^3 s^{-1} (1961 to 2011, ref. BMLFUW (2011)). Runoff events >30 m^3 s^{-1} occurred 14 times in the observation period (2011: 3, 2012: 4, 2013: 7) and, like precipitation events, cumulated in summer (May-Sep: 8, Oct-Nov: 2, Dec-Feb: 4). The largest flood in the monitoring period reached a maximum hourly discharge of 104 m^3 s^{-1} (2 June 2013) and was one of the strongest recorded floods in that region with an estimated return period of around 100 years (Eybl *et al.*, 2013).

Variability in sediment flux I: total sediment composition

Sediment flux in Lake Mondsee was trapped over the period from January 2011 to December 2013 at two different sites within the lake basin (Fig. 1), one located in a position proximal to the inflow of the main tributary river (distance: 900 m) and one in a distal position (2800 m). The total sediment flux including both, lake internal (autochthonous) and external (allochthonous) components, was by median 4 g m^{-2} d^{-1} (prox.: 4·2 g m^{-2} d^{-1}, dist.: 3·6 g m^{-2} d^{-1}) and exhibited (i) a seasonal variability with higher mean flux rates in summer (5 to 6 g m^{-2} d^{-1} in May-Sep) and lower mean flux rates in autumn and winter (1 to 1·5 g m^{-2} d^{-1} in Oct-Jan) and (ii) short-term peaks of up to 758 g m^{-2} d^{-1} in the proximal and up to 59 g m^{-2} d^{-1} in the distal lake basin (Fig. 3A and B).

The seasonal variability of the trapped sediment flux is also expressed by distinct changes in sediment composition (Fig. 3C and D). Calcite contents varied from 20 to 40% (October to April) to 60 to 95% (May to September) with maximum values between July and August (Fig. 3C). The higher calcite contents in summer reflect biochemical precipitation of calcite in epilimnic waters which is a typical seasonal process in mid-latitude hardwater lakes (Koschel *et al.*, 1983). The contents of organic matter in trapped sediment, calculated from the measured TOC contents, varied between 5 and 25% and exhibited a clear maximum in April and a second, lower maximum of 15 to 20% in late autumn (October-December) when bulk sediment fluxes were lowest (Fig. 3B and D). The maximum in April was predominantly made up of diatom frustules, whereas in autumn amorphous organic matter and diatoms were abundant as revealed by smear slide investigations. Between October and March and within peak sediment flux samples the sediment was mainly composed of dolomitic and siliciclastic matter. The observed seasonal variations in sediment flux lead to the formation of characteristic diatom, calcite and mixed sub-layers preserved and described in the varved sediment record of Lake Mondsee (Lauterbach *et al.*, 2011; Swierczynski *et al.*, 2012).

Variability in sediment flux II: spatial distribution

We characterized the spatial sediment distribution in the lake water body by both its vertical and horizontal

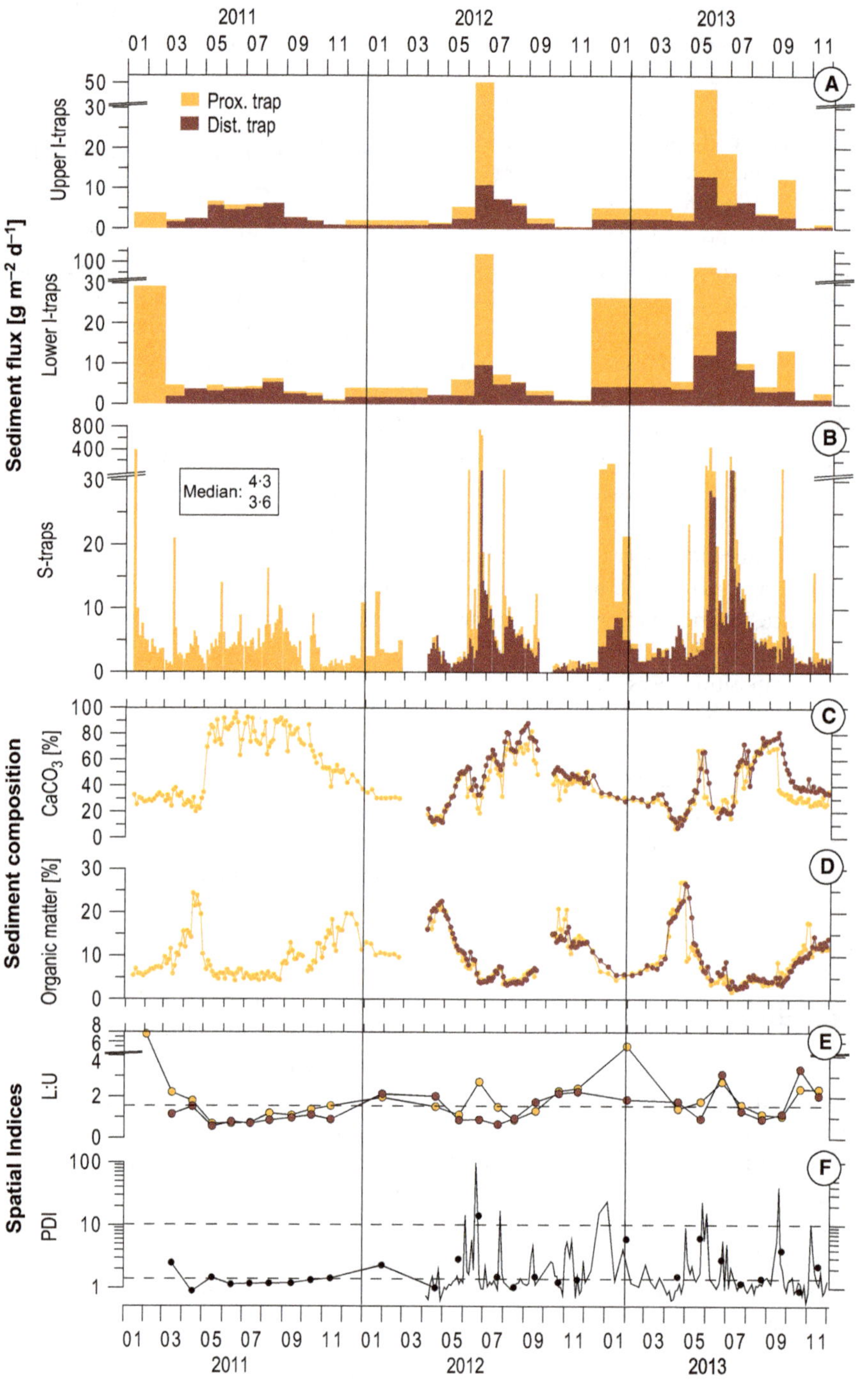

Fig. 3. Sediment fluxes in Lake Mondsee at the proximal and distal location (01/2011 to 11/2013): (A) monthly values in integrating traps in the upper (14 m water depth) and lower water column (3 m above lake floor, prox.: 53 m, dist.: 60 m); (B) 3 to 12 days values in sequential traps in the lower water column and contents of (C) calcite and (D) organic matter (Prox.: 2011 to 2013, dist.: 2012 to 2013). Spatial sediment flux patterns are expressed by ratios of sediment flux between: (E) lower and upper traps (L : U) and (F) proximal and distal traps (PDI), dots in (F) indicate monthly values calculated from I-traps.

variability in order to decipher pathways and mechanisms of lake internal sediment transport. The vertical variability is expressed by the flux ratio of the lower and upper traps (L : U, ref. Cockburn & Lamoureux (2008)) calculated for the 28 I-trap samples representing monthly means (Fig. 3E). From these samples, 51% (prox.: 13 samples, dist.: 16 samples) ranged between 0·6 and 1·5 indicating comparable flux rates throughout the water column and 38% of the values ranged between 1·5 and 2·5 (prox.: 12 samples, dist.: 9 samples) representing a mixture of vertical and lateral sediment fluxes. L : U values >2·5 rarely occurred (11%, prox.: 4 samples, dist.: 2 samples), reflecting sediment flux predominately in the lower water column mainly driven by hyperpycnal underflows (Cockburn & Lamoureux, 2008). It has to be considered that the vertical sediment flux ratios likely represent minimum estimates of sedimentation during times of underflows, as the traps were designed to capture mainly sediment that settled vertically through the water column and were allocated 3 m above the lake bottom likely leading to an underestimation of material transported by underflows. The vertical sediment distribution indicates seasonally changing sediment pathways. In summer (May-Sep), the L : U ratio was <1·5 for 90% of all values indicating predominately downward sediment flux from the upper water column. In fall to spring the L : U ratio was >1·5 for 90% of all values pointing to the contribution of lateral sediment transport from the shoreline and/or the tributary streams during winter.

The lateral sediment distribution is expressed by the flux ratio of the proximal and the distal site (Proximal-Distal Index [PDI], ref. Lamoureux (1999)). PDI values were calculated for the lowermost I-traps and sequential traps (Fig. 3F) revealing 64% of the values ranging between 0·6 and 1·5 representing a uniform sedimentation pattern and 36% of the values at the proximal site exceeding those at the distal site by factors >1·5 proving sediment input from the main tributary river and decreasing sediment deposition towards distal direction through settling of particles. In few cases (6%) the PDI values exceed 10 indicating localized sediment fluxes to the proximal lake basin.

To identify events with strongly increased sedimentation in the sediment flux time series, we defined high sediment flux values as peaks which (i) exceeded the median sediment flux of 4 g m^{-2} d^{-1} and (ii) showed an increase to the previous sample of more than 50%. This resulted in 32 sediment flux peaks occurring during 35 months of observation at the proximal trap location (2011 to 2013) and 14 peaks during 20 months of observation at the distal trap location (2012 to 2013) (Fig. 4). 12 of these 14 peaks occurred synchronously at both locations or

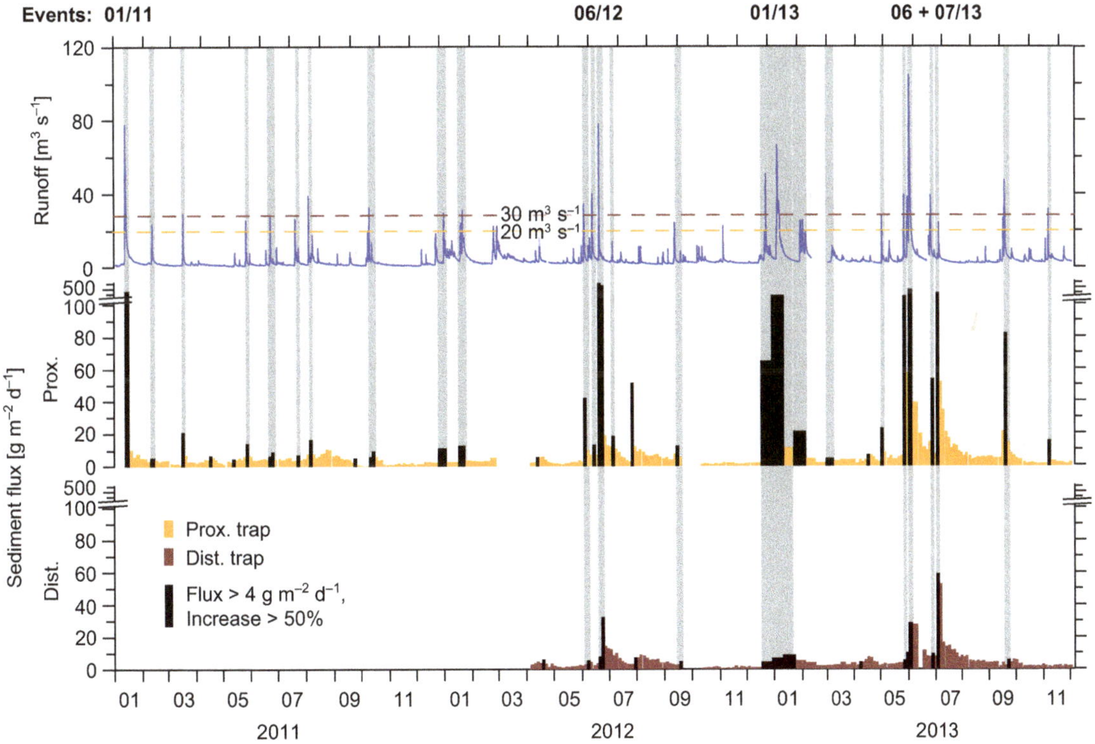

Fig. 4. Comparison of trapped sediment flux in Lake Mondsee with runoff in the Griesler Ache River. Black coloured samples mark peaks in sediment flux as defined by exceeding the median sediment flux (4 g m^{-2} d^{-1}) and an increase to the previous sample of 50%. Grey bars mark sediment flux peaks coincided with elevated runoff. Dashed lines indicate thresholds for triggering peaks in sediment flux: 20 m^3 s^{-1} in the proximal and 30 m^3 s^{-1} in the distal lake basin. The five runoff events that triggered the highest sediment flux to the lake basin (>100 g m^{-2} d^{-1} at the proximal trap) are named at the top.

delayed by one sample at the distal location. Four peaks occurred only in one part of the basin; two at the proximal and two at the distal trap. For peak sediment flux events at both trap sites the spatial distribution of detrital material is variable expressed in PDI values ranging between 2 and 25.

Sediment flux versus runoff

The 3-year time series of trapped sediment flux was compared to runoff data from the outlet gauge of the main tributary river, the Griesler Ache, in order to test if a relation exists between river floods and the spatio-temporal sediment distribution within the lake (Fig. 4). Most of the defined peaks in sediment flux (prox.: 26 of 32; flux: 5 to 758 g m^{-2} d^{-1}; dist.: 11 of 14; 4 to 59 g m^{-2} d^{-1}) coincided with elevated river runoff ranging from 10 to 104 m^3 s^{-1}. For one peak in sediment flux (July 2012), no discharge data were available so that this event was excluded from further analyses. Five peaks of the proximal (16% of all sediment flux peaks) and two peaks of the distal trap (15%) did not relate to elevated runoff. Interestingly, peaks not related to runoff events occurred independently either at the proximal or at the distal site (Fig. 4).

Comparing flood magnitudes in terms of their hourly peak river discharge reveals that 10% of low magnitude floods in the range 10 to 20 m^3 s^{-1} triggered a measurable peak in sediment flux in the proximal lake basin, whereas for floods >20 m^3 s^{-1} already 96% resulted in a distinct sediment flux at this location (Fig. 4). At the distal location elevated sediment flux was never observed for floods <20 m^3 s^{-1}, but 80% of floods >30 m^3 s^{-1} resulted in a peak. Two peaks in sediment flux at the distal location were related to floods between 20 and 30 m^3 s^{-1}. The sediment flux peaks in the proximal lake basin occurred within 3 days after the discharge peak, i.e. within the same sampling interval (Fig. 5). In the distal lake basin the period of highest sediment flux mostly expanded over two sampling intervals (3 to 6 days).

Besides the coincidence of flood occurrence and enhanced sediment flux, we compared the amount of sediment deposited over the whole flooding period with peak runoff values and observed an exponential relation (Fig. 6A). The correlation coefficient decreases from the proximal ($r^2_{prox.}$ = 0·62) to the distal lake basin ($r^2_{dist.}$ = 0·31) and is strongly affected by one event (07/13, Fig. 4), when a very high sediment deposition (prox.: 1258 g m^{-2}, dist: 468 g m^{-2}) but only a low maximum discharge were recorded (24 m^3 s^{-1}). If this event is

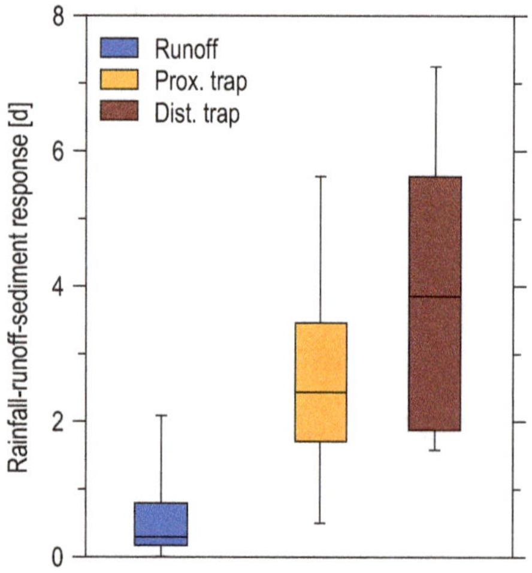

Fig. 5. Time lag between peaks in precipitation and runoff as well as between peaks in runoff and sediment flux at the proximal and distal traps (end of sampling interval) of (prox.: n = 25, dist.: n = 7).

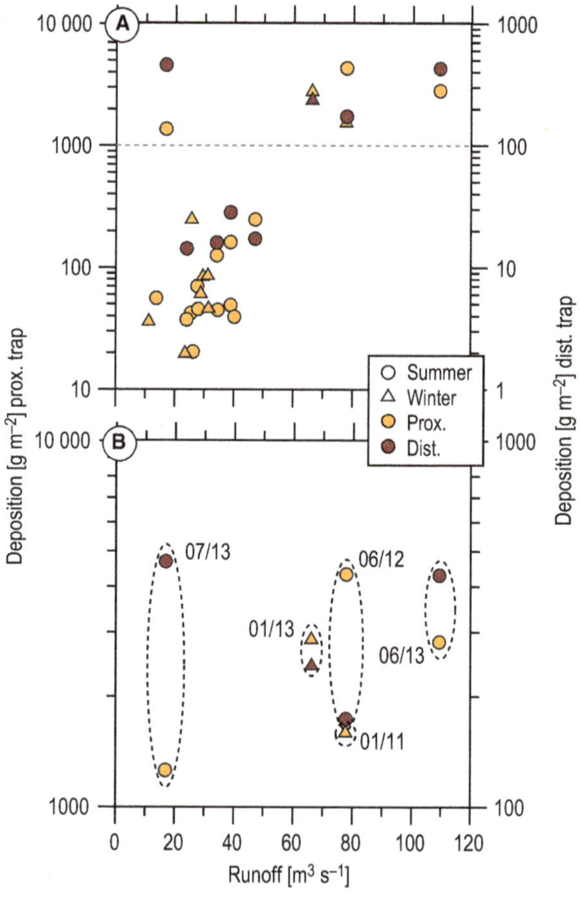

Fig. 6. (A) Correlation of peak runoff (hourly values) and sediment deposition during the monitoring period 2011 to 2013 applying an exponential regression. Statistics: Prox.: n = 25, r^2 = 0·76, P < 0·01, Dist.: n = 7, r^2 = 0·80, P < 0·01; 07/13 event excluded from statistics. (B) Enlarged view of plot (A) for the five strongest sediment flux events.

excluded, the correlation becomes better and more similar at both sites ($r^2_{prox.} = 0{\cdot}76$, $r^2_{dist.} = 0{\cdot}80$).

Hydro-sedimentary dynamics during major floods

In total, 26 floods resulting in enhanced sediment flux to the lake floor were recorded in the 3-year monitoring period. For five of these floods the sediment yield exceeded 1000 g m^{-2} at the proximal site (1258 to 4320 g m^{-2}) that were triggered by precipitation of more than 40 mm d^{-1} and resulting runoff peaks ranging from 24 to 104 m^3 s^{-1} (Table 1). At the distal site the sediment yield was significantly lower (174 to 468 g m^{-2}).

Three of the five floods with highest sediment yields occurred in summer and two in winter (Table 1). The 01/11 flood lasted for 3 days with a maximum precipitation of more than 65 mm d^{-1} in all three gauged catchments (Fig. 2). The 01/13 event was characterized by a persistent wet phase over 2 months with two main precipitation periods of 3 and 4 days, respectively, and maxima of 57 to 83 mm d^{-1} (Fig. 8). Whereas the two winter floods were both triggered by precipitation events that covered the whole catchment area, the three strongest summer floods were triggered by regional and local scale rainfall events (Fig. 2). The latter were convective events of high intensity and short duration (<1 day) affecting only parts of the catchment, like the 06/12 event in the Fischbach sub-catchment (max. rainfall at station SWB: 65 mm d^{-1}) and the 07/13 event in the Wangauer Ache catchment (Oberwang: 48 mm d^{-1}). In contrast, the 06/13 event was triggered by a series of synoptic scale low pressure systems (Blöschl *et al.*, 2013), resulting in strong precipitation over the whole catchment area lasting for 2 weeks (Table 1).

The stream-flow rose within 1 to 10 h after the main precipitation events (Figs 7 and 8). In January 2011, the flood hydrograph peaked at 78 m^3 s^{-1} and was larger than 10 m^3 s^{-1} for 3 days. The two runoff peaks between December 2012 and January 2013 were characterized by maximum discharges of 50 and 66 m^3 s^{-1} and durations of 3 and 7 days. The local rainfall events in June 2012 and July 2013 triggered peak flows of 78 m^3 s^{-1} during the 06/12 event and 24 m^3 s^{-1} during the 07/13 event. The comparably low peak flow in the Griesler Ache during the 07/13 event is due to the local character of the precipitation event which mainly covered the catchment of the Wangauer Ache which is not part of our

Table 1. Hydro-sedimentary data on the five largest sediment transfer events to Lake Mondsee within the monitoring period 2011 to 2013. Note that the 07/13 event is a very rare local event transporting sediments from a different sub-catchment (Wangauer Ache) that is not included in the monitoring program.

	01/11	06/12	01/13	06/13	07/13
Catchment data					
Precipitation					
Affected catchment area	Regional	Fischbach	Regional	Regional	Wangauer A.
Date	10 to 15 Jan	20 Jun	9 Dec to 12 Jan	19 May to 06 Jun	03 to 05 Jul
Days >10 mm	2	1	3+3+4	1+2+4	2
Maximum [mm d^{-1}]	71	65	83	90	48
Sum [mm]	104	82	350	320	75
Discharge Griesler Ache					
Days >10 m^3 s^{-1}	3	1	3+7	1+8	1
Max. hourly discharge [m^3 s^{-1}]	78	78	66	104	24
Max. daily discharge [m^3 s^{-1}]	56	21	43	79	11
Max. turbidity [NTU]	No data	1730	1078	1410	No data
Max. measured SSC [g L^{-1}]	No data	19	1	62	No data
Max. rated SSC [g L^{-1}]	No data	20	2	103	No data
Lake data					
Prox. Trap					
Max. sediment flux [g m^{-2} d^{-1}]	391	758	163	452	54
Sediment deposition [g m^{-2}]	1603	4320	2880	2825	1258
D : $Q_{prox.}$	No data	0·1	0·3	0·2	< 0·1
L : $U_{prox.}$	8	3	5	2	3
Dist. Trap					
Max. sediment flux [g m^{-2} d^{-1}]	No data	32	9	29	16
Sediment deposition [g m^{-2}]	No data	174	245	430	468
D : $Q_{dist.}$	No data	0·1	0·2	0·5	0·1
L : $U_{dist.}$	No data	1	2	1	3
PDI	No data	25	12	7	3

monitoring network (Table 1, Fig. 1). The June 2013 flood was exceptional for the entire observation period and reached by far the highest maximum discharge values ($104\ m^3\ s^{-1}$) and the longest duration (9 days $>10\ m^3\ s^{-1}$). The flood resulted in a lake level rise of more than 1·5 m and flooding of the city of Mondsee.

Suspended sediment concentration in the river (SSC), as recorded by turbidity measurements and automatic water samples, reached highest values during the 06/13 event ($SSC_{max} = 61\ g\ L^{-1}$ in samples, $103\ g\ L^{-1}$ rated from turbidity) and was lowest during the 01/13 event ($SSC_{max} = 1{\cdot}1\ g\ L^{-1}$ in samples, $1{\cdot}6\ g\ L^{-1}$ rated from turbidity). During all events, SSC increased with increasing discharge and reached maxima during the rising limb of the flood hydrograph and already declined 1 to 2 h before the flood peak (Figs 7 and 8).

The sediment depositions in the lake occurred with time lags of 1 to 4 days after the peak runoff at the proximal location and of up to 10 days at the distal trap (Figs 7 and 8). The total sediment deposition in the proximal basin was highest during the 06/12 event ($4320\ g\ m^{-2}$) and lowest during the 07/13 event ($1258\ g\ m^{-2}$), whereas at the distal site the opposite has been observed, i.e. highest sediment deposition during the 07/13 event ($468\ g\ m^{-2}$) and lowest during the 06/12 event ($174\ g\ m^{-2}$).

The sources of detrital sediment are determined by their mineralogical composition (Table 2). The sediments originating from the Flysch catchment predominately consist of siliciclastic material made up of quartz, mica and feldspars, whereas the Northern Calcareous Alps are predominantly composed of dolomite (Figure S1). This allows us to use the dolomite/quartz ratio (D : Q) to distinguish between the two main sediment source areas (Table 2). Riverbed material of the main tributaries Griesler and Wangauer Ache exhibit similar D : Q ratios of 0·2. This value is in accordance with most values measured in sediment trap samples in the lake, ranging between 0·2 and 0·5, indicating these catchments as main sources for detrital sediments. The lowest D : Q values <0·2 are measured for the Fischbach bed load reflecting a clear dominance of Flysch sediments in this sub-catchment that exclusively drains a Flysch area (Figure S1). Similar values are measured in sediments trapped after the local 06/12 and 07/13 events. The highest D : Q values >1·0 were measured for the bed load of the Kienbach creek reflecting a dominance of material originating from the Northern Calcareous Alps. Sediments trapped in the distal trap after the 06/13 flood exhibit D : Q values between 0·5 and 1·0 pointing to sediment transport from both, the Flysch and the Limestone catchments.

Sediment trap samples of the five events with highest sediment yields were further analysed for their grain size distribution (Fig. 9). The resulting distributions exhibit unimodal (prox.: 4 samples, dist.: 2 samples) and bimodal patterns (prox.: 1 samples, dist.: 4 samples) with maxima around 10 to 60 μm for unimodal and at 2 to 4 μm and 10 to 60 μm for bimodal functions, respectively. Thus, samples with a bimodal grain size distribution contain a higher portion of fine silt and clay particles. Fine-grained particles were generally more abundant in samples trapped (i) in summer, (ii) after the main sediment flux event and (iii) in the distal lake basin. The only exception is the June 2013 flood event (Fig. 9C), when trapped samples in the distal lake basin exhibited a unimodal function and a coarser maximum (36 μm) than in the proximal lake basin (24 μm).

DISCUSSION

Detrital layers in lake sediments commonly are interpreted as flood recorders. However, the potential of these geoarchives for extending instrumental flood records back in time is still not fully exploited due to our limited knowledge about (i) how the amount and spatial distribution of detrital sediment are related to flood parameters and (ii) to which extent this relation is affected by local factors in the catchment and the lake. These are considered as potential bias for establishing palaeoflood time series from lake sediment records (e.g. Schiefer *et al.*, 2006, 2011; Dugan *et al.*, 2009).

The 3-year monitoring at Lake Mondsee provides a comprehensive set of hydro-sedimentary data from the lake and its catchment comprising 26 floods with very different magnitudes (return periods ranging from ≪1 year to 100 years), seasonal occurrence and sediment response (4 to $758\ g\ m^{-2}\ d^{-1}$). This reveals new insights into processes of flood layer deposition with unprecedented detail including a variety of influencing factors like peak runoff and season, sediment availability or the type of precipitation event giving the basis for an improved evaluation of the depositional record of Lake Mondsee. Our monitoring data have been related to the available sub-recent depositional record covering the time period from 1976 to 2005 (Fig. 11) (Kämpf *et al.*, 2014) since it was not yet possible to obtain undisturbed sediment cores of the last 3 years due to the too high water content of the topmost sediments.

Parameters controlling flood-related sediment flux

Runoff magnitude

A general link between runoff events in the Griesler Ache and sediment flux into the lake is evidenced by the measured sediment response to 26 flood events. Moreover, we

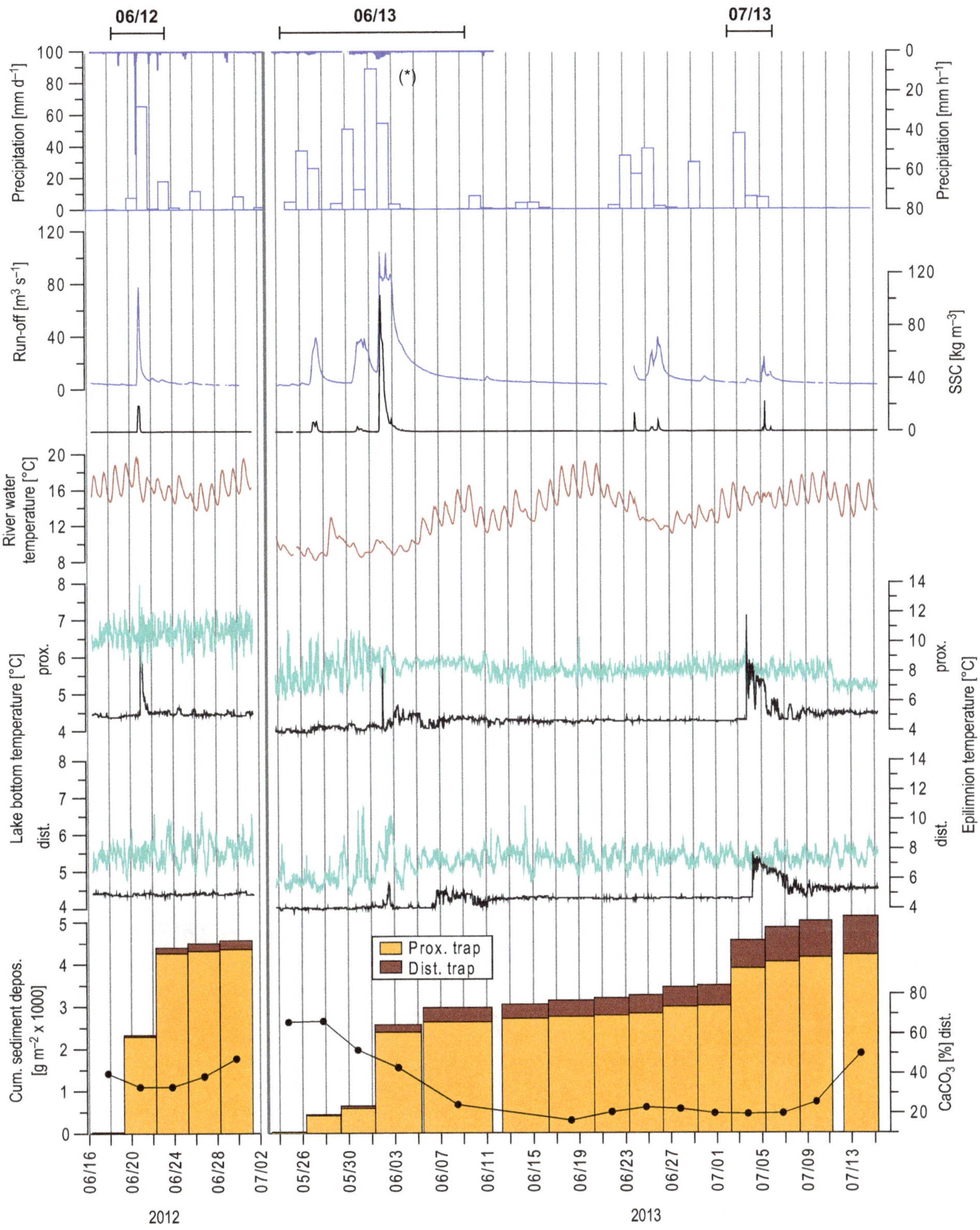

Fig. 7. Precipitation, runoff, river sediment concentration, river and lake water temperature and trapped sediment deposition 3 m above the lake bottom at the proximal and distal trap location during the largest summer floods. (*) Daily precipitation data from 2013 measured at the Oberwang gauging station. The distance between vertical grey lines represents 2 days.

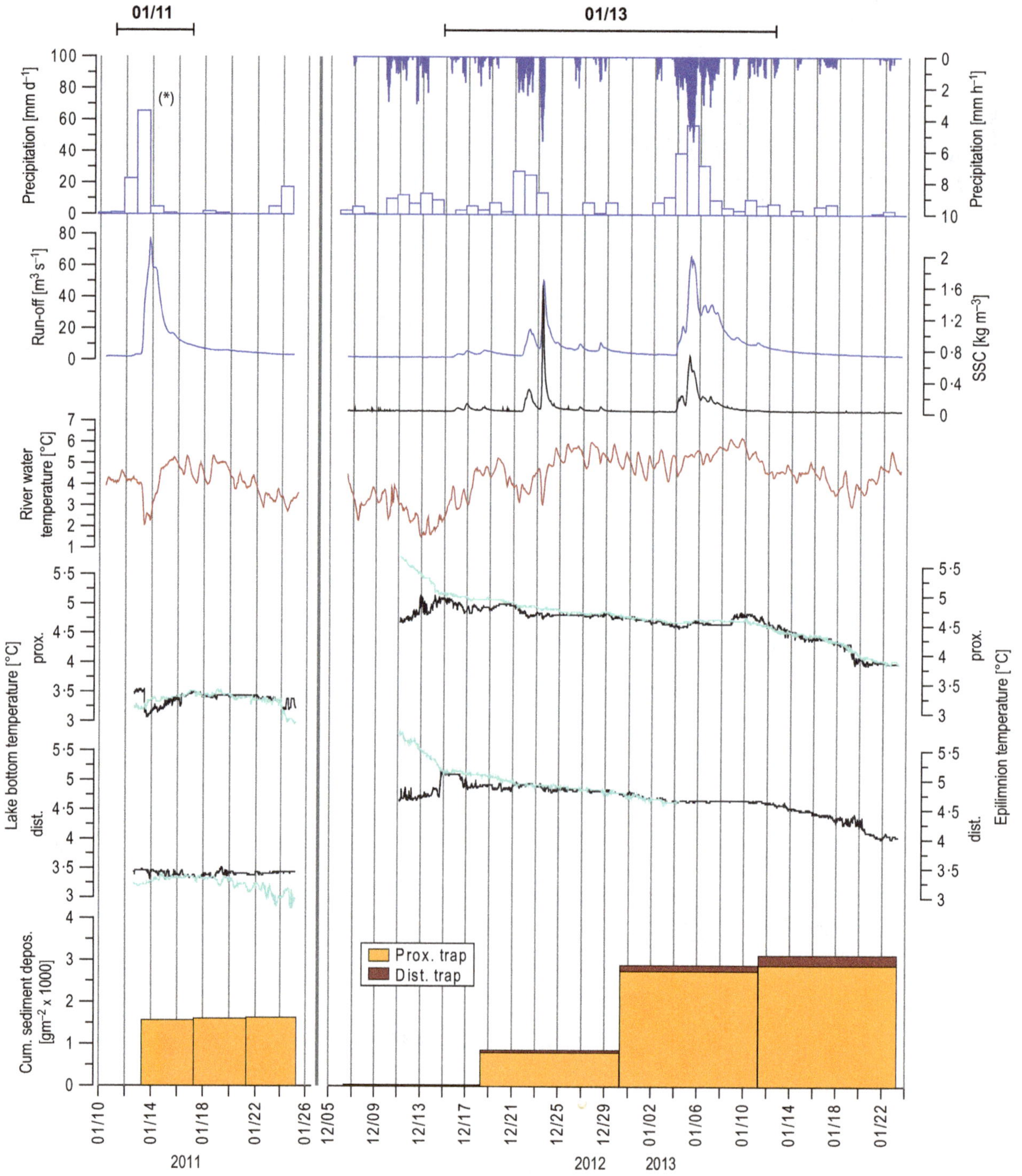

Fig. 8. Precipitation, run-off, river sediment concentration, river and lake water temperature and trapped sediment deposition 3 m above the lake bottom at the proximal and distal trap location during the largest winter floods. (*) Precipitation data from the 01/11 event measured at Thalgauberg gauging station. The distance between vertical grey lines represents 2 days.

found different empiric runoff thresholds for both sites, above which suspended sediment is transported to the respective location: 20 $m^3 s^{-1}$ for the proximal and 30 $m^3 s^{-1}$ for the distal site (Fig. 4). Lower amplitude floods caused a measurable sediment flux only in two exceptional cases.

In addition to the coincidence of peaks in the runoff and sediment trap time series, we observed a significant

Table 2. Mineralogical composition [%] of sediment trap samples obtained during major flood events (I = integral trap), river bed material and two detrital layers in the sub-recent sediment record. The ratio of dolomite and quartz contents (D : Q) was used to discriminate the Flysch (quartz-rich) and limestone (dolomite-rich) dominated sub-catchments (see also Figure S1).

Sample	Calcite	Chlorite	Dolomite	Kaolinite	Illite	Orthoclase	Albite	Quartz	Smectite	D : Q
Sediment traps										
13/6 prox.	26	4	6	2	10	3	4	28	16	0·2
13/6 prox._I_upper	32	4	5		9	2	3	27	18	0·2
13/6 prox._I_lower	27	4	6		10	3	5	29	17	0·2
13/7 prox.	25	5	2	4	12	3	4	31	15	0·1
13/1 prox.	25	3	9	2	7	3	7	35	10	0·3
12/6 prox.	21	5	4	3	11	3	5	31	17	0·1
13/6 dist.	43	4	18		10	2	2	20		0·9
13/6 dist._I_upper	41	3	4		9	2	2	20	20	0·2
13/6 dist._I_lower	38	4	9	3	8	2	2	19	15	0·5
13/7 dist.	23	5	2	3	12	2	2	26	25	0·1
13/1 dist.	33	4	6	3	10	2	3	24	16	0·2
12/6 dist.	35	4	2	4	12	3	2	22	18	0·1
River bed										
Griesler Ache	32		11			5	5	48		0·2
Kienbach	8		90					2		37·6
Wangauer Ache	57		6				3	33		0·2
Fischbach	19		4			6	7	64		0·1
Detrital layers										
DL 1986 dist.	18	3	64		2			8	5	7·7
DL 2002 prox.	21	5	2	4	13	3	4	28	20	0·1

exponential relation between peak runoff and the amount of trapped sediment in the proximal ($r^2 = 0{\cdot}76$) and distal lake basin ($r^2 = 0{\cdot}80$) if the local 07/13 event with predominant sediment flux from the ungauged secondary stream Wangauer Ache is excluded (Fig. 6A).

Hitherto, we only discussed the relation between runoff and sediment transport. More important for interpreting the depositional record, however, is an assessment of the amount of sediment flux that is necessary to produce a recognizable detrital layer in the sediment record. We can calculate this minimum sediment yield from the thickness of the finest detected detrital layers found in the sediments (0·2 mm; Kämpf *et al.*, 2014) and a mean dry density of 1·5 g cm^{-3}. The resulting minimum sediment yield of approximately 300 g m^{-2} was exceeded by four of the 26 observed floods at the proximal site that were triggered by runoff peaks ranging from 66 to 104 m^3 s^{-1} and precipitation >65 mm d^{-1} (Fig. 4, Table 1). At the distal site, where the long flood record has been established (Swierczynski *et al.*, 2013), only the strongest of these four events likely resulted in sufficient sediment transport to form a detrital layer (June 2013, max. runoff: 104 m^3 s^{-1}). Hence, based on the amount of sediment deposition we predict the formation of one flood triggered detrital layer at the distal location (June 2013) and four layers at the proximal site (January 2011, June 2012, January, June 2013). Another event in July 2013 that also supplied sediment amounts >300 g m^{-2} to the proximal and distal locations (Fig. 7) is not included in the further discussion since this event was caused by local precipitation in the Wangauer Ache catchment, which is not part of our monitoring network (Fig. 1).

The flood discharge values, which according to the observational data should have caused detrital layer formation, are for both locations in the same range as empirical discharge thresholds for detrital layers revealed from the depositional record (Kämpf *et al.*, 2014). These are >60 m^3 s^{-1} for the proximal site (observation: 66 m^3 s^{-1}, 78 m^3 s^{-1}, 104 m^3 s^{-1}) and >80 m^3 s^{-1} for the distal site (observation: 104 m^3 s^{-1}). The good agreement between these independently obtained data further supports the existence of discharge thresholds for flood layer deposition and makes us confident that these thresholds can be reasonable well determined.

Despite the existence of thresholds in discharge for detrital layer formation we do not observe a correlation between runoff and sediment yield for the strongest floods (Fig. 6B). This suggests that the availability of fine-grained sediment in the rivers and variable sediment distribution within the lake basin also play a crucial role for detrital layer formation and will be discussed in the following.

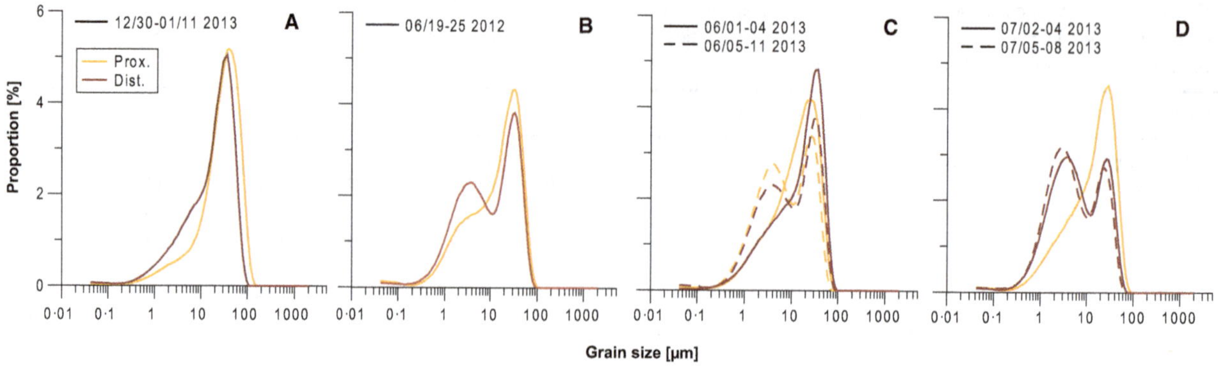

Fig. 9. Grain size distribution of trapped sediment samples of the four largest sediment transfer events: (A) 01/13, (B) 06/12, (C) 06/13, (D) 07/13. Straight-lined plots indicate sediment flux peak samples; dashed plots indicate samples after the sediment flux peaks.

Sediment availability

The measured variations in suspended sediment concentration in the streams provide evidence for the sediment availability in the catchment as an important factor (Fig. 10). Fine sediments are mainly derived from the channel bank of the streams but sometimes also from patches of local erosion events, e.g. landslides, as revealed by field observations after the 06/13 flood (Figure S3). The sediment input to the streams further depends on the season since measured sediment concentrations were much lower during the winter flood (01/13: max. 1 g L^{-1}) compared to summer floods (20 to 103 g L^{-1}). Although only two strong winter floods are observed, we assume that this seasonal difference is generally valid and can be explained by reduced soil erosion in winter due to snow cover and frozen ground (Dugan *et al.*, 2009).

Sediment availability is further changing within the course of a flood as demonstrated by the relation of river SSC and discharge (Fig. 10). Clockwise hysteresis functions indicate higher sediment concentrations during the rising limbs of the flood hydrographs proving river transport capacity as the main driver for riverine sediment transport in the initial phase of a flood when fine sediments are sufficiently available. In the later flood stage when sediments have been washed out and less fine material is available the riverine sediment transport is reduced as has been observed in various rivers (e.g. Forbes & Lamoureux, 2005; López-Tarazón *et al.*, 2010). This effect is also obvious from lower SSC values during the higher second flood peak of the 01/13 event which followed 13 days after a lower first runoff peak, which, in contrast, resulted in higher SSC values. For floods of only short duration, like the 06/12 event that lasted only for 18 hrs, river SSC followed the flood hydrograph during the rising and falling limb for the main flooding interval (>40 m^3 s^{-1}, 5 h) suggesting no limitations in sediment availability during such short floods.

Although the interpretation of the hysteresis plot of the strongest event (06/13) is limited because runoff during peak flow conditions (>100 m^3 s^{-1}) was likely underestimated due to river bank overflow, this example demonstrates that extreme high SSC values (>100 g L^{-1}) not necessarily result in the highest suspended sediment transport into the lake. If a certain threshold is reached and the floodplain becomes flooded, the flow velocity decreases and suspended sediment accumulates in the floodplain (Figure S3). This process is known as catchment storage (e.g. Orwin *et al.*, 2010) and is observed for the very strong 06/13 flood which led to lower amounts of trapped sediments than, for example, the 06/12 flood with much lower SSC values (max. 20 g L^{-1}).

Contrary with factors reducing sediment input to the lake, we also found few cases when sediment deposition can be significantly increased by additional sediment releases from local sources. Evidence for mixing of flood-triggered sediments from different sources is revealed for the distal location which is located ca. 800 m in front of the secondary Kienbach creek inflow (Fig. 1). The development of an underflow originating from the Kienbach and its impact on sediment deposition at the distal site can be demonstrated for the June 2013 flood. Besides tracing slumps in the head-catchment and the deposition of an approximately 0·5 m thick sediment fan of sand and gravel in the delta area of the Kienbach creek (Figure S3), we measured a 0·2 K temperature increase at the lake bottom at the distal site even 1 h before the temperature rise of the deep water at the proximal site (Fig. 7). This underflow did not reach the proximal site and thus reflects a local event with short travel time reaching the distal site before the underflow triggered by the main tributary (Griesler Ache). The Kienbach creek as sediment source is confirmed by coarser grain sizes (Fig. 9) and predominant dolomitic composition of the trapped sediments (Table 2) which reflects the Kienbach catchment geology of the Northern Calcareous Alps. In the

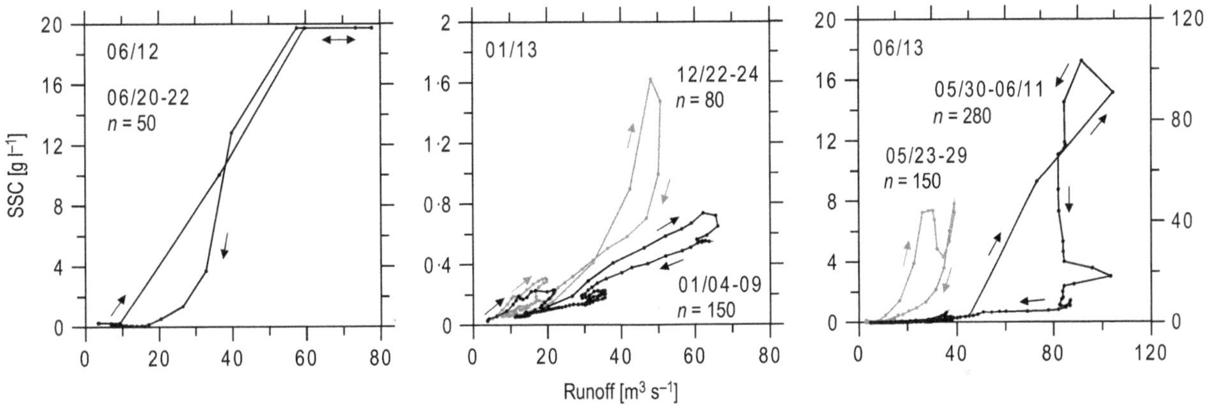

Fig. 10. River runoff – suspended sediment concentration (SSC) hysteresis plots for the three strongest recorded flood events in the Griesler Ache river (06/12, 01/13, 06/13). Note the different scaling of the y-axes. Dots mark single measurements in 1-hr intervals.

depositional record, local sediment flux from the Kienbach sub-catchment to the distal location has been suggested for four flood layers and for one debris flow layer (DL 16 in Fig. 11) deposited between 1976 and 2005 (Swierczynski *et al.*, 2009; Kämpf *et al.*, 2014).

In summary, the occurrence and thickness of detrital layers depends also on timing and preconditions of floods that might led to either an over- or an underestimation of the number of palaeofloods in the depositional record. On the one hand, detrital layers can be disproportionally thin or even miss when a flood followed a previous flood after a short time not sufficient to refill the riverbed with sediment or when catchment erosion is reduced (Czymzik *et al.*, 2010; Kämpf *et al.*, 2012b). On the other hand, additional sediment release from local sediment sources can result in the formation of non-flood triggered detrital layers or increase flood layer thickness (Girardclos *et al.*, 2007; Dugan *et al.*, 2009; Swierczynski *et al.*, 2009; Czymzik *et al.*, 2010). However, the potential bias by such local layers can be reduced by detailed micro-facies and geochemical analyses of individual detrital layers (Mangili *et al.*, 2005; Swierczynski *et al.*, 2009) and tracing detrital layers in multiple sediment cores (Lamoureux, 1999; Schiefer *et al.*, 2006, 2011; Kämpf *et al.*, 2012b, 2014).

Lake internal sediment distribution

The different spatial sediment distribution in the lake basin of individual events (PDI = 2 to 25) is related to variations in the formation of hyperpycnal underflows and mesopycnal interflows (Sturm & Matter, 1978; Chapron *et al.*, 2005). Sudden temperature increases of 2 to 3·5 K recorded in the deep water column during the largest summer floods proved the occurrence of underflows (Fig. 7). In winter, temperature fluctuations were less distinct or even negative (−0·4 K to +0·1 K in Fig. 8) due to the lower river water temperature. Underflows developed during the four strongest floods, but reached the distal site only during the 06/13 event 7 h after passing the proximal site. From this time lag, a minimum current speed of 9 cm s^{-1} can be calculated which is rather low compared to larger lakes where much stronger underflows were measured like, for example, lakes Walensee (20 to 50 cm s^{-1}), Geneva (30 to 90 cm s^{-1}) and Constance (30 to 130 cm s^{-1}) (Lambert *et al.*, 1976; Lambert & Giovanoli, 1988). The resulting lower sediment transport capacity of underflows in Lake Mondsee is proven also by the measured vertical differences in sediment flux in the proximal ($L : U_{prox.} = 2$ to 8) and distal locations ($L : U_{dist.} = 1$ to 2) indicating underflows as the predominant transport mechanism in the proximal and interflows in the distal lake basin.

The rather low number of underflows reaching the distal site compared to observations from other lakes (Schiefer *et al.*, 2006; Cockburn & Lamoureux, 2008; Jenny *et al.*, 2014) can be explained by the specific basin morphometry of Lake Mondsee (Fig. 1). At the point where the main tributary, the Griesler Ache, enters the lake, the elongated shape of the basin turns from NW-SE to N-S direction and later back to the original NW-SE direction. Due to this pronounced kink, the inflowing waters from the Griesler Ache are not straightaway directed towards the outflow but deflected first to the South and then to the East. Consequently, the flow velocity is expected to decrease, resulting in a lower sediment transport capacity. Lakes with frequent occurrence of underflows commonly have one large tributary stream and a simple elongated basin shape with straight flow direction from the main tributary inflow towards the outflow (Lambert *et al.*, 1976; Best *et al.*, 2005; Crookshanks & Gilbert, 2008; Jenny *et al.*, 2014).

The observed processes of detrital sediment distribution within Lake Mondsee provide explanations for the

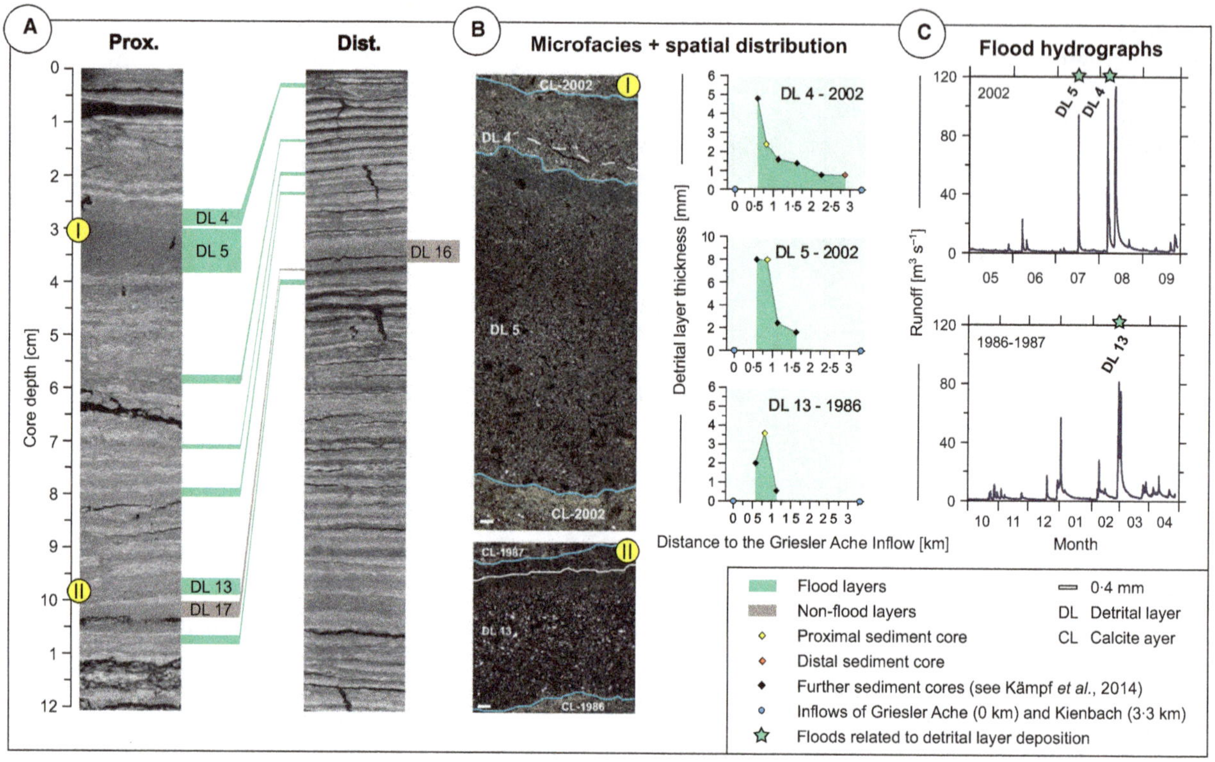

Fig. 11. Detrital layers in the varved sediment record of Lake Mondsee: (A) Thin section scans (polarized light) of the upper 12 cm of sediment cores close to the proximal and distal trap location (Fig. 1) and correlation of detrital layers, (B) microfacies of selected detrital layers (microscopic images under polarized light, 20× magnification) showing graded layers in the proximal core and different spatial distribution patterns, (C) seasonal flood hydrographs (hourly values) of the Griesler Ache River coincided to detrital layer deposition according to the seasonal resolution of the varved detrital layer record: summer = May-Sep and winter = Oct-Apr.

occurrence or absence of detrital layers in the depositional record obtained at the distal location (Swierczynski *et al.*, 2013). There, flood layers are mainly transported by interflows and appear as very fine, only microscopically detectable (0·2 to 0·8 mm), non-graded silt/clay layers, whereas graded layers triggered by underflows only occur at the proximal site in an area within 1·5 km around the river mouths (Fig. 11). Importantly, interflows are favoured in the summer season when the water column is stratified and a pycnocline develops along which sediment is transported. In contrast, even high amplitude winter floods supplied only small sediment amounts to the distal lake basin like the January 2013 flood; the sediment deposition (245 g m^{-2}) most likely was not sufficient to form a discernible detrital layer. This explains why the flood record from the distal site has been regarded as summer time series (Swierczynski *et al.*, 2012). To obtain a flood record including both summer and winter floods a coring site at the proximal position should be selected.

Another factor controlling the spatial distribution of detrital sediment within the lake basin is the flood duration. This can be demonstrated for sediment deposition during the short 06/12 event (74 mm rainfall in 4 h, Table 1) with a focal area at the proximal location since the under- and interflows broke down quickly when river discharge dropped and thus did not reach the distal basin as reflected by low sedimentation rates in the distal trap (Fig. 7). In contrast, the 06/13 flood lasted for 9 days resulting in strong and long lasting interflows as reflected by elevated sediment deposition at the distal trap over 10 days (Fig. 7).

Reconstructing palaeoflood time series in particular from distal sediment records, therefore, needs to concern not only about peak discharges but also the duration of floods. Strong but short floods might not reach the distal location but result in thick detrital layers in proximal cores (e.g. DL 5 in Fig. 11; associated peak runoff: 100 $m^3\ s^{-1}$, daily mean: 24 $m^3\ s^{-1}$). For the depositional record this means that (i) daily instead of hourly runoff values are more appropriate for defining discharge thresholds for detrital layer formation and (ii) distal coring locations are especially suitable for reconstructing large-scale rather than local flood events.

Implications for the use of detrital layers as flood proxies

Our monitoring data provide clues for both, (i) reliable interpretation of detrital layers as flood proxies and (ii) selecting suitable coring locations for palaeoflood reconstruction.

The monitoring data reveals that thresholds in discharge exist, above which flood layer formation becomes very likely. Obviously, these thresholds depend on the coring location and most likely vary also for different lakes (Czymzik *et al.*, 2010; Kämpf *et al.*, 2012b; Corella *et al.*, 2014; Jenny *et al.*, 2014). However, the monitoring and its comparison with the sediment record demonstrate that these thresholds can be empirically determined and thus provide information about the minimum amplitude of floods recorded in depositional records.

Reconstructing individual flood amplitudes from detrital layer thickness, however, still remains challenging. For Lake Mondsee with a complex lake basin morphometry, a vegetated catchment and sediments predominantly produced in the water column through biological productivity, this approach failed because several other factors including sediment availability, local sediment supply and lake internal sediment distribution significantly bias magnitude estimates of palaeofloods. In contrast, for small proglacial and nival lakes with one main tributary inflow and predominantly clastic-detrital sedimentation a correlation between peak runoff and sediment flux has been demonstrated (Desloges & Gilbert, 1994; Chutko & Lamoureux, 2008; Schiefer *et al.*, 2011). More lakes must be investigated to find out if there is a systematic difference between lakes with different settings and sediment types.

Detrital layer successions in depositional records should not per se be considered as complete flood time series. We have demonstrated that there might be additional layers due to local sediment transport but also 'missing' layers, i.e. floods that did not result in detrital layer formation. This can be either due to random or systematic processes, like the lack of winter flood layers at the distal location in Lake Mondsee. Both, additional, non-flood triggered, and missing layers represent a potential and assumedly site specific bias in flood reconstruction. However, we have also demonstrated that it is possible to reduce the potential bias through (i) detailed micro-facies and geochemical analyses and (ii) a careful choice of coring sites ideally integrating proximal and distal locations (Czymzik *et al.*, 2010; Schiefer *et al.*, 2011; Jenny *et al.*, 2014; Kämpf *et al.*, 2014).

Although flood reconstructions are conducted to a wide range of lakes, varved sediment records provide particularly suitable archives for process studies to develop quantitative flood proxies in terms of their hydrological characteristics.

CONCLUSIONS

The 3-year integrated lake and catchment flood monitoring at Lake Mondsee revealed detailed insights into the complex chain of processes leading to flood layer formation. In particular, reasons for variable sediment flux at the lake floor could be identified including hydrological factors such as runoff magnitude and duration, local geomorphological factors influencing sediment availability in the catchment and factors controlling the spatial distribution of detrital material within the lake. This knowledge has implications for the long flood record established at the distal lake basin of Lake Mondsee:

1. Threshold processes in the runoff – sediment flux relation define flood magnitudes above which suspended matter is transported and detrital layers are formed. The long Lake Mondsee flood layer chronology records all floods exceeding 80 m^3 s^{-1} river discharge and lasting for at least 2 days.
2. The depositional record at the distal site represents mainly summer floods due to the seasonally favoured development of mesopycnal interflows, the main transport agent for lake internal sediment distribution. Winter flood layers, deposited by hyperpycnal underflows, can be only found at core locations close to the river inflow.
3. Reconstruction of flood amplitudes by layer thickness is limited because the sediment yield after floods is not linearly related to runoff but is additionally affected by various other geomorphological and lake internal factors.
4. The potential bias of detrital layers triggered by local erosion events rather than by high magnitude floods can be reduced by detailed micro-facies and high-resolution geochemical analyses.

It has been shown that comparison of sediment monitoring with sub-recent detrital layer records provides valuable information for an advanced interpretation of detrital layers as flood recorders and for systematic selection of the most suitable coring location within a lake basin. Even if our dataset obviously is to a certain degree site specific for Lake Mondsee, we expect the fundamental mechanisms and controlling processes as valid also for many other lake sites.

ACKNOWLEDGEMENTS

This study contributes to the Potsdam Research Cluster for Georisk Analysis, Environmental Change and Sustainability (PROGRESS) part A.3 'Extreme events in geoarchives' funded by the German Federal Ministry for Education and Research (BMBF). We are especially grateful to Prof. Rainer Kurmayer and Prof. Thomas Weisse (both University of Innsbruck, Research Institute for Limnology Mondsee) for their support and scientific collaboration as well as to Kurt Mayrhofer (University of Innsbruck, Research Institute for Limnology Mondsee), Richard Niederreiter (UWITEC) and numerous students for technical support and help during field work. We further thank Sebastian Lorenz (University of Greifswald) for his help during grain size measurements, Petra Meyer (GFZ Potsdam) for elemental analyses as well as Georg Schettler and Jens Mingram (both GFZ Potsdam) for useful advices regarding the monitoring set-up and lab analyses. Andreas Hendrich (GFZ Potsdam) is acknowledged for his help with illustrating the figures. We thank two anonymous reviewers for constructive comments which helped to improve the manuscript.

References

Arnaud, F., Revel, M., Chapron, E., Desmet, M. and Tribovillard, N. (2005) 7200 years of Rhône river flooding activity in Lake Le Bourget, France: a high-resolution sediment record of NW Alps hydrology. Holocene, **15**, 420–428.

Best, J.L., Kostaschuk, R.A., Peakall, J., Villard, P.V. and Franklin, M. (2005) Whole flow field dynamics and velocity pulsing within natural sediment-laden underflows. Geology, **33**, 765.

Blöschl, G., Nester, T., Komma, J., Parajka, J. and Perdigão, R.A.P. (2013) The June 2013 flood in the Upper Danube basin, and comparisons with the 2002, 1954 and 1899 floods. Hydrol. Earth Syst. Sci., **17**, 5197–5212.

BMLFUW (2011) Hydrographisches Jahrbuch von Österreich 2011. BMLFUW, Vienna, 967 pp.

Chapron, E., Arnaud, F., Noël, H., Revel, M., Desmet, M. and Perdereau, L. (2005) Rhone River flood deposits in Lake Le Bourget: a proxy for Holocene environmental changes in the NW Alps, France. Boreas, **34**, 404–416.

Chutko, K.J. and Lamoureux, S.F. (2008) Identification of coherent links between interannual sedimentary structures and daily meteorological observations in Arctic proglacial lacustrine varves: potentials and limitations. Can. J. Earth Sci., **45**, 1–13.

Cockburn, J.M.H. and Lamoureux, S.F. (2008) Inflow and lake controls on short-term mass accumulation and sedimentary particle size in a High Arctic lake: implications for interpreting varved lacustrine sedimentary records. J. Paleolimnol., **40**, 923–942.

Corella, J.P., Benito, G., Rodriguez-Lloveras, X., Brauer, A. and Valero-Garcés, B.L. (2014) Annually-resolved lake record of extreme hydro-meteorological events since AD 1347 in NE Iberian Peninsula. Quatern. Sci. Rev., **93**, 77–90.

Crookshanks, S. and Gilbert, R. (2008) Continuous, diurnally fluctuating turbidity currents in Kluane Lake, Yukon Territory. Can. J. Earth Sci., **45**, 1123–1138.

Czymzik, M., Dulski, P., Plessen, B., von Grafenstein, U., Naumann, R. and Brauer, A. (2010) A 450 year record of spring-summer flood layers in annually laminated sediments from Lake Ammersee (southern Germany). Water Resour. Res., **46**, W11528.

Czymzik, M., Brauer, A., Dulski, P., Plessen, B., Naumann, R., von Grafenstein, U. and Scheffler, R. (2013) Orbital and solar forcing of shifts in Mid- to Late Holocene flood intensity from varved sediments of pre-alpine Lake Ammersee (southern Germany). Quatern. Sci. Rev., **61**, 96–110.

Desloges, J.R. and Gilbert, R. (1994) Sediment source and hydroclimatic inferences from glacial lake sediments: the postglacial sedimentary record of Lillooet Lake, British Columbia. J. Hydrol., **159**, 375–393.

Dokulil, M. and Skolaut, C. (1986) Succession of phytoplankton in a deep stratifying lake: Mondsee, Austria. Hydrobiologica, **138**, 9–24.

Dugan, H.A., Lamoureux, S.F., Lafrenière, M.J. and Lewis, T. (2009) Hydrological and sediment yield response to summer rainfall in a small high Arctic watershed. Hydrol. Process., **23**, 1514–1526.

Eybl, J., Godina, R., Lalk, P., Lorenz, P., Müller, G., Pavlik, H., Weilguni, V. and Heilig, M. (2013) Hochwasser im Juni 2013 – Die hydrographische Analyse. BMLFUW, Vienna, Austria.

Forbes, A. and Lamoureux, S. (2005) Climatic controls on streamflow and suspended sediment transport in three large Middle Arctic Catchments, Boothia Peninsula, Nunavut, Canada. Arct. Antarct. Alpine Res., **37**, 304–315.

Francus, P., Bradley, R.S., Abbott, M.B., Patridge, W. and Keimig, F. (2002) Paleoclimate studies of minerogenic sediments using annually resolved textural parameters. Geophys. Res. Lett., **29**, 1998.

Gilbert, R., Crookshanks, S., Hodder, K.R., Spagnol, J. and Stull, R.B. (2006) The record of an extreme flood in the sediments of Montane Lillooet Lake, British Columbia: implications for paleoenvironmental assessment. J. Paleolimnol., **35**, 737–745.

Girardclos, S., Schmidt, O.T., Sturm, M., Ariztegui, D., Pugin, A. and Anselmetti, F.S. (2007) The 1996 AD delta collapse and large turbidite in Lake Brienz. Mar. Geol., **241**, 137–154.

Glur, L., Wirth, S.B., Büntgen, U., Gilli, A., Haug, G.H., Schär, C., Beer, J. and Anselmetti, F.S. (2013) Frequent floods in the European Alps coincide with cooler periods of the past 2500 years. Sci. Rep., **3**, 2770.

Hsü, K.J. and Kelts, K. (1985) Swiss lakes as a geological laboratory. Naturwissenschaften, **72**, 315–321.

**Jenny, J.-P., Wilhelm, B., Arnaud, F., Sabatier, P., Giguet Covex, C., Mélo, A., Fanget, B., Malet, E., Ployon, E. and

Perga, M.E. (2014) A 4D sedimentological approach to reconstructing the flood frequency and intensity of the Rhône River (Lake Bourget, NW European Alps). J. Paleolimnol., **51**, 469–483.

Kämpf, L., Brauer, A., Dulski, P., Feger, K., Jacob, F. and Klemt, E. (2012a) Sediment imprint of the severe 2002 summer flood in the Lehnmühle reservoir, eastern Erzgebirge (Germany). E&G Quatern. Sci. J., **61**, 3–15.

Kämpf, L., Brauer, A., Dulski, P., Lami, A., Marchetto, A., Gerli, S., Ambrosetti, W. and Guilizzoni, P. (2012b) Detrital layers marking flood events in recent sediments of Lago Maggiore (N. Italy) and their comparison with instrumental data. Freshwat. Biol., **57**, 2076–2090.

Kämpf, L., Brauer, A., Swierczynski, T., Czymzik, M., Mueller, P. and Dulski, P. (2014) Processes of flood-triggered detrital layer deposition in the varved Lake Mondsee sediment record revealed by a dual calibration approach. J. Quatern. Sci., **29**, 475–486.

Koschel, R., Brenndorf, J., Proft, G. and Recknagel, R. (1983) Calcite precipitation as a natural mechanism of eutrophication. Arch. Hydrobiol., **98**, 380–408.

Lambert, A. and Giovanoli, F. (1988) Records of riverbome turbidity currents and indications of slope failures in the Rhone delta of Lake Geneva. Limnol. Oceanogr., **33**, 458–468.

Lambert, A., Kelts, K. and Marshall, N.F. (1976) Measurements of density underflows from Walensee, Switzerland. Sedimentology, **23**, 87–105.

Lamoureux, S. (1999) Spatial and interannual variations in sedimentation patterns recorded in nonglacial varved sediments from the Canadian High Arctic. J. Paleolimnol., **21**, 73–84.

Lamoureux, S.F., Stewart, K.A., Forbes, A.C. and Fortin, D. (2006) Multidecadal variations and decline in spring discharge in the Canadian middle Arctic since 1550 AD. Geophys. Res. Lett., **33**, L02403.

Lapointe, F., Francus, P., Lamoureux, S.F., Saïd, M. and Cuven, S. (2012) 1750 years of large rainfall events inferred from particle size at East Lake, Cape Bounty, Melville Island, Canada. J. Paleolimnol., **48**, 159–173.

Lauterbach, S., Brauer, A., Andersen, N., Danielopol, D.L., Dulski, P., Hüls, M., Milecka, K., Namiotko, T., Obremska, M., von Grafenstein, U. and Participants, D. (2011) Environmental responses to Lateglacial climatic fluctuations recorded in the sediments of pre-Alpine Lake Mondsee (northeastern Alps). J. Quatern. Sci., **26**, 253–267.

Lewis, J. (1996) Turbidity-controlled suspended sediment sampling for runoff-event load estimation. Water Resour. Res., **32**, 2299–2310.

Lewis, J. and Eads, R. (2009) Implementation Guide for Turbidity Threshold Sampling: Principles, Procedures, and Analysis. USDA.

López-Tarazón, J.A., Batalla, R.J., Vericat, D. and Balasch, J.C. (2010) Rainfall, runoff and sediment transport relations in a mesoscale mountainous catchment: The River Isábena (Ebro basin). Catena, **82**, 23–34.

Mangili, C., Brauer, A., Moscariello, A. and Naumann, R. (2005) Microfacies of detrital event layers deposited in Quaternary varved lake sediments of the Pianico-Sellere Basin (northern Italy). Sedimentology, **52**, 927–943.

Meyers, P.A. and Teranes, J.L. (2001) Sediment organic matter. In: Tracking Environmental Change Using Lake Sediments. Volume 2: Physical and Geochemical Methods (Eds W.M. Last and J.P. Smol), pp. 239–269. Kluwer Academic Publishers, Dordrecht, The Netherlands.

Mueller, P., Thoss, H., Kaempf, L. and Güntner, A. (2013) A buoy for continuous monitoring of suspended sediment dynamics. Sensors, **13**, 13779–13801.

Orwin, J.F., Lamoureux, S.F., Warburton, J. and Beylich, A. (2010) A framework for characterizing fluvial sediment fluxes from source to sink in cold environments. Geogr. Ann., **92**, 155–176.

Schiefer, E., Menounos, B. and Slaymaker, O. (2006) Extreme sediment delivery events recorded in the contemporary sediment record of a montane lake, southern Coast Mountains, British Columbia. Can. J. Earth Sci., **43**, 1777–1790.

Schiefer, E., Gilbert, R. and Hassan, M.A. (2011) A lake sediment-based proxy of floods in the Rocky Mountain Front Ranges, Canada. J. Paleolimnol., **45**, 137–149.

Schlolaut, G., Brauer, A., Marshall, M.H., Nakagawa, T., Staff, R.A., Bronk Ramsey, C., Lamb, H.F., Bryant, C.L., Naumann, R., Dulski, P., Brock, F., Yokoyama, Y., Tada, R. and Haraguchi, T. (2014) Event layers in the Japanese Lake Suigetsu "SG06" sediment core: description, interpretation and climatic implications. Quatern. Sci. Rev., **83**, 157–170.

Siegenthaler, C. and Sturm, M. (1991) Die Häufigkeit von Ablagerungen extremer Reuss-Hochwasser : Die Sedimentationsgeschichte im Urnersee seit dem Mittelalter. Mitt. Bundesamt für Wasserwirtschaft, **4**, 127–139.

Støren, E.N., Dahl, S.O., Nesje, A. and Paasche, Ø. (2010) Identifying the sedimentary imprint of high-frequency Holocene river floods in lake sediments: development and application of a new method. Quatern. Sci. Rev., **29**, 3021–3033.

Sturm, M. and Matter, A. (1978) Turbidites and varves in Lake Brienz (Switzerland): deposition of clastic detritus by density currents. Int. Assoc. Sedimentol. Spec. Publ., **2**, 147–178.

Swierczynski, T., Lauterbach, S., Dulski, P. and Brauer, A. (2009) Die Sedimentablagerungen des Mondsees (Oberösterreich) als ein Archiv extremer Abflussereignisse der letzten 100 Jahre. In: Klimawandel in Österreich – Die letzten 20.000 Jahre…und ein Blick voraus (Eds R. Schmidt, C. Matulla and R. Psenner), pp. 115–126. Innsbruck University Press, Innsbruck, Austria.

Swierczynski, T., Brauer, A., Lauterbach, S., Martín-Puertas, C., Dulski, P., von Grafenstein, U. and Rohr, C. (2012)

A 1600 yr seasonally resolved record of decadal-scale flood variability from the Austrian Pre-Alps. Geology, **40**, 1047–1050.

Swierczynski, T., Lauterbach, S., Dulski, P., Delgado, J., Merz, B. and Brauer, A. (2013) Mid- to late Holocene flood frequency changes in the northeastern Alps as recorded in varved sediments of Lake Mondsee (Upper Austria). Quatern. Sci. Rev., **80**, 78–90.

Van Husen, D. (1989) Blatt 65–Mondsee. Geol. Karte der Republik Österreich 150 000

Wilhelm, B., Arnaud, F., Sabatier, P., Magand, O., Chapron, E., Courp, T., Tachikawa, K., Fanget, B., Malet, E., Pignol, C., Bard, E. and Delannoy, J.J. (2013) Palaeoflood activity and climate change over the last 1400 years recorded by lake sediments in the north-west European Alps. J. Quatern. Sci., **28**, 189–199.

Wirth, S.B., Gilli, A., Simonneau, A., Ariztegui, D., Vannière, B., Glur, L., Chapron, E., Magny, M. and Anselmetti, F.S. (2013a) A 2000 year long seasonal record of floods in the southern European Alps. Geophys. Res. Lett., **40**, 4025–4029.

Wirth, S.B., Glur, L., Gilli, A. and Anselmetti, F.S. (2013b) Holocene flood frequency across the Central Alps – solar forcing and evidence for variations in North Atlantic atmospheric circulation. Quatern. Sci. Rev., **80**, 112–128.

Supporting Information

Additional Supporting Information may be found in the online version of this article:

Figure S1. Mineralogical composition of river bed samples (blue arrow) and detrital layers in sediment cores (red arrow).

Figure S2. Calibration of turbidity of the Griesler Ache River recorded at St. Lorenz gauging station against suspended sediment concentration (SSC) derived from automatically taken water samples for the three flood events 06/12, 01/13 and 06/13.

Figure S3. Photographs documenting fluvial and in-lake sediment transport: (A) 06/13 event: channel bank erosion at Fischbach (SWB station), (B) 06/13 event: slump in the Kienbach valley, (C) 06/13 event: deposition of a ca. 0·5 m thick fan of sand and gravel in the Kienbach delta (A to C: photographs taken on 12 June 2013), (D) 01/11 event: Griesler Ache at St. Lorenz gauging station (14 January 2011 09:40 a.m.; runoff = 48 $m^3 s^{-1}$), (E) 06/13 event: Griesler Ache close to St. Lorenz gauging station (02 June 2013 06:00 p.m.; runoff = 83 $m^3 s^{-1}$, SSC = 7 $g L^{-1}$), (C) 06/12 event: Griesler Ache inflow into Lake Mondsee; (21 June 2012 12:10 p.m., runoff = 10 $m^3 s^{-1}$, SSC = 0·1 $g L^{-1}$).

7

Diagenetic incorporation of Sr into aragonitic bivalve shells: implications for chronostratigraphic and palaeoenvironmental interpretations

MARIA C. MARCANO*, TRACY D. FRANK†, SAMUEL B. MUKASA‡, KYGER C LOHMANN* and MARCO TAVIANI§,¶

*Department of Earth and Environmental Sciences, University of Michigan, 2534 C. C. Little Building, 1100 North University Ave., Ann Arbor, MI 48109-1005, USA

†Department of Earth and Atmospheric Sciences, University of Nebraska-Lincoln, 214 Bessey Hall, Lincoln, NE 68588-0340, USA

‡College of Engineering and Physical Sciences, University of New Hampshire, Kingsbury Hall W289, 33 Academic Way, Durham, NH 03824-3591, USA

§Istituto di Scienze Marine (ISMAR) – CNR, Via Gobetti 101, 40129, Bologna Italy

¶Biology Department, Woods Hole Oceanographic Institution, 266 Woods Hole Road, Woods Hole, MA 02543, USA

Keywords
Aragonite, bivalvia, diagenesis, Miocene, $^{18}O/^{16}O$, porewater, $^{87}Sr/^{86}Sr$.

ABSTRACT

Aragonite is easily altered during diagenesis, therefore presumed pristine when present. In effect, beyond polymorphic transformation to calcite, alteration paths of aragonite remain poorly understood despite heavy reliance on such material to produce palaeoenvironmental and chronostratigraphic interpretations. Previous work on core material from Southern McMurdo Sound, Antarctica, showed that unlike their calcitic counterparts, seemingly unaltered aragonite shell fragments invariably produced older than expected $^{87}Sr/^{86}Sr$ ages. In this study, we pursued additional analyses of these aragonite shells and of the porewater of the core to understand this discrepancy. Aragonite mineralogy was reconfirmed and elemental mapping of shell fragments revealed growth lines within the middle layer suggestive of good preservation. The outer layer, however, showed anomalously high Sr concentrations (average $4{\cdot}5 \pm 0{\cdot}6$ mole% $SrCO_3$; *ca* 25 mmol mol^{-1} Sr/Ca) and was depleted in ^{18}O and ^{13}C compared to the middle layer, both features inconsistent with pristine material. The $\delta^{18}O$ values and Sr concentrations of the porewater were used to model outer layer compositions reasonably well. Coincidentally, porewater Sr isotope composition was in general agreement with the age model of the core only at the aragonite-bearing interval suggesting that Sr-isotopic disequilibrium between porewater and the carbonates was the rule rather than the exception in the core. The Sr isotope compositions of the aragonite shells are most likely the result of early diagenesis as suggested by the inconsistent O and C isotope compositions between shell layers and the anomalously high Sr concentrations. We conclude that knowledge of Sr concentration and distribution in shells is critical to determine the viability of Sr stratigraphy and the scale at which it may be applied. Reliance on traditional indicators of lack of alteration, such as cathodoluminescence, Mn-Fe concentration, and the presence of labile mineralogies to assert chronostratigraphic and palaeoenvironmental questions may produce erroneous conclusions due to obscurely altered material.

INTRODUCTION

The $^{87}Sr/^{86}Sr$ ratios of unaltered marine biogenic carbonates are routinely compared to the well-established secular variation of Sr isotopic composition of sea water (Hess *et al.*, 1986; Howarth & McArthur, 1997; McArthur *et al.*, 2001) to provide chronostratigraphic control. Because the residence time of Sr in sea water is considerably longer than the mixing time of the oceans and because Sr is not fractionated by near-surface physicochemical processes, Sr isotopic composition of sea water at any one time is homogeneous and thus reflects the balance between global rock distribution, rock-type proportion and weathering intensity (Graustein, 1989; Faure & Mensing, 2005). Given that carbonate secreting organisms do not fractionate Sr when they incorporate it in their skeletons, the $^{87}Sr/^{86}Sr$ ratio of the open oceanic sea water can be reconstructed through geological time using unaltered biogenic carbonate material. In this context, well-preserved samples can also be used to estimate Sr sources and to model past mixing and weathering rates, although these relationships are not straightforward and numerous components of the climate system appear to play different roles of variable importance at different points in time. For example, modelling suggests that most of the increase in sea water $^{87}Sr/^{86}Sr$ since the mid Pliocene can be explained solely by the relative rise in phyllosilicates weathering prompted by global cooling (Li *et al.*, 2007), while Kashiwagi *et al.* (2008) conclude that sea water $^{87}Sr/^{86}Sr$ cannot be used as a direct proxy for silicate weathering and atmospheric CO_2 decrease during the Cenozoic.

Strontium stratigraphy is not applicable to intervals where plateaus in the secular variation of radiogenic Sr prevent time discrimination, or to areas where large freshwater inputs can deviate the biogenic carbonate signal from the global averages (Ingram & Depaolo, 1993; Israelson & Buchardt, 1999; Sessa *et al.*, 2012). However, Sr isotope composition from well-preserved biogenic carbonates of neritic and bathyal marine environments, where conventional chronostratigraphic methods are problematic due to frequent hiatuses and reworking, can be helpful in establishing time constraints. That was an immediate analytical target offered by the several macrofossil-bearing intervals of core AND-2A recovered by the ANDRILL (ANtarctic Geological DRILLing) program during its second field campaign (i.e., the Southern McMurdo Sound (SMS) project) during the austral summer of 2007 (Fig. 1).

The multinational ANDRILL program is a joint effort to drill the margin of Antarctica in search of Cenozoic stratigraphic records to study variation in ice sheets and obtain a better understanding of polar climate evolution (Harwood *et al.*, 2005, 2006, 2009). Specific goals for the SMS project included: (i) improve chronostratigraphic control; (ii) document melt-water discharge events from the Dry Valleys of the TransAntarctic Mountains and (iii) evaluate the persistence of polar conditions in Antarctica over the past 15 Myr. Extracting this critical information on continental evolution from marginal sediments, however, is not without complications. The biogenic-rich intervals of the core were sampled in part to assess the aforementioned objectives using Sr and O isotopes.

The Sr isotope compositions of pristine marine samples are considered to accurately reflect precipitation age. Previous analyses of venerid and pectinid shells from the macrofossil-rich intervals of core AND-2A (Taviani *et al.*, 2009) were suggestive of minimal to no diagenetic alteration. The Sr isotope compositions of unaltered calcitic pectinid fragments provided additional chronostratigraphic control to the core (Acton *et al.*, 2008). In contrast with calcite fragments, Sr isotope compositions of the aragonitic venerid shells were lower than expected in all samples resulting in older than reasonable Sr-isotopic ages, while Sr concentrations of the aragonite shells were higher than those observed in modern bivalves (Marcano *et al.*, 2009).

Given the significance of chronostratigraphic and palaeoenvironmental interpretations reliant on the Sr content of carbonate samples, and considering that the aragonite anomalous Sr isotope compositions of samples from core AND-2A were not unequivocally explained using basic isotopic mass balance calculations, these samples were further explored here using more detailed techniques and evaluated against new Sr data from the porewater of the core. Unlike biogenic carbonates, interstitial fluids in AND-2A appeared to be highly modified as suggested by initial chemical analyses of porewater and fracture filling cements (Gui, 2009; Frank & Gui, 2010). Here, the Sr concentrations and isotope compositions of AND-2A porewater were explored to better understand the results from shell studies and more broadly assess the use of $^{87}Sr/^{86}Sr$ as a definite chronostratigraphic indicator from carbonates whose alteration has been ruled out using limited criteria.

Previous work: Sr compositions of biogenic material from other Antarctic cores

Several studies from previous Antarctic coring efforts produced Sr isotope ages consistent with independently determined age estimates. The Sr isotope ratios measured in aragonitic bivalves recovered from DSDP Site 270 in the southeastern portion of the Ross Sea and from CIROS-1, off Ferrar Glacier in McMurdo Sound (Barrera, 1989), produced ages consistent with biostratigraphic estimates. Similarly, several aragonitic bivalve samples from the Cape Roberts Project (CRP) cores, drilled about 65 km north of CIROS-1, produced Sr ages consistent with Ar isotope ages and diatom-based biostratigraphy (Lavelle, 1998, 2000, 2001). Most of these ages (8 out of 11) remain valid even after the age model of CRP-2/2A was astrochronologically adjusted across the Oligocene-Miocene boundary (Naish *et al.*, 2008).

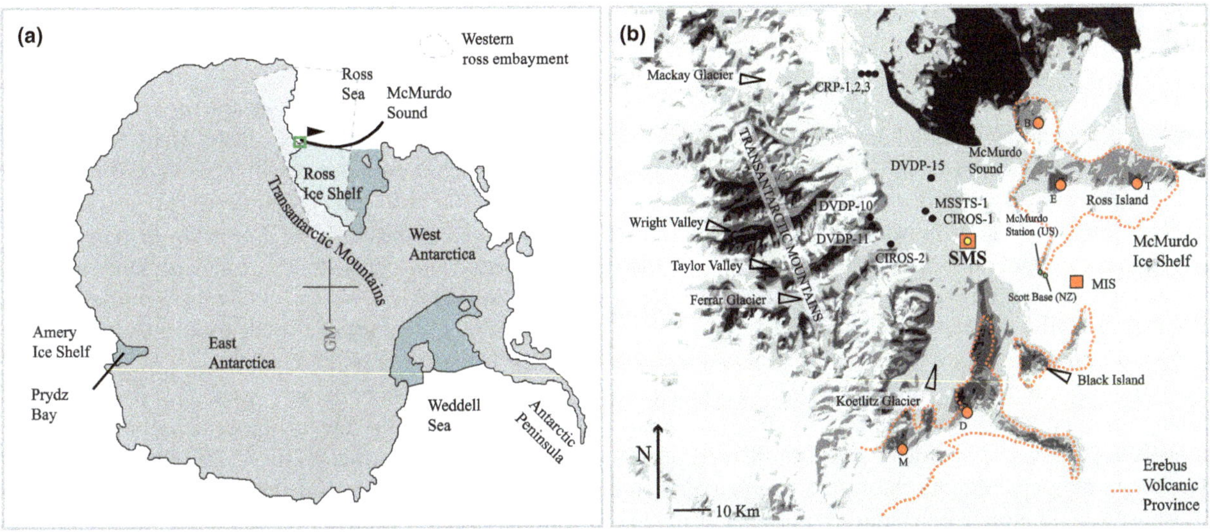

Fig. 1. Location map. (A) McMurdo Sound at the edge of the Ross Ice Shelf, Antarctica. (B) ANDRILL SMS (Southern McMurdo Sound) and MIS (McMurdo Ice Shelf) drill-holes location in Southern McMurdo Sound (squares). CRP – Cape Roberts Project; DVDP – Dry Valley Drilling Project; MSSTS – McMurdo Sound Sediment and Tectonic Study; CIROS – Cenozoic Investigations of the Ross Sea. Dotted line demarks the Erebus volcanic province; B – Mount Bird; E – Mount Erebus; T – Mount Terror; M – Mount Morning; D – Mount Discovery. Modified from Harwood *et al.* (2005).

Other studies have produced mixed results, with Sr isotope compositions inconsistent with stratigraphic age. For example, shells from the first ANDRILL core recovered in McMurdo Sound (AND-1B, MIS – McMurdo Ice Sheet Project, Naish *et al.*, 2007) yielded somewhat ambiguous Sr isotope ages (Wilson *et al.*, 2007). Within the uncertainty of shell fragment identification, the bivalves appeared to be consistently associated with lower than expected Sr isotope values, while other macrofossils (mainly foraminifera and cirripeda) produced ages closer to the age model of the core. Mineralogy of bivalve fragments was not specified, and therefore it is uncertain whether the older ages were associated with aragonitic bivalves only, as was the case in the AND-2A core. Wilson *et al.* (2007) concluded that the most likely cause of the age discrepancy was contamination by matrix sediments. However, results continued to be equivocal after several cycles of HCl etching of the shell fragments, suggesting that ambiguous Sr isotope compositions were probably intrinsic to the shell and not a product of contamination. Partial equilibration with porewaters was proposed by Wilson *et al.* (2007) as an alternate cause of the anomalous isotopic compositions, but this hypothesis was not tested.

The Sr isotope compositions of bivalve shells positively identified as aragonite reported from ODP Site 739 in Prydz Bay, Antarctica, also produced unexpected ages (Thierstein *et al.*, 1991). In this case, the Sr isotope ages were younger (latest Early Oligocene-earliest Miocene) than the aragonite shell-bearing stratigraphic units (uppermost Eocene-lowermost Oligocene). Porewaters from the same depths as the shell-bearing diamictites were analysed for Sr concentration and isotopic composition. The Sr concentrations were between 1·5 and 2 times higher than that of modern sea water, and Sr isotope ratios were among the most radiogenic ever recorded by the ODP and DSDP. Thierstein *et al.* (1991) calculated that measured aragonite Sr isotope compositions would require either a 90% contribution from freshwater to the depositional fluid or 10% to 17% post-depositional incorporation of Sr from the extant porewater. They also calculated that porewater compositions could be reasonably derived from altered continental detritus.

Approaches used in the studies mentioned above were not consistent. For example, identification, description and criteria to rule out alteration or transport varied widely. This may in part explain the disparity of results. In any case, the apparent pristine aragonite of the macrofossil shells remained puzzling and required further investigation.

MATERIALS AND METHODS

Previous analyses to identify and discriminate alteration effects on these and other selected macrofossil fragments are described in Marcano *et al.* (2009). Here, we focus on AND-2A venerid aragonitic fragments recovered between 429·28 m and 430·51 m below seafloor (mbsf) identified as in-house samples number 9, 10, 11-1, 15-2 and 15-3. These fragments belong to a venerid bivalve recently shown to be a species new to Science and described by Beu & Taviani (2014) as *Retrotapes andrillorum*. The outer and middle layer could be clearly recognized in most fragments. Marcano *et al.* (2009) sampled the outer and middle layers separately for chemical analyses, but relatively large sample sizes were required for conventional powder X-ray diffraction technique. Results could not account for the presence of small, localzed alteration products or resolve differences among layers.

Powder X-ray microdiffraction (PXRD) was used in this study to reconfirm mineralogy without sample homogenization. PXRD patterns were collected on a Rigaku R-Axis Spider diffractometer with an imaging plate detector using graphite monochromated Cu-Kα radiation (1·5406 Å) at ambient temperature. Whole fragments of venerid shell exhibiting both outer and middle layers were used. The sample was mounted on a cryoloop with paratone N oil for analysis, and the diffractogram obtained from sections of samples where both layers occupied approximate equal volumes. To avoid preferred orientation, images were collected for 5 min while rotating the sample about the φ-axis at $10°s^{-1}$ and oscillating ω between 120° and 180° at $1°s^{-1}$, with ψ set at 45° (see additional online material for an illustration of the sample orientation). These were integrated with a 0·05° step size with the AreaMax2 software package.

The aragonite shell fragments were further sampled from the outer and middle layer to measure oxygen and carbon isotope composition and supplement previous published values (Marcano *et al.*, 2009). Separates were roasted in vacuo, at 200°C, to eliminate volatile contaminants. Oxygen and carbon isotope ratios were determined using an automated Kiel IV device coupled to a triple-collector gas source Finnigan MAT 253 isotope-ratio mass spectrometer and reported against the VPDB (Vienna PeeDee Belemnite) standard. Standard deviations for both carbon and oxygen are equal to or better than 0·1‰.

Electron microprobe elemental mapping was done to localize areas of anomalous Sr concentration in the shell. The X-ray mapping was carried out on a cross section of a shell fragment (sample 10) using wavelength-dispersive (WLD) spectrometry in a Cameca SX-100 electron probe microanalyser (EPMA). Data were collected from three spectrometers with a detection limit of 270 p.p.m. To avoid precision limitations imposed by the analytical time necessary to quantify the entire Sr map, concentrations were measured along three transects, each 270 μm in length along the cross-section of the fragment. Beam diameter was adjusted to 5, 2 and <0·5 μm with each step size varying from 5 to 2 μm, producing a total of 245 measurements. Secondary electron images were also acquired from the mapped areas and for the rest of the shell fragments, which showed cleaned, unaltered prismatic material.

Porewater chemistry was measured on-ice and discussed elsewhere (Panter *et al.*, 2008). Here, Sr concentrations of porewaters were determined on a Perkin Elmer Optima 3300 DV inductively coupled plasma-optical emission spectrometer (ICP-OES) using 10-point calibration curves. One High-Purity® standard solution (Trace Metals in Drinking Water) and one in-house standard indicated that accuracy of the chemical analyses was ±5% RSD or better for Sr and Ca.

The Sr concentrations were used to determine initial porewater sample size for Sr isotope determinations. After complete porewater evaporation 2·5N HCl was added to the residues. Strontium was then separated using column chromatography (Mukasa *et al.*, 1991). Samples were dried to a solid, treated with a drop of 14N HNO_3, re-dried, and loaded on a single Re filament. The Sr was loaded with 0·1 vol% H_3PO_4 and $TaClO_4$ solution. The $^{87}Sr/^{86}Sr$ measurements were done on a VG Sector multi-collector thermal ionization mass spectrometer (TIMS). The Sr isotope composition was corrected for mass fractionation using $^{86}Sr/^{88}Sr = 0{\cdot}1194$. The repeated analyses of NBS-987 standard ($n = 3$) gave an average $^{87}Sr/^{86}Sr = 0{\cdot}710252 \pm 10 \times 10^{-6}$. Total blanks averaged 0·35 ng for Sr, which are negligible (Table 1).

All analyses described above were carried out in the Chemistry Department and the Earth and Environmental Sciences Department of the University of Michigan in Ann Arbor, Michigan.

RESULTS

Powder patterns from the PXRD patterns were processed in Jade Plus3 to calculate peak positions and intensities. The Jade software package developed by Materials Data Inc. (MDI) supports and provides access to comprehensive XRD databases such as the Powder Diffraction Files (PDF 2 and 4), produced by the International Center for Diffraction Data (ICDD). The suggested match was exclusively aragonite, with both layers contributing equally to the result.

Electron microprobe elemental mapping identified areas of anomalous Sr concentration in the shell (Fig. 2; see additional online material for a complete list of the results). The backscattered electron (BSE) image suggested that the mineralogy of the middle and outer layers of the venerid fragment was uniform (i.e., no high-Sr alteration phase). Instead of occurring in isolated phases, Sr appeared highly and evenly concentrated in the prismatic outer layer of the shell and along well-defined lines in the crystallographic homogeneous middle layer. Samples

Table 1. Sr and Ca concentrations and Sr isotope compositions of AND-2A porewater.

In-house ID		Depth (mbsf)	p.p.m.* Ca	Sr	mmol mol^{-1} Sr/Ca	$^{87}Sr/^{86}Sr$	Error
PW	1	9·67 to 9·72	590·707	10·014	7·75	0·70807	$1{\cdot}1 \times 10^{-5}$
PW	2-1	30·09 to 30·15	477·209	4·172	4·00	0·70987	$1{\cdot}0 \times 10^{-5}$
PW	3-1	37·41 to 37·46	342·930	3·041	4·06	0·70996	$1{\cdot}0 \times 10^{-5}$
PW	4	43·72 to 43·77	187·016	1·624	3·97	0·71003	$1{\cdot}1 \times 10^{-5}$
PW	5	51·30 to 51·35	174·668	2·092	5·48	0·71018	$1{\cdot}7 \times 10^{-5}$
PW	6	57·21 to 57·26	181·939	1·495	3·76	0·71012	$1{\cdot}5 \times 10^{-5}$
PW	7	62·66 to 62·71	289·175	5·026	7·95	—	—
PW	8	81·03 to 81·08	—	—	—	0·71022	$1{\cdot}0 \times 10^{-5}$
PW	9	91·97 to 93·03	—	—	—	0·71017	$1{\cdot}0 \times 10^{-5}$
PW	10-1	92·97 to 93·03	268·230	3·465	5·91	0·71009	$1{\cdot}3 \times 10^{-5}$
PW	11	116·22 to 116·27	805·600	10·752	6·11	0·71001	$1{\cdot}0 \times 10^{-5}$
PW	12	155·76 to 155·81	—	—	—	0·70963	$1{\cdot}0 \times 10^{-5}$
PW	13	235·66 to 235·76	—	—	—	0·70889	$1{\cdot}0 \times 10^{-5}$
PW	14-1	336·18 to 336·28	1973·065	31·512	7·31	0·70872	$1{\cdot}2 \times 10^{-5}$
PW	15	353·53 to 353·63	2017·073	35·442	8·04	0·70870	$1{\cdot}1 \times 10^{-5}$
PW	16	545·01 to 545·11	1604·868	49·121	14·00	0·70842	$1{\cdot}0 \times 10^{-5}$
PW	17	619·35 to 619·45	3154·357	61·999	8·99	0·70855	$1{\cdot}0 \times 10^{-5}$
PW	18	779·69 to 779·79	4136·569	84·366	9·33	0·70692	$1{\cdot}0 \times 10^{-5}$
PW	19	109·84 to 809·94	—	—	—	0·70711	$1{\cdot}0 \times 10^{-5}$
PW	20	963·44 to 963·54	3558·582	81·286	10·45	0·70802	$1{\cdot}0 \times 10^{-5}$

*Accuracy better than ±5% for Sr and Ca concentrations.

strictly within the outer prismatic layer averaged 2·3 ± 0·4 wt% Sr (4·5 ± 0·6 mole% $SrCO_3$) (n = 15, error is ± 2σ), a concentration about five times that of the middle layer lines (0·5 ± 0·2 wt% Sr or 1·0 ± 0·4 mole% $SrCO_3$, n = 43, error is ±2σ). These high-Sr lines in turn had about twice the background Sr concentration of the middle layer (0·3 ± 0·1wt%, 0·5 ± 0·2 mole% $SrCO_3$, n = 184, error is ±2σ).

The Sr^{2+} concentration of porewater varied from a minimum of 1·5 to 84·4 p.p.m.; while Ca^{2+} varied from 175 to 4,137 p.p.m. (Table 1). Behaviour of Sr^{2+} and Ca^{2+} downcore was very similar decreasing in the upper 60 m to their respective minima and increasing thereafter. The Sr^{2+} and Ca^{2+} concentrations diverged between *ca* 336 mbsf and 545 mbsf, where Ca^{2+} decreased while Sr continued to increase. In the uppermost 60 m of the core, Sr^{2+}/Ca^{2+} varied from 7·75 mmol mol^{-1} at *ca* 10 mbsf to its minimum (*ca* 4 mmol mol^{-1}) and back to *ca* 8 mmol mol^{-1}. From about 93 mbsf to the bottom of the core, porewater Sr^{2+}/Ca^{2+} steadily increased to a maximum of *ca* 10 mmol mol^{-1}, with only one sample off the trend at *ca* 545 mbsf (Fig. 3).

Porewater Sr isotope compositions in the upper *ca* 200 m of the core showed large deviations from the age model of the core (Fig. 4; Table 1). Close to the top of the core, porewater $^{87}Sr/^{86}Sr$ was well below modern sea water values (0·70807 ± 15). From this depth to *ca* 235 mbsf, Sr isotope compositions increased to their maximum (0·71022 ± 15 at 73·18 mbsf) and then decreased to values close to those predicted by the age model of the core (0·70889 ± 15 at 235·71 mbsf) where they remained relatively stable to a depth of *ca* 620 mbsf. Porewater $^{87}Sr/^{86}Sr$ dropped to its lowest values at 779·74 mbsf (0·70692 ± 15) and finally increased to the deepest porewater measured at *ca* 960 mbsf to values still below those corresponding to the age model (0·70802 ± 15).

Supplementary stable isotopes samples produced average $\delta^{18}O$ of −5·7‰ from the outer layer and −2·1‰ from the middle layer, while $\delta^{13}C$ averages are −4·5‰ and −0·2‰ from outer and middle layer samples, respectively (Fig. 5). Aragonite $\delta^{18}O$ and $\delta^{13}C$ are highly correlated.

INTERPRETATION AND DISCUSSION

Strontium in aragonite bivalve shells

Thin bivalve shells are considered by some to be among the least reliable materials to preserve sea water Sr isotope ratios (Smalley *et al.*, 1994). However, because aragonite is easily altered during early diagenesis (Bathurst, 1975; Brand & Veizer, 1980), bivalve shells that retain their original aragonite mineralogy are presumed pristine and thus expected to preserve their original chemical composition. Lack of pervasive alteration of AND-2A venerid

Fig. 2. Sr compositional map over background BSE image (visible at the edges) of in-house sample number 10 from 430·49 to 430·51 mbsf. Scale bar applies to both images. Two of the three measured compositional profiles are superimposed on the elemental map. Measurement locations are shown as points along the abscissa. Profile A: beam diameter 2 µm, sample spacing 2 µm. Profile B: beam diameter 0·5 µm, sample spacing 5 µm. The representative SEM image overlay in the upper-left has a different scale.

shells was previously inferred based on the absence of cathodoluminescence, high Sr and low Fe-Mn concentrations (Marcano *et al.*, 2009). In the present study, the mineralogy of both layers of the same bivalve samples was reconfirmed by microdiffraction.

The Sr concentration contrast observed within the middle layer of the venerid fragments from AND-2A (Fig. 2) may be reasonably explained as the result of Sr concentration fluctuating in response to seasonal changes, which either influence Sr availability or the physiology of the bivalve that in turns controls Sr distribution. For example, in a number of previous studies, Sr incorporation into bivalves has been variably correlated to temperature, and also to growth rate and maturity (Dodd, 1965; Stecher *et al.*, 1996; Hart & Blusztajn, 1998; Dutton *et al.*, 2002; Gillikin *et al.*, 2005; Freitas *et al.*, 2006; Surge & Walker, 2006; Elliot *et al.*, 2009). However, no unique relationship has been established, and as suggested by some, controls on Sr incorporation into bivalve shells may even be species specific (Gillikin *et al.*, 2005; Bailey & Lear, 2006). Regardless of the cause, controlling factors of Sr incorporation into skeletal aragonite vary as the carbonate is incrementally added to the shell, thereby producing a record of growth such as that observed in Figure 2.

However, suggestive of primary precipitation the observed growth lines are, their Sr concentrations are anomalously high compared to modern bivalves, in particular those of the outer layer. Modern molluscs in general discriminate against skeletal Sr incorporation, and their Sr content is usually less than 0·77 mole% $SrCO_3$ (Kinsman, 1969; Veizer, 1983; Morse & MacKenzie, 1990). Modern aragonite cements, on the other hand, contain between 1·2 mole% $SrCO_3$ and 2·3 mole% $SrCO_3$, although inorganic aragonite with up to 14 mole % $SrCO_3$ has been reported from hot spring deposits (Morse & MacKenzie, 1990). Unusually high Sr contents in AND-2A bivalves suggest that recrystallization of the

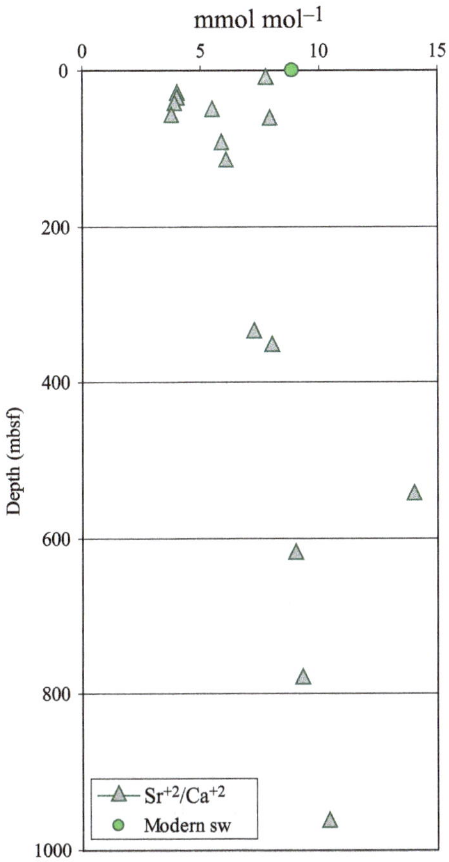

Fig. 3. Sr^{2+}/Ca^{2+} variation with depth. Below *ca* 93 mbsf, Sr^{2+}/Ca^{2+} increases moderately to the core bottom with one sample at *ca* 545 mbsf off the trend.

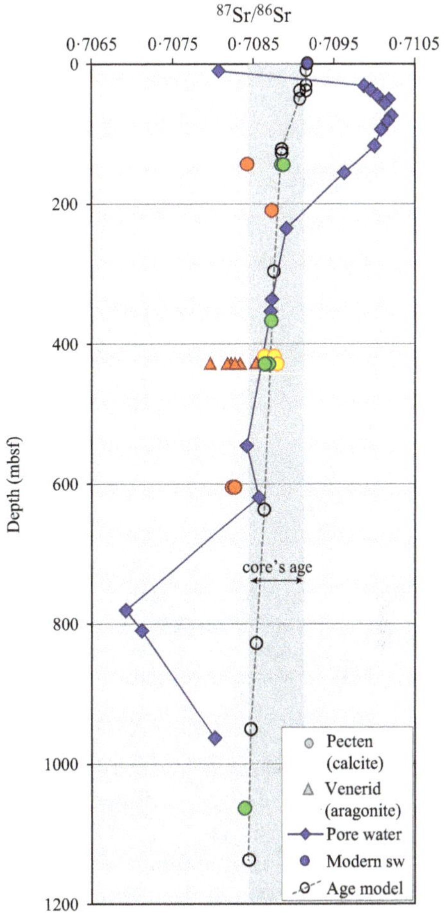

Fig. 4. Sr isotope compositions of porewater (blue diamonds) and previously analysed carbonate samples (Marcano *et al.*, 2009). Circles: pectinid samples (calcite). Triangles: venerid samples (aragonite). Green symbols are samples with no clear indications of alteration; yellow symbols are samples marginally altered; red symbols are clearly altered samples. Black open circles: AND-2A age model data. Overall age range is indicated by the grey vertical band. Modern sea water $^{87}Sr/^{86}Sr$ value indicated by blue circle at depth 0 mbsf. The large and predominant discrepancy between the Sr isotope values of the samples and the porewater is indicative of disequilibrium and suggestive of allochtonous fluids. Limited concurrences are better interpreted as coincidental.

outer layer, a neomorphic process in which the mineral persists after reaction, probably occurred (Folk, 1965; Bathurst, 1975). Therefore, the prismatic structure observed in the Sr-rich outer layer of AND-2A venerid samples, observed also in modern aragonite bivalves, is not sufficient evidence to reject the potential presence of a secondary carbonate because the crystal form can persist through mineralogical stabilization. Crystal structure preservation occurs when diagenetic reactions take place through migrating solution films, which allows chemical changes to occur at the crystal boundaries without developing porosity (Kinsman, 1969; Maliva *et al.*, 2000).

Authigenic aragonite has not been described in AND-2A, but general conditions favourable for carbonate precipitation from the porewater (i.e., very high alkalinity) exist throughout the length of the core. Also, Wada & Okada (1989) described aragonite cements at a variety of depths in the CIROS-1 core, which was drilled less than 30 km away from the locality of this study (Fig. 1). Detailed description of the porewater chemistry and the diagenetic conditions are published elsewhere (Panter *et al.*, 2008; Frank & Gui, 2010). Knowledge of the specific saturation state of the pore fluids, however, is not useful given the disequilibrium between porewater and shell aragonite shown by Sr isotopes and the localized nature of the recrystallization process.

As pointed out by Marcano *et al.* (2009), calculated non-thermodynamic Sr partition coefficients (K_{Sr}) for aragonite between 0·4 and 2 are necessary to obtain the measured Sr concentrations in AND-2A shells if these were precipitating from normal modern sea water. A simple non-thermodynamic calculation using a Sr partition

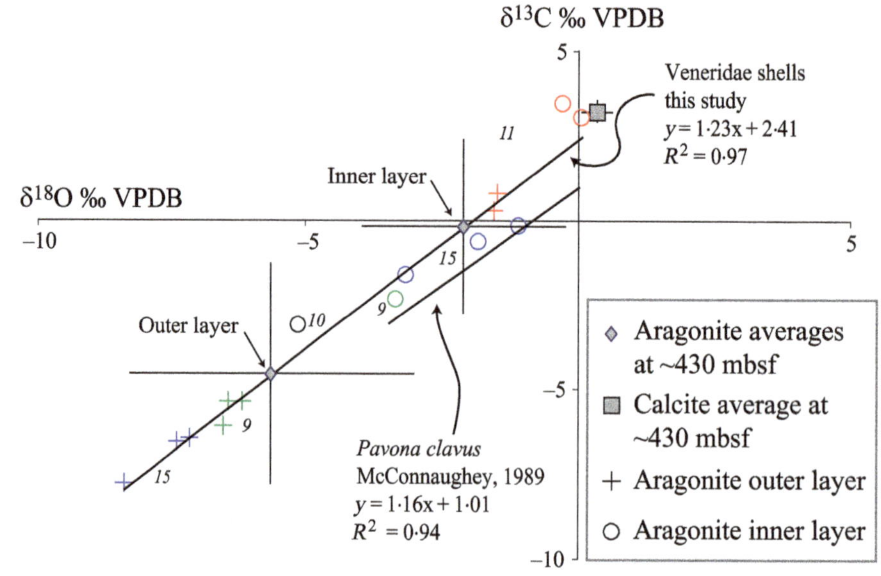

Fig. 5. Oxygen and carbon isotope compositions of AND-2A bivalve samples. Open squares: pectinid samples (calcite). Open circles and crosses: middle and outer layer samples, respectively (aragonite). In-house sample identification number in italics: 15 blue symbols, 11 red symbols, 10 black symbols, 9 green symbols. Error of averages is 2σ. Linear fit was calculated to all aragonite data. Tropical coral *Pavona clavus* data from McConnaughey (1989a,b) plotted for comparison.

coefficient of 1·13 (calculated by interpolating to 24°C the values of the distribution coefficient calculated by Kinsman & Holland (1969) at 16°C and 80°C), suggests that the Sr concentrations of aragonite cement precipitating from modern porewater at the level of interest, where Sr^{2+}/Ca^{2+} could potentially vary between 8 mmol mol^{-1} and 14 mmol mol^{-1} (Fig. 3), should be between 9 mmol mol^{-1} and 16 mmol mol^{-1}. Although narrower, this range is in good agreement with the 4 to 18 mmol mol^{-1} Sr/Ca measured in bulk samples of the AND-2A aragonite shell fragments (Marcano *et al.*, 2009). There are limitations associated with this calculation, in particular when considering the high ionic strength of AND-2A porewater. Nevertheless, higher aragonite Sr concentrations can certainly be modelled using the Sr in porewater instead of sea water.

Inorganic precipitation of aragonite has been used to calculate the temperature-dependent partition of Sr in aragonite. These experiments have produced relatively consistent K_{Sr} values close to 1 for temperatures <100°C (Kinsman & Holland, 1969; Dietzel *et al.*, 2003; Gaetani & Cohen, 2006). All of these calculations imply that the Sr/Ca in aragonite should be very close to the Sr^{2+}/Ca^{2+} in the fluid from which it precipitates. However, Bathurst (1975) in analysing the calculations of Kinsman & Holland (1969), noted the large uncertainties involved (>3500 p.p.m. variation in Sr concentration of inorganic aragonite precipitating from a known fluid at a fixed temperature). Moreover, Gaetani & Cohen (2006) pointed out the large discrepancy between the values based on experimental precipitation and their theoretical calculations. Their theoretical approaches return considerably lower distribution coefficients, which are directly instead of inversely correlated to temperature. The authors suggest that aragonites that conform to a partition coefficient close to 1 are actually strongly enriched in Sr relative to the expected equilibrium concentrations, which should be between 0·5 mmol mol^{-1} and 1 mmol mol^{-1} for temperatures from 15 to 75°C. This disagreement suggests that the incorporation of Sr in aragonite is a complex process not yet fully understood and probably more so in biogenic aragonite.

Alteration scenarios

Although recrystallization without mineralogical change probably occurred, a satisfactory diagenetic process to explain it and further physical evidence of its occurrence are both difficult to produce. The original aragonitic outer shell material must have been dissolved to once more precipitate as aragonite. Carbonate cement compositions are in part controlled by the mineralogy of the particles present in the sediment (Walter, 1986), and although in this case cementation *sensu stricto* did not occur, it illustrates the plausibility of primary mineralogy controlling diagenetic phases. Unless conditions that inhibit calcite precipitation exist, the more thermodynamically stable calcite crystals will tend to precipitate from fluids saturated with respect to aragonite. Aragonite and high-magnesian calcite, the common skeletal carbonate materials to precipitate in shallow environments, are in metastable equilibrium with sea water and will tend to recrystallize to calcite in the early diagenetic environment (inversion of Folk, 1965; polymorphic transformation of Bathurst, 1975). Although the presence of ions in the pore fluids influences mineral equilibria, and the free energies of formation of pure aragonite and pure calcite are close enough that small changes in the fluid can impact their equilibria (Bathurst, 1975; Morse & MacKenzie, 1990), it is difficult to explain fluid changes across migrating films so that aragonite is dissolved and reprecipitated as the film advances. Even if kinetic processes supersede thermodynamic controls of solution-precipitation (as is common in natural carbonates), it is unclear why diagenetic stabilization should increase Sr concentration. Diagnostic physical products of recrystallization without mineralogical change were absent in petrographic and SEM preliminary observations of AND-2A samples.

Studies on solid-state diffusion of Cd^{2+} on calcite demonstrated that this mechanism is capable of incorporating cations into the bulk of the solid and likely to play a role in the uptake of trace metals by other mineral groups as well (Stipp *et al.*, 1992). Although diffusion in carbonate minerals has been regarded relevant only at high temperature and pressure, it may be an important factor in standard conditions, highlighting the potential uncertainties associated to techniques based on the assumption of minerals functioning as perfectly closed system (Stipp *et al.*, 1998).

AND-2A aragonite shells are characterized by low Mn and Fe concentrations, an otherwise good indicator for lack of extensive diagenetic alteration. However, the low Mn and Fe concentrations may be a reflection of their absence in the diagenetic environment, probably an indication of conditions sufficiently oxidizing to prevent reduction of oxides and hydroxides. Abundant iron-rich oxidizing phases and the near absence of organic matter support this possibility. In addition, Mn and Fe share structural affinity to the trigonal calcite, not the orthogonal aragonite. This alone may have prevented their incorporation into the secondary carbonate, even if present in the sediment after the death of the bivalves. Slow recrystallization may have also prevented inclusion of Mn and Fe in aragonite (Morse & MacKenzie, 1990).

If Sr concentrations of the aragonite shell can be achieved as described, and recrystallization occurred and was limited largely to the outer layer, crystal structure may have played a role in promoting chemical changes. Solubility can be influenced by crystal properties and mineral structure differences typically exist between bivalve layers and also between growth lines within a single layer. According to Walter & Morse (1985), thermodynamically more stable carbonate phases can in some cases dissolve faster than less stable ones due to differences in microstructural complexity. Consequently, the original contrast in crystalline structure is likely to translate into a stability gradient between the bivalve layers, which will result in selective or localized diagenetic stabilization.

Because the alteration scenarios discussed above are based mainly on the excessive Sr concentration of the outer layer and its contrast with the middle layers, which are not by themselves diagenetic indicators, it is necessary to examine them as potential primary features. While some modern aragonite bivalves show Sr enrichment in the outer layer (Shirai *et al.*, 2014), not all do. Elliot *et al.* (2009) obtained similar average Sr/Ca from different layers of modern *Tridacna gigas* specimens, although the middle layers showed higher variability. Neither average nor variability differences could be distinguished between the Sr/Ca ratios from the prismatic outer layer and the crossed-lamellar middle layer of *Mercenaria campechiensis* from the Gulf Coast of Florida (salinity between 19 and 36), although Sr concentration behaviour with respect to $\delta^{18}O$ differed between layers (Surge & Walker, 2006). Takesue *et al.* (2008) found significant differences in several cation-to-Ca ratios as a function of the aragonite crystal structure in the estuarine bivalve *Corbula amurensis*, with the notable exceptions of Sr and Na. The shell used to perform the latter analysis grew in waters with a maximum salinity of 28·5. Interestingly, Foster *et al.* (2009) studying the marine cold water bivalve *Arctica islandica* found significant Sr/Ca differences between samples from the axis of maximum growth and those from transects parallel to it, as well as between and within aragonite from the umbo and the outer shell layer. They concluded that changes in shell architecture were a likely candidate to control Sr incorporation through Sr distribution coefficient changes associated with differences in crystal growth. Their analyses were not designed to sample different crystal structures, but the differences they observed keep open the possibility of structure mediated Sr incorporation in some aragonitic bivalves, perhaps an indirect consequence of contrasting growth rates. Still, in the studies mentioned above, Sr/Ca in the aragonite shells remains below *ca* 3 mmol mol^{-1}, while the outer layer of AND-2A sample 10, for example, averages 25 mmol mol^{-1}. This mechanism is most likely unable to concentrate Sr at the levels found in this outer layer.

In summary, a primary origin of aragonitic shells from AND-2A may be suggested by the presence of primary growth lines in the shells and the lack of Sr-isotopic equilibrium between the shells and the ambient porewater. The difficulties associated with physically documenting and chemically explaining potential diagenetic changes in the outer layer of AND-2A *Retrotapes* and the uncertainties in the partition coefficient of Sr in aragonite could be used to further the argument of lack of alteration. Nevertheless, the extreme Sr concentration of the outer layer is difficult to reconcile with precipitation by a living bivalve and also with the fresh water input that would be required to explain the more negative $\delta^{18}O$ measured in the outer layer, which will be discussed later. A primary origin is therefore unlikely.

Porewater Sr isotope compositions

Excluding a relatively small deviation at 545·06 mbsf, porewater Sr isotope compositions are within the age model of the core and range between *ca* 336 and 620 mbsf (Fig. 4). This section includes the aragonite-bearing interval at *ca* 430 mbsf. At all other depths porewater Sr isotope compositions diverge substantially from the age model of the core. Unlike porewater, Sr-isotopic ratios of unaltered calcite shell fragments were in agreement with the age model in the aragonite-bearing interval (366·83 mbsf) and also above and below (144·05 mbsf and 1063·72 mbsf; Marcano *et al.*, 2009). Although porewater may appear to have maintained its original Sr isotope composition at the aragonite-bearing level, this is probably not the case.

Carbonate Sr isotopes require high water-to-rock ratios (on the order of 10^3) to equilibrate with fluids unless these are brines. In that case equilibration can occur at considerably lower water-to-rock ratios (*ca* 10), similar to those required to equilibrate oxygen during freshwater diagenesis (Banner & Hanson, 1990). AND-2A porewater is highly saline, increasing linearly at a rate of about 30 per 100 m depth and stabilizing below *ca* 500 mbsf to salinities between 150 and 200 in the practical salinity scale (Frank & Gui, 2010). Alkalinity is also high (maximum *ca* 55 meq kg^{-1}).

For brines in particular, appropriate chemical models to describe mineral solubility in subsurface conditions are lacking, and non-equilibrium processes are probably the norm (Morse & MacKenzie, 1990). Nevertheless, given the observed carbonate-porewater isotopic contrasts, the highly modified chemistry of AND-2A porewater and the age of the stratigraphic column, late modification or substitution of the porewater probably occurred in this area, which is located only a few tens of kilometres from the coast. Porewater compositions of the DSDP Sites 270 to 273 in the continental shelf of the Ross Sea, for example, are very different from AND-2A fluids. Maximum alkalinity reported is 25 meq kg^{-1}, although it remains mostly below 10 meq kg^{-1}, while salinity drops with depth from normal marine to *ca* 27 (Mann & Gieskes, 1975). They also reported a slight Sr concentration increase downcore possibly associated with aragonite dissolution. Sr concentration in AND-2A decreases in the uppermost part of the core and begins to increase steadily below *ca* 120 mbsf. The unusual Sr isotopic composition of AND-2A porewater is highlighted by the Elderfield & Gieskes (1982) study, which summarized Sr concentration and isotopic composition trends from 37 DSDP holes with latitudinal distribution between 0·5° and 69·9° (Median = 31·53° lat). In contrast to AND-2A, overall porewater isotopic compositions in all analysed drill-holes decrease with depth from normal or close to normal modern marine values in the upper 10 mbsf, to a minimum of 0·70490. Strontium isotope compositions significantly greater than modern sea water were not measured in any of the more than 160 porewater data sets included in this study.

The extant porewater Sr isotope compositions appear not to be just the direct result of reactions with volcanic glass. Throughout AND-2A volcanic alteration products are found in close proximity to unaltered glass (Fielding *et al.*, 2008). Given the reactive nature of volcanic material, the presence of unaltered grains may in part indicate that some of the alteration products originated at the sedimentary source instead of the subsurface. The Sr isotope compositions of porewater are inadequate to explain the anomalous Sr isotope compositions in AND-2A aragonites and give support to a parautochtonous origin of the porewater. What is more, these observations highlight the potential for invalid interpretations if stabilization with the current porewater had occurred. In such a case, the Sr isotope composition of the water fortuitously matching the age model of the core would have not raised further suspicions about the integrity of the aragonite shells. Consequently, ruling out alteration using only basic criteria such as mineralogy, crystal structure, and Mn and Fe content may be misleading.

Oxygen and Carbon isotopic compositions of aragonite bivalve shells

Calcite pectinids and aragonite venerids at *ca* 430 mbsf have contrasting oxygen and carbon isotope compositions (Fig. 5). While all measured shell calcites at *ca* 430 mbsf ($n = 3$) produced very similar $\delta^{18}O$ and $\delta^{13}C$ values, aragonites did not. Calculated equilibrium $\delta^{18}O$ values for aragonite (between $-10{\cdot}3‰$ and $-11{\cdot}0‰$) using the oxygen isotope composition of the porewater (about -10 ‰; Frank & Gui, 2010) and the borehole temperatures at the level of the aragonite-bearing interval (between 21°C and 24°C; Wonik *et al.*, 2008) are at least 2‰ lower than the lowest $\delta^{18}O$ measured in the venerid shell outer layer ($-8{\cdot}4‰$) and up to 9·5‰ lower than the most positive shell $\delta^{18}O$ value (−1·5‰). These estimates were calculated using the Grossman & Ku (1986) corrected palaeotemperature equation, which is in good agreement with relationships based on theoretical estimates of water-aragonite oxygen fractionation at equilibrium (Kim *et al.*, 2007).

The modelled $\delta^{18}O$ values show variable levels of disequilibrium between the aragonitic shells and the porewater for the conditions of the core today. However, using lower temperatures, it is possible to replicate the measured aragonite $\delta^{18}O$ values. This scenario requires alteration to occur earlier in the burial history of the samples without further re-equilibration. In the upper 50 m of the core, temperatures increase from *ca* 0 to 5°C, while water $\delta^{18}O$ decreases from modern sea water values of *ca* −1‰ to about −10‰ (Wonik *et al.*, 2008; Frank & Gui, 2010). Aragonite $\delta^{18}O$ precipitating in this interval would have compositions between *ca* +3·5‰ and −6‰. Under stable conditions with respect to temperature and porewater $\delta^{18}O$, maximum alteration could have occurred when the sample was *ca* 200 mbsf. This is a good indication that parts of the shell could have been subjected to early alteration. The moderate Sr^{2+}/Ca^{2+} increase at depth (Fig. 3) suggests that dissolution of Sr-rich phases is occurring, although this normal diagenetic trend may not be locally dominant as appears to be the case at the aragonite-bearing interval.

The carbon isotope composition of the shell is a function of the isotopic composition of the bicarbonate from which it forms, which in turn reflects the incorporation of organic and inorganic carbon, as well as metabolic fractionation. Variations in carbon isotopic composition indicate, in part, changes in productivity that are a function of depth or microhabitat conditions (Grossman & Ku, 1986; Hoefs, 1997), and these are difficult to predict or estimate. Although metabolic CO_2 appears to have a limited influence on molluscan carbonate (McConnaughey & Gillikin, 2008; Beirne *et al.*, 2012), it is unlikely that C can be incorporated into biogenic carbonate exclusively reflecting hydrological conditions (i.e., without vital effects). The $\delta^{13}C$ difference between calcite and aragonite was argued to reflect in part contrasting life styles between the pectinid and venerid bivalves (Marcano *et al.*, 2009).

Considerable differences exist in both oxygen and carbon isotope values between outer and middle layers. Values in outer layers are more negative than those from corresponding middle layers. Bivalve shell layers are formed from different pallial fluid sources along different areas of the mantle (Moore, 1969; McConnaughey & Gillikin, 2008), and adjacent points from different layers do not necessarily form at the same time. This discrepancy may be exacerbated by rapid growth. Primary intra-shell contrast in $\delta^{13}C$ is common, but intra-shell contrasts in $\delta^{18}O$, although reported (Elliot *et al.*, 2009), are normally absent (Surge & Walker, 2006). While $\delta^{18}O$ comparisons between layers in most studies are based on long, usually multiyear data from complete valves, the millimetre-sized fragments analysed here provide data from a single point and this may help explain the contrast. However, the $\delta^{18}O$ difference between the layers can be better explained as the product of localized alteration.

The high correlation between $\delta^{18}O$ and $\delta^{13}C$ values and its similarity with such relationships observed in corals (Gonzalez & Lohmann, 1985; McConnaughey, 1989a, b) could be argued to represent disequilibrium precipitation in contrast with the equilibrium fractionation that characterizes most modern molluscs (Epstein *et al.*, 1953; Grossman & Ku, 1986). A fairly good correlation between $\delta^{18}O$ and $\delta^{13}C$ values has been reported for the modern venerid *M. mercenaria* (Elliot *et al.*, 2003), and departures from established oxygen isotope equilibrium fractionation of aragonite exist in other clams (Carré *et al.*, 2005). These observations suggest that disequilibrium precipitation in aragonite bivalves is possible. Kinetic effects probably associated with periods of rapid growth are the main cause of disequilibrium and likely a consequence of the seasonal growth in bivalves. Although the similarity between the slopes of linear fits to the venerid data and to coral data is intriguing (Fig. 5), the possibility of primary precipitation out of equilibrium cannot be tested, and as was mentioned above, the more negative values of the outer layer are not explained using the primary precipitation argument. Thus, the high $\delta^{18}O$ – $\delta^{13}C$ correlation remains unexplained.

CONCLUSION

Previous work had shown that aragonite bivalve fragments recovered from ANDRILL AND-2A core in Southern McMurdo Sound, Antarctica, passed generally accepted criteria for unaltered aragonite and showed Sr concentration contrast along growth lines of the middle layer. The Sr concentrations in the outer layer, however, were anomalously high, and Sr isotope compositions of all subsamples were less radiogenic than expected (Marcano *et al.*, 2009). Within the uncertainties associated with the incorporation of Sr into aragonite, the observed high-Sr concentration of the shells was modelled here using AND-2A porewater. Alteration was also suggested by the persistent depletion in ^{18}O of the outer layers compared to the middle layers, and the successful calculation of shell isotopic values using core temperatures and porewater $\delta^{18}O$. The Sr isotope compositions of the porewater and the aragonite shells differed. AND-2A porewater was highly modified and in general disequilibrium with the core carbonates. In addition, Sr isotope compositions of the porewater deviated from that of the Ross Sea shelf and from *ca* 160 other porewater compositions reported by Elderfield & Gieskes (1982). The Sr concentration at the levels reported here (i.e., Sr/Ca as high as 25 mmol mol^{-1}), although not easily explained chemically or supported petrographically as a product of alteration, is in strong disagreement with modern primary shell precipitation. The presence of aragonite with no visual indications and only partial evidence of chemical alteration have been also reported for other molluscs, mostly on the nacreous layer of inoceramids, ammonoids and gastropods (Dauphin *et al.*, 2007; Cochran *et al.*, 2010; Wierzbowski *et al.*, 2012). Together, these observations confirm that the anomalous Sr concentrations and isotope compositions observed in the aragonite bivalves of AND-2A, equally expressed along diachronous layers of the shell, are likely the result of early diagenesis. Changes may have been selective, affecting differently the distinct crystal structures of the shell, which are variably susceptible to alteration. Evaluating mineralogy, cathodoluminescence and abundance of minor elements typically associated with diagenesis (i.e., Mn, Fe) is not enough to rule out alteration in Cenozoic Antarctica biogenic aragonites. Knowledge of Sr concentration and distribution in the shell is critical to determining the viability of Sr stratigraphy and the scale at which it may be applicable.

ACKNOWLEDGEMENTS

We are very thankful to Carl Henderson, Lora Wingate, Anja Schleicher, Antek G. Wong-Foy, Sara Worsham and Lindsay Shuller for technical assistance. Many thanks also to Adam Matzger, Rod Ewing, Marcus Johnson, Jamie Gleason and Glenn Gaetani for sharing their expertise. NSF grant #EAR-991135 helped cover part of the analytical costs. This article commits to the GEOSMART PNRA Project and it is Ismar-Bologna scientific contribution no. 1862. The ANDRILL project is a multinational collaboration between the Antarctic programmes of Germany, Italy, New Zealand and the United States. Antarctica New Zealand is the project operator and developed the drilling system in collaboration with A. Pyne. Antarctica New Zealand supported the drilling team at Scott Base; Raytheon Polar Services Corporation supported the science team at McMurdo Station and the Crary Science and Engineering Laboratory. The ANDRILL Science Management Office at the University of Nebraska-Lincoln provided science planning and operational support. The scientific studies are jointly supported by the US National Science Foundation, the New Zealand Foundation for Research Science and Technology and the Royal Society of New Zealand Marsden Fund, the Italian Antarctic Research Programme, the German Research Foundation and the Alfred Wegener Institute for Polar and Marine Research. We recognize and appreciate the valuable input from Adrian Immenhauser and two anonymous reviewers, which help improve the original manuscript.

References

Acton, G., **Crampton, J.**, **Di Vincenzo, G.**, **Fielding, C.G.**, **Florindo, F.**, **Hannah, M.J.**, **Harwood, D.M.**, **Ishman, S.E.**, **Johnson, K.**, **Jovane, L.**, **Levy, R.H.**, **Lum, B.**, **Marcano, M.C.**, **Mukasa, S.B.**, **Ohneiser, C.**, **Olney, M.**, **Riesselman, C.**, **Sagnotti, L.**, **Stefano, C.**, **Strada, E.**, **Taviani, M.**, **Tuzzi, E.**, **Verosub, K.L.**, **Wilson, G.S.** and **Zattin, M.** (2008) Preliminary Integrated chronostratigraphy of the AND-2A core, ANDRILL Southern McMurdo Sound Project Antarctica. In: Studies from the ANDRILL, Southern McMurdo Sound Project, Antarctica (Eds D.M. Harwood, F. Florindo, F. Talarico and R.H. Levy), Terra Antarct., **15**, 212–220.

Bailey, T.R. and **Lear, C.H.** (2006) Testing the effect of carbonate saturation of the Sr/Ca of biogenic aragonite: a case study from the River Ehen, Cumbria, UK. Geochem. Geophys. Geosyst., **7**, Q03019, doi:10.1029/2005GC001084.

Banner, J.L. and **Hanson, G.N.** (1990) Calculation of simultaneous isotopic and trace element variations during water-rock interaction with applications to carbonate diagenesis. Geochim. Cosmochim. Acta, **54**, 3123–3137.

Barrera, E. (1989) Strontium isotope ages. In: Antarctic Cenozoic History from the CIROS-1 drillhole, McMurdo Sound, Antarctica (Ed. B. , P. J), DSIR Bull., **245**, 151–152.

Bathurst, R.G.C. (1975) Carbonate Sediments and Their Diagenesis. Elsevier, Amsterdam, 593 pp.

Beirne, E.C., **Wanamaker, A.L.**, **Jr** and **Feindel, S.C.** (2012) Experimental validation of environmental controls on the $\delta^{13}C$ of *Arctica islandica* (ocean quahog) shell carbonate. Geochim. Cosmochim. Acta, **84**, 395–409.

Beu, A. and **Taviani, M.** (2014) Early Miocene Mollusca from McMurdo Sound, Antarctica (ANDRILL 2A drill core), with a review of Antarctic Oligocene and Neogene Pectinidae (Bivalvia). Palaeontology, **57**, 299–342.

Brand, U. and **Veizer, J.** (1980) Chemical diagenesis of a multicomponent carbonate system 1: trace elements. J. Sed. Petrol., **50**, 1219–1236.

Carré, M., **Bentaleb, I.**, **Blamart, D.**, **Ogle, N.**, **Cardenas, F.**, **Zevallos, S.**, **Kalin, R.M.**, **Ortilieb, L.** and **Fontugne, M.** (2005) Stable isotopes and sclerochronology of the bivalve *Mesodesma donacium*: potential application to Peruvian paleoceanographic reconstructions. Palaeogeogr. Palaeoclimatol. Palaeoecol., **228**, 4–25.

Cochran, J.K., **Kallenberg, K.**, **Landman, N.H.**, **Harries, P.J.**, **Weinreb, D.**, **Turekian, K.K.**, **Beck, A.J.** and **Cobban, W.A.** (2010) Effect of diagenesis on the Sr, O, and C isotope composition of late Cretaceous mollusks from the western interior seaway of North America. Am. J. Sci., **310**, 69–88.

Dauphin, Y., **Williams, C.T.** and **Barskov, I.S.** (2007) Aragonitic rostra of the Turonian belemnitid *Goniocamax*: arguments from diagenesis. Acta Palaeontol. Pol., **52**, 85–97.

Dietzel, M., **Gussone, N.** and **Eisenhauer, A.** (2003) Co-precipitation of Sr^{2+} and Ba^{2+} with aragonite by membrane diffusion of CO_2 between 10 and 50°C. Chem. Geol., **203**, 139–151.

Dodd, R.D. (1965) Environmental control of strontium and magnesium in *Mytilus*. Geochim. Cosmochim. Acta, **29**, 385–398.

Dutton, A., **Lohmann, K.C.** and **Zinsmeister, W.J.** (2002) Stable isotope and minor elements proxies for Eocene climate of Seymour Island, Antarctica. Palaeoceanography, **17**, 1016.

Elderfield, H. and **Gieskes, J.M.** (1982) Sr isotopes in interstitial waters of Deep Sea Drilling Project cores. Nature, **300**, 493–497.

Elliot, M.B., **deMenocal, B.K.L.** and **Howe, S.S.** (2003) Environmental controls on the stable isotopic composition of *Mercenaria mercenaria*. Potential application to paleoenvironmental studies. Geochem. Geophys. Geosyst., **4**, 1056.

Elliot, M., **Welsh, K.**, **Chicott, C.**, **McCulloch, M.**, **Chappell, J.** and **Ayling, B.** (2009) Profiles of trace elements and stable isotopes derived from giant long-lived *Tridacta gigas* bivalves: potential applications in paleoclimate studies. Palaeogeogr. Palaeoclimatol. Palaeoecol., **280**, 132–142.

Epstein, S., **Buchsbaum, R.**, **Lowenstam, H.A.** and **Urey, H.C.** (1953) Revised carbonate-water isotopic temperature scale. Geol. Soc. Am. Bull., **64**, 1316–1326.

Faure, G. and **Mensing, T.** (2005) Isotopes. Principles and Applications. John Wiley & Sons, New Jersey, 897 pp.

Fielding, C.G., **Atkins, C.B.**, **Basset, K.N.**, **Browne, G.H.**, **Dunbar, G.B.**, **Field, B.D.**, **Frank, T.D.**, **Krissek, L.A.**, **Panter, K.S.**, **Passchier, S.**, **Pekar, S.F.**, **Sandroni, S.** and **Talarico, F.** (2008) Sedimentology and stratigraphy of the AND-2A core, ANDRILL Southern McMurdo Sound Project, Antarctica. In: Studies from the ANDRILL, Southern McMurdo Sound Project, Antarctica (Eds D.M. Harwood, F. Florindo, F. Talarico and R.H. Levy), Terra Antarct., 15, 77–112.

Folk, R. (1965) Some aspects of recrystallization in ancient limestones. SEPM Spec. Publ., **13**, 14–48.

Foster, L.C., **Allison, N.**, **Finch, A.A.** and **Anderson, C.** (2009) Strontium distribution in the shell of the aragonite bivalve *Arctica islandica*. Geochem. Geophys. Geosyst., **10**, Q03003, doi:10.1029/2007GC001915.

Frank, T.D., **Gui, Z.** and the ANDRILL SMS Science Team (2010) Cryogenic origin for brine in the subsurface of southern McMurdo Sound, Antarctica. Geology, **38**, 587–590.

Freitas, P.S., **Clarke, L.J.**, **Kennedy, H.**, **Richardson, C.A.** and **Abrantes, F.** (2006) Environmental and biological controls on elemental (Mg/Ca, Sr/Ca and Mn/Ca) ratios in shells of the king scallop *Pecten maximus*. Geochim. Cosmochim. Acta, **70**, 5119–5133.

Gaetani, G.A. and **Cohen, A.L.** (2006) Element partitioning during precipitation of aragonite from seawater: a framework for understanding paleoproxies. Geochim. Cosmochim. Acta, **70**, 4617–4634.

Gillikin, D.P., **Lorrain, A.**, **Navez, J.**, **Taylor, J.W.**, **André, L.**, **Keppens, E.**, **Baeyens, W.** and **Dehairs, F.** (2005) Strong biological controls on Sr/Ca rations in aragonitic marine bivalve shells. Geochem. Geophys. Geosyst., **6**, Q05009, doi:10.1029/2004GC000874.

Gonzalez, L. and **Lohmann, K.C.** (1985) Carbon and oxygen isotopic composition of Holocene reefal carbonates. Geology, **13**, 811–814.

Graustein, W.C. (1989) $^{87}Sr/^{86}Sr$ ratios measure the sources and flow of strontium in terrestrial ecosystems. In: Ecological Studies (Eds P.W. Rundel, J.R. Ehleringer and K.A. Nagy), **68**, 491–512.

Grossman, E.L. and **Ku, T.-L.** (1986) Oxygen and carbon isotope fractionation in biogenic aragonite; temperature effects. Chem. Geol., **59**, 59–74.

Gui, Z. (2009) Origin of brines in Neogene sediments of the Ross Sea, Antarctica: AND-2A Core, ANDRILL, Southern McMurdo Sound Project. MS Thesis. University of Nebraska, Lincoln, 51 pp.

Hart, S.R. and **Blusztajn, J.** (1998) Clams as recorders of ocean ridge volcanism and hydrothermal vent field activity. Science, **280**, 883–886.

Harwood, D.M., **Florindo, F.**, **Levy, R.H.**, **Fielding, C.G.**, **Pekar, S.F.** and **Speece, M.A.** (2005) ANDRILL Southern McMurdo Sound Project Scientific Prospectus. ANDRILL SMO Contribution No. 5, Lincoln, 29 p.

Harwood, D.M., **Levy, R.H.**, **Cowie, J.**, **Florindo, F.**, **Naish, T.**, **Powell, R.** and **Pyne, A.** (2006) Deep drilling with the ANDRILL program in Antarctica. Sci. Drilling, **3**, 43–45.

Harwood, D.M., **Florindo, F.**, **Talarico, F.**, **Levy, R.H.**, **Kuhn, G.**, **Naish, T.**, **Niessen, F.**, **Powell, R.**, **Pyne, A.** and **Wilson, G.S.** (2009) Antarctic drilling recovers stratigraphic records from the continental margin. EOS Trans. Am. Geophys. Union, **90**, 90–91.

Hess, J., **Bender, M.** and **Schilling, J.-G.** (1986) Evolution of the ratio of strontium-87 to strontium-86 in seawater from cretaceous to present. Science, **231**, 979–984.

Hoefs, J. (1997) Stable Isotope Geochemistry. Springer, Berlin, 201 pp.

Howarth, R.J. and **McArthur, J.M.** (1997) Statistics for strontium isotope stratigraphy. A robust LOWESS fit to marine Sr-isotope curve for 0–206 Ma, with look-up tables for the derivation of numerical age. J. Geol., **105**, 441–456.

Ingram, B.L. and **Depaolo, D.J.** (1993) A 4300-year strontium isotope record of estuarine palaeosalinity in San-Francisco Bay, California. Earth Planet. Sci. Lett., **119**, 103–119.

Israelson, C. and **Buchardt, B.** (1999) Strontium and oxygen isotopic composition of East Greenland rivers and surface waters: implication for palaeoenvironmental interpretation. Palaeogeogr. Palaeoclimatol. Palaeoecol., **153**, 93–104.

Kashiwagi, H., Ogawa, Y. and **Shikazono, N.** (2008) Relationship between weathering, mountain uplift, and climate during the Cenozoic as deduced from the global carbon-strontium cycle model. Palaeogeogr. Palaeoclimatol. Palaeoecol., **270**, 139–149.

Kim, S.-T., O'Neil, J.R., Hillaire-Marcel, C. and **Mucci, A.** (2007) Oxygen isotope fractionation between synthetic aragonite and water: influence of temperature and Mg^{2+} concentration. Geochim. Cosmochim. Acta, **71**, 4704–4715.

Kinsman, D. (1969) Interpretations of Sr^{2+} concentrations in carbonate minerals and rocks. J. Sed. Petrol., **39**, 486–508.

Kinsman, D. and **Holland, H.D.** (1969) The co-precipitations of cations with $CaCO_3$ – IV. The co-precipitation of Sr^{2+} with aragonite between 16° and 96°C. Geochim. Cosmochim. Acta, **33**, 1–17.

Lavelle, M. (1998) Strontium-isotope stratigraphy of the CRP-1 drillhole, Ross Sea, Antarctica. Terra Antarct., **5**, 691–696.

Lavelle, M. (2000) Strontium isotope stratigraphy and age model for CRP-2/2A, Victoria Land Basin, Antarctica. Terra Antarct., **7**, 611–619.

Lavelle, M. (2001) Strontium isotope stratigraphy for CRP-3, Victoria Land Basin, Antarctica. Terra Antarct., **8**, 593–597.

Li, G.J., Chen, J., Ji, J., Liu, L., Yang, J. and **Sheng, X.** (2007) Global cooling forced increase in marine strontium isotopic ratios: importance of mica weathering and a kinetic approach. Earth Planet. Sci. Lett., **254**, 303–312.

Maliva, R.G., Missimer, T.M. and **Dickson, J.A.D.** (2000) Skeletal aragonite neomorphism in Plio-Pleistocene sandy limestones and sandstones, Hollywood, Florida, USA. Sed. Geol., **136**, 147–154.

Mann, R., Gieskes J.M. (1975) Interstitial waters studies, Leg 28. In: Initial Reports of the Deep Sea Drilling Project (Eds D. E. Hayes, L.A. Frakes, P.J. Barrett, D.A. Burns, P.-H. Chen, A.B. Ford, A.G. Kaneps, E. M. Kemp, D.W. McCollum, D.J. W. Piper, R. E. Wall, and P.N. Webb), **28**, 805–814.

Marcano, M.C., Mukasa, S., Lohmann, K.C., Stefano, C., Taviani, M. and **Andronikov, A.** (2009) Chronostratigraphic and paleoenvironmental constraints derived from the $^{87}Sr/^{86}Sr$ and $\delta^{18}O$ signal of Miocene bivalves, Southern McMurdo Sound, Antarctica. Global Planet. Change, **69**, 124–132.

McArthur, J.M., Howarth, R.J. and **Bailey, T.R.** (2001) Strontium isotope stratigraphy: LOWESS Version 3. Best-fit line to the marine Sr-isotope curve for 0 to 509 Ma and accompanying look-up table for deriving numerical age. Look-up table Version 4:08/04. J. Geol., **109**, 155–169.

McConnaughey, T. (1989a) ^{13}C and ^{18}O isotopic disequilibrium in biological carbonates: I. Patterns. Geochim. Cosmochim. Acta, **53**, 151–162.

McConnaughey, T. (1989b) ^{13}C and ^{18}O isotopic disequilibrium in biological carbonates: II. In vitro simulation of kinetic isotope effects. Geochim. Cosmochim. Acta, **53**, 163–171.

McConnaughey, T. and **Gillikin, D.P.** (2008) Carbon isotopes in mollusks shell carbonates. Geo-Mar. Lett., **28**, 287–299.

Moore, R.C. (1969) Treatise on invertebrate paleontology. Part N. Mollusca 6 Bivalvia, Vols 1, 2. No. 3. The Geological Society of America, Laurence, 951 pp.

Morse, J.W. and **MacKenzie, F.T.** (1990) Geochemistry of Sedimentary Carbonates. Elsevier, Amsterdam, 707 pp.

Mukasa, S.B., Shervais, J.W., Wilshier, H.G. and **Nielson, J.E.** (1991) Intrinsic Nd, Pb, and Sr isotopic heterogeneities exhibited by the Lherz Peridotite massif, French Pyrenees. J. Petrol., Special Volume **2**, 117–134.

Naish, T.R., Powell, R., Levy, R.H., Henrys, S., Krissek, L.A., Niessen, F., Pompilio, M., Scherer, R., Wilson, G.S. and **Team, t.A.-M.S.** (2007) Synthesis of the initial scientific results of the MIS project (AND-1B core), Victoria Land Basin, Antarctica. Terra Antarct., **14**, 317–327.

Naish, T.R., Wilson, G.S., Dunbar, G.B. and **Barrett, P.J.** (2008) Constraining the amplitude of Late Oligocene bathymetric changes in western Ross Sea during orbitally-induced oscillation in the East Antarctic Ice Sheet: (2) implication for global sea-level changes. Palaeogeogr. Palaeoclimatol. Palaeoecol., **260**, 66–76.

Panter, K.S., Talarico, F., Basset, K.N., Del Carlo, P., Field, B.D., Frank, T.D., Hoffmann, S., Kuhn, G., Sandroni, S., Taviani, M., Bracciali, L., Cornamusini, G., von Eynatten, H., Rocchi, S. and **Team, t.A.-S.S.** (2008) Petrology and geochemistry of the AND-2A core, ANDRILL Southern McMurdo Sound Project, Antarctica. In: Studies from the ANDRILL, Southern McMurdo Sound Project, Antarctica (Eds D.M. Harwood, F. Florindo, F. Talarico and R.H. Levy), Terra Antarct., **15**, 147–192.

Sessa, J.A., Ivany, L.C., Schlossnagle, T.H., Samson, S.D. and **Schellenberg, S.A.** (2012) The fidelity of oxygen and strontium isotope values from shallow shelf settings: implications for temperature and age reconstructions. Palaeogeogr. Palaeoclimatol. Palaeoecol., **342**, 27–39.

Shirai, K., Schöne, B.R., Miyaji, T., Radarmacher, P., Krause, R.A., Jr and **Tanabe, K.** (2014) Assessment of the mechanism of elemental incorporation into bivalve shells (*Arctica islandica*) based on elemental distribution at the microstructural scale. Geochim. Cosmochim. Acta, **126**, 307–320.

Smalley, C., Higgins, A.C., Howarth, R.J., Nicholson, H., Jones, C.E., Swinburne, N.H.M. and **Bessa, J.** (1994) Seawater Sr isotope variations through time: a procedure for constructing a reference curve to date and correlate marine sedimentary rocks. Geology, **22**, 431–434.

Stecher, H.A., Krantz, D.E., Lord, C.J., Luther, G.W. and **Bock, K.W.** (1996) Profiles of strontium and barium in

Mercenaria mercenaria and *Spisula solidissima* shells. Geochim. Cosmochim. Acta, **60**, 3445–3456.

Stipp, S.L.S., **Hochella, M.F., Jr**, **Parks, G.A.** and **Leckie, J.O.** (1992) Cd^{+2} uptake by calcite, solid-state diffusion, and the formation of solid-solution: interface processes observed with near-surface sensitive techniques (XPS, LEED, and AES). Geochim. Cosmochim. Acta, **56**, 1941–1954.

Stipp, S.L.S., **Konnerup-Madsen, J.**, **Franzreb, K.**, **Kulik, A.** and **Mathieu, H.J.** (1998) Spontaneous movement of ions through calcite at standard temperature and pressure. Nature, **396**, 356–359.

Surge, D. and **Walker, K.J.** (2006) Geochemical variation in microstructural shell layers of the southern quahog (*Mercenaria campechiensis*): implications for reconstructing seasonality. Palaeogeogr. Palaeoclimatol. Palaeoecol., **237**, 182–190.

Takesue, R.K., **Bacon, C.R.** and **Thompson, J.K.** (2008) Influences of organic matter and calcification rate on trace elements in aragonitic estuarine bivalves shells. Geochim. Cosmochim. Acta, **72**, 5431–5445.

Taviani, M., **Hannah, M.**, **Harwood, D.M.**, **Ishman, S.E.**, **Johnson, K.**, **Olney, M.**, **Riesselman, C.**, **Tuzzi, E.**, **Beu, A.G.**, **Blair, S.**, **Cantarelli, V.**, **Ceregato, A.**, **Corrado, S.**, **Mohr, B.**, **Nielson, S.H.H.**, **Persico, D.**, **Petrushak, S.**, **Raine, J.I.**, **Warny, S.** and the ANDRILL-SMS Science Team (2009) Palaeontological characterisation and analysis of the AND-2A core, ANDRILL Southern McMurdo Sound Project, Antarctica. Terra Antarct., **15**, 113–146.

Thierstein, H.R., **MacDougall, J.D.**, **Martin, E.E.**, **Larsen, B.**, **Barron, J.A.** and **Baldauf, J.** (1991) Age determinations of Paleogene diamictites from Prydz Bay (Site 739), Antarctica, using Sr isotopes of mollusks and biostratigraphy of microfossils (diatoms and coccoliths). In: Proceedings of the Ocean Drilling Program, Scientific Results (Eds J. Barron, B. Larsen, J.Q. Baldauf, C. Alibert, S. Berkowitz, J.-P. Caulet, S. Chambers, A. Cooper, R. Cranston, W. Dorn, W. Ehrmann, R. Fox, G. Fryxell, M. Hambrey, B. Huber, C. Jenkins, S.-H. Kang, B. Keating, K. Mehl, I.I. Noh , Q. Oilier, A. Pittenger, H. Sakai, C. Schroder, A. Solheim, D. Stockwell, H. Thierstein, B. Tocher, B. Turner, and W. Wei), **119**, 742–745.

Veizer, J. (1983) Trace elements and isotopes in sedimentary carbonates. Rev. Mineral., **11**, 265–299.

Wada, H. and **Okada, H.** (1989) Carbonate isotopes. In: Antarctic Cenozoic History from the CIROS-1 Drillhole, McMurdo Sound, Antarctica (Ed. P.J. Barrett), DSIR, **245**, 195–200.

Walter, L.M. (1986) Relative efficiency of carbonate dissolution and precipitation during diagenesis, a progress report of the role of solution chemistry. In: Special Publication Society of Economic Paleontologists and Mineralogists (Ed. D.L. Gautier), SEPM Spec. Publ., **38**, 1–11.

Walter, L.M. and **Morse, J.W.** (1985) The dissolution kinetics of shallow marine carbonates in seawater: a laboratory study. Geochim. Cosmochim. Acta, **49**, 1503–1513.

Wierzbowski, H., **Anczkiewicz, R.**, **Bazarnik, J.** and **Pawlak, J.** (2012) Strontium isotope variations in Middle Jurassic (Labe Bajocian-Callovian) seawater: implications for Earth's tectonic activity and marine environments. Chem. Geol., **334**, 171–181.

Wilson, G.S., **Levy, R.H.**, **Browne, G.H.**, **Cody, R.**, **Dunbar, N.**, **Florindo, F.**, **Henrys, S.**, **Graham, I.**, **McIntosh, W.C.**, **McKay, R.M.**, **Naish, T.**, **Ohneiser, C.**, **Powell, R.**, **Ross, J.**, **Sagnotti, L.**, **Scherer, R.**, **Sjunneskog, C.**, **Strong, C.P.**, **Taviani, M.**, **Winter, D.** and **Team, t.A.-M.S.** (2007) Preliminary integrated chronostratigraphy of the AND-1B core, ANDRILL McMurdo Ice Shelf Project, Antarctica. Terra Antarct., **14**, 297–316.

Wonik, T., **Grelle, T.**, **Hardwerger, D.**, **Jarrard, R.D.**, **McKee, A.**, **Patterson, T.**, **Paulsen, T.**, **Pierdominici, S.**, **Schmitt, D.R.**, **Schroder, H.**, **Speece, M.A.**, **Wilson, T.** and **Team, t.A.-S.S.** (2008) Downhole measurements in the AND-2A borehole, ANDRILL Southern McMurdo Sound Project, Antarctica. In: Studies from the ANDRILL, Southern McMurdo Sound Project, Antarctica (Eds D.M. Harwood, F. Florindo, F. Talarico and R.H. Levy), Terra Antarct., **15**, 57–68.

8

Microbial sedimentary imprint on the deep Dead Sea sediment

CAMILLE THOMAS*,[1], YAEL EBERT†, YAEL KIRO‡, MORDECHAI STEIN[§], DANIEL ARIZTEGUI* and THE DSDDP SCIENTIFIC TEAM

*Department of Earth Sciences, University of Geneva, rue des Maraichers 13, Geneva CH 1205, Switzerland (E-mail: camille.thomas@univ-savoie.fr)
†Institute of Earth Sciences, Hebrew University of Jerusalem, Edmond J. Safra Campus, Givat Ram, Jerusalem, IL 91904, USA
‡Lamont-Doherty Earth Observatory, Columbia University, 61 Rt. 9W, Palisades, NY 10964, USA
§Geological Survey of Israel, 30 Malkhe Israel St., Jerusalem, IL 95501, USA

Keywords
EPS, geomicrobiology, hypersaline, iron-sulphur mineralization.

[1]Present address: UMR 42 CARRTEL, Alpine Research Center on Lake Food Webs, University of Savoie Mont-Blanc, 73376 Le Bourget du Lac, France

ABSTRACT

A study of an International Continental Drilling Program core recovered from the middle of the modern Dead Sea has identified microbial traces within this subsurface hypersaline environment. A comparison with an active microbial mat exhibiting similar evaporative processes characterized iron-sulphur mineralization and exopolymeric substances resulting from microbial activity. Exopolymeric substances were identified in the drilled sediment but unlike other hypersaline environments, it appears that they have a limited effect on the precipitation of calcium carbonate in the sedimentary column. Sulphate reduction, however, plays a role in all types of evaporative facies, leading to the formation of diagenetic iron sulphides in glacial and interglacial intervals. Their synthesis seems to occur under progressive sulphidation that generally stops at greigite because of incomplete sulphate reduction. The latter may be caused by a lack of suitable organic matter in this hypersaline, hence energy-demanding, environment. Pyrite may be found in periods of high lake productivity, when more labile organic matter is available. The carbon and sulphur cycles are thus influenced by microbial activity in the Dead Sea environment and this influence results in diagenetic transformations in the deep sediment.

INTRODUCTION

The Dead Sea is one of the most saline lakes in the world. In addition, it also has extremely high levels of divalent cations Ca^{2+} and Mg^{2+}, making it even harder for life to cope with its chemistry (Nissenbaum, 1975). While few microbes have adapted to such environments, successful colonists include *Archaea* members of the extreme halophilic class *Halobacteria*, as well as a few halophilic *Bacteria* (Bodaker *et al.*, 2010; Rhodes *et al.*, 2012).

In such an environment, salinity gradients encourage and support life. Submarine freshwater springs host numerous and diverse microbial mats benefiting from local dilution of the brine (Ionescu *et al.*, 2012). In the recent past, rainy winters have also brought enough freshwater to the lake to dilute the shallow layers by 70%, enabling blooms of the halophilic algae *Dunaliella parva* (Oren & Shilo, 1982; Oren *et al.*, 1995). The highly labile organic matter produced by the autotrophic eukaryotes would have been immediately degraded by blooms of *Halobacteria* (Oren, 1983).

The microbial influence on the Dead Sea subsurface is less well-known. The recovery of a 457 m long core from the deepest part of the lake via the International Continental Drilling Program (ICDP)-funded Dead Sea Deep Drilling Project (DSDDP) has shown a potential for climatic and palaeoenvironmental reconstructions (Lazar *et al.*, 2014; Neugebauer *et al.*, 2014; Torfstein *et al.*, 2015). In order to constrain the impact of microbial activity on this pristine archive, a geomicrobiological investigation was performed. Initial results revealed the importance of assessing the role of microbial activity in the early diagenesis of the Dead Sea sediment, highlighting differential distribution of microbes with diverse functional potential along the core (Thomas *et al.*, 2014; Ariztegui *et al.*, 2015). The DSDDP project also provides a unique opportunity to explore the diagenetic processes occurring in the sediment of the Dead Sea before they become exposed and fully desiccated.

Here, we investigate the potential traces of past or current microbial activity in the Dead Sea depositional record, and qualify early diagenesis in its deep sediment. In order to do so, the sedimentology and mineralogy of an active microbial mat from the western shore of the Dead Sea have been examined and identified microbial traces were searched for in the cored material.

Hypersaline environments often allow for the growth of microbial mats as their salinity prevents the development of grazing communities (Des Marais, 1995). Being often close to saturation with respect to halite, gypsum and calcium carbonates, it can be difficult to differentiate abiotic mineralization from organomineralization. The role of EPS as a matrix for mineral nucleation, particle binding or cation concentration is a cornerstone of the calcium carbonate mineralization in hypersaline sediments (Dupraz *et al.*, 2004; Dupraz & Visscher, 2005). We address EPS recognition and its role in the Dead Sea realm, which is known to precipitate aragonite abiotically from the lake water column (Stein *et al.*, 1997). In addition, the iron-sulphur mineralization previously recognized as diagenetic phases in palaeomagnetic studies (Ron *et al.*, 2006) is further investigated. Studies of the subsurface waters of the Dead Sea Basin have highlighted major redox transformations affecting carbon, iron and sulphur among other elements (Avrahamov *et al.*, 2010, 2014; Kiro *et al.*, 2013). It is therefore necessary to assess the extent of such changes and whether they reach the deepest region of the Dead Sea Basin, or if they are constrained to the aquifer/brine interface at its shoulders.

GEOLOGICAL AND LIMNOLOGICAL SETTING

The Dead Sea is located on the lowest spot on continental Earth, at a current elevation of 428·9 m b.s.l (2015). Today's water salinity is among the highest on Earth for a lake of this size, reaching 348 g L^{-1} of total dissolved salts (Oren & Gunde-Cimerman, 2012). Intensive use of catchment area water for irrigation, and potash mining have led to a decrease in water level over the last 40 years at an average rate of 0·7 m per year (Abu Ghazleh *et al.*, 2009; Tahal Group and the Geological Survey of Israel, 2011). Rare freshwater inputs bringing nutrients and carbonate ions to the lake are the Jordan River, and in rainy periods, flash floods fed by canyons (wadis) cutting through the Judean Mountains and the Jordanian Plateaus.

Before 1979, sufficiently rainy conditions had allowed a diluted epilimnion to establish through the mixing of freshwater with the calcium-chlorine brine of the Dead Sea. Anoxic conditions developed in the resulting hypolimnion, leading to high hydrogen sulphide concentration and the occurrence of iron sulphide explained by bacterial sulphate reduction activity (Nissenbaum, 1975; Nishri & Stiller, 1984). These conditions disappeared quickly from the water column after complete overturn in the winter of 1978–1979 (Steinhorn *et al.*, 1979). Exceptionally rainy winters in 1980 and 1991 led to the renewed formation of temporal meromictic conditions. The diluted epilimnion constituted a less extreme environment in which the halophilic Green Algae *Dunaliella* could develop and bloom, imparting a red colour to the lake (Oren & Shilo, 1982; Oren *et al.*, 1995).

An array of evidence suggests the presence of relatively similar conditions supporting the influence of life on the geochemical record of the Lake Lisan sediment. For example, within the Lisan Formation in the Perazim Valley (west of the Dead Sea southern basin), a general increase in the $\delta^{13}C$ of the aragonite laminae deposited during the last glacial period has been interpreted as a result of increasing autotrophic activity in the palaeo-epilimnion (Kolodny *et al.*, 2005).

Sulphur isotopes from gypsum intervals also support the occurrence of active sulphate reduction in the water column and bottom sediment of the lake (Torfstein *et al.*, 2005, 2008). Negative $\delta^{34}S$ values from disseminated gypsum and thin gypsum laminae of the Lisan Formation (from $-26‰$ to $1‰$) are interpreted as being a result of bacterial sulphate reduction in the anoxic hypersaline palaeo-hypolimnion. Sulphur concretions from the Masada section also bear lighter $\delta^{34}S$ isotopes than the surrounding gypsum values (27 to 29 ‰ lower) and iron and manganese concentration gradients were interpreted to result from incomplete microbial sulphate reduction terminating at S^0 (Bishop *et al.*, 2013).

The two limnological regimes of the lake are reflected in the nature of the evaporitic minerals precipitating in both situations, constituting a good proxy for changes in the limnological regime. During meromictic conditions, aragonite precipitation dominates (Stein *et al.*, 1997). This is emphasized in outcrops and core material by alternating laminae of aragonite and detrital marls (AAD), indicative of glacial periods (Stein *et al.*, 1997; Neugebauer *et al.*, 2014). During periods of increasing aridity, the lake becomes holomictic and gypsum precipitation occurs, followed by halite deposition. This type of deposit, intercalated by detrital sediment is largely dominant in interglacial periods. Since carbon and sulphur cycles vary greatly from one situation to another, our analyses are linked to these different types of facies in order to tackle the microbial impact on the Dead Sea sediment.

MATERIAL

A microbial mat was collected from an ephemeral saline pond on the western shore of the Dead Sea, near the En Qedem spring site (Fig. 1B). Salinity of the pond has reached 36·9%, with concentrations of major elements as follows: Na^+ = 2·12 M, Ca^{2+} = 0·58 M, Mg^{2+} = 1·59 M and Cl^- = 6·11 M and SO_4^{2-} = 18·9 mM, slightly above those of the current Dead Sea water (Ionescu *et al.*, 2012). The upper part of the mat consists of a gypsum and halite-rich biofilm (Fig. 1E) coloured by red pigments originating mainly from the carotenoids of *Dunaliella* cells, which have developed in the diluted water of the ephemeral pond (Fig. 1D). Microbial communities were analysed (Thomas *et al.*, 2015) and are mainly composed of heterotrophic *Archaea* of the *Halobacteria* class, while almost no bacterial DNA could be sequenced. Aragonite stellate clusters are also present in the mat as a result of aragonite needles precipitating from the pond water (Fig. 2A). Below this EPS-rich red layer (Fig. 2B and C), alternating laminae of pure aragonite (white), mixed aragonite and detrital material (light grey), and detrital sediment only (dark grey) are found (Fig. 2D). They form subcentimetre alternations of typical uncompacted AAD. Based on the extracted 16S rRNA gene sequences, the aragonite lamina host a similar archaeal community to the immediately overlying mat crust (Thomas *et al.*, 2015). However, the bacterial community seems more diverse, with halophilic fermenters such as *KBI Candidate Division* and *Halanaerobiaceae*, and rare *Desulfohalobium*-related organisms (halophilic sulphate reducer).

A cube of sediment was cut using sterile implements, immediately wrapped in Parafilm® M and stored at −4°C. Subsampling was done using a sterile scalpel and picks for various microscopy analyses in the geomicrobiology laboratory at the Department of Earth Sciences, University of Geneva.

The core material was taken from ICDP hole 5017-1A drilled during the DSDDP expedition. This core was retrieved in November-December 2010 from the middle of the Dead Sea (N 31°30′28·98″, E 35°28′15·60″) at a depth of 297 m. The core material was sampled either onshore from core-catchers at a specially tailored geomicrobiology laboratory, at Ein Gedi, Israel, or during the core-opening parties at GFZ Potsdam, Germany in June, 2011. In all cases, the outermost parts of the samples were discharged in order to avoid oxidation effects and contamination with drilling fluids.

METHODS

Scanning electron microscope

Samples were examined on smear slides after ethanol drying by use of an optical microscope. Halite samples were

Fig. 1. Overview of sampling sites and material. (A) Map of the Levantine region with Dead Sea location. (B) Location of drilling site 5017-1A (green square) and of the Qedem area, where the microbial mat was sampled (red square). (C) View of the ICDP-drilling platform at site 5017-1A. (D) Hypersaline pond north of Ein Qedem, where the microbial mat sample was taken. The pool is *ca* 15 m long. (E) Photograph of the mat (red, orange and green in colour) with underlying laminated sediments (grey to white).

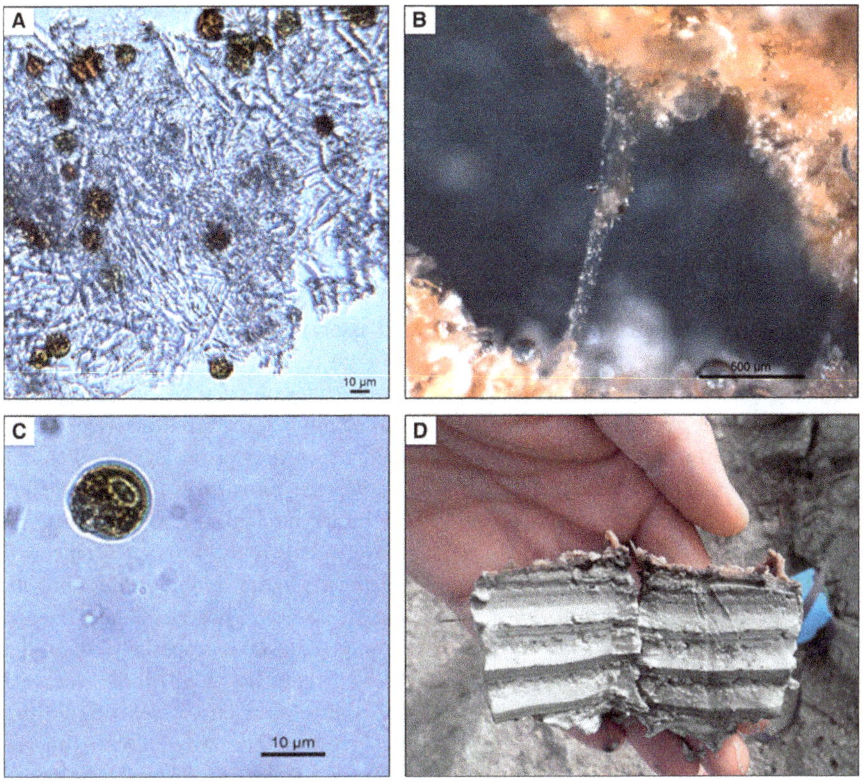

Fig. 2. (A) Aragonite needles and Dunaliella (orange circles) in the water of the ephemeral pond in Ein Qedem. (B) Photograph of EPS thread within a cavity of the gypsum-halite mat crust. (C) Close-up view of Dunaliella algae. Note that the nucleus is visible. (D) Section of the mat and underlying sediment exhibiting the AAD behaviour typical of the Lisan Formation.

already dry and were only polished with sand paper and clean alcohol. Scanning electron microscopy (SEM) was performed after ethanol and air drying, and critical point drying on a Jeol® JSM-7001 FA at the University of Geneva. Samples were then mounted on an aluminium stub with double-sided conductive carbon tape and gold-coated (15 nm) by low vacuum sputter coating.

Cryo-SEM was performed at C-SEM, in Neuchâtel, Switzerland on a Philips XL30 ESEM –FEG equipped with a Gatan Alto cryo-transfer system. The microscope was operated at 20 kV under a pressure of less than 10^{-4} mbar. The sample was freeze-dried in liquid nitrogen, transferred to the cryo-chamber stabilized at −140°C, where it was sectioned perpendicularly to the mat and sediment laminae using a blade, sublimated and sputtered with platinum. The sectioned surface was then observed.

Polished halite sections were examined using an Environmental Scanning Electron Microscope (ESEM) FEI Quanta 200 at the Scientific and Technological Centers, Universitat de Barcelona (CCiTUB). Chemical identification was performed by EDS.

Energy dispersive X-ray spectrometry (EDS) and back-scattered electron microscopy were used for chemical analysis on SEM, ESEM and cryo-SEM, allowing semi-quantitative determination of mineral chemistry.

X-ray fluorescence

Mapping of different elements within a section of microbial mat was performed using micro X-ray fluorescence (µXRF) at the University of Geneva under an EAGLE µProbe machine with a rhodium tube at a voltage of 40 kV and a current of 420 µA with a 50 µm spot. Mineral distribution was then interpreted based on each image and relative colour intensity using the software of the Adobe® Collection Suite and the Vision32® software version 4·953.

Elemental composition of core 5017-1A was measured using XRF at GFZ Potsdam, Germany on an ITRAX corescanner of COX Analytical Systems. A chromium tube was used at 30 kV and 30 mA. These data only allow for qualitative interpretations as they originate from raw counts (in counts per seconds) with sufficient counting statistics from un-calibrated quick core section scans as detailed by Neugebauer *et al.* (2014).

Electron microprobe

Magnetic fractions were collected as follows: sediment samples were extracted using plastic cubes, gathered using a magnetic probe and mounted in epoxy. The samples were analysed using a JEOL JX8230 superprobe EPMA with EDS and four wavelength-dispersive spectrometers (WDS) for microanalysis at the Institute of Earth Sciences, Hebrew University of Jerusalem. Both back-scattered electron (BSE) and secondary electron images were taken. Beam conditions for EDS analysis were set to 15 keV and 15 nA, and for WDS mapping 15 keV and 50nA. All phases were analysed for titanium, iron, sulphur, and oxygen using silicate and oxide standards. Data were processed with a PRZ correction procedure. The WDS measurement error is 0·5%. The difference between greigite and pyrite is bigger, than the error. Beam precision is 0·2 μm, allowing small crystals to be measured (as compared with 1 μm for EDS). In addition to the standard mineral correction, a matrix correction for the epoxy was also performed.

RESULTS

X-ray microfluorescence scanning on the mat

The facies of the microbial mat and its underlying sediment are similar to those defined in the cores (Neugebauer *et al.*, 2014). Halite mixed with gypsum is analogous to that of interglacial periods, except that it is completely embedded in EPS. Aragonitic laminae alternating with detritus show identical distribution to those of the AAD facies. The chemical compositions of the laminae are highlighted by key elements (Fig. 3). Chlorine is present in high and homogenous concentrations in the sediment and halite precipitated on the surface of the section (topography image) as the sample dried while being mapped at high resolution. Sulphur is concentrated in two laminae, and correlates with the strongest calcium peaks, indicating the presence of gypsum minerals. As with chlorine, the distribution of calcium (green) within the sediment is relatively homogeneous as it is highly concentrated in the Dead Sea water. Higher fluorescence of calcium corresponds to gypsum minerals. Aragonite is best mapped with strontium, showing the presence of aragonitic precipitation in light grey to white laminae of the mat (light blue in μXRF). It is intercalated by detrital input mapped under iron bands (light yellow). Diffuse detritus is also present below this band, imparting a greyish colour to the aragonite laminae. The 'Fe × S' mapping allows the areas of greatest iron sulphide mineralization to be highlighted. Although some background noise is observed due to the superposition of a gypsum and detritus signature, higher fluorescence identifies iron sulphide minerals, as pyrite, greigite or other transition phases and is represented in blue in the sketch (Fig. 3). The main iron-sulphur phases are observed at the margins of the detrital lamina, within the aragonite laminae, rather than within the iron-rich detritus lamina itself.

X-ray fluorescence scanning in the core

In the core, detailed XRF profiles highlight the lithologies and associated facies (further detailed in Neugebauer *et al.*, 2014). The 'Fe × S' curve is superimposed over the sulphur profile of core 1A, in order to emphasize areas of iron-sulphur mineralization (Fig. 4). Sulphur peaks are mainly seen in gypsum levels and sulphur concretion-rich intervals. In some cases (e.g. 100 m, 110 m, 160 m, 340–350 mblf), the high 'Fe × S' signatures of sulphur concretions overlap with the signal generated by iron-sulphur minerals as figure resolution does not allow for differentiation on a centimetre scale. Higher 'Fe × S' peaks are found between 0 and 80 m, especially within the 45 to 75 m interval where the 'Fe × S' profile does not always correlate with the sulphur curve, hence indicating other signatures than gypsum and native sulphur concretions (i.e. iron sulphide minerals). In general, the intensity is lower for 'Fe × S' profiles below 100 m, regardless of measured sulphur fluorescence. Examples of the morphologies and size of iron sulphide minerals are presented on the right side of Fig. 4, together with the corresponding EDS spectra.

Morphology of iron-sulphur phases

The chemistry and morphology of iron-sulphur minerals varies independently of depth. Euhedral minerals forming an octahedron have been found at a depth of 65·38 m (Fig. 5A) in the 'laminated detritus' facies (*ld*, Neugebauer *et al.*, 2014). Micron-scale spheroids were also found in such facies and also in gypsum intervals at 90·64 m (Fig. 5B), while rough-surface spherulites ca. 5 μm in diameter were retrieved at 337·62 m in the AAD (Fig. 5C).

Euhedral pyrite, typically 5 to 20 μm in diameter, occurs at the boundary between the detrital and aragonitic layers of the microbial mat (Fig. 5A). Core examples are much smaller. They are generally individual euhedrons of 5 μm maximum, sometimes presenting overgrowth (Fig. 5B). True pyrite framboids were observed in the mat (Fig. 5C) and consist of small octahedrons. In the core, framboids are rare and of limited size. They are formed of aggregated micro-spheroids (Fig. 5D) or crystals (Fig. 5E and F). This form may be a precursor phase of micron-scale spherules or larger rough-surface spherulites (Fig. 5G and H, respectively).

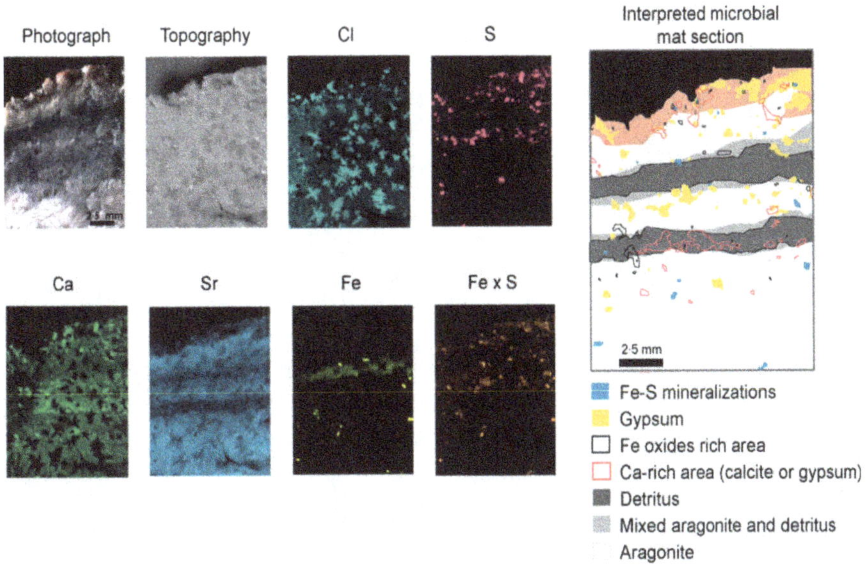

Fig. 3. Micro-XRF mapping of a microbial mat and the laminated sections of modern sediments along the Dead Sea shore. The dominant mineral is authigenic aragonite (Sr-mapping) with some gypsum (S and Ca). The darker layer is detrital material (calcite and clays). Disseminated Fe-sulphide is outlined by the 'Fe × S' mapping, and is principally found in aragonitic layers. Star shaped crystals are the result of secondary precipitation caused by desiccation of the samples during measurement. See Methods for details on the interpreted sketch. Scale bar is the same for each scanned image and is 2·5 mm.

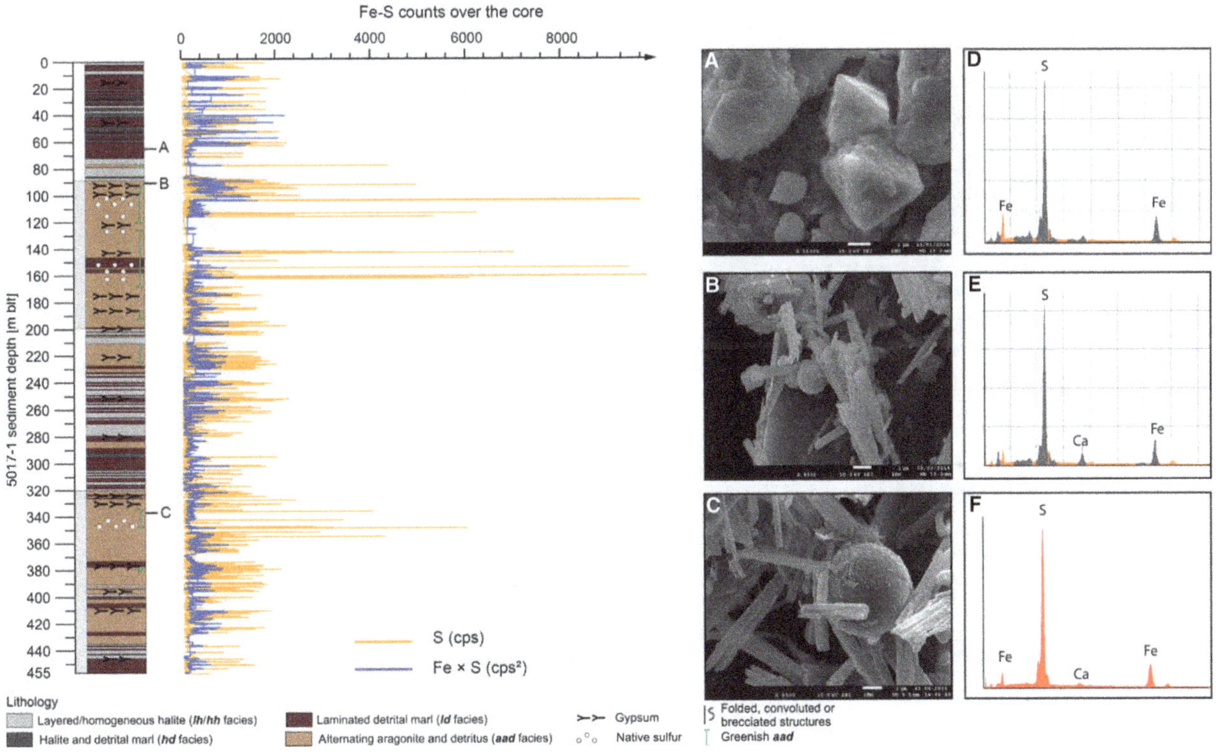

Fig. 4. Interpreted lithological profile of core 1A taken from Neugebauer *et al.* (2014) and corresponding S and Fe × S profiles as given by XRF scanning. On the right hand, SEM photographs of (A) micron-sized euhedral Fe-S mineralization from core 1-A-32 at 65·38 m. (B) Fe-S spheroid at 90,64 m, core 1-A-43. (C) rough- surface spherulite from core 1-A-131 at 337·62 m with the corresponding EDS spectra (D, E and F respectively). Depth in metres below lake floor.

Fig. 5. Iron-sulphur mineralization morphologies in the Dead Sea realm. (A) Euhedral pyrite from a white aragonitic lamina of the mat. Aragonite needles are visible in the background. (B) Similar euhedral morphology of Fe-S mineral in core 1-A-32, at 65·38 mblf. (C) Cryo-SEM picture of a framboidal pyrite embedded in EPS at the mat-sediment interface of the mat sample. (D) Agglomerate of four framboidal mineralizations in the core (65·38 mblf). (E) Back-scattered electron imaging of a Fe-S rough-surface spherulite being formed in core 1- A-30, at 59·29 mblf. (F) Close-up view of (E) in secondary electron imaging, emphasizing the rough droplet like topography of the mineralization. (G) Micron-scale Fe-S spheroid at 90·64 mblf, core 1-A-43. (H) Rough surface spherulite of micron scale at the tip of aragonitic stellate cluster in core 1-A-131, at 337·62 mblf. Pictures were all taken under secondary electron imaging in a regular SEM, except when specifically stated.

Chemistry of iron-sulphur phases

In the core, iron-sulphur minerals rarely reach the iron-sulphur ratio characteristic of pyrite (S/Fe = 2; Table 1 and Fig. 6) except for the largest examples (> 5 µm), which approach true pyritic compositions (FeS_2). They are either euhedral minerals in the mat (Fig. 5A), or rough-surface spherulites at 350 m in the core (Fig. 4C).

In some dark, high magnetic susceptibility laminae associated with AAD (Fig. 7A), iron sulphide rims with a composition of greigite are observed around some titano-magnetite grains (Fig. 8). The titanomagnetite appears partially dissolved and iron-sulphur minerals also form on cracks. The composition of the iron-sulphur rim varies from greigite to pyrite depending on depth. Such rims are found both in the Holocene interglacial sediment and in the Lake Lisan-associated sediment (last glacial period). Greigite is also observed infilling chambers of organisms (e.g. foraminifera) in the Holocene AAD sediment (Fig. 9). Measurements of the atomic ratio of these intra-skeletal iron-sulphur minerals revealed S:Fe values between 1·32 and 1·35, characteristic of greigite (Fig. 9 and Table 1). Except in the mat, EDS and WDS measurements show rather continuous S/Fe ratios, with higher values (pyrite-like) attributed to deep AAD samples (140–151 m, 337 m) and mackinawite to greigite chemistry for others (Table 1 and Fig. 6).

Although not fully characterized, iron-sulphur mineral phases also occur around gypsum in detrital-rich halite intervals (Fig. 10 and Table 1). Most of the other minerals, generally in the form of spherules or octahedrons, have a similar chemistry to greigite (Fe_3S_4). Few iron-sulphur precipitates lie below the mackinawite line (Fe_8S_9), and they are sub-micron in size. The resolution for EDS spot analysis is superior to 1 µm in length and width. Data from smaller minerals should be considered with caution and have been highlighted in Fig. 6.

Microscopic analysis of EPS

Exopolymeric substances have been clearly observed in the mat, with the naked eye, as well as with a regular microscope (Fig. 2) and with cryo-SEM. A cerebroid structure under cryo-SEM facilitates EPS identification in the mat (Fig. 11A) together with a transparent film-like texture in close-up view, frequently with embedded aragonite stellate clusters (Fig. 11B). This texture differs from the characteristic honeycomb morphology generally observed in cryo-SEM sections (Defarge *et al.*, 1996; Dupraz *et al.*, 2004). This difference is interpreted as being a result of the extreme salinity of the Dead Sea.

In regular SEM mode, the EPS are very smooth and appear transparent, often covering the sediment (Fig. 11C). The EPS in the core have a filmy texture, sometimes folding in on itself, or torn apart during preparation (Fig. 11D), possibly presenting voids and more rugged surfaces. Aragonitic stellate clusters or individual needles are often found embedded in these EPS.

Cryo-SEM observations of the mat sample allowed sharp, freshly precipitated aragonite stellate clusters to be distinguished embedded in the cerebroid texture typical of the EPS (Fig. 12A). The thin needles constituting the stellate cluster are individually covered by the biofilm in close-up view (Fig. 12B). Rough surfaces on the torn EPS are formed by nanoglobules (Fig. 12C). They are also visible between aragonitic needles, apparently entangled in this biological structure (Fig. 12D). The needles here are very thin and elongated, without sharp tips nor smooth surface. In the core, sharp and thin aragonite stellate clusters, as well as fragments of thick ones may emerge or pierce through the EPS (Fig. 12E). They also are found aggregated on its surface (Fig. 12F), or in between EPS and gypsum minerals, where the EPS host numerous iron sulphide minerals, as demonstrated by back-scattered imaging (Fig. 12G). At depths greater than 20 mblf (Table 1), EPS-like structures could not be identified although smooth film-like textures were found associated with aragonitic stellate clusters (Fig. 12H). Here, they correspond to halite minerals, demonstrating the complexity of untangling biotic features from abiotic ones in the Dead Sea sediment.

DISCUSSION

Indicators of microbial activity in the Dead Sea sediment

EPS

Examples of EPS are well represented within the microbial mat (Fig. 2B and D). Their recognition is generally difficult in the wet Dead Sea sediment. Numerous poorly investigated precipitates, occurring in the Dead Sea naturally or upon desiccation before SEM observation, may take on a smooth sheet-like structure (Fig. 11H) similar to documented textures for EPS (Spadafora *et al.*, 2010). Chemical analysis of hypothesized EPS is not equivocal as X-ray penetration of EDS generally exceeds the EPS thickness. Loading of these structures with calcium, magnesium and chloride also tends to mask its carbon and oxygen content (Fig. S1 of Supplementary Material). Morphological characterization of EPS, therefore, was often the most reliable solution and analysis of chemical composition and morphology under cryo-SEM mode allowed for conclusive identification.

In the core, EPS are also observed. However, their observation is limited mainly to the top of the core

Table 1. Summary of the samples observed and measured under scanning electron microscope (SEM), environmental scanning electron microscope (ESEM) and electron microprobe (EPMA). Changes in shade mark changes in sample (hence in depth).

Core (5017-A)	Depth (mblf)	Lithology	Type of analysis	S/Fe ratio	EPS observed
1-1	0	lh/hh	ESEM	?	n/a
1-1	0,24	lh/hh	SEM	–	Yes
2-2	2,74	aad	EPMA	1,36	Yes
2-2	2,74	aad	EPMA	1,32	
2-2	2,74	aad	EPMA	1,30	
2-2	2,74	aad	EPMA	1,36	
2-2	2,74	aad	EPMA	1,30	
2-2	2,74	aad	EPMA	1,31	
2-cc	3,29	lh/hh	SEM	?	No
3-cc	6,34	lh/hh	SEM	?	No
5-cc	9,39	lh/hh	SEM	?	No
13-cc	23,59	ld	SEM	?	Yes
29-cc	57,26	gy	SEM	1,25	No
29-cc	57,26	gy	SEM	0,69	
30-cc	59,29	ld	SEM	1,08	No
30-cc	59,29	ld	SEM	1,30	
30-cc	59,29	ld	SEM	1,33	
30-cc	59,29	ld	SEM	1,33	
30-cc	59,29	ld	SEM	1,38	
30-cc	59,29	ld	SEM	0,79	
30-cc	59,29	ld	SEM	1,08	
30-cc	59,29	ld	SEM	0,86	
30-cc	59,29	ld	SEM	1,30	
30-cc	59,29	ld	SEM	1,33	
30-cc	59,29	ld	SEM	1,50	
30-cc	59,29	ld	SEM	1,38	
30-cc	59,29	ld	SEM	1,08	
30-cc	59,29	ld	SEM	0,79	
32-cc	65,38	aad	SEM	0,59	No
32-cc	65,38	aad	SEM	1,23	
32-cc	65,38	aad	SEM	1,42	
32-cc	65,38	aad	SEM	1,17	
32-cc	65,38	aad	SEM	1,46	
32-cc	65,38	aad	SEM	1,23	
32-cc	65,38	aad	SEM	1,61	
32-cc	65,38	aad	SEM	1,37	
32-cc	65,38	aad	SEM	1,33	
32-cc	65,38	aad	SEM	1,46	
32-cc	65,38	aad	SEM	1,38	
32-cc	65,38	aad	SEM	1,40	
32-cc	65,38	aad	SEM	1,35	
32-cc	65,38	aad	SEM	1,45	
32-cc	65,38	aad	SEM	1,69	
32-cc	65,38	aad	SEM	1,68	
32-cc	65,38	aad	SEM	1,47	
32-cc	65,38	aad	SEM	1,22	
32-cc	65,38	aad	SEM	1,45	
32-cc	65,38	aad	SEM	1,14	
32-cc	65,38	aad	SEM	1,66	
32-cc	65,38	aad	SEM	0,95	
32-cc	65,38	aad	SEM	1,46	
32-cc	65,38	aad	SEM	1,47	

(Continued)

Table 1. Continued.

Core (5017-A)	Depth (mblf)	Lithology	Type of analysis	S/Fe ratio	EPS observed
32-cc	65,38	aad	SEM	1,19	
32-cc	65,38	aad	SEM	0,98	
32-cc	65,38	aad	SEM	1,72	
33-cc	67,38	ld	SEM	1,24	No
33-cc	67,38	ld	SEM	1,45	
33-cc	67,38	ld	SEM	1,22	
33-cc	67,38	ld	SEM	1,57	
33-cc	67,38	ld	SEM	0,97	
33-cc	67,38	ld	SEM	1,47	
33-cc	67,38	ld	SEM	1,59	
33-cc	67,38	ld	SEM	1,40	
33-cc	67,38	ld	SEM	1,63	
33-cc	67,38	ld	SEM	1,36	
33-cc	67,38	ld	SEM	1,06	
33-cc	67,38	ld	SEM	1,40	
33-cc	67,38	ld	SEM	1,48	
33-cc	67,38	ld	SEM	1,64	
33-cc	67,38	ld	SEM	1,49	
33-cc	67,38	ld	SEM	1,25	
33-cc	67,38	ld	SEM	1,06	
33-cc	67,38	ld	SEM	1,14	
33-cc	67,38	ld	SEM	1,31	
33-cc	67,38	ld	SEM	1,33	
36-1	73,425	lh/hh	ESEM	–	n/a
37-2	75,425	lh/hh	ESEM	?	n/a
37-cc	75,58	ld	SEM	1,59	no
37-cc	75,58	ld	SEM	1,00	
37-cc	75,58	ld	SEM	0,56	
37-cc	75,58	ld	SEM	1,51	
37-cc	75,58	ld	SEM	1,22	
37-cc	75,58	ld	SEM	0,72	
37-cc	75,58	ld	SEM	0,42	
40-1	82,8	lh/hh	ESEM	?	n/a
41-1	85,445	lh/hh	ESEM	–	n:a
43-cc	90,64	gy	SEM	1,45	No
43-cc	90,64	gy	SEM	1,62	
52-cc	118,08	ld	SEM	1,39	No
52-cc	118,08	ld	SEM	1,45	
53-cc	121,13	ld	SEM	–	No
60-3	142,88	ld	EPMA	2,19	n/a
60-3	142,88	ld	EPMA	1,21	n/a
61-1	144,08	ld	EPMA	1,94	n/a
61-1	144,08	ld	EPMA	1,98	
61-1	144,08	ld	EPMA	1,88	
61-1	144,08	ld	EPMA	1,75	
61-1	144,08	ld	EPMA	1,66	
61-1	144,08	ld	EPMA	1,73	
61-3	146,13	aad	EPMA	1,85	n/a
61-3	146,13	aad	EPMA	2,25	
61-3	146,13	aad	EPMA	2,29	
61-3	146,13	aad	EPMA	2,24	
61-3	146,13	aad	EPMA	1,98	
61-3	146,13	aad	EPMA	1,97	
61-3	146,13	aad	EPMA	2,31	
61-3	146,13	aad	EPMA	2,35	

(Continued)

Table 1. Continued.

Core (5017-A)	Depth (mblf)	Lithology	Type of analysis	S/Fe ratio	EPS observed
61-3	146,13	aad	EPMA	2,30	
61-3	146,13	aad	EPMA	2,11	
63-3	151,47	ld	EPMA	1,83	n/a
63-3	151,47	ld	EPMA	2,01	
63-3	151,47	ld	EPMA	1,95	
63-3	151,47	ld	EPMA	1,85	
63-3	151,47	ld	EPMA	0,94	
63-3	151,47	ld	EPMA	1,82	
63-3	151,47	ld	EPMA	2,03	
63-3	151,47	ld	EPMA	2,01	
80-1	199,175	lh/hh	ESEM	–	n/a
81-2	202,935	lh/hh	ESEM	?	n/a
88-2	224,6	ld	SEM	?	no
91-2	233,38	lh/hh	ESEM	?	n/a
91-2	233,52	lh/hh	ESEM	?	n/a
93-1	236,2	lh/hh	ESEM	–	n/a
93-1	237,03	lh/hh	ESEM	?	n/a
95-1	241,41	lh/hh	ESEM	–	n/a
97-2	248,24	lh/hh	ESEM	?	n/a
97-2	248,43	lh/hh	ESEM	?	n/a
98-2	250,665	lh/hh	ESEM	–	n/a
98-3	253,052	lh/hh	ESEM	–	n/a
100-2	254,82	lh/hh	ESEM	–	n/a
100-cc	256,955	lh/hh	ESEM	?	n/a
103_3	263,383	lh/hh	ESEM	–	n/a
104-2	266,5	lh/hh	ESEM	?	n/a
106-2	272,21	lh/hh	ESEM	–	n/a
108-2	278,534	lh/hh	ESEM	?	n/a
125-1	319,865	lh/hh	ESEM	?	n/a
125-1	319,93	lh/hh	ESEM	?	n/a
131-cc	337,62	aad	SEM	1,77	no
131-cc	337,62	aad	SEM	1,93	
174-1	437,59	lh/hh	ESEM	–	n/a
174-1	438,84	lh/hh	ESEM	?	n/a
176-2	445,35	lh/hh	ESEM	?	n/a
176-2	445,7	lh/hh	ESEM	?	n/a
Mat layer	**Depth (mblf)**	**Lithology**	**Type of analysis**	**S/Fe ratio**	**EPS observed**
Mat surface	0,01	ha-gy in EPS	SEM	?	Yes
Mat detrital	0,06	aad	SEM	2,04	Yes
Mat aragonite	0,16	aad	SEM	2,03	No
Mat aragonite	0,16	aad	SEM	1,98	No

S/Fe ratios were calculated after EDS measurements for SEM and WDS measurements for EPMA. They were not calculated on ESEM but are annotated '?' when iron sulphides were observed, and 'no' when no iron sulphide phases could be found. EPS presence/absence is given for the observed samples on SEM. They were not looked for using ESEM and EPMA (n/a). Depth is in metres below lake floor. Lithology codes are described in Neugebauer *et al.* (2014). Lh/hh layered and homogenous halite, hd halite and detrital marl, aad alternating aragonite and detritus, ld laminated detrital marl, gy gypsum laminae or layers. For comparison, stoichiometric ratios of iron sulphide minerals are: $S/Fe_{pyrite} = 2$; $S/Fe_{greigite} = 1{,}33$; $S/Fe_{mackinawite} = 0{,}89$.

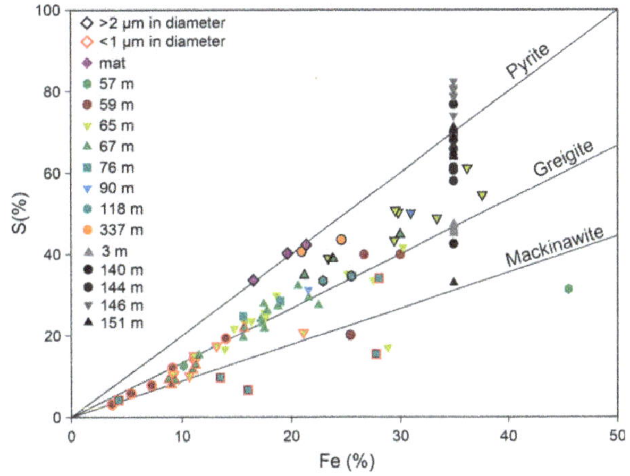

Fig. 6. Iron-sulphide mineral compositions taken from EDS measurements (colour) and EPMA (black and grey). The microprobe ratios have been normalized to a given iron percentage (35%) for better understanding. EDS measurements from small minerals (<1 μm) are circled in red to stress the bias induced by this type of measurements on micron-sized objects (see Methods). Pyrite line as y = 1/2(x), greigite line as y = 3/4(x) and mackinawite line as y = 9/8(x).

(Table 1) with minimal recognition below that. Extraction of EPS was attempted but did not work as a large quantity of salt was carried along in the process, inducing a large bias analytically. The failure to identify EPS below 23 m suggests either (i) that they were missed, (ii) degradation with time in the deeper sediment or (iii) difference/absence of production in the first place. The finding of EPS in the upper part of the core suggests that microbial activity has occurred at one point in the sediment. Their disappearance deeper in the core may also support anaerobic heterotrophy.

Protection from the high calcium and magnesium divalent cation concentrations is probably the main reason for EPS production in the Dead Sea. The bonding capacity of EPS towards divalent cations (Kawaguchi & Decho, 2002; Dupraz & Visscher, 2005) could lower *in situ* concentrations of these cations, which is the principal reason the Dead Sea is inimical to life (Nissenbaum, 1975; Bodaker *et al.*, 2010). The EDS measurements on EPS (Fig. S1 of supplementary material) show concentrations of calcium and magnesium in the EPS matrix; however, such measurements must be interpreted very carefully given analytical inaccuracy on such thin structures.

Iron sulphide minerals

In the microbial mat, several types of iron-sulphur minerals were observed. Framboidal pyrite was only located close to the mat while large euhedral pyrite occurred in the underlying aragonitic laminae (Fig. 3). In the core, however, although euhedral minerals were observed, they

Fig. 7. Photographs of the cores. (A)Black layer exhibiting peaks in magnetic susceptibility and greigite occurrences near AAD intervals. (B) Isolated S^0 concretions in reworked intervals. (C) S^0 concretions in disturbed and undisturbed AAD.

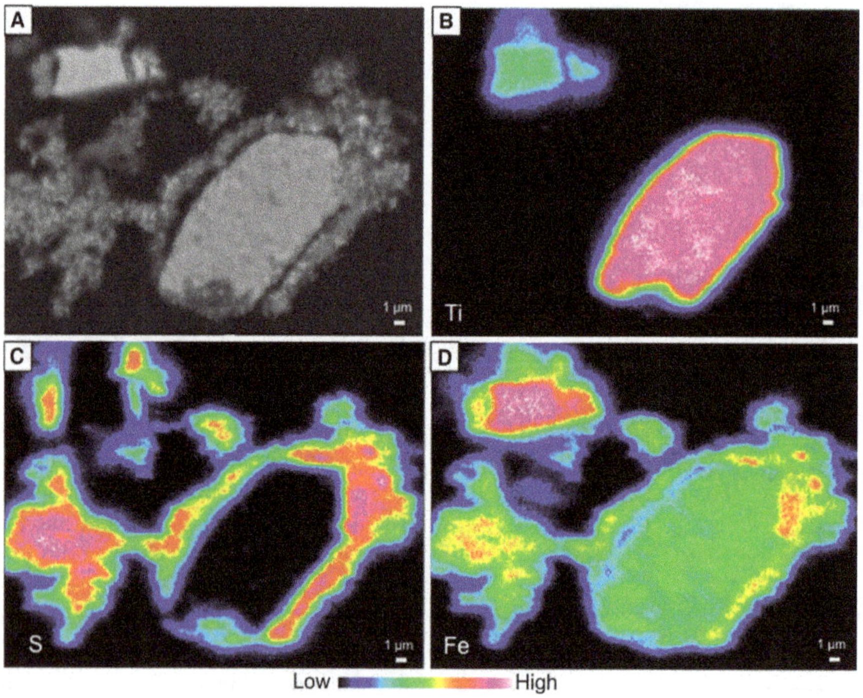

Fig. 8. Greigite rim around titanomagnetite in Holocene AAD sediment. (A) Back-scattered electron image. (B) WDS mapping of relative distribution of titanium highlights the titanomagnetite mineralogy. An outward depletion is observed at its rim. (C) WDS mapping of relative distribution of sulphur highlights the presence of Fe-S around the mineral (D) WDS mapping of relative distribution of iron shows enrichment in the Fe-S rim when compared to the internal chemistry of the titanomagnetite.

were generally small in size (<2 μm diameter). Their chemistry is also significantly different, with Fe/S ratios often limited to that of mackinawite or greigite (Fig. 6), and much less pyrite.

Linking morphology with formation processes is still a matter of debate. The composition and morphology of iron sulphide minerals results from numerous parameters such as the variety of reactive iron available, the

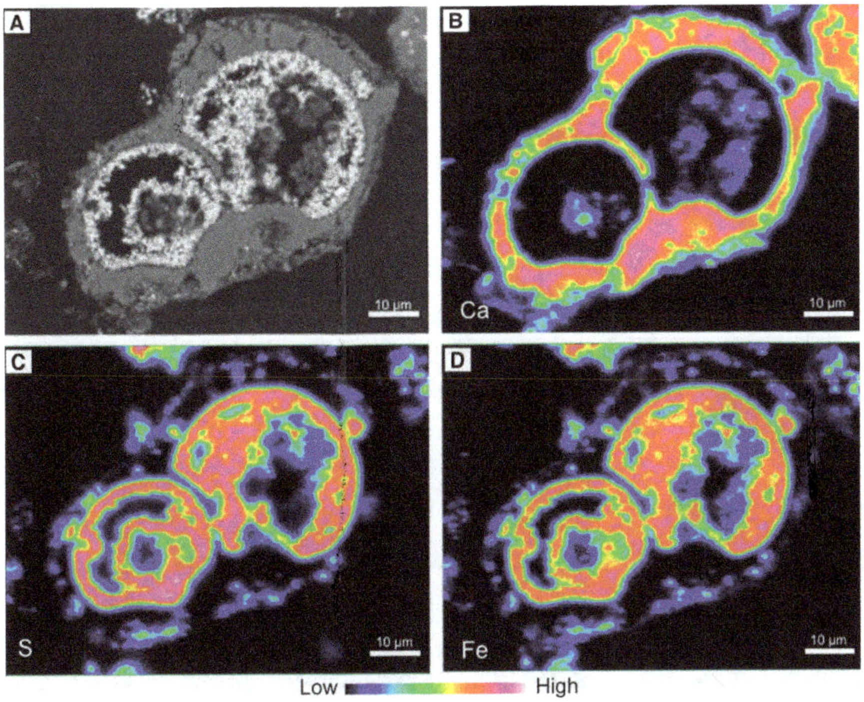

Fig. 9. Greigite and Fe-S infill within foraminifera. (A) Back-scattered electron imaging showing differences between the foraminifera lodges and its infill. (B) WDS mapping of relative distribution of Ca. (C) WDS mapping of relative distribution of S. (D) WDS mapping of relative distribution of Fe. Note the connection between the two chambers seen in (A). (C) sulphur and (D) iron concentrations mark the Fe-S mineral infilling, probably replacing the organic matter within the organism shell.

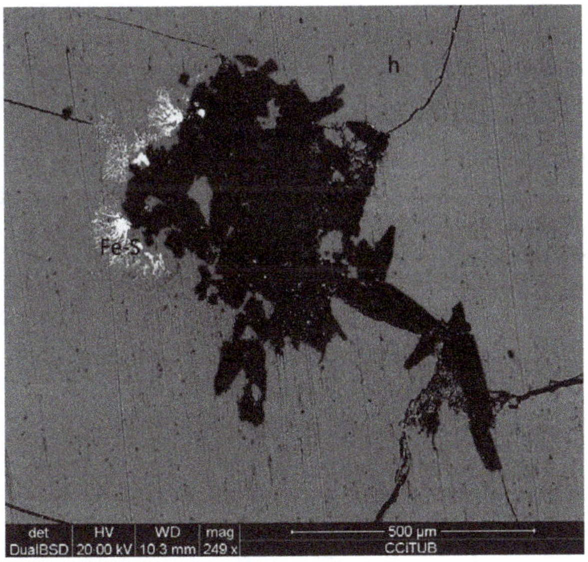

Fig. 10. Back-scattered electron photograph from SEM of a polished halite sample. Fe-S mineralizations develop around gypsum (g) inclusions within a halite (h) crystal.

saturation rate, the concentration of reduced sulphur and stability of redox conditions (Berner, 1984; Wang & Morse, 1996; Wilkin & Barnes, 1997; Wilkin & Arthur, 2001). A combination of observations from field and laboratory models attempting to mimic sedimentary processes have allowed characterization of framboidal, pentagonal, octahedral, tetrahedral, dodecahedral and icosahedral pyrite forms originating from biogenic factors (Astafieva *et al.*, 2005). This type of geometry encompasses all euhedral morphologies described above for the Dead Sea core and mat. Framboidal pyrites are formed of these euhedral structures as well as from micro-spheroids (Wilkin & Barnes, 1997), as observed in the core (Figs 4B and 5D), and have been demonstrated to be biomarkers in the sediment (Popa *et al.*, 2004; Vuillemin *et al.*, 2013).

Sulphate reduction has been indirectly documented in several studies of the Lake Lisan sediment (Gavrieli *et al.*, 2001; Torfstein *et al.*, 2005) and the stratified Dead Sea water column (Nissenbaum, 1975; Nissenbaum & Kaplan, 1976). Although less studied, microbial reduction of iron is also suggested as a source of ferrous iron in the sediment (Nishri & Stiller, 1984; Gavrieli *et al.*, 2001). Based on these observations, the presence of iron-sulphur minerals in the core of the Dead Sea argue for similar processes (microbial iron and sulphate reduction) in the deep Dead Sea Basin.

In addition, sulphur concretions were localized in the whole core (Neugebauer *et al.*, 2014). It is hard to explain their distribution as is seems that they occur both in well

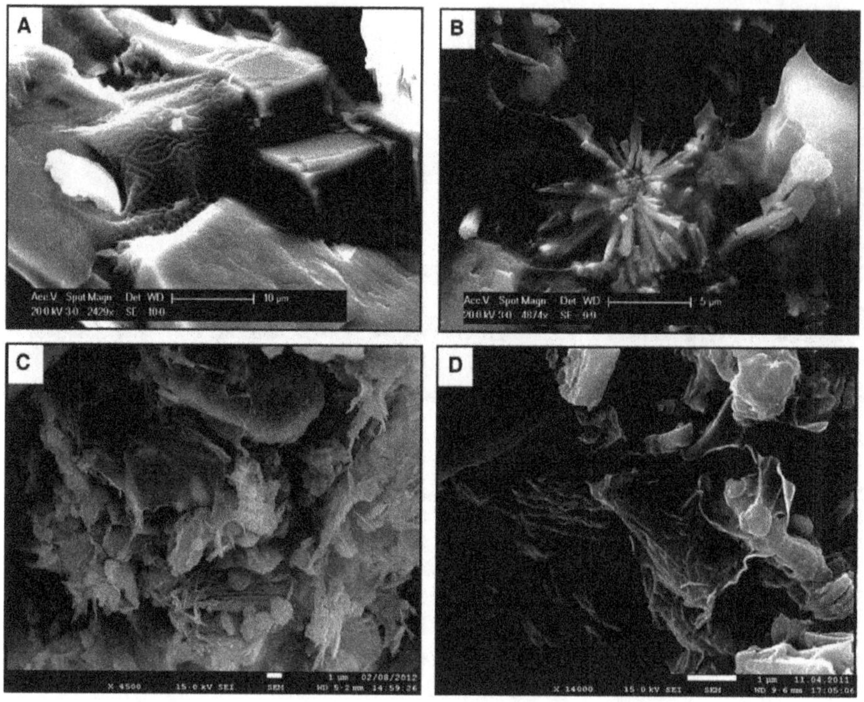

Fig. 11. Photographs of EPS morphologies in the mat (A) Cryo-SEM picture of EPS sheet in between gypsum crystals. Note the wrinkly cerebroid structure clearly defining the biofilm. (B) Cryo-SEM picture of an aragonite stellate cluster embedded in sheet-like EPS. (C) Regular SEM picture of sheet-like EPS covering detrital sediment. Globules are clay minerals, needles are aragonite. (D) Regular SEM picture of sheet-like EPS, torn and folded as it can occur in the core, here at 23·59 mblf in core 1-A-13.

preserved and in seismically disturbed AAD, as well as in mass wasting deposits (Fig. 7B and C). Different pathways can explain their formation, from incomplete sulphate reduction as suggested by Bishop *et al.* (2013) to oxidation of reduced sulphur species by turbidity events. Both pathways support the role of microbial sulphate reduction in the accumulation of sulphide or S^0 in the sediment of the Dead Sea Basin. The fact that the chemistry of iron sulphides does not systematically reach that of pyrite, and allows coexistence of greigite and pyrite, raises questions regarding the process of pyritization in this sedimentary environment. This issue is addressed in the next paragraphs.

Organic matter production and sulphate reduction

Microbial communities inhabiting the gypsum sediments are mainly composed of highly adapted halophilic *Archaea*, relying mostly on fermentation of varied organic substrates (Thomas *et al.*, 2014). Alternatively, communities inhabiting the AAD seem to rely on the degradation of biomass derived from moderate halophiles living within the stratified water column (Ariztegui *et al.*, 2015) and EPS production could be related to a stress response of these moderate halophiles. It could also originate from sulphate reducing microbes inhabiting these sediments, as they have been recognized to be important EPS producers (Bosak & Newman, 2005; Braissant *et al.*, 2007). Degradation of these EPS by fermenters or sulphate reducing bacteria may release organic molecules such as aminated compounds (Mishra & Jha, 2009). Such compounds would be subsequently available for osmoprotectant synthesis or uptake by community members utilizing a 'low-salt strategy' (Roesser & Müller, 2001). They could also support other communities such as methanogens as non-competitive substrate (Thomas *et al.*, 2014), enabling more activity and organic matter degradation.

In the microbial mat

The quality of organic matter is critical for microbial activity in deep lacustrine sediments (Glombitza *et al.*, 2013; Ariztegui *et al.*, 2015). In the Dead Sea, energy requirements for osmotic adaptation select for metabolisms yielding the largest amount of ATP for a given quantity of substrate (Oren, 2001). In such competition, labile substrates and nutrient availability have a major impact. Water column communities and shallow sedimentary ones, like those observed in the microbial mat,

Fig. 12. Photographs of EPS structures in the mat and the core. (A) Cryo-SEM back-scattered electron photograph of an aragonite stellate cluster embedded in wrinkly EPS. Close-up view (B) shows the tight relationship between aragonitic thin needles in the EPS matrix. (C) Nanoglobules on the surface of torn and folded EPS. In the back, EPS display vacuolar textures. (D) Aragonitic needles covered with EPS (as seen by torn and vacuolar textures) and with nanoglobule-like textures at their tips. Needles are thin, poorly defined and not smoothened here. (E-F) Aragonite stellate clusters embedded in EPS form core 1-A-2 at 2·74 mblf and 3·22 mblf. (G) Back-scattered electron imaging of Fe-S mineralizations embedded in EPS, in between aragonitic agglomerates (top) and gypsum mineral (bottom) from 2·74 mblf. (H) Aragonite stellate cluster at the surface of halite smooth surface at depth 337·62 mblf, in core 1-A-131. A-D from the mat, the rest from the core. All pictures taken in secondary electron imaging unless stated otherwise.

for example, will benefit from fresher and more labile organic matter and will subsequently be able to perform metabolic activity deeper ones cannot afford.

The mat microcosm benefits both from the presence of a salinity gradient and from fresh organic matter. Among others, *Dunaliella* algae probably developed and survived after fresh water inputs to a small, ephemeral pond (rain events occurred in December 2010; Fig. 2A and C). Mixing with Dead Sea water, dissolution of surrounding salt and evaporative process in such a shallow pond would quickly raise elemental concentrations to those observed at sampling time. The decay of these abundant autotrophs (Fig. 2A) provides highly labile organic matter to the shallow sedimentary community, allowing mat development. Such high-quality organic matter would encourage high rates of sulphate reduction, facilitating the formation of framboidal pyrite in its vicinity, as described by Passier *et al.* (1997) in the Mediterranean Basin. Indeed, framboids are only observed very close to the gypsum crust within the microbial mat, while large euhedral pyrite dominates in the aragonite layers.

Both microtextures have been shown to result from differential rates of pyritization (Wilkin & Barnes, 1996). In their demonstration of pyritization via iron loss from iron monosulphides, the authors show that fast pyritization leads to framboidal pyrite while euhedral crystals result from slow initial pyritization. Intense sulphate reduction in the microbial mat could allow framboidal pyrite to develop. Excess HS^- could diffuse away and react with iron monosulphide to form the large euhedral pyrites present in the aragonite lamina, at the interface with detrital layers (Fig. 3). Sulphur isotope measurements on iron-sulphur phases would allow testing for the formation of these minerals in open versus closed systems. The liberation of reduced iron derived from the neighbouring detritus layer would be sufficient to support the process of pyritization.

In the deep sediment

In the core, very few actual framboids were observed. Iron-sulphur minerals often occurred as micron-sized individual spherules, only rarely aggregating into framboids. At given intervals, euhedral iron-sulphur minerals similar (in morphology) to those observed in the mat, but often smaller in size, could be observed.

The chemistry of these iron-sulphur minerals extends from iron monosulphides to pyrite (Fig. 6 and Table 1). The smallest iron-sulphur phases have an iron-sulphur ratio between mackinawite and greigite. Iron monosulphides have been acknowledged as important precursors in the formation of pyrite (Berner, 1970; Canfield *et al.*, 1992; Wilkin & Barnes, 1996). Pyritization has been observed to occur through different pathways including: (i) the polysulphide pathway consisting of adding S^0 to the iron-sulphur precursor (Rickard, 1975; Luther, 1991); (ii) the ferrous iron loss pathway (Wilkin & Barnes, 1996) where oxidants such as O_2, nitrate, Fe(III), Mn(IV) or organic matter allow the release of a Fe^{2+} ion or (iii) the H_2S pathway consisting of a solid state reaction (Drobner *et al.*, 1990; Rickard & Luther, 1997; Pósfai *et al.*, 1998). Transformation of mackinawite to greigite, and greigite to pyrite necessitate similar reactants, although kinetic limitations highlighted by Hunger & Benning (2007) prevent the full transformation of mackinawite to pyrite in sulphur limiting conditions.

The data presented here are consistent with a pyritization process in which S^0 is added to iron-sulphur precursors and suggests a transitional path from mackinawite to greigite and eventually to pyrite in the Dead Sea core, with limiting S^0. Although no general rate can be observed since a clear correlation is not seen between pyrite content and depth, it seems that progressive sulphidation as suggested by Sweeney & Kaplan (1973) and Schoonen & Barnes (1991) allows pyrite formation in the Dead Sea. This sulphidation is most probably the result of microbial sulphate reduction in the Dead Sea sediment. Such a process has been identified in the present Dead Sea (Häusler *et al.*, 2014) where it can benefit from very sharp salinity gradients formed, for example, by underwater fresh water springs.

Sulphate reduction rates in the sediment are, however, supposed to be very slow because of the high salinity and low input of fresh carbon (Häusler *et al.*, 2014). It means that the production of sulphide from an allegedly infinite sulphate source in most of the deep Dead Sea sediment (as observed from gypsum saturation), is limited. Recently, Bishop *et al.* (2013) suggested that sulphate reduction in the Dead Sea sediment does not even proceed all the way to hydrogen sulphide, because of the lack of an electron donor. Such hypotheses are based on the sulphur isotope composition of native sulphur concretions found in the Lisan Formation. Bishop *et al.* (2013) argue that the absence of pyrite in the sulphate-rich Dead Sea sediment is linked to this incomplete sulphate reduction terminating at S^0, thus allowing mackinawite and greigite, and sometimes pyrite, to coexist, supporting observations from Hunger & Benning (2007) in limiting sulphur conditions.

Changes in limnological regimes results in switches in microbial processes

Microbes in stratified versus mixed water column in the Dead Sea

Microbial life in the Dead Sea is limited by the amount of energy required to equilibrate with a highly concentrated external medium. As such, metabolic pathways are selected for their efficiency (Oren, 2010). In such a setting, the lake biosphere benefits from freshwater inputs in different ways: (i) a lowering of the salinity of the upper water column, forming a salinity gradient within the lake, (ii) input of normally life limiting nutrients and (iii) creation or input of fresh and labile organic matter (Oren, 1983; Oren *et al.*, 1995).

Periods of lake stratification are thus characterized by an upper layer hosting primary production and a hypolimnion where anaerobic degraders, such as sulphate reducing bacteria or fermenters, can degrade the produced organic matter (Nissenbaum & Kaplan, 1976; Torfstein *et al.*, 2005; Ariztegui *et al.*, 2015). This is reflected by microbial communities in the resulting sedimentary facies (Thomas *et al.*, 2015). On the other hand, during holomictic periods (generally dominant when evaporation largely exceeds precipitation), the absence of water dilution, in conjunction with nutrient starvation and lack of labile organic matter, only allows extreme halophiles such as archaeal members of the *Halobacteria* class to develop and subsist in the water column and in the subsequent sediment (Bodaker *et al.*, 2010; Thomas *et al.*, 2015).

The data presented here agree with such inferences. The presence of framboidal and euhedral pyrite (Fig. 5A and C) in the microbial mat can be explained by the complete sulphate reduction encouraged by the presence of a salinity gradient in the pond and fresh organic material provided by *Dunaliella* algae. Furthermore, it is proposed that pyrite occurs only when enough fresh organic matter is transported to the sediment, with potentially sufficient dilution of the hypolimnion (as suggested by Lazar *et al.*, 2014). Such events could have occurred at the climax of glacial periods in the Dead Sea Basin, for example, during the Lisan period, as observed in the core (Table 1 and Fig. 6). Pyrite could therefore be used as a productivity proxy in the deep Dead Sea sediment. Differential occurrences of iron-sulphur minerals, in particular greigite, in halite and AAD facies also support major variations in the iron, sulphur and carbon cycles in each limnological setting.

Cycling of iron, sulphur and carbon in the dynamic Dead Sea

Examination of changes in major and minor elements in the water column of the Dead Sea during its latest turnover (winter 1978–1979) revealed the impact of lake stratification on the sulphur and iron cycles. Nishri & Stiller (1984) estimated that all dissolved iron in the lake's water column had oxidized within a year after the complete penetration of oxidant waters to the depth of the lake. Iron-sulphur minerals, which used to be observed precipitating directly in those bottom waters, were also fully replaced by iron oxy-hydroxides. Neev & Emery (1967) demonstrated the direct precipitation of iron-sulphur minerals on ropes hanging in the hypolimnion of the Dead Sea before its complete mixing in 1978. Measurement of ferrous iron concentrations in the lake at this time supported a diffusion of dissolved reduced iron from the sediment (Nishri & Stiller, 1984). Analysis of groundwater present in the Dead Sea shore subsurface supports the slow but efficient reduction of iron in the hypersaline subsurface (Kiro *et al.*, 2013). Syngenetic iron-sulphur precipitation is thus possible in the Dead Sea hypolimnion, and probably consumes the available ferrous iron in this water until HS^- or Fe^{2+} is no longer available. The subsequent particulate iron is eventually incorporated within the sediment.

Although available iron reacts with sulphide, shallow sediments only harbour greigite phases, and rarely show pyrite chemistry. Greigite is, for example, seen infilling the chambers of allochtonous foraminifera (Fig. 9) that could originate from ponds around the lake (Almogi-Labin *et al.*, 1992). The organic matter they provide would be readily available for sulphate reducing organisms in the lake. Ferrous iron ions in the water column may allow precipitation of iron sulphides. Iron sulphides within the sediment, however, would enter a zone where fresh organic matter and thus HS^- becomes less available. Similarly, ferrous iron may become less available. Rims of greigite around titano-magnetite (Fig. 8) would support such limitations. Canfield *et al.* (1992) demonstrated that although iron sources are available in the sediment, the reactivity of iron is the key factor in subsurface environments. Potentially, iron oxides form a source of reactive iron in this setting, allowing diagenetic iron sulphide development in the sediment at relatively slow rates. Eventually, growth may continue in the deeper sediment, if enough sulphide accumulates and is allowed to react with iron.

In a fully mixed and halite-precipitating water column, the microbial cycles start only in the shallow sediment, and are relatively slow as no salinity gradient is expected in the deep subsurface. Microbial reduction is thought to be limited and diagenetic transformation occurs at relatively low and steady rates, similar to what could be observed in the deeply buried AAD, but with less labile organic matter available.

Interactions and redox gradients are probably very different from conditions in the shallow parts of the lake. Most information available on the cycling of iron and sulphur in the Dead Sea has been derived from environments where groundwater interacts with the deep brine of the core (Kiro *et al.*, 2013; Avrahamov *et al.*, 2014). The deep sediment experiences almost no interaction with diluted groundwater and mainly interacts with brine. It is thus likely that the relatively active processes observed in the shores of the Dead Sea, like microbial sulphate reduction, methanogenesis, anaeorobic oxidation of methane and iron reduction and oxidation (Ionescu *et al.*, 2012; Avrahamov *et al.*, 2014), are extremely limited in the deep sediment. Cycles of iron, sulphur and carbon must then be seen very differently in the deeper part of the lake. New work, potentially arising from the study of DSDDP cores, will allow better characterization of the processes at stake in this very peculiar subsurface environment.

Microbial effects on early diagenesis in the deep Dead Sea sediment

Aragonite precipitation in the Dead Sea: abiotic or microbially influenced?

Precipitation of aragonite occurs in the Dead Sea when magnesium/calcium-rich brines of the hypolimnion encounter carbonate-loaded fresher water of the epilimnion. This is thought to occur either through simple addition of these carbonates (Barkan *et al.*, 2001), or through alkalinity rise during CO_2 escape caused by evaporation, or by CO_2 consumption by photosynthetic blooming organisms (Neev & Emery, 1967; Begin *et al.*, 1974; Kolodny *et al.*, 2005). Carbonate ions thus seem to be limiting the precipitation of aragonite in the Dead Sea. As a result, their production in the sediment or at the water sediment interface could be a potential trigger in the precipitation of aragonite, especially since associated metabolic processes promote a rise in alkalinity (Dupraz & Visscher, 2005).

Dupraz *et al.* (2009) have listed potential mechanisms for enhancing calcium carbonate precipitation. Among them, complete sulphate reduction would permit the release of EPS-bound Ca^{2+} as well as carbonate ions by degradation of EPS, raising alkalinity and the $CaCO_3$ saturation index (Visscher *et al.*, 1998, 2000). Effects of sulphate reduction on alkalinity have been lately discussed by Meister (2013) and, in particular, the fact that the production of one H^+ per mole of reduced sulphate also leads to a pH decrease, hence to more carbonate dissolution than precipitation. Such a model seems to be dependent on the type of organic matter used as electron donor (Gallagher *et al.*, 2014).

In the case of the Dead Sea, degradation of EPS and the release of bonded calcium is not required, as suggested in the process of biologically induced mineralization (Dupraz *et al.*, 2009), since calcium is already present in the porewater at extreme concentration. The use of formate or hydrogen as an electron donor for SRB could thus also promote carbonate precipitation in the form of aragonite (Gallagher *et al.*, 2012), and is known to occur at high salinity for various species (Oren, 2010).

The presence of nanospheres in the mat (Fig. 11C), and what seems to resemble the transformation of the organic matrix to calcium carbonate minerals (Fig. 11D), may thus be influenced by the microbial production of carbonate ions in the sediment, from the available allochthonous or autochthonous organic matter. These nanoglobules have been recognized to be the initiating stage of calcium carbonate precipitation in numerous EPS-rich environments (Benzerara *et al.*, 2006; Bontognali *et al.*, 2008; Spadafora *et al.*, 2010), or to be preferentially associated with sulphate reducing bacteria (Aloisi *et al.*, 2006). Eventually, they lead to the formation of stellate aragonite clusters, embedded in, or originating from, the biofilm sheet remains (Fig. 11D–F) as, for example, those observed in the marine stromatolites of Highborne Cay, Bahamas (Reid *et al.*, 2000). The final aragonite morphology is similar to other aragonite precipitates (needles or stellate clusters) in the Dead Sea. Clustering into stars may occur through EPS related nucleation, but also via nucleation on the surface of detrital minerals or bacterial surfaces (Spadafora *et al.*, 2010). No direct morphological characteristic can thus be identified as the only result of the influence of EPS on aragonite precipitation.

While potentially observable in the mat (but not recorded in the bulk aragonite carbon isotopes; unpublished data), microbially influenced aragonite precipitation is unlikely to occur in the core. First of all, no visible proof was identified. An absence of nanoglobules or of conclusive aragonite-EPS interactions and morphologies tends to discriminate against such pathway. Second, this pathway would depend on the complete degradation of organic matter through microbial activity, which is not supported by the data presented here. Instead, this study supports either a slow and/or incomplete sulphate reduction, as previously demonstrated in the Dead Sea environment (Bishop *et al.*, 2013; Häusler *et al.*, 2014), which would have a limited effect on the alkalinity of the sedimentary micro-environment (Dupraz & Visscher, 2005). The impact of calcium carbonate generated via sulphate reduction on the Dead Sea environment should then be minor compared to the huge amount of aragonite needles precipitating in the water column: 15 mol m^{-2} y^{-1} in Lake Lisan (Stein *et al.*, 1997), and 1·4 mol m^{-2} y^{-1} after heavy flooding during winter, 1992 (Barkan *et al.*, 2001). Stable isotopic measurements of bulk aragonitic carbon did not indicate specifically microbially induced fractionation, although some measured values are relatively low (unpublished), suggesting that the impact of such a pathway on the total aragonitic pool is minimal. However, it is relevant in order to better understand future diagenetic processes affecting porosity and organic matter degradation in hypersaline sediments.

Iron-sulphur minerals as diagenetic imprint of the deep Dead Sea sediment

This study supports diagenetic development of greigite in the deep sediment. This process had already been highlighted by Ron *et al.* (2006) and Frank *et al.* (2007). Iron sulphide rims regularly observed around titano-magnetite in AAD intervals (Figs 7A and 8) suggest rapid reaction of soluble sulphide with dissolved iron originating from the detrital iron fraction. This would undoubtedly affect the magnetic signal of the Dead Sea sediment, as suggested by Rowan & Roberts (2006). Iron-sulphur precipitates have been observed in the halite sections of the core too, when gypsum is an associated evaporative phase (Fig. 10). The detrital component seems to be the source of iron there, and iron-sulphur forms around the gypsum minerals. Different behaviours are thus observed, whether the focus is on sediments derived from arid or more humid periods. This suggests that reduction may also occur in holomictic settings, while it has so far been described mainly from the sediments of Lake Lisan or in general in a stratified lake (Nissenbaum, 1975; Torfstein *et al.*, 2008). This early diagenesis may, however, be erased by a second stage, if sediments are exposed in outcrop and desiccate. For example, neither greigite nor pyrite has been observed in the Masada sections of Lake Lisan (Torfstein *et al.*, 2005; Bishop *et al.*, 2013) and halite intervals dissolve when exposed.

The iron and sulphur cycles are greatly influenced by microbes in the Dead Sea subsurface with phase changes from gypsum sulphate to HS^-, S^0, and iron-sulphur mineralization. The quality of the available organic carbon is critical to the formation of these phases. It also appears that the processes vary depending on whether halite-rich intervals precipitating during arid periods are examined, or AAD intervals characteristic of more humid episodes. Similar to the effects on microbial communities (Ariztegui *et al.*, 2015; Thomas *et al.*, 2015), freshwater inputs largely affect the precipitation and diagenetic processes in the deep Dead Sea Basin.

CONCLUSION

The Dead Sea is a unique system that can hardly be compared to any other hypersaline systems. It possesses some hypersaline biological system characteristics such as specifically adapted microbial communities. However, the latter develop at much slower rates, and with different limiting conditions (low sulphide, high calcium and magnesium concentrations). The drilled record sheds light on the evolution of deep sediments before they undergo dehydration or dissolutive transformation. It implies diagenetic transformations occurring in the deeper part of the basin as follow:

- The formation of iron-sulphur mineralization emphasizes the existence of a microbial sulphur cycle influencing mineral precipitation and organic matter preservation.
- Iron-sulphur minerals are relatively small and often close to greigite in composition, supporting a slow or incomplete microbial sulphate reduction in the Dead Sea sediment, whether arid or more humid conditions prevail. On the other hand, the occurrence of authigenic pyrite may be an important marker of lake productivity.
- The iron cycle is also influenced by secondary dissolution and re-precipitation of magnetic fractions.
- Comparison with an active microbial mat of the Dead Sea shore allowed exopolymeric substances to be detected in the core sediments. Identified mainly based on morphology, they have the potential to serve as templates for aragonite precipitation at depth, implying that aragonite may not only form in the water column under conditions previously envisaged.

Together, these findings support the existence of living microbes in the sediment of the Dead Sea, and remain the only proof for their metabolic activity. Finally, these results have strong implications regarding the fate of organic matter and on processes involved in mineralization and porosity development from the perspective of diagenetic evolution of hypersaline subsurface systems. They also support the microbial impact on the sedimentological and geochemical record of the deep Dead Sea sediment. Such observations must be taken into account in future palaeoenvironmental reconstructions.

ACKNOWLEDGEMENTS

This research was funded by the Swiss National Science Foundation (projects 200021-132529 and 200020-149221/1) for CT and by The Dead Sea Drill Excellence Center of the Israel Science Foundation (grant #1736/11) for YK and YE. This work has been realized in close collaboration with members of the DSDDP Scientific Team, for which a list can be found at www.icdp-online.org. The authors thank the DSDDP technical team in Israel and Potsdam, and the ICDP engineers that have helped in the set-up of the material and the analysis at GFZ Potsdam during core-opening. N. Waldmann and A. Vuillemin are deeply acknowledged for direction and assistance on sampling. P. Dulski and his lab are thanked for XRF core scanning, C. Dupraz, M. Dadras and M. Leboeuf for

Cryo-SEM assistance and advice and J. Garcia-Veigas at the CCiT lab facilities in the University of Barcelona for ESEM. We also thank A. Martignier and R. Martini at the University of Geneva and M. Pacton for their precious SEM expertise. The authors also wish to thank P. Visscher for fruitful conversation and two anonymous reviewers who greatly improved this article. The authors declare no conflict of interest.

References

Abu Ghazleh, S., **Hartmann, J.**, **Jansen, N.** and **Kempe, S.** (2009) Water input requirements of the rapidly shrinking Dead Sea. *Naturwissenschaften*, **96**, 637–643.

Almogi-Labin, A., **Perelis-Grossovicz, L.** and **Raab, M.** (1992) Living Ammonia from a hypersaline inland pool, Dead Sea area, Israel. *J. Foraminifer. Res.*, **22**, 257–266.

Aloisi, G., **Gloter, A.**, **Kruger, M.**, **Wallmann, K.**, **Guyot, F.** and **Zuddas, P.** (2006) Nucleation of calcium carbonate on bacterial nanoglobules. *Geology*, **34**, 1017–1020.

Ariztegui, D., **Thomas, C.** and **Vuillemin, A.** (2015) Present and future of subsurface biosphere studies in lacustrine sediments through scientific drilling. *Int. J. Earth Sci.*, **104**, 1655–1665.

Astafieva, M.M., **Rozanov, A.Y.** and **Hoover, R.** (2005) Framboids: their structure and origin. *Paleontol. J.*, **39**, 457.

Avrahamov, N., **Yechieli, Y.**, **Lazar, B.**, **Lewenberg, O.**, **Boaretto, E.** and **Sivan, O.** (2010) Characterization and dating of saline groundwater in the Dead Sea area. *Radiocarbon*, **52**, 1123–1140.

Avrahamov, N., **Antler, G.**, **Yechieli, Y.**, **Gavrieli, I.**, **Joye, S.B.**, **Saxton, M.**, **Turchyn, A.V.** and **Sivan, O.** (2014) Anaerobic oxidation of methane by sulfate in hypersaline groundwater of the Dead Sea aquifer. *Geobiology*, **12**, 511–528.

Barkan, E., **Luz, B.** and **Lazar, B.** (2001) Dynamics of the carbon dioxide system in the Dead Sea. *Geochim. Cosmochim. Acta*, **65**, 355–368.

Begin, Z.B., **Ehrlich, A.** and **Nathan, Y.** (1974) Lake Lisan: the Pleistocene precursor of the Dead Sea. *Geol. Surv. Isr. Bull.*, **63**, 11–30.

Benzerara, K., **Menguy, N.**, **López-García, P.**, **Yoon, T.-H.**, **Kazmierczak, J.**, **Tyliszczak, T.**, **Guyot, F.** and **Brown, G.E.** (2006) Nanoscale detection of organic signatures in carbonate microbialites. *Proc. Natl Acad. Sci. USA*, **103**, 9440–9445.

Berner, R.A. (1970) Sedimentary pyrite formation. *Am. J. Sci.*, **268**, 1–23.

Berner, R.A. (1984) Sedimentary pyrite formation: an update. *Geochim. Cosmochim. Acta*, **48**, 605–615.

Bishop, T., **Turchyn, A.V.** and **Sivan, O.** (2013) Fire and brimstone: the microbially mediated formation of elemental sulfur nodules from an isotope and major element study in the paleo-Dead Sea. *PLoS ONE*, **8**, e75883.

Bodaker, I., **Sharon, I.**, **Suzuki, M.T.**, **Feingersch, R.**, **Shmoish, M.**, **Andreishcheva, E.**, **Sogin, M.L.**, **Rosenberg, M.**, **Maguire, M.E.**, **Belkin, S.**, **Oren, A.** and **Béjà, O.** (2010) Comparative community genomics in the Dead Sea: an increasingly extreme environment. *ISME J.*, **4**, 399–407.

Bontognali, T.R.R., **Vasconcelos, C.**, **Warthmann, R.J.**, **Dupraz, C.**, **Bernasconi, S.M.** and **McKenzie, J.A.** (2008) Microbes produce nanobacteria-like structures, avoiding cell entombment. *Geology*, **36**, 663–666.

Bosak, T. and **Newman, D.K.** (2005) Microbial kinetic controls on calcite morphology in supersaturated solutions. *J. Sed. Res.*, **75**, 190.

Braissant, O., **Decho, A.W.**, **Dupraz, C.**, **Glunk, C.**, **Przekop, K.M.** and **Visscher, P.T.** (2007) Exopolymeric substances of sulfate-reducing bacteria: interactions with calcium at alkaline pH and implication for formation of carbonate minerals. *Geobiology*, **5**, 401–411.

Canfield, D.E., **Raiswell, R.** and **Bottrell, S.H.** (1992) The reactivity of sedimentary iron minerals toward sulfide. *Am. J. Sci.*, **292**, 659–683.

Defarge, C., **Trichet, J.**, **Jaunet, A.M.**, **Robert, M.**, **Tribble, J.** and **Sansone, F.J.** (1996) Texture of microbial sediments revealed by cryo-scanning electron microscopy. *J. Sed. Res.*, **66**, 935.

Des Marais, D.J. (1995) The biogeochemistry of hypersaline microbial mats. In: *Advances in Microbial Ecology* (Ed. J.G. Jones), pp. 251–274. Plenum, Springer US, New York.

Drobner, E., **Huber, H.**, **Wächtershäuser, G.**, **Rose, D.** and **Stetter, K.O.** (1990) Pyrite formation linked with hydrogen evolution under anaerobic conditions. *Nature*, **346**, 742–744.

Dupraz, C. and **Visscher, P.T.** (2005) Microbial lithification in marine stromatolites and hypersaline mats. *Trends Microbiol.*, **13**, 429–438.

Dupraz, C., **Visscher, P.T.**, **Baumgartner, L.K.** and **Reid, R.P.** (2004) Microbe-mineral interactions: early carbonate precipitation in a hypersaline lake (Eleuthera Island, Bahamas). *Sedimentology*, **51**, 745–765.

Dupraz, C., **Reid, R.P.**, **Braissant, O.**, **Decho, A.W.**, **Norman, R.S.** and **Visscher, P.T.** (2009) Processes of carbonate precipitation in modern microbial mats. *Earth-Sci. Rev.*, **96**, 141–162.

Frank, U., **Nowaczyk, N.R.** and **Negendank, J.F.W.** (2007) Palaeomagnetism of greigite bearing sediments from the Dead Sea, Israel. *Geophys. J. Int.*, **168**, 904–920.

Gallagher, K.L., **Kading, T.J.**, **Braissant, O.**, **Dupraz, C.** and **Visscher, P.T.** (2012) Inside the alkalinity engine: the role of electron donors in the organomineralization potential of sulfate-reducing bacteria. *Geobiology*, **10**, 518–530.

Gallagher, K., **Dupraz, C.** and **Visscher, P.** (2014) Two opposing effects of sulfate reduction on carbonate precipitation in normal, marine, hypersaline and alkaline environments: COMMENT. *Geology*, **42**, 313–314.

Gavrieli, I., **Yechieli, Y.**, **Halicz, L.**, **Spiro, B.**, **Bein, A.** and **Efron, D.** (2001) The sulfur system in anoxic subsurface brines and its implication in brine evolutionary pathways: the Ca-chloride brines in the Dead Sea area. *Earth Planet. Sci. Lett.*, **186**, 199–213.

Glombitza, C., Stockhecke, M., Schubert, C.J., Vetter, A. and **Kallmeyer, J.** (2013) Sulfate reduction controlled by organic matter availability in deep sediment cores from the saline, alkaline Lake Van (Eastern Anatolia, Turkey). *Front. Microbiol.*, **4**, 1–12.

Häusler, S., Weber, M., Siebert, C., Holtappels, M., Noriega-Ortega, B.E., De Beer, D. and **Ionescu, D.** (2014) Sulfate reduction and sulfide oxidation in extremely steep salinity gradients formed by freshwater springs emerging into the Dead Sea. *FEMS Microbiol. Ecol.*, **90**, 956–969.

Hunger, S. and **Benning, L.G.** (2007) Greigite: a true intermediate on the polysulfide pathway to pyrite. *Geochem. Trans.*, **8**, 1.

Ionescu, D., Siebert, C., Polerecky, L., Munwes, Y.Y., Lott, C., Häusler, S., Bižić-Ionescu, M., Quast, C., Peplies, J., Glöckner, F.O., Ramette, A., Rödiger, T., Dittmar, T., Oren, A., Geyer, S., Stärk, H.J., Sauter, M., Licha, T., Laronne, J.B. and **de Beer, D.** (2012) Microbial and chemical characterization of underwater fresh water springs in the Dead Sea. *PLoS ONE*, **7**, e38319.

Kawaguchi, T. and **Decho, A.W.** (2002) Isolation and biochemical characterization of extracellular polymeric secretions (EPS) from modern soft marine stromatolites (Bahamas) and its inhibitory effect on $CaCO_3$ precipitation. *Prep. Biochem. Biotechnol.*, **32**, 51–63.

Kiro, Y., Weinstein, Y., Starinsky, A. and **Yechieli, Y.** (2013) Groundwater ages and reaction rates during seawater circulation in the Dead Sea aquifer. *Geochim. Cosmochim. Acta*, **122**, 17–35.

Kolodny, Y., Stein, M. and **Machlus, M.** (2005) Sea-rain-lake relation in the Last Glacial East Mediterranean revealed by $\delta^{18}O$-$\delta^{13}C$ in Lake Lisan aragonites. *Geochim. Cosmochim. Acta*, **69**, 4045–4060.

Lazar, B., Sivan, O., Yechieli, Y., Levy, E.J., Antler, G., Gavrieli, I. and **Stein, M.** (2014) Long-term freshening of the Dead Sea brine revealed by porewater Cl^- and $\delta^{18}O$ in ICDP Dead Sea deep-drill. *Earth Planet. Sci. Lett.*, **400**, 94–101.

Luther, G.W. (1991) Pyrite synthesis via polysulfide compounds. *Geochim. Cosmochim. Acta*, **55**, 2839–2849.

Meister, P. (2013) Two opposing effects of sulfate reduction on carbonate precipitation in normal marine, hypersaline, and alkaline environments. *Geology*, **41**, 499–502.

Mishra, A. and **Jha, B.** (2009) Isolation and characterization of extracellular polymeric substances from micro-algae Dunaliellasalina under salt stress. *Bioresour. Technol.*, **100**, 3382–3386.

Neev, D. and **Emery, K.** (1967) The Dead Sea, depositional processes and environments of evaporites. *Geol. Surv. Isr. Bull.*, **41**, 147.

Neugebauer, I., Brauer, A., Schwab, M., Waldmann, N., Enzel, Y., Kitagawa, H., Torfstein, A., Frank, U., Dulski, P., Agnon, A., Ariztegui, D., Ben-Avraham, Z., Goldstein, S.L., Stein, M. and **DSDDP Scientific Party** (2014) Lithology of the long sediment record recovered by the ICDP Dead Sea Deep Drilling Project (DSDDP). *Quatern. Sci. Rev.*, **102**, 149–165.

Nishri, A. and **Stiller, M.** (1984) Iron in the Dead Sea. *Earth Planet. Sci. Lett.*, **71**, 405–414.

Nissenbaum, A. (1975) The microbiology and biogeochemistry of the Dead Sea. *Microb. Ecol.*, **2**, 139–161.

Nissenbaum, A. and **Kaplan, I.R.** (1976) Sulfur and carbon isotopic evidence for biogeochemical processes in the Dead Sea. In: *Environmental Biogeochemistry* (Ed. J. Nriagu), pp. 309–325. Ann Arbor Science Publishers, Ann Arbor.

Oren, A. (1983) Population dynamics of halobacteria in the Dead Sea water column. *Limnol. Oceanogr.*, **28**, 1094–1103.

Oren, A. (2001) The bioenergetic basis for the decrease in metabolic diversity at increasing salt concentrations: implications for the functioning of salt lake ecosystems. *Hydrobiologia*, **466**, 61–72.

Oren, A. (2010) Thermodynamic limits to microbial life at high salt concentrations. *Environ. Microbiol.*, **13**, 1908–1923.

Oren, A. and **Gunde-Cimerman, N.** (2012) Fungal Life in the Dead Sea. In: *Progress in Molecular and Subcellular Biology* (Ed. C. Raghukumar), pp. 115–132. Springer, Heidelberg.

Oren, A. and **Shilo, M.** (1982) Population dynamics of Dunaliella parva in the Dead Sea. *Limnol. Oceanogr.*, **27**, 201–211.

Oren, A., Gurevich, P., Anati, D., Barkan, E. and **Luz, B.** (1995) A bloom of Dunaliella parva in the Dead Sea in 1992: biological and biogeochemical aspects. *Hydrobiologia*, **297**, 173–185.

Passier, H., Middelburg, J., de Lange, G.J. and **Böttcher, M.** (1997) Pyrite contents, microtextures, and sulfur isotopes in relation to formation of the youngest eastern Mediterranean sapropel. *Geology*, **25**, 519–522.

Popa, R., Kinkle, B.K. and **Badescu, A.** (2004) Pyrite framboids as biomarkers for iron-sulfur systems. *Geomicrobiol J.*, **21**, 193–206.

Pósfai, M., Buseck, P.R., Bazylinski, D.A. and **Frankel, R.B.** (1998) Reaction sequence of iron sulfide minerals in bacteria and their use as biomarkers. *Science*, **280**, 880–883.

Reid, R.P., Visscher, P.T., Decho, A.W., Stolz, J.F., Bebout, B.M., Dupraz, C., Macintyre, I.G., Paerl, H.W., Pinckney, J.L., Prufert-Bebout, L., Steppe, T.F. and **DesMarais, D.J.** (2000) The role of microbes in accretion, lamination and early lithification of modern marine stromatolites. *Nature*, **406**, 989–992.

Rhodes, M.E., Oren, A. and **House, C.H.** (2012) Dynamics and persistence of Dead Sea microbial populations as shown by high-throughput sequencing of rRNA. *Appl. Environ. Microbiol.*, **78**, 2489–2492.

Rickard, D. (1975) Kinetics and mechanism of pyrite formation at low temperatures. *Am. J. Sci.*, **275**, 636–652.

Rickard, D. and **Luther, G.W.** (1997) Kinetics of pyrite formation by the H_2S oxidation of iron (II) monosulfide in aqueous solutions between 25 and 125°C: the mechanism. *Geochim. Cosmochim. Acta*, **61**, 135–147.

Roesser, M. and **Müller, V.** (2001) Osmoadaptation in bacteria and archaea: common principles and differences. *Environ. Microbiol.*, **3**, 743–754.

Ron, H., Nowaczyk, N.R., Frank, U., Marco, S. and **McWilliams, M.O.** (2006) Magnetic properties of Lake Lisan and Holocene Dead Sea sediments and the fidelity of chemical and detrital remanent magnetization. *New Front. Dead Sea Paleoenviron. Res.*, **Special Pa**, 171–182.

Rowan, C.J. and **Roberts, A.P.** (2006) Magnetite dissolution, diachronous greigite formation, and secondary magnetizations from pyrite oxidation: Unravelling complex magnetizations in Neogene marine sediments from New Zealand. *Earth Planet. Sci. Lett.*, **241**, 119–137.

Schoonen, M.A.A. and **Barnes, H.L.** (1991) Reactions forming pyrite and marcasite from solution: I. Nucleation of FeS_2 below 100 C. *Geochim. Cosmochim. Acta*, **55**, 1495–1504.

Spadafora, A., Perri, E., Mckenzie, J.A. and **Vasconcelos, C.** (2010) Microbial biomineralization processes forming modern Ca: Mg carbonate stromatolites. *Sedimentology*, **57**, 27–40.

Stein, M., Starinsky, I.A., Katz, I.A., Goldstein, J.S.L., Machlus, M. and **Schramm, A.** (1997) Strontium isotopic, chemical, and sedimentological evidence for the evolution of Lake Lisan and the Dead Sea. *Geochim. Cosmochim. Acta*, **61**, 3975–3992.

Steinhorn, I., Assaf, G., Gat, J.R., Nishri, A., Nissenbaum, A., Stiller, M., Beyth, M., Neev, D., Garber, R., Friedman, G.M. and **Weiss, W.** (1979) The Dead Sea: deepening of the mixolimnion signifies the overture to overturn of the water column. *Sci.*, **206**, 55–57.

Sweeney, R.E. and **Kaplan, I.R.** (1973) Pyrite Framboid Formation; Laboratory Synthesis and Marine Sediments. *Econ. Geol.*, **68**, 618–634.

Tahal Group and the Geological Survey of Israel (2011). Red Sea to Dead Sea Water Conveyance (RSDSC) Study Program: Dead Sea Study.

Thomas, C., Ionescu, D., Ariztegui, D. and the DSDDP Scientific Party (2014) Archaeal populations in two distinct sedimentary facies of the subsurface of the Dead Sea. *Mar. Genomics*, **17**, 53–62.

Thomas, C., Ionescu, D., Ariztegui, D. and the DSDDP Scientific Team (2015) Impact of paleoclimate on the distribution of microbial communities in the subsurface sediment of the Dead Sea. *Geobiology*, **13**, 546–561.

Torfstein, A., Gavrieli, I. and **Stein, M.** (2005) The sources and evolution of sulfur in the hypersaline Lake Lisan (paleo-Dead Sea). *Earth Planet. Sci. Lett.*, **236**, 61–77.

Torfstein, A., Gavrieli, I., Katz, A., Kolodny, Y. and **Stein, M.** (2008) Gypsum as a monitor of the paleo-limnological–hydrological conditions in Lake Lisan and the Dead Sea. *Geochim. Cosmochim. Acta*, **72**, 2491–2509.

Torfstein, A., Goldstein, S.L., Kushnir, Y., Enzel, Y., Haug, G. and **Stein, M.** (2015) Dead Sea drawdown and monsoonal impacts in the Levant during the last interglacial. *Earth Planet. Sci. Lett.*, **412**, 235–244.

Visscher, P.T., Reid, R.P., Bebout, B.M., Hoeft, S.E., Macintyre, L.G. and **Thompson, J.A.** (1998) Formation of lithified micritic laminae in modem marine stromatolites (Bahamas): the role of sulfur cycling. *Am. Mineral.*, **83**, 1482–1493.

Visscher, P.T., Reid, R.P. and **Bebout, B.M.** (2000) Microscale observations of sulfate reduction: correlation of microbial activity with lithified micritic laminae in modern marine stromatolites. *Geology*, **28**, 919–922.

Vuillemin, A., Ariztegui, D., Coninck, A., Lücke, A., Mayr, C. and **Schubert, C.** (2013) Origin and significance of diagenetic concretions in sediments of Laguna Potrok Aike, southern Argentina. *J. Paleolimnol.*, **50**, 275–291.

Wang, Q. and **Morse, J.W.** (1996) Pyrite formation under conditions approximating those in anoxic sediments I. Pathway and morphology. *Mar. Chem.*, **52**, 99–121.

Wilkin, R.T. and **Arthur, M.A.** (2001) Variations in pyrite texture, sulfur isotope composition, and iron systematics in the Black Sea: evidence for Late Pleistocene to Holocene excursions of the O_2–H_2S redox transition. *Geochim. Cosmochim. Acta*, **65**, 1399–1416.

Wilkin, R.T. and **Barnes, H.L.** (1996) Pyrite formation by reactions of iron monosulfides with dissolved inorganic and organic sulfur species. *Geochim. Cosmochim. Acta*, **60**, 4167–4179.

Wilkin, R.T. and **Barnes, H.L.** (1997) Formation processes of framboidal pyrite. *Geochim. Cosmochim. Acta*, **61**, 323–339.

9

Evolution of the Northern Tethyan Helvetic Platform during the late Berriasian and early Valanginian

CHLOE MORALES*[,1], JORGE E. SPANGENBERG†, ANNIE ARNAUD-VANNEAU‡, THIERRY ADATTE† and KARL B. FÖLLMI†

*Institute of Earth Sciences, University of Lausanne, Géopolis, 1015, Lausanne, Switzerland
†Institute of Earth Surface Dynamics, University of Lausanne, Géopolis, 1015, Lausanne, Switzerland
‡Association Dolomieu, 6 Chemin des Grenouilles, 38700, La Tronche, France

Keywords
Berriasian, environmental change, Helvetic Platform, Valanginian.

[1]Present address: Marine Palynology Group, Institute of Earth Sciences, University of Utrecht, Van Unnikgebouw, Heidelberglaan 24, 3584 CS, Utrecht, the Netherlands

ABSTRACT

The Early Cretaceous period is characterized by widespread carbonate production in tropical and subtropical epicontinental seas, which was modulated by changes in sea-level, detrital and nutrient fluxes, and the global carbon cycle. As a result, carbonate platforms were sensitive recorders of environmental change, which often anticipated global environmental perturbations. A good example is provided by the northern Tethyan carbonate platform, which is presently preserved in the central European Helvetic Alps. There, the latest early to late Valanginian Weissert episode of global change, which is defined by the first important positive shift in $\delta^{13}C$ records of the Cretaceous, is expressed by a prolonged, stepwise drowning phase. In this contribution, a detailed reconstruction of palaeoenvironmental change before and during the Weissert episode is provided based on three representative sections of the Helvetic platform. The sections are placed along a deepening transect and correlated by means of ammonite and microfossil biostratigraphy, sequence stratigraphy and $\delta^{13}C$ chemostratigraphy. In a first phase of palaeoenvironmental change during the latest Berriasian, photozoan carbonate production was stopped by a major and hitherto undetected drowning episode, which was followed by a phase of renewed carbonate production by heterozoan biota. This phase was linked to major sea-level rise, a change to a more humid climate and strong regional subsidence associated with tectonic block tilting. During the Valanginian, the circulation of nutrient-enriched sea waters prevented a return to oligotrophic conditions and two further drowning episodes occurred, which are both documented by condensed phosphate-rich beds and dated as middle early Valanginian and late Valanginian to early Hauterivian. The exact causes of the three-step deterioration in carbonate production are not established but a link to episodic volcanic activity is likely, eventually related to the formation of the Paranà-Etendeka large igneous province.

INTRODUCTION

The Early Cretaceous was characterized by generally high pCO_2 levels and correspondingly reinforced greenhouse conditions, which were favourable to important carbonate production in tropical and subtropical shallow-water settings (Hay, 2008). During this time interval, carbonate deposition was modulated and in certain regions episodically interrupted by global sea-level change, changes in the global carbon cycle and corresponding pH and pCO_2 in sea water, which were often related to phases of major volcanic activity, and changes in detrital and nutrient flux rates. The resulting carbonate deposits represent first-order archives of palaeoclimate and palaeoenvironmental change, which recorded and often anticipated global perturbations (Föllmi *et al.*, 1994, 2006; Weissert *et al.*, 1998; Huck *et al.*, 2011).

During the Early Cretaceous, a first episode of major palaeoenvironmental change occurred during the latest early to the late Valanginian (Weissert episode), which is defined by the first important carbon-isotope excursion (CIE) of the Cretaceous. The mechanisms leading to this episode are likely to be sought in increased volcanic and hydrothermal activity, which – depending on which time scale is used – may be attributed to the formation of the Paranà-Etendeka Large Igneous Province (Lini *et al.*, 1992; Erba *et al.*, 2004; Martinez *et al.*, 2015). The Weissert episode is preceded by a phase of significant palaeoenvironmental and palaeoclimate change, which occurred during the late Berriasian and early Valanginian. In the western European domain (England, Germany, France and Switzerland), a phase of enhanced humidity began during the late Berriasian, and probably reached a maximum in the earliest Valanginian (Hallam *et al.*, 1991; Schnyder *et al.*, 2005; Föllmi, 2012; Morales *et al.*, 2013). A general increase in marine nutrient levels in ocean basins is recorded from the latest Berriasian to the late Valanginian (Föllmi, 1995; Duchamp-Alphonse *et al.*, 2007) and on northern Tethyan platforms (Föllmi *et al.*, 2007; Morales *et al.*, 2013), which interfered with the evolution of carbonate platforms at that time. In addition, carbonate production was impacted by sea-level variations, the occurrence and timing of which are not well-constrained and still under debate (Schlager, 1981; Haq *et al.*, 1987; Hardenbol *et al.*, 1998; Gréselle & Pittet, 2010; Haq, 2014). Finally, and on a more regional scale, the change in platform morphology from a distally steepened ramp to a swell-dominated ramp and the disappearance of a barrier close to the Berriasian-Valanginian boundary documented in the Jura region probably influenced the distribution of continental fluxes on the northern Tethyan shelf and in the adjacent basin (Morales *et al.*, 2013).

The Helvetic platform succession, which is presently preserved in the northern part of the central European Alps, documents the afore-mentioned palaeoenvironmental changes in great detail. A change from photozoan to heterozoan carbonate production has been observed near the Berriasian-Valanginian boundary (Ischi, 1978; Burger, 1985; Wyssling, 1986; Mohr, 1992; Föllmi *et al.*, 1994, 2007), and two condensed phosphatic and glauconitic-rich layers of early Valanginian (Büls Beds), and late early Valanginian to Hauterivian age (Gemsmättli Bed), highlight the occurrence of two successive incipient drowning phases (Haldimann, 1977; Wyssling, 1986; Kuhn, 1996). While the presence of ammonites within these two beds permits the two drowning episodes to be accurately dated, the stratigraphy of the platform carbonates deposited prior to the formation of these two condensed beds is less well-constrained, impeding its correlation with the general record of environmental change during the late Berriasian and earliest Valanginian.

In this contribution, a detailed study of three representative sections along a proximal-distal transect through the upper Berriasian and Valanginian Helvetic succession is presented. An improved stratigraphic framework based on ammonite, benthic foraminiferal and calpionellid biostratigraphy is established and combined with $\delta^{13}C$ bulk-rock chemostratigraphy and sequence stratigraphy. The evolution in facies and microfacies associated with variations in mineralogical and phosphorus contents are used to examine changes in accommodation, ecology, nutrients and climate. The goal thereby is to trace the onset and evolution of the impact of palaeoenvironmental change prior to and during the Weissert episode on the north-western Tethyan carbonate platform. In a second step, the Helvetic sedimentary succession is compared with sediments preserved in the Provence, Pyrenean and Jura Platforms, which allows a regional view of the palaeoenvironmental changes that occurred along the northern Tethyan margin to be developed. Finally, the potential influence of intense volcanic activity on palaeoecological and environmental changes is evaluated.

GEOLOGICAL SETTING AND DESCRIPTION OF STUDIED SECTIONS

The succession of Early Cretaceous platform deposits (Funk *et al.*, 1993; Föllmi *et al.*, 2006, 2007) starts with the Zementstein Formation, which is characterized by dark and monotonous marly carbonate deposits and dated from the *Berriasella jacobi* and *Subthurmannia occitanica* zones by ammonites and calpionellids (early Berriasian; Mohr, 1992). It is overlain by the Öhrli Formation, which documents the development of a photozoan carbonate platform with the predominant deposition of oolitic and bioclastic sediments, containing a rich and diverse fauna of benthic foraminifera, corals, green algae and echinoderms (Burger, 1985, 1986; Mohr, 1992). The Öhrli Formation includes two marly and calcareous intervals – the Lower Öhrli Marl Member, Lower Öhrli Limestone Member, Upper Öhrli Marl Member and Upper Öhrli Limestone Member, respectively. The age of the Öhrli Formation is poorly constrained. A maximum age is provided by calpionellids and ammonites found at its base, which indicate an early Berriasian age (*S. occitanica* zone) (Mohr, 1992). The Öhrli Formation passes laterally into the monotonous hemipelagic marly succession of the Palfris Formation (Burger, 1985, 1986; Wyssling, 1986).

Both formations are overlain by the marly and sand-rich Vitznau Formation, which was attributed to the early Valanginian based on palynomorphs (Pantic & Burger, 1981). The overlaying Betlis Formation is rich in echinoderms and bryozoans and marks the development of a heterozoan platform. The occurrence of the ammonite

Thurmanniceras thurmanni s.l. (Wyssling, 1986) indicates a *T. pertransiens* age. Distal occurrences of the Betlis Formation include the condensed and phosphatic Büls Bed, which was dated as late *T. pertransiens* – early *B. campylotoxus* zone (Kuhn, 1996). The overlying condensed and phosphatic Gemsmättli Bed and its lateral equivalent, the sandy Pygurus Bed, provides a minimum age corresponding to the *S. verrucosum* zone (Wyssling, 1986; Kuhn, 1996; Föllmi *et al.*, 2007).

Three sections were selected, which represent inner shelf, barrier and outer shelf settings along a proximal-distal transect (Säntis, Dräckloch, Vitznau; Fig. 1). The Säntis section (47°15′3″N; 9°19′16″E) is located at the foot of the Säntis Mountain close to Schwägalp (canton Appenzell Ausserrhoden). It belongs to the Säntis-Drusberg nappe. The measured succession is approximately 105 m thick and includes the Upper Öhrli Limestone Member, Vitznau Formation, Betlis Formation, and the Gemsmättli Bed (Fig. 2). The Dräckloch section (46°58′27″N, 8°56′7″E) is part of the Gassenstock Mountain (Bös Fulen summit), situated 5 km south of the village of Richisau on the border between the cantons of Glarus and Schwyz. The outcrop belongs to the Axen nappe. The measured section is 400 m thick and starts with the Öhrli Formation (Fig. 3A, Lower Öhrli Limestone Member, Upper Öhrli Marl Member and Upper Öhrli Limestone Member), and goes on with the Vitznau and Betlis formations (Fig. 3B). The Betlis Formation is overlain by the sandy, coarse-grained sediments of the Pygurus Bed. The Vitznau section (47°0′25″N; 8°30′23″E) is situated close to the village of Vitznau (canton Lucerne). Tectonically it belongs to the 'Randkette' attached to the Wildhorn nappe. The measured section is approximately 60 m thick and includes the Palfris Formation, the Upper Öhrli Limestone Member, and the Vitznau and Betlis formations (Fig. 4).

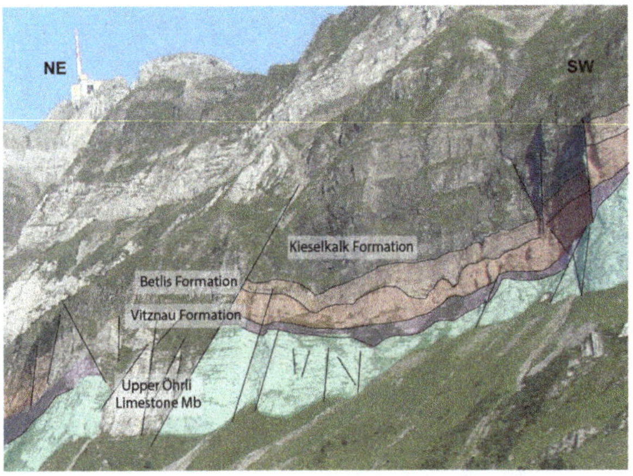

Fig. 2. Panorama of the north-western slope of the Säntis Mountain, where the Lower Cretaceous succession is duplicated. The Vitznau Formation is 20 m thick (photo location: 47°15′15″N 9°18′60″E). Turquoise, purple and red shadings correspond to prominent photozoan, recessive heterozoan and prominent heterozoan deposits, respectively. Black borders highlight contacts between formations. Subvertical black lines correspond to fault zones.

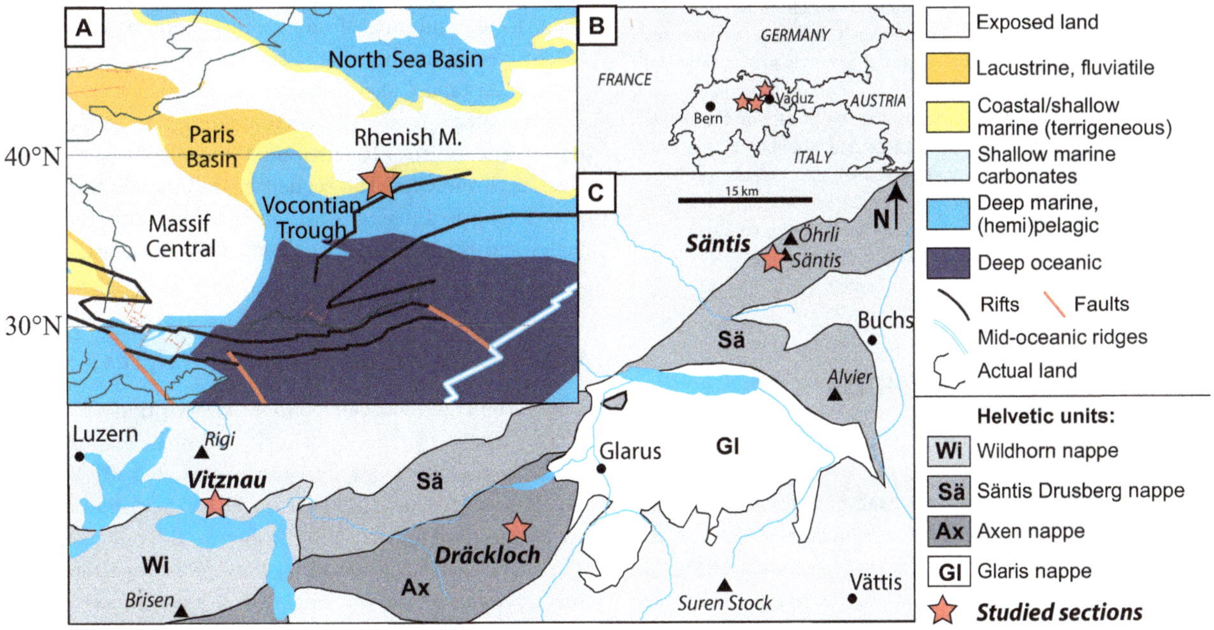

Fig. 1. Location of the sections. (A) Palaeogeographic map (compiled after Dercourt *et al.*, 2000, http://jan.ucc.nau.edu/~rcb7/, and Stampfli & Hochard, 2009); (B) Map of Switzerland; (C) Tectonic map.

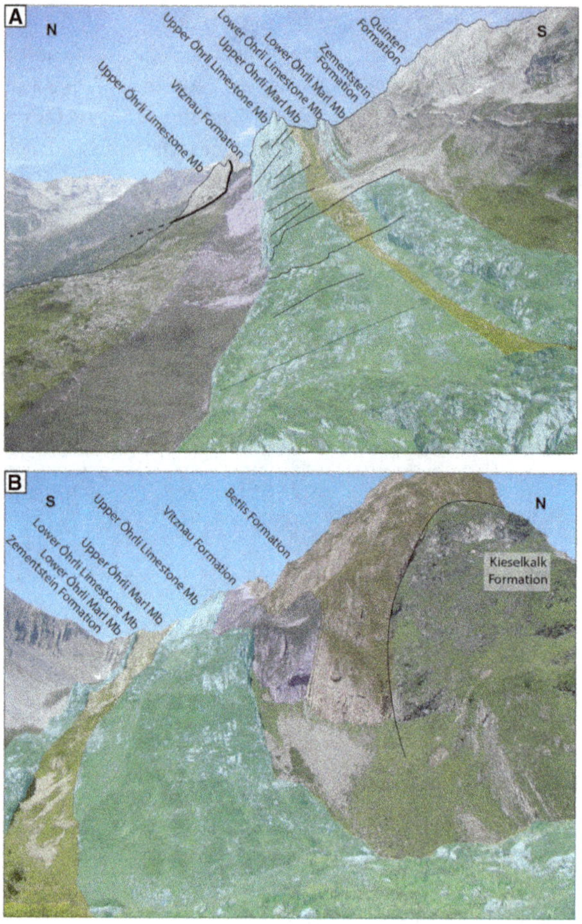

Fig. 3. Panoramas of the Dräckloch area where the sedimentological succession is inverted. Photo location: 46°58′35″N 8°56′08″E. Turquoise, yellow, purple and red shadings correspond to prominent photozoan, recessive photozoan, recessive heterozoan and prominent heterozoan deposits, respectively. Black lines correspond to fault zones. (A) View to the east, showing the lower part of the section. The repetition of the Upper Öhrli Limestone Member in the north direction is linked to a thrust (bold black line), which settles on top of the Vitznau Formation. For scale, the Vitznau Formation is 55 m thick. (B) View to the west, showing the upper part of the succession in a basinward direction. The contact between the Valanginian Betlis and the Hauterivian Kieselkalk formations is faulted.

MATERIAL AND METHODS

Biostratigraphy

In total, 334 thin sections were examined for their benthic foraminiferal and calpionellid assemblages (152 for the Säntis section, 134 for the Dräckloch section and 48 for the Vitznau section). The stratigraphic range of marker benthic foraminifera species was attributed following Darsac (1983), Boisseau (1987), and Blanc (1996), and of calpionellids following Remane (1963, 1985), Remane *et al.* (1986, 1998), Blanc (1996), and Blau & Grün (1997). In addition, an ammonite was identified from the Vitznau section.

Fig. 4. Panorama of the Vitznau section (photo taken from the cable car between Vitznau and Hinterbergen, 47°00′49″N 8°29′54″E). Yellow, turquoise, purple and red shadings correspond to recessive heterozoan, prominent photozoan, recessive heterozoan and prominent heterozoan deposits, respectively. The lateral continuity of the Upper Öhrli Limestone Member is not ensured in this outcrop. Black lines correspond to fault zones. For scale, the Vitznau Formation is 33 m thick.

Microfacies and sequence stratigraphy

The outcrops, samples and thin sections were analysed for their facies and microfacies, which were described and interpreted following the classification of Arnaud-Vanneau & Arnaud (2005). Twelve facies zones were thereby differentiated, from F1 corresponding to hemipelagic environments, to F10 attributed to shallow subtidal to tidal environments. The sequence stratigraphic framework was established using field observations and trends in facies and microfacies (Vail *et al.*, 1987; Catuneanu *et al.*, 2009).

Carbon and oxygen isotope analyses

The carbon and oxygen isotope composition ($\delta^{13}C$ and $\delta^{18}O$ values) of 266 bulk-rock carbonate samples (69 from Säntis, 145 from Dräckloch and 52 from Vitznau) was determined using a Thermo Fisher Scientific carbonate-preparation device and GasBench II connected to a Thermo Fisher Scientific Delta Plus XL isotope ratio mass spectrometer, which was operated in a continuous helium flow mode. The CO_2 was extracted by reaction with crystalline orthophosphoric acid of pro-analysis quality shifted to liquid state through heating at 70°C. Ratios of carbon and oxygen isotopes are reported in the delta (δ) notation as the per mil (‰) deviation relative to the Vienna–Pee Dee belemnite standard. Replicate analyses demonstrated an analytical reproducibility for the international calcite standard NBS-19 and the laboratory standards Carrara Marble of better than ±0·05‰ for $\delta^{13}C$ and ±0·1‰ for $\delta^{18}O$.

Phosphorus content

Total phosphorus (P) contents were measured on 320 bulk-rock samples (129 from the Säntis section, 143 from the Dräckloch section and 48 from the Vitznau section) following the procedure described in Bodin *et al.* (2006). The concentration of P was obtained in ppm by calibration with known standard solutions using a UV/Vis photospectrometer (Perking Elmer UV/Vis Photospectrometer Lambda 10, $\lambda = 865$ nm) with a mean precision of 5%.

Bulk-rock mineralogy

The bulk-rock mineralogy was determined with a Thermo scientific ARL X'TRA IP2500 X-ray diffractometer using a semi-quantification method using external standards and following the procedures of Kübler (1983, 1987) and Adatte *et al.* (1996). The precision was 5 to 10% for phyllosilicates and 5% for grain minerals. A total of 113 samples were run for the section at Säntis, 147 for the section at Dräckloch, and 47 for the section at Vitznau. Relative contents of phyllosilicates, quartz, K-feldspar, Na-plagioclase, calcite, dolomite, pyrite, goethite and ankerite were determined. Variations in K-feldspar, Na-plagioclase, dolomite, pyrite, goethite and ankerite proportions were not significant enough to be shown in this publication. They were nevertheless taken into account in the calculation of the percentages of the other minerals.

DATA DOCUMENTATION AND INTERPRETATION

Lithostratigraphy and microfacies

Säntis section

The lower part of the section (the first 53 m) corresponds to the upper part of the Öhrli Formation (Figs 2 and 5). This unit includes abundant large benthic foraminifera accompanied by green algae, rudists, gastropods, calcareous sponges, corals, bivalves and echinoderms. Its microfacies ranges from F2 to F10 (Table 1), and is largely dominated by shallow-water limestone deposited at either side of a shoal. Three dissolution levels were observed within the Upper Öhrli Limestone Member. The first is located 20 m above the base of the section, where dissolution features infilled by mud (Fig. 6) affect 2 m of the underlying sediment. A second level occurs at 26 m above the base of the section, where muddy cavity infillings are present (Figs 5 and 7A). The Öhrli Formation terminates with facies characterized by enhanced microbial activity (intense micritization of clasts) and an abundance of gastropods and thin miliolids (F10). Its top surface is a complex surface marked by dissolution vugs, macroscopic borings and infillings by mudstone of the overlying Vitznau Formation (Fig. 6). These borings are also present in the overlying bed, indicating the superposition of two hardgrounds. The three dissolution levels are interpreted as epikarstic layers.

The base of the Vitznau Formation (between 53 to 57 m above section base) is characterized by a microfacies rich in crinoids and bryozoans with an important degree of sedimentary reworking (F4/FT). This microfacies type is replaced at 57 m above the base of the section by peloidal, crinoid-rich microfacies showing a relatively scarce and poorly diversified fauna of bivalves, bryozoans and circalittoral foraminifera (F3). Based on these observations, the Vitznau Formation indicates a change towards heterozoan carbonate production.

At the base of the Betlis Formation (69 to 72 m above the base of the section), oncoids and reworked bioclasts were observed (F4/T). Numerous chert nodules occur between 67 and 86·5 m above the base of the section. An erosive surface was identified at 86·5 m. The upper part of the Betlis Formation (from 96 to 107 m above the base of the section) shows an important increase in detrital quartz, with the occurrence of well-rounded and broken quartz grains (0·2 to 1 mm). The top of the Betlis Formation is characterized by a hardground with borings infilled by phosphatic and glauconitic sediment of the overlying condensed Gemsmättli Bed (Fig. 8B).

Dräckloch section

The Dräckloch section (Figs 3 and 9) starts with the Lower Öhrli Limestone Member, which shows a microfacies rich in quartz, circalittoral foraminifera, sponge spicules, echinoderms and containing sparse larger agglutinated foraminifera (F2/3) typical of platform slope deposits. This member evolves (22 to 55 m above the base of the section) towards coarser grainstone containing large benthic foraminifera, echinoderms and small ooids, together with numerous large rounded mud intraclasts and bioclasts (shallow-water photozoan facies F5/6). The few samples collected from the marly upper part of the Lower Öhrli Limestone Member and the Upper Öhrli Marl Member (55 to 130 m above section base) show a return to facies F2/3.

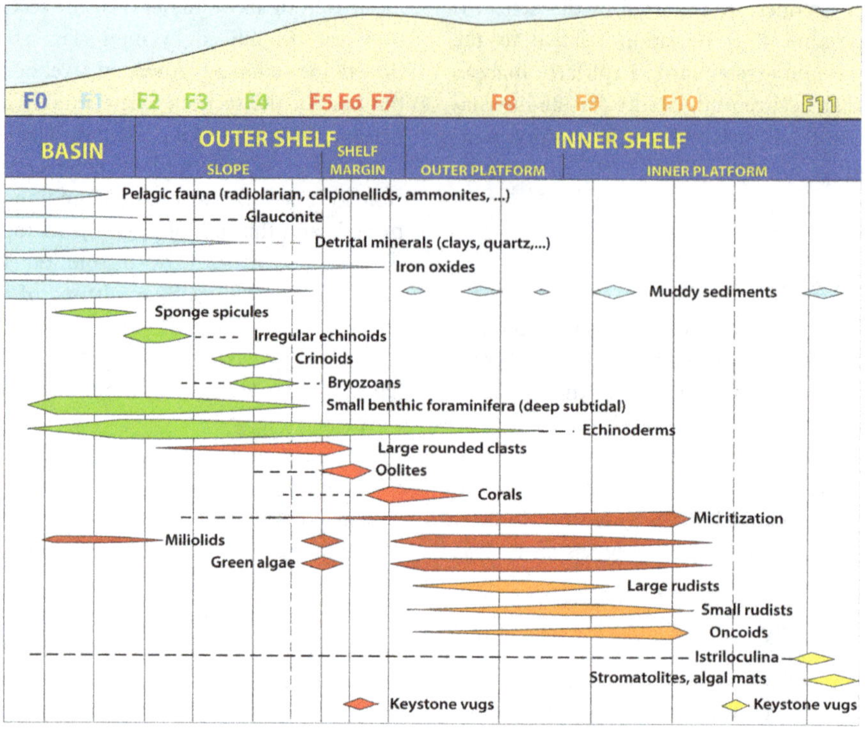

Fig. 5. Ideal distribution of faunas along a shelf-basin transect and resulting distribution of microfacies. Classification after Arnaud-Vanneau & Arnaud (2005). F0: Biomicrite with radiolaria and calpionellids. F1: Wackestone with sponge spicules. F2: Wackestone with *spatangidae*. F3: Grainstone/packstone with rounded echinoderm debris and small foraminifera. F4: Grainstone/packstone with crinoids and bryozoans. F5: Grainstone with large rounded debris. F6: Grainstone with oolites. F7: Grainstone/packstone with corals; Boundstone. F8: Wackestone/packstone with large foraminifera, sometimes accompanied by large rudists. F9: Packstone/grainstone with *Miliolidae* and rudists. F10: Packstone/grainstone with oncoids and *Bacinella.* F11: Mudstone with bird's eyes; emersion facies.

The microfacies of the overlaying Upper Öhrli Limestone Member shows an increasing proportion of large benthic foraminifera, echinoderms and green algae (F6/7) between 130 and 174 m above the base of the section. At 174 m, a rapid change towards facies F3 is noted, and an evolution towards coarser grainstone (up to F8) follows to a point 246 m above the base of the section. Fringing cements and dissolution vugs with mud and microsparite infillings (Fig. 8C and D) were observed between 236 and 246 m above the base of the section. From 246 to 260 m above section base, a change in biota occurs: corals and calcareous sponge debris become dominant and benthic foraminifera nearly disappear with the exception of *Andersenolina* (Fig. 8E). The presence of dolomitic extraclasts indicative of a confined and very shallow environment mixed with mud of outer shelf origin (containing *Lenticulina*), hermatypic and ahermatypic coral debris, rudists and bryozoans (Fig. 7H) suggest strong sedimentary reworking (F5/FT). The thickness of this interval (246 to 260 m above section base) could not be evaluated because of the presence of a fault zone (indicated in Figs 3 and 10). Nevertheless, the floro-faunal microfacies associations are similar below and above this faulted zone, indicating a certain continuity in facies. The top of the Upper Öhrli Limestone Member shows borings infilled by a complex succession of partly phosphatized sediments including mud with small peloids, and pyrite (Fig. 10A and B), suggesting a hardground (Fig. 11A) and condensation. The uppermost bed of the Upper Öhrli Limestone Member shows evidence of sediment reworking with the occurrence of detrital mud pebbles with dolomite extraclasts (FT). Bladed circumgranular cements are observed in this bed (Fig. 8F), pointing to early diagenetic cementation in the vadose zone.

The overlying Vitznau Formation starts with a layer containing abundant reworked bryozoans, corals, *Gryphaea* and serpulids floating in a clayey matrix (FT, Figs 10C and 12C), which documents a change towards heterozoan carbonate production. A second hardground with borings infilled with pyrite is observed at the top of this interval. The lower part of the Vitznau Formation at 246 to 300 m above the base of the section shows limestone-marl alternations containing abundant *Gryphaea* in life position or weakly transported, some brachiopods,

Table 1. Facies description and their palaeoenvironmental significance (adapted from Arnaud-Vanneau & Arnaud, 2005)

Facies type		Description	Location
F1	Wackestone with sponge spicules	Marl and bluish-grey argillaceous limestone containing significant amount of clay and silted quartz. Sponge spicules and ostracods are found in abundance. The fauna also includes irregular sea urchins, Gryphaea, and small agglutinated foraminifera.	Hemipelagic facies
F2	Wackestone with irregular sea urchins and small benthic foraminifers	Marly limestone including irregular sea urchins, small agglutinated foraminifera and peloids in abundance, and sponge spicules, ostracods and serpulids in a lesser extent.	Lower offshore, quiet deep subtidal
F3	Packstone-grainstone with echinoderm fragments and small benthic foraminifers	Well sorted and slightly more calcareous (less detrital grains) than F2 with sparse macrofauna. Preponderancy of echinoderm fragments, peloids and small benthic foraminifers.	Lower and upper offshore, deep subtidal
F4	Packstone-grainstone with bryozoans and crinoids	Hydrodynamic facies often cross-bedded with coarser grains predominantly constituted of bryozoans and crinoid debris. Scarce presence of reworked large benthic foraminifers such as *Lenticulina* or *Andersenolina*	Lower and upper offshore, deep subtidal, below the photic zone
F5	Grainstone with large rounded bioclasts	Limestone deposited under high energy, often showing oblique and cross stratifications. Contain abundant rounded large-sized benthic foraminifera, mixed with diverse clasts (echinoderms, bivalves, gastropods, green algae, calcareous sponges, etc.)	Upper offshore and shoreface, shallow subtidal to tidal, photic zone
F6	Grainstone with oolites	High energy limestone with oblique and cross stratifications, characterized by the abundance of oolites	Shoreface, shallow subtidal to tidal, photic zone
F7	Grainstone and boundstone with hermatypic corals	High energy limestone with rounded skeletal debris, often containing oolites	Edge of the platform: photic zone, subtidal to tidal
F8	Packstone-wackestone with large benthic foraminifera and rudists	Coarse limestone with abundant and diversified fauna of large benthic foraminifera, but also rudists, gastropods, calcareous sponges, green algae and a high percentage of micritized bioclasts, as evidence of microbial activity, echinoderms are still present	External part of the inner platform
F9	Packstone-wackestone with miliolidae	Massive calcareous meter-thick beds with micritized debris and miliolids. Absence of echinoderms	Inner platform, shallow subtidal
F10	Packestone-grainstone with oncolites and Bacinella	Massive calcareous meter-thick beds with micritized debris, oncolites and Bacinella. Absence of echinoderms	Inner platform, shallow subtidal to tidal
FT	Facies of transgression	High degree of reworking, leading to the mixing of biota living in different environments	Initiation of a relative sea-level rise

circalittoral foraminifera and sponge spicules (deep subtidal facies F2/F3). Limestone layers are absent in the upper part of the Vitznau Formation (300 to 315 m above section base) and *Gryphaea* is less abundant.

The Betlis Formation begins 315 m above the base of the section with a layer containing ooids, abundant crinoids, sparse bryozoans, small benthic foraminifera, sponges and gastropods, together with reworked mud pebbles from the Vitznau Formation (FT). From 315 to 370 m, a peloidal microfacies containing crinoids and bryozoans is present (F3 to F4). Similarly to the Säntis section, numerous chert layers and nodules are present in the interval between 315 and 357 m (Fig. 11B). At 370 m, a level containing crinoids, bivalves, small benthic foraminifera, *Lenticulina* and ostracods shows important reworking (FT). The Pygurus Member on top of the Betlis Formation includes millimetre-sized, rounded and fractured quartz grains, and bioturbation is important.

Vitznau section

The Vitznau section (Fig. 13) starts with the upper part of the Palfris Formation, in which muddy facies containing sparse bioclasts of echinoderms and reworked circalittoral foraminifera (F1/F2) evolve towards a grainstone facies with peloids, sparse bryozoans and small benthic foraminifera (F3). In the overlying Upper Öhrli Limestone Member, 12·5 to 23·4 m above the base of the section, a change towards coarser grainstone with mud pebbles, echinoderms, sparse ooids, large benthic foraminifera, corals and sponges (F5) is observed. The last 1·6 m of the Öhrli Formation shows a complex succession of erosive surfaces and hardgrounds. A first irregular erosive surface associated with limestone lag pebbles (Fig. 11C) occurs at 23·4 m above section base. The overlying calcareous layer is perforated and the borings are infilled by marls of the covering layer (Fig. 11D). This

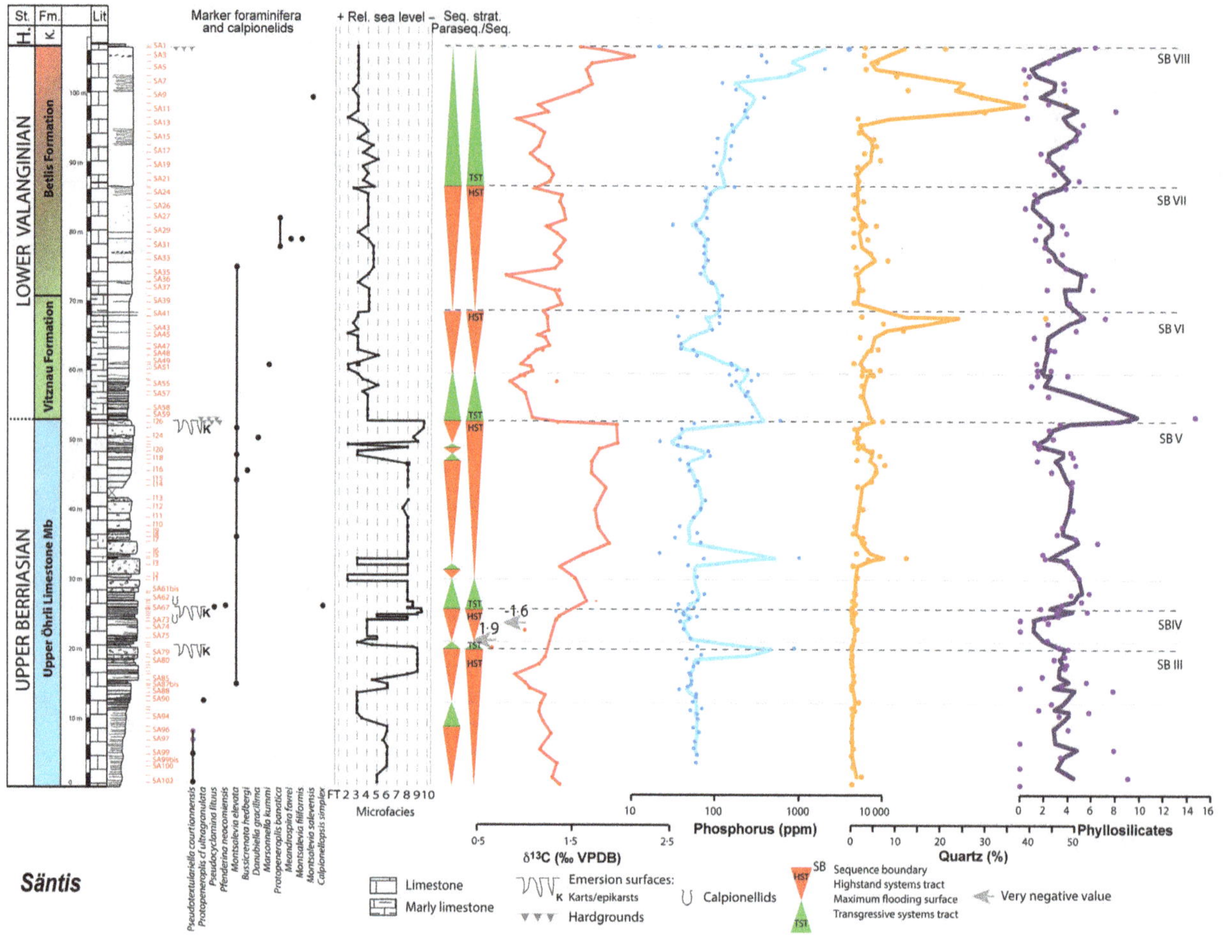

Fig. 6. Sedimentology, geochemistry and mineralogy of the Säntis section. Microfacies classification after Arnaud-Vanneau & Arnaud (2005). Five sequence boundaries are identified. An abrupt change of carbonate production and an important hiatus are highlighted at the Berriasian-Valanginian boundary. Generally higher $\delta^{13}C$ values, and low P are observed within Berriasian photozoan limestone, whereas lower $\delta^{13}C$ values, and higher P characterise Valanginian heterozoan limestone. In the top part of the section, late Valanginian platform drowning is materialized by a condensed phosphatic layer infilling perforations. The Weissert episode is associated with peaks in P and quartz contents. Numerical data are included in Appendix.

thin marly layer is rich in reworked and pyritized extraclasts (Fig. 10D). The next two layers have an erosive base (Fig. 11E). Their microfacies consists of peloids, echinoderms and small benthic foraminifera with sparse reworked platform debris (benthic foraminifera, ooids and extraclasts) (F3). In the uppermost part of the Upper Öhrli Limestone Member, the microfacies additionally contains calcareous sponges, corals and ooids (F6). A

Fig. 7. Marker benthic foraminifera identified in the studied sections. Scale 1 is used for *Pseudotextulariella courtionnensis* sp. and *Pseudotextulariella courtionnensis* (upper lower – lower upper Berriasian, images A and B, and C to J respectively), scale 2 for *Montsalevia elevata* (upper Berriasian, K to R), *Montsalevia salevensis* (lower Valanginian, S), and *Pfenderina neocomiensis* (upper Berriasian – Valanginian, T). (A) Dräckloch, Upper Öhrli Marls Member (GAS70); (B) Dräckloch, Upper Öhrli Limestone Member (GAS111); (C) Säntis, Upper Öhrli Limestone Member (SA99); (D) Dräckloch, Upper Öhrli Limestone Member (GAS109); (E) Dräckloch, Upper Öhrli Limestone Member (GAS111); (F) Dräckloch, Upper Öhrli Limestone Member (GAS135); (G) Dräckloch, Upper Öhrli Limestone Member (GAS139); (H) Dräckloch, Upper Öhrli Limestone Member (GAS143); (I) Dräckloch, Upper Öhrli Limestone Member (GAS155); (J) Vitznau, Upper Öhrli Limestone Member (Vz40); (K) Säntis, Upper Öhrli Limestone Member (I15); (L) Säntis, Upper Öhrli Limestone Member (SA87); (M) Dräckloch, Upper Öhrli Limestone Member (GAS109); (N) Dräckloch, Upper Öhrli Limestone Member (GAS125); (O) Dräckloch, Upper Öhrli Limestone Member (GAS127); (P) Dräckloch, Upper Öhrli Limestone Member (GAS153); (Q) Vitznau, Upper Öhrli Limestone Member (Vz40); (R) Vitznau, Upper Öhrli Limestone Member (Vz43); (S) Säntis, Betlis Formation (SA9); and (T) Säntis, Upper Öhrli Limestone Member (SA66) (GASV8), Dräckloch.

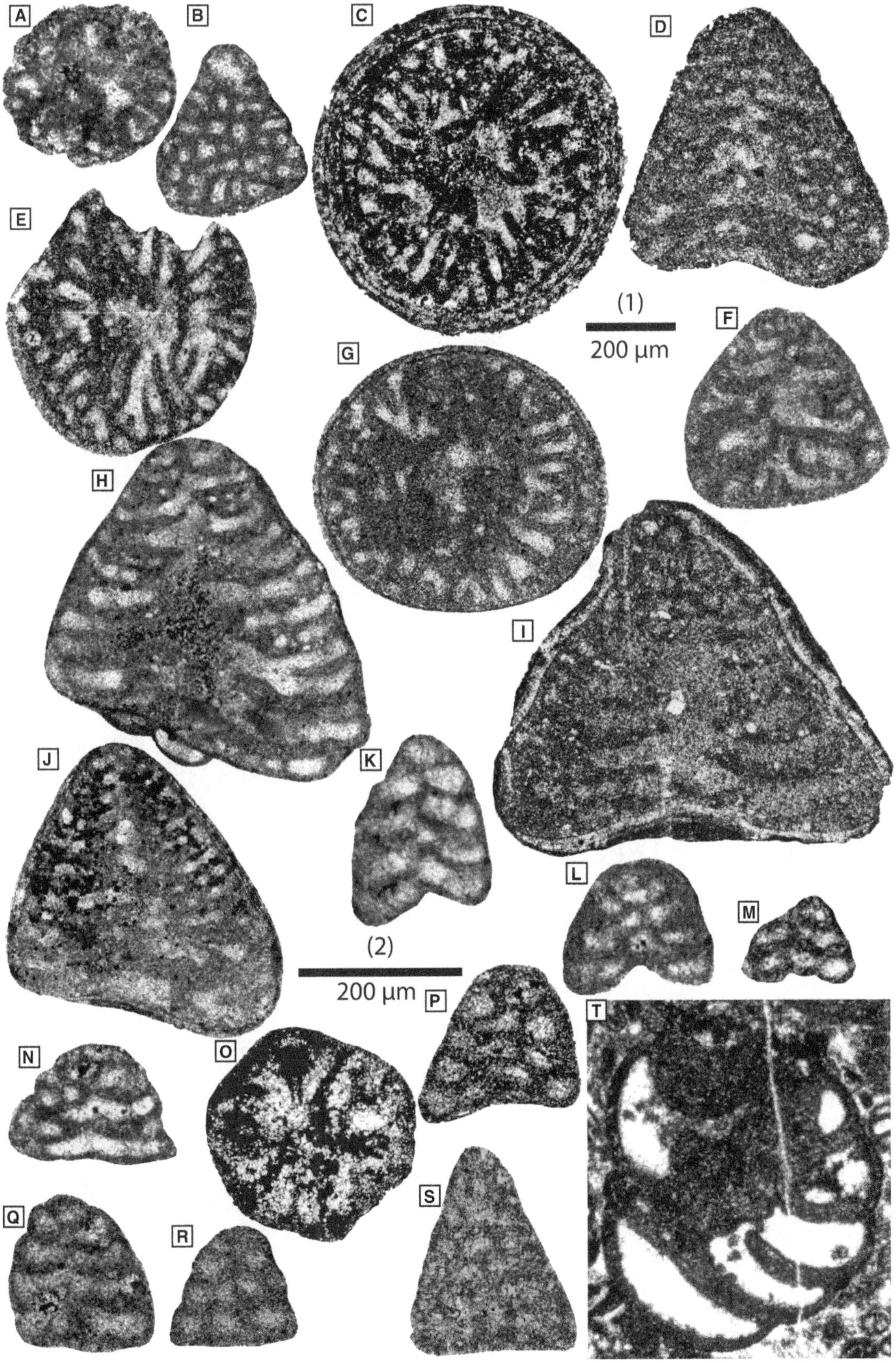
A
B
C
D
E
(1)
200 μm
F
G
H
I
J
K
L
M
(2)
200 μm
P
T
N
O
S
Q
R

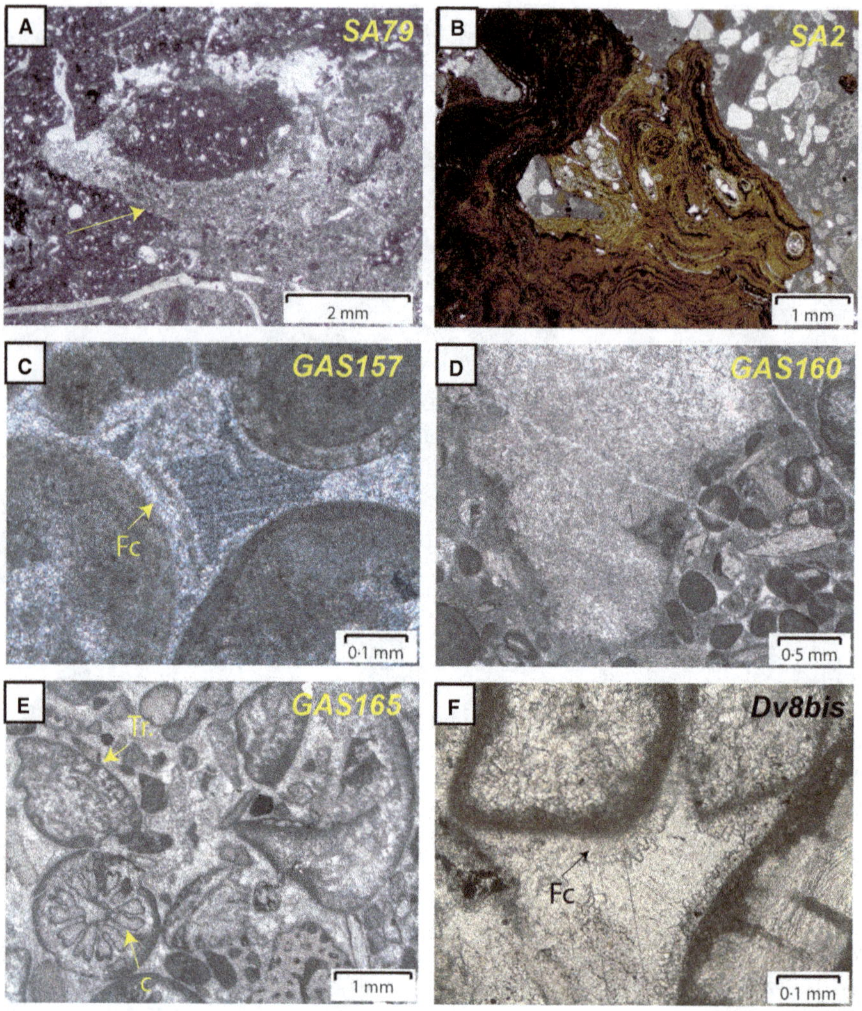

Fig. 8. (A) Epikarst with early infillings (yellow arrow), Upper Öhrli Limestone Member (SA79, transmitted light), Säntis; (B) Phosphatic crust (Ph) and quartz grains, uppermost Betlis Formation perforated and infilled by sediments of the overlying Gemsmättli Bed (SA2, polarized and transmitted light, respectively), Säntis; Dräckloch; (C) Fringing cements (Fc), Upper Öhrli Limestone Member (GAS157, polarized light), Dräckloch; (D) Dissolution vug with early infillings, Upper Öhrli Limestone Member (GAS160, transmitted light), Dräckloch; (E) Bioclastic grainstone with ahermatypic coral (c) and *Trocholina* foraminifera (Tr.), uppermost Upper Öhrli Limestone Member (GAS165, transmitted light), Dräckloch; (F) Fringing cements (Fc), uppermost Upper Öhrli Limestone Member (Dv8bis, transmitted light), Dräckloch.

hardground is observed on top of the Upper Öhrli Limestone Member, whose borings are often filled with pyrite (Fig. 10E). A further marine hardground occurs on top of the first marl-limestone alternation of the Vitznau Formation. The corresponding microfacies shows intense reworking with ostracods, microbial mats, bryozoans, broken serpulids, micritized crinoids and corals, together with mud pebbles rich in quartz (FT), up to 26·3 m above the base of the section (Fig. 10F). From 26·3 to 56·8 m, the microfacies consists of monotonous wackestone containing ostracods, sponge spicules and serpulids (F1/F2). Numerous *Gryphaea* were observed in life position or reworked.

At 56·8 m above the base of the section, an erosive bank rich in bivalves is present (Fig. 11F), which is followed by a recessive marly interval of 1 m. Finally, field observations indicate that the Betlis Formation is composed of a peloidal echinodermal carbonate (F3). The lack of samples in this interval is due to difficult access in steep terrain and the presence of vegetation hiding the transition between the Vitznau and Betlis formations.

Carbon and oxygen isotope data

Oxygen isotope values (see supplementary data) oscillate between -5 and $-2‰$ in the Säntis section (with a mean

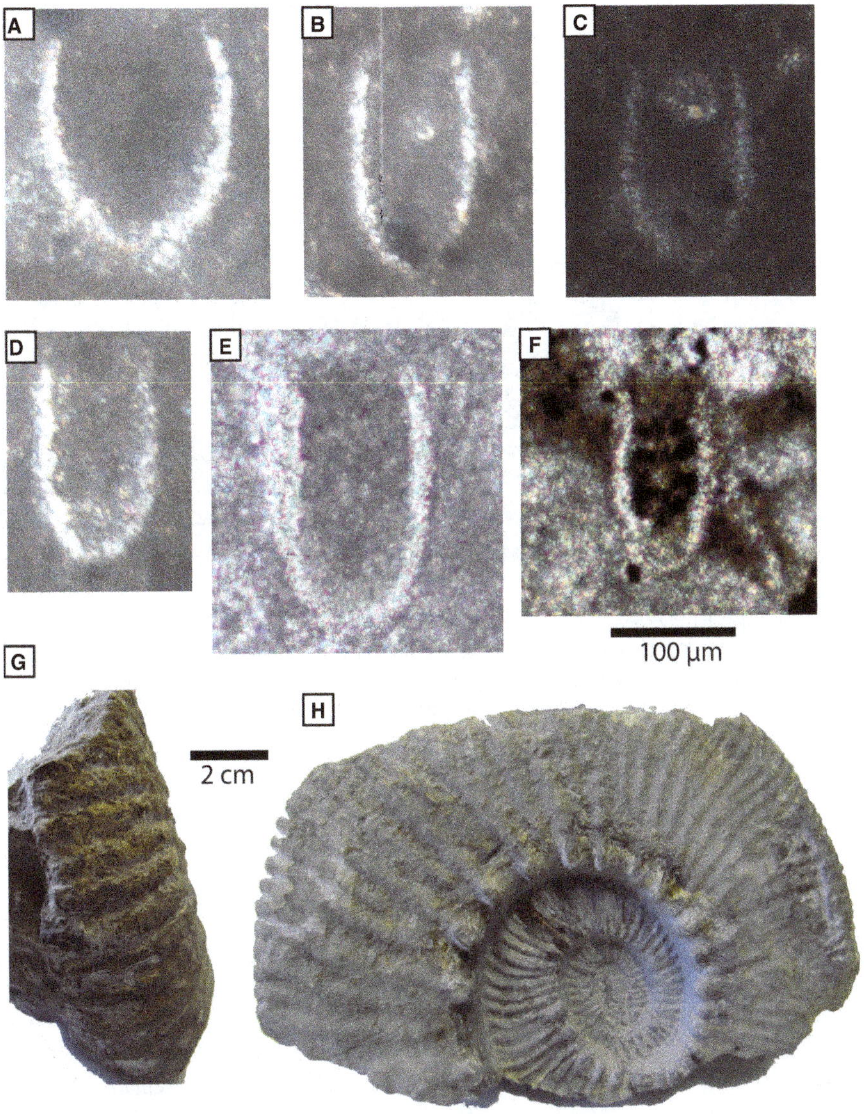

Fig. 9. Marker calpionellids and ammonite. (A) *Remaniella filipescui*, calp. zones B to D3, Upper Öhrli Limestone Member (Vz45), Vitznau; (B) *Calpionellopsis oblonga*, calp. zones D2 to D3, Upper Öhrli Limestone Member (Vz45), Vitznau; (C) *Calpionellopsis oblonga*, calp. zones D2 to D3, Upper Öhrli Limestone Member (Vz45), Vitznau; (D) *Calpionellopsis simplex*, calp. zones D2 to D3, Upper Öhrli Limestone Member (Vz45), Vitznau; (E) *Calpionellopsis simplex*, calp. zones D2 to D3, Upper Öhrli Limestone Member (SA66), Säntis; (F) possible *Tintinnopsella longa*, calp. zones D1 to E2, Vitznau Formation; (G and H) *Thurmanniceras thurmanni* s.str. (upper Berriasian, *T. otopeta* ammonite zone), ventral and umbilical views, respectively.

of −3·5‰ and a standard deviation of 0·4‰), and −6 and −2‰ in the Dräckloch and Vitznau sections (with means of −3·7 and −3·5‰ and standard deviations of 0·8 and 0·7‰, respectively). The $\delta^{18}O$ values reflect significant diagenetic overprint (Choquette & James, 1987) and are not further discussed here.

The $\delta^{13}C$ long-term trends are similar between the three sections (Figs 5, 9, 10 and 12): relatively heavy but variable values (with mean values of 1·3, 1·1 and 0·9‰, and standard deviations of 0·7, 0·6 and 0·4‰, in the Säntis, Dräckloch and Vitznau sections, respectively) are observed in the Öhrli Formation and its distal equivalent, the Palfris Formation. In contrast, lower $\delta^{13}C$ values (with mean values of 1·3, 0·5 and 0·3‰, and standard deviations of 0·2, 0·4 and 0·2‰, in the Säntis, Dräckloch and Vitznau sections, respectively) are recorded in the Vitznau and the Betlis formations. The Säntis section is the only section that records a positive $\delta^{13}C$ shift (of 1·2‰) at the top of the Betlis Formation. The decrease in $\delta^{13}C$ values (0·7‰) is abrupt and

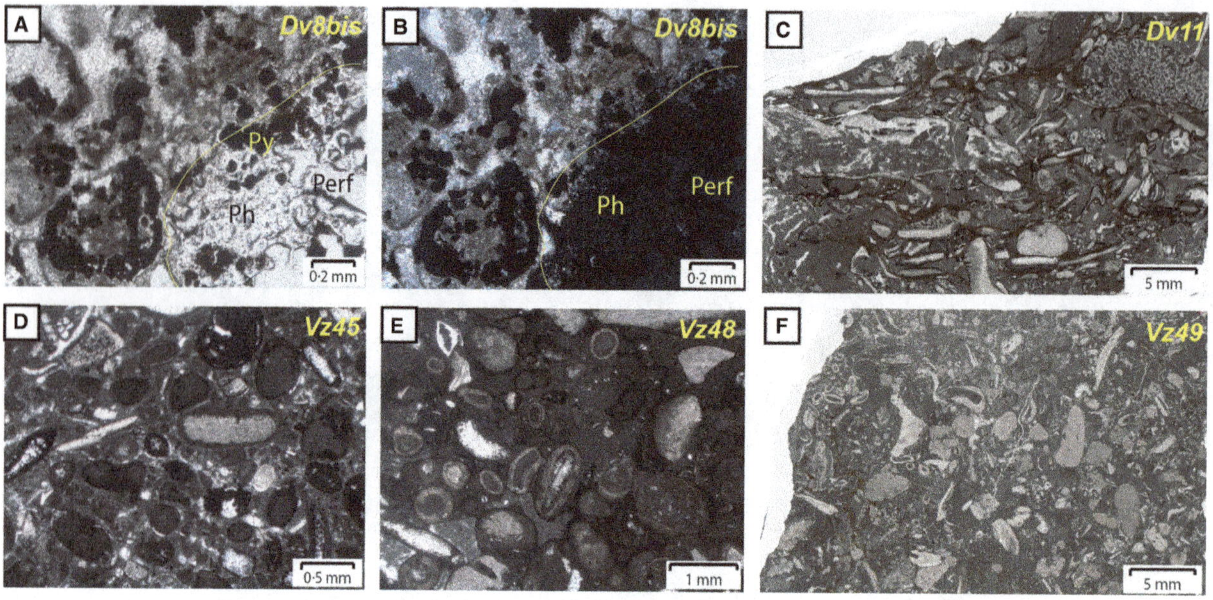

Fig. 10. (A and B) Boring (Bor) filled by phosphate (Ph) and pyrite (Py), uppermost Upper Öhrli Limestone Member (Dv8bis, transmitted and polarized light, respectively), Dräckloch; (C) Reworked sediment with a dolomitized sponge, worn corals, bryozoans, pebbles with serpulids, ostracods and *Lenticulina* in a muddy matrix, lowermost Vitznau Formation (Dv11, transmitted light), Dräckloch; (D) Reworked sediment with micritized and partly pyritized platform clasts in a mud matrix, uppermost Upper Öhrli Limestone Member (Vz45, transmitted light), Vitznau; (E) Level of intense sediment reworking with extraclasts of oolitic and crinoidal sediments in a mud matrix, uppermost Upper Öhrli Limestone Member (Vz48, transmitted light), Vitznau; (F) Reworked bryozoans, crinoids, coral debris, serpulids and ostracods, lowermost Vitznau Formation (Vz49, transmitted light), Vitznau.

occurs at the limit between the Upper Öhrli Limestone Member and the Vitznau Formation in the Säntis and Vitznau sections, and within the upper part of the Upper Öhrli Limestone Member in the Dräckloch section.

The circulation of meteoric or altered marine pore waters tends to decrease the oxygen and carbon-isotopic values in carbonates during diagenesis (Choquette & James, 1987). As such, a minor overprint of $\delta^{13}C$ values is to be expected. However, the $\delta^{13}C$ vs. $\delta^{18}O$ plot (Fig. 14) exhibits a relatively poor correlation coefficient ($R^2 < 0{\cdot}4$). In general, carbon-isotope records in neritic carbonates have been shown to be only marginally affected by burial diagenesis, and have been used as a reliable stratigraphic tool (Ferreri *et al.*, 1997; Hennig, 2003; Föllmi *et al.*, 2006; Weissert *et al.*, 2008). Exceptions are, however, possible in the presence of emersion surfaces and depending on mineralogy (for instance aragonite versus calcite; Ferreri *et al.*, 1997; Swart & Eberli, 2005; Weissert *et al.*, 2008). The consistent trends between the sections, as well as the comparable $\delta^{13}C$ trends and value ranges with correlated sections in the Jura Mountains and the Vocontian Basin (La Chambotte, Juracime, Montclus; Morales *et al.*, 2013), suggest that the $\delta^{13}C$ records are rather well preserved.

Phosphorus content as a nutrient tracer

Low P contents were measured in the Lower Öhrli Limestone Member, followed by higher values in the Upper Öhrli Marl Member (100 and 250 ppm on average in the Dräckloch section). In the Palfris Formation of the Vitznau section, P values oscillate around 200 ppm (Fig. 13). A decrease in P concentrations was measured in the overlying Upper Öhrli Limestone Member with mean values of 60, 80 and 70 ppm in the Säntis (Fig. 6, with the exception of two samples with values close to 1000 ppm), Dräckloch and Vitznau sections (Fig. 12), respectively. In the Öhrli Formation, variations in P concentrations are therefore associated with lithological changes. Therefore, the level of nutrients in the sea water was significantly controlled by depth.

Phosphorus contents abruptly increase at the base of the Vitznau Formation in the three sections. In this formation, mean values of 110, 250 and 300 ppm were measured in the Säntis, Dräckloch and Vitznau sections, respectively, which are similar to those in the Betlis Formation. In the Säntis and Dräckloch sections, the uppermost part of the Betlis Formation shows a progressive increase in P values up to 3700 and 2750 ppm, respectively. The increase in *P* levels in the Vitznau and Betlis

Fig. 11. Outcrop photographs of key facies. (A) Borings with pyrite infillings, top of the Upper Öhrli Limestone Member (Dv8bis), Dräckloch; (B) Chert nodules (arrows), Betlis Formation (Be15), Dräckloch; (C) Vertical plane section of an erosion surface with lag pebbles (dotted line), Upper Öhrli Limestone Member (between Vz44and Vz45), Vitznau; (D) Calcareous bed with borings infilled by the overlying marly level, Upper Öhrli Limestone Member (Vz45), Vitznau; (E) Erosion surface (arrow), Upper Öhrli Limestone Member (between Vz46 and Vz46bis), Vitznau; (F) Erosion base (dotted line) of a level rich in bivalve shells, upper part of the Vitznau Formation (Vb11), Vitznau.

formations is associated with the proliferation of suspension-feeding organisms (essentially crinoids and bryozoans), leading to the production of heterozoan carbonates. The relatively high P contents observed in the Betlis Formation indicate that depth was not the only factor controlling the nutrient level, and that additional sources were involved.

Bulk-rock mineralogy as a proxy for detrital input

The phyllosilicate and quartz contents increase in the Lower Öhrli Limestone Member and Upper Öhrli Marl Member in the Dräckloch section (from 5 to 35% and from 10 to 25%, respectively), as well as in the Palfris Formation in the Vitznau section (up to 20% and 25%, respectively). They decrease below the detection limit of the XRD (<5%) in the overlying Upper Öhrli Limestone Member in all three sections. Therefore, these trends show a good correlation with lithological changes and are comparable to trends in P values.

In the Vitznau Formation, the phyllosilicate and quartz contents increase to 17% and 18%, respectively in the Dräckloch section, and up to 32 and 47%, respectively in the Vitznau section, indicating a strong increase in detrital material. In the Säntis section, they remain below the detection limit (<5%), perhaps linked to the more proximal location of this section. In the Betlis Formation, the

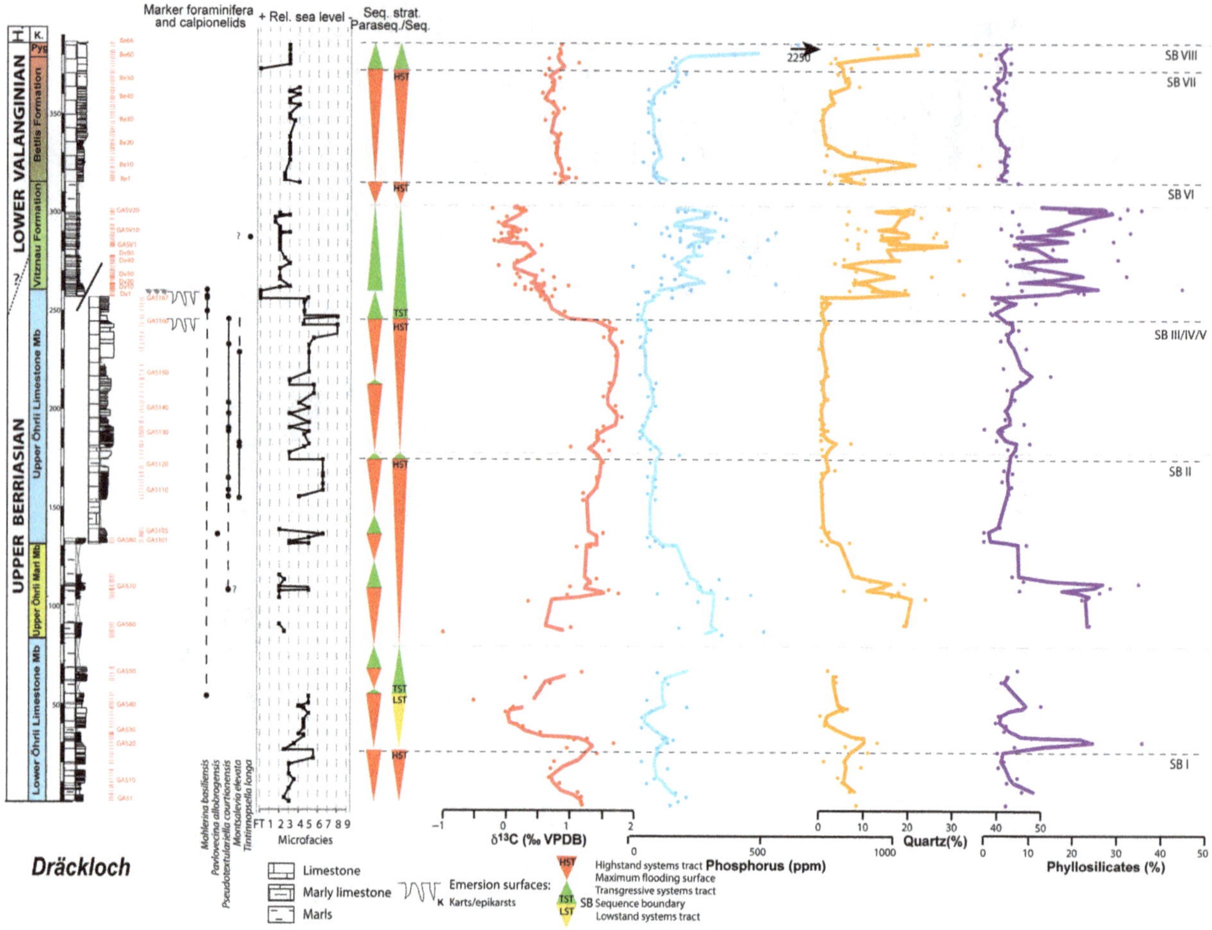

Fig. 12. Stratigraphy, sedimentology, geochemistry and mineralogy of the Dräckloch section. Microfacies classification after Arnaud-Vanneau & Arnaud (2005). Five sequence boundaries are identified. An abrupt change in carbonate production and important hiatuses are highlighted through the Berriasian-Valanginian boundary. Highly reworked deposits on top of an epikarst, hardgrounds and an abrupt negative shift in $\delta^{13}C$ values witness the time gap. Note that the Weissert carbon-isotope excursion (CIE) is not preserved at the top of the section (the contact between the Pygurus Member and the Kieselkalk Formation is faulted), but peaks in P and quartz contents witness the ongoing drowning. Numerical data are included in Appendix.

phyllosilicate content falls below 10% in all three sections, but a different behaviour for the quartz content is observed. In the Dräckloch section, relatively high values occur at the base of the formation (36% at section meter 326). In the Säntis and Dräckloch sections, the quartz content falls below 5% and rises in the uppermost part of the formation to 44% and 36%, respectively. This goes along with the presence of large (>1 mm) rounded quartz grains and a concomitant increase in *P* values shortly before the drowning of the Weissert episode.

Age control and correlation of sections

A combination of lithological, biological and sequence stratigraphic tools were used to correlate the sections. Stratigraphic sequences were determined on the base of key surfaces (sequence boundaries, transgressive and maximum flooding surfaces). Sequence boundaries in shallow-water deposits can be recognized by the presence of unconformities associated with erosion and subaerial exposure (Vail *et al.*, 1987). Transgressive surfaces are often erosive and overlaid by reworked deposits, and maximum flooding surfaces are indicated by a maximum of accommodation space. In certain cases, accommodation maxima (between the transgressive and highstand systems tracts) correspond to a transitional interval termed maximum flooding zone (mfz). Parasequences were interpolated based on (micro-)facies and lithological observations, and are indicated in the corresponding figures. However, given the locally unequal sampling resolution due to limited access and fault zones, the determination of parasequences should only be considered as an

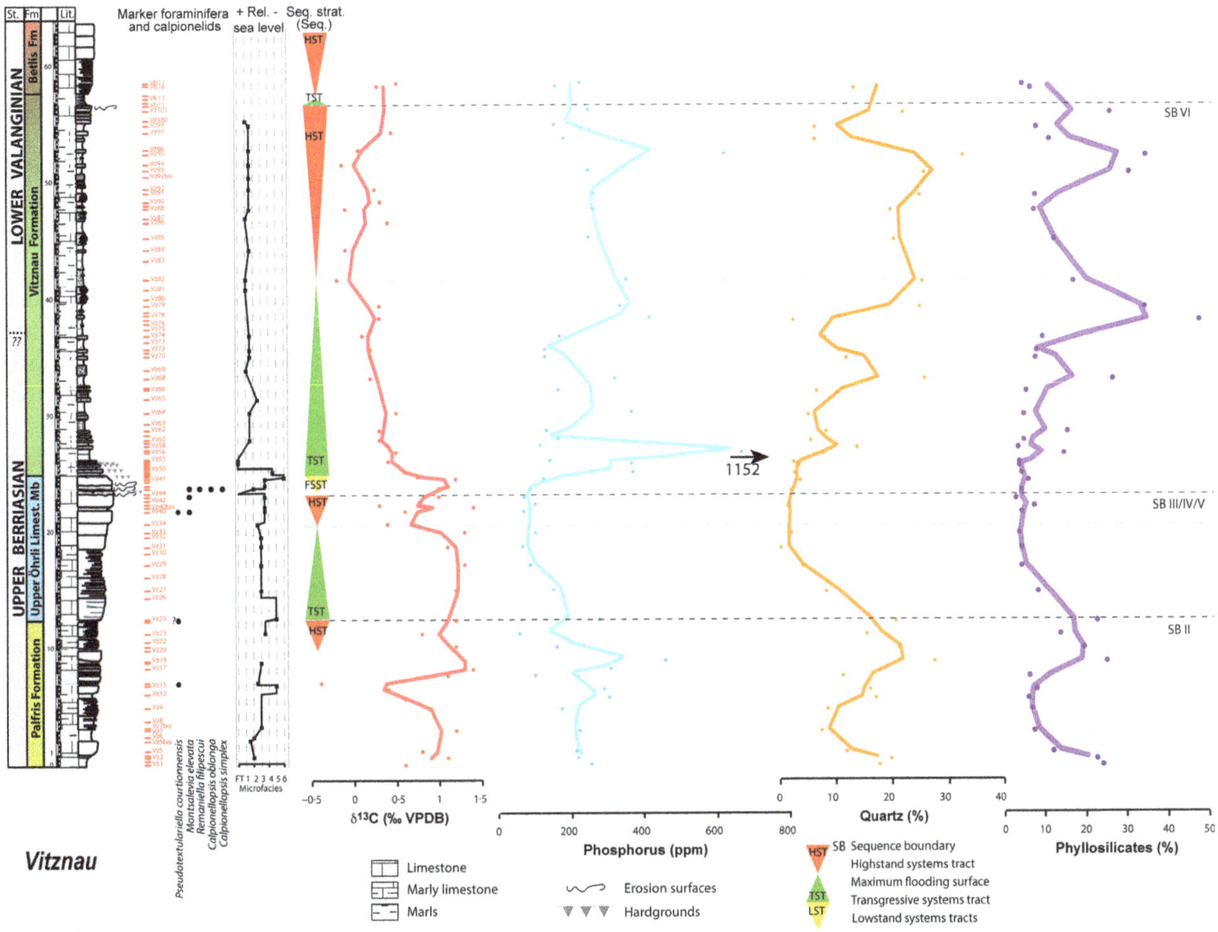

Fig. 13. Stratigraphy, sedimentology, geochemistry and mineralogy of the Vitznau section. Microfacies classification after Arnaud-Vanneau & Arnaud (2005). Two sequence boundaries are identified. A succession of truncation surfaces associated with reworked platform deposits is observed in the upper Berriasian and interpreted as a FSST. A succession of four hardgrounds is associated with the relative sea-level rise following the FSST. The abrupt negative shift in $\delta^{13}C$ values and peak in phosphorus contents associated with the hardground highlight the time gap. With the transgression, higher phosphorus, quartz and phyllosilicate contents are observed. Numerical data are included in Appendix.

approximation. The studied sections include a total of seven sequences, labelled from I to VII.

The age of the sequences was determined from their content in stratigraphic marker species (benthic foraminifera, calpionellids, ammonites), which allowed the sections to be correlated (Table 2). The stratigraphic distributions of marker benthic foraminifera appear to be consistent with those in various western Tethyan regions (Fig. 15): the Jura Mountains (Darsac, 1983; Blanc, 1996; Morales *et al.*, 2013), Pyreneans (Peybernès & Combes, 1994) and Provence (Virgone & Masse, 1996; Virgone, 1997). The distribution of *Pseudotextulariella courtionensis* is particularly notable, since this foraminifer was identified in all three Helvetic sections. This species has a relatively wide geographic distribution since it has been recorded in the French Alps (Arnaud-Vanneau & Darsac, 1984), the French and Swiss Jura Mountains (Darsac, 1983; Boisseau, 1987; Pasquier, 1995; Blanc, 1996), the Pyrenees (Peybernès & Combes, 1994), Provence (Virgone & Masse, 1996; Virgone, 1997), and in the Helvetic Alps (Pasquier, 1995). Its biostratigraphic range probably extends back to the late early Berriasian (calpionellid zone C) and covers the early late Berriasian (*S. boissieri* ammonite zone, *Malbosiceras paramimounum* ammonite subzone, which corresponds to calpionellid zone D1). This was established in the Jura Mountains (Darsac, 1983; Boisseau, 1987; Blanc, 1996), and a compatible age range is noted in the Pyrenees and Provence (Peybernès & Combes, 1994; Virgone & Masse, 1996; Virgone, 1997).

Pseudotextulariella courtionensis may be associated with another large foraminifer, *Pavlovecina allobrogensis*, which has a shorter biostratigraphic range. *Pavlovecina allobrogensis* is found in a 1 to 2 m thick marker horizon in the Jura Mountains (Darsac, 1983; Boisseau, 1987; Adatte,

1988; Blanc, 1996), and indicates an early late Berriasian age (early *S. boissieri* ammonite zone, *M. paramimounum* ammonite subzone; early calpionellid zone D1; Blanc, 1996).

The first appearance of *Montsalevia elevata* and *Pfenderina neocomiensis* and the last appearance of *Pseudotextulariella courtionensis* mark the later part of the late Berriasian (Darsac, 1983; Boisseau, 1987; Adatte, 1988; Blanc, 1996). *Protopeneroplis Banatica* is found in latest Berriasian – earliest Valanginian deposits (Blanc, 1996). The early Valanginian is characterized by the first appearance of *Montsalevia salevensis* (Darsac, 1983; Boisseau, 1987; Blanc, 1996).

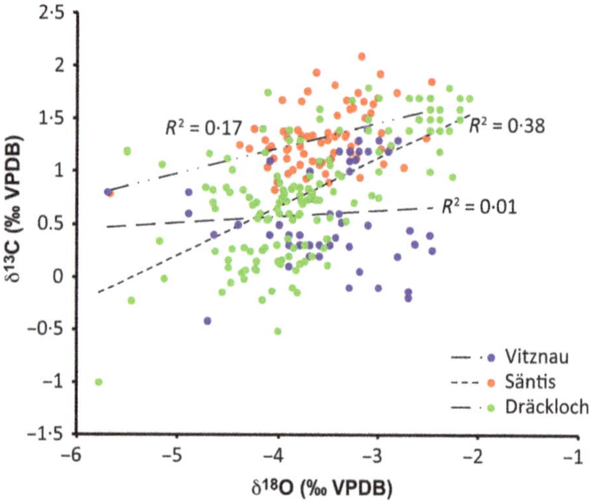

Fig. 14. Carbon versus oxygen isotope crossplot. The absence of correlation between $\delta^{13}C$ and $\delta^{18}O$ values shows that diagenesis did not significantly affect the $\delta^{13}C$ values.

In addition, calpionellids were identified in the different sections. In the Vitznau section, a Berriasian age is assigned to the upper part of the Upper Öhrli Limestone Member by the presence of *Remaniella filipescui* (Fig. 9), which extends from calpionellid zones B to D3 (Blanc, 1996). The presence of *Calpionellopsis* in the upper part of the Upper Öhrli Limestone Member in the sections at Säntis and Vitznau (Fig. 9) indicates calpionellid zone D2/D3 (Remane, 1963, 1985; Remane *et al.*, 1986, 1998).

Finally, an ammonite was found in the scree, immediately below the section of Vitznau (Fig. 9). The specimen is a *Thurmanniceras thurmanni* s.str. and indicates the early *Thurmanniceras otopeta* ammonite zone corresponding to the latest Berriasian (Blanc *et al.*, 1992).

The first sequence (sequence I), is marked by the joint presence of *Pseudotextulariella courtionensis* and *Montsalevia elevata*, and as such is of late early – early late Berriasian age. Sequence I is well-documented in the Dräckloch section (Fig. 15). Since this section is composed of outer shelf to outer platform deposits, sequence boundaries are not necessarily defined by emersion surfaces, and lowstand systems tracts (LST) can be preserved. There, sequence I starts 25 m above the base of the section, where the first sequence boundary (SB I) is placed

Table 2. Age attribution of stratigraphic sequences based on their marker fauna content. Biostratigraphies are based on the distribution of benthic foraminifera (principally established by Darsac, 1983 and Blanc, 1996), calpionellids (Blanc, 1996; Remane *et al.*, 1998) and ammonites (Blanc *et al.*, 1992). See the corresponding text for additional details

Stage	Calpionellid zonation	Sequence stratigraphic units	Marker fauna: Benthic foraminifera	Marker fauna: Calpionellids	Marker fauna: Ammonites
Lower VALANGINIAN	E	Sequence VII	FO *Montsalevia salevensis*		
Transition		Sequence VI	FO *Meandrospira favrei* FO *Protopeneroplis banatica* FO *Montsalevia filiformis* LO *Montsalevia elevata*		
	D3?	Sequence V			*Thurmanniceras thurmanni* s.str ?
Upper BERRIASIAN	D2/D3	Sequence IV Sequence III	FO *Pfenderina neocomiensis*	*Remaniella filipescui* *Calpionellopsis simplex* *Calpionellopsis oblonga*	
	D1	Sequence II	LO *Pseudotextulariella courtionnensis*		
Transition	C/D1	Sequence I	FO *Montsalevia elevata* LO *Pavlovecina allobrogensis* FO *Pavlovecina allobrogensis* FO *Pseudotextulariella courtionnensis*		

The question mark is related to the uncertainty of the sedimentary interval to which the ammonite belong.

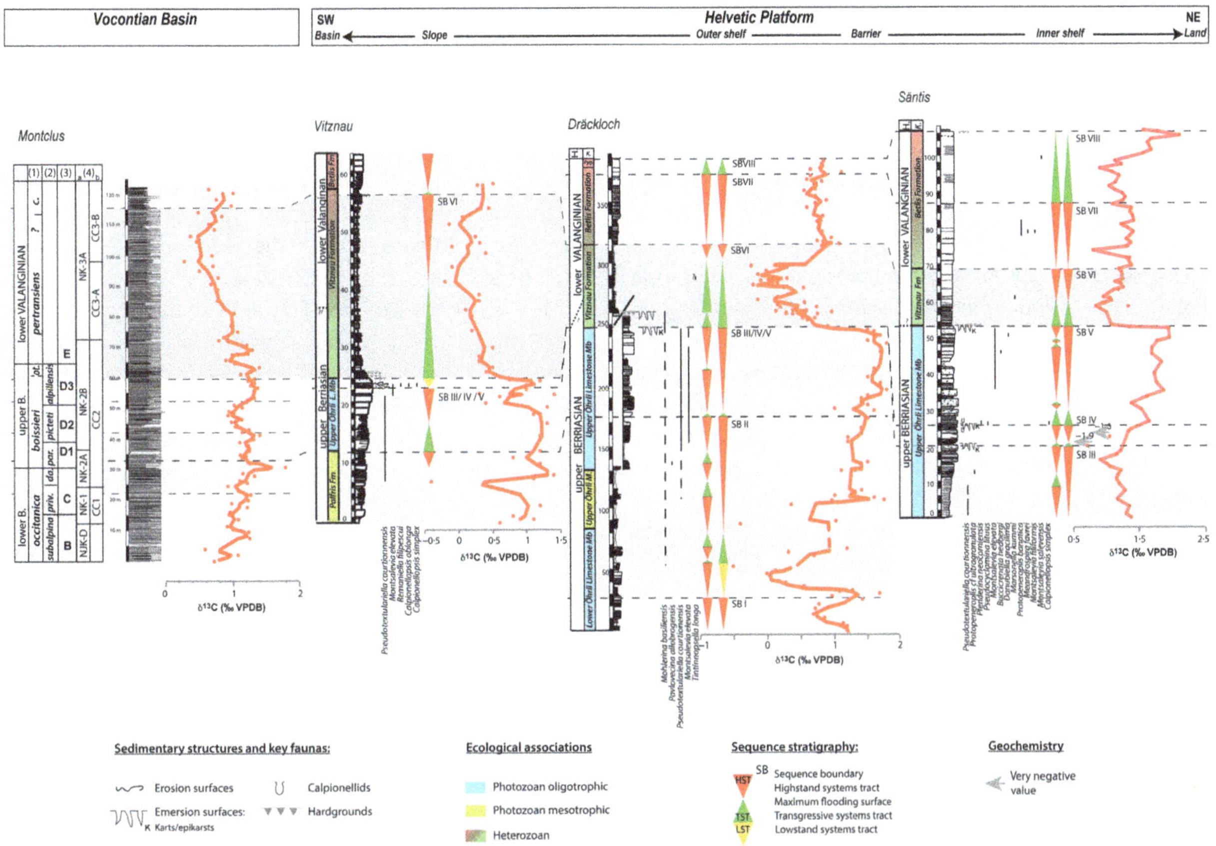

Fig. 15. Correlation of the sections across a NE/SW proximal-distal transect according to biostratigraphy and sequence stratigraphy; and comparison with the Vocontian Basin based on biostratigraphy and chemostratigraphy (Montclus, Morales *et al.*, 2013).

at a change from facies F5/6 to F2/3. The overlying shallowing-upward interval thickens towards the west (e.g. towards the basin, Fig. 3) and is interpreted as a LST. The deepening upward trend towards facies F2 constitutes the transgressive systems tract (TST), and the mfz is placed in the more recessive layers covered by vegetation. The HST is well developed and documents the progressive installation of the photozoan platform. In the more distal Vitznau section, *Pseudotextulariella courtionensis* is the only stratigraphic marker found in the first sequence (from 0 to 12 m above section base). Since this foraminifer is also present in the overlying sequence and since *Pavlovecina allobrogensis* is generally found in shallow-water deposits, this first sequence is attributed to sequence I.

The following sequence (sequence II) contains also *Pseudotextulariella courtionensis* and *Montsalevia elevata*, but *Pavlovecina allobrogensis* is absent. This association is documented in the three sections. In the Säntis succession, sequence II is present at the base of the section and continues 20 m up section. In the Dräckloch section, SB II is placed at an abrupt change in facies from F5/F8 to F3 at 175 m above section base. Sequence II shows a shallowing-upward trend towards facies F8 up to 246 m above the base of the section, corresponding to a HST. In the Vitznau section, SB II is placed at the base of a more prominent calcareous bed showing reworked platform clasts (F5). The following interval with a deepening trend from facies F5 to F3 is interpreted as a TST. The mfs is placed where the most distal facies (F2/3) is observed 20·6 m above the base of the section.

Sequence III is marked by the disappearance of *Pseudotextulariella courtionensis*. Sequence IV shows the first appearance of *Pfenderina neocomiensis* and is the last sequence observed in the photozoan platform succession of the Upper Öhrli Limestone Member. In the Säntis section, SB III corresponds to the first epikarstic level. Above, the facies shows a rather abrupt deepening to facies F3 corresponding to a TST, which is followed by a shallowing-upward trend to facies F9/10, typifying the HST. SBIV is placed 26 m above the base of the section, where a second epikarst level is observed. The upper part of the Upper Öhrli Limestone Member is dominated by facies F8, but shows abrupt changes to facies F2 and F3.

The presence of parasequence boundaries, local tectonic activity, and/or the transfer by storms (washover) may explain the occurrence of such outer platform deposits in the external lagoon.

In the Dräckloch section, fringing cements and dissolution vugs indicative of conditions close to emersion are observed between 236 and 246 m above the base of the section. SB III is placed at 246 m, where the $\delta^{13}C$ record shows an abrupt shift to lighter values, indicating the presence of an important hiatus. Above this level (from 246 to 260 m above the base of the section), shallow-water organisms still largely dominate carbonates, but an abrupt change in the carbonate fabric is observed with a clear dominance of corals and calcareous sponges, and the near disappearance of benthic foraminifera (except for *Trocholinas* which are found in abundance, and *Mohlerina basiliensis*). Sequences III and IV are therefore not observed in the Dräckloch section. This implies that sequence boundaries III, IV and V are combined, and a significant part of the late Berriasian is missing.

In the Vitznau section, an important erosional surface associated with lag pebbles marks a sequence boundary at 23·5 m above section base. Between 23·5 and 26·3 m above the base of the section, a succession of erosional surfaces is present. The top of this interval shows the reworking of partly lithified sediments rich in ooids and bioclasts. Given the relatively distal position of the Vitznau section, this interval is interpreted as a falling-stage systems tract (FSST). The marker calpionellid *Calpionellopsis simplex* was found together with *Calpionellopsis oblonga* and *Remaniella filipescui* (Fig. 9), and with the benthic foraminifera *Montsalevia elevata* (Fig. 7). *Pseudotextulariella courtionensis*, which is abundant in the underlying sequence, is absent from these reworked deposits. Consequently, this FFST may belong either to the combination of sequences III, IV and V, or to sequence V alone.

Sequence V is marked by an increase in detrital minerals, a change towards heterozoan carbonate production, and a lower diversity and abundance in benthic foraminifera (large specimens are no longer observed). In the Säntis section, SB V is documented by an epikarst overlapped by a hardground at 53 m above the base of the section, indicating that this sequence boundary is combined with the transgressive surface. The occurrence of a second and similar hardground in the immediately overlying bed, and of an interval containing reworked extraclasts of corals and calcareous sponges together with intraclasts of heterozoan organisms (58 m above section base) witness an important transgression (TST). A mfz is then placed where the deepest microfacies was observed (F2, outer shelf).

In the Dräckloch section, SB V is placed where dissolution vugs indicate conditions close to the emersion. Sequence V starts with the uppermost beds of the Upper Öhrli Limestone Member, where important sediment reworking is observed. This facies is interpreted as a lag on top of the transgressive surface, and the overlying interval as a TST. The phosphatic and pyritic infillings of the hardground on top of the Upper Öhrli Limestone Member, as well as the presence of a second hardground on top of a reworked layer in the Vitznau Formation confirms the occurrence of major relative sea-level rise during sequence V. The mfs of this sequence is placed within the more recessive part of the Vitznau Formation, which is covered by vegetation (between 309 and 315 m above section base).

In the Vitznau section, this transgression of major amplitude is equally documented by two hardgrounds and sediment reworking, located on top of the FFST (26·3 m above section base). The mfs of this sequence is also placed in the more recessive layers of the Vitznau Formation (facies F1). The ammonite *Thurmanniceras thurmanni* s.str. (Fig. 9) indicating the early part of the *T. otopeta* zone (latest Berriasian) was found in the scree. Ahermatypic corals were found in the host rock of the ammonite, which were also observed in the Dräckloch section in the early TST of sequence V (lower part of the Vitznau Formation). Stable isotope and P analyses performed on the host rock of the ammonite indicate that its origin is from the Vitznau Formation (values of 0·20‰ $\delta^{13}C$, −2·70‰ $\delta^{18}O$, 137 ppm P are only found at 33 m). If the determination and the position of the ammonite are correct (*Thurmanniceras thurmanni* s.l. has an extended range into the *T. pertransiens* ammonite zone; Wippich, 2003; Bujtor, 2013), the lower part of the Vitznau Formation at Vitznau (e.g. early TST of sequence V) has a latest Berriasian age.

Sequence VI shows the development of heterozoan carbonates deposits and corresponds to the uppermost part of the Vitznau Formation in the Vitznau section, and to the lower part of the Betlis Formation in the Vitznau, Dräckloch and Säntis sections. In the three sections, its sequence boundary is mingled with the transgressive surface, and overlain by lag deposits. SB VI is placed at 69 m above the base of the Säntis section, at 315 m above the base of the Dräckloch section, and at 56·8 m above the base of the Vitznau section where erosive banks showing intense sediment reworking are observed. Because no significant change in facies occurs within sequence VI in the Säntis and Dräckloch sections, the mfs probably coincides with the sequence boundary. In the Vitznau section, the mfz is placed within the overlying marly interval, which probably corresponds to a deeper depositional environment as it shows the most recessive layers of the sequence. There, only the lower part of the Betlis Formation is accessible, and interpreted as the base of a HST. Marker benthic foraminifera indicate an age close to the Berriasian-Valanginian boundary for sequence VI (Table 2).

Sequence VII corresponds to the top of the heterozoan Betlis Formation. In the Säntis section, SB VII is placed at 86·6 m above the base of the section, where an erosive surface associated with reworked oncoids was observed. The mfs is then placed within the hardground associated with the Gemsmättli Bed at the top of the Betlis Formation. In the Dräckloch section, SB VII is noted at 370 m above the base of the section with a level of important reworking. The uppermost part of this interval corresponds to the Pygurus Member, which is part of the Pygurus-Gemsmättli complex documenting the major drowning phase of the Valanginian Helvetic platform (Haldimann, 1977; Wyssling, 1986; Kuhn, 1996; Föllmi *et al.*, 2006, 2007). This sequence is thereby characterized by significant detrital input and ends with a major condensation phase. The duration of the condensation of this level (more than 3 Myr, from the late Valanginian to the early Hauterivian) implies the presence of several sequence boundaries within this level (Godet, 2013), therefore highlighting a complex sequence stratigraphic surface (SB VIII).

DISCUSSION

Global sea-level change during the late Berriasian – early Valanginian

The stratigraphic distribution of marker foraminifer and calpionellid species allows for a correlation of the Helvetic platform with other shallow-water deposits in the north-western Tethyan area (Fig. 16), and therefore permits differentiation between local and global factors controlling sedimentation. Faunal associations characteristic of sequences I and II were identified in the Pierre Châtel and Vions formations (Jura Mountains; Darsac, 1983; Morales *et al.*, 2013), in the Calcaire Blanc Inférieur Formation and in the Membre Marneux Inférieur of the Marnes Vertes Infracrétacées Formation (Provence; Virgone, 1997), and the Calcaires à Trocholines et Dasycladacées (Pyrenees; Peybernès & Combes, 1994). Similarly, marker species of sequences III and IV were found in the top of the Vions Formation and the Lower Member of the Chambotte Formation (Jura Mountains), the Membre Carbonaté of the Marnes Vertes Infracrétacées (Provence) and the Calcaires Roux (Pyrenees). These sequences were correlated following the scheme proposed in Fig. 15. The sedimentary successions show comparable changes in accommodation space, which are interpreted to express north-western and northern Tethyan sea-level variations during the late early and the late Berriasian.

An important transgression is recorded through the Berriasian-Valanginian boundary. This sea-level event is marked by the general deposition of more open marine marly facies in the Helvetic region (Vitznau Formation, sequence V), the Jura Mountains (Guiers Member of the Chambotte Formation), Provence (Membre Marneux Supérieur of the Marnes Vertes Infracrétacées), and the Pyrenees (uppermost part of the Calcaires Roux Formation). This transgression is also associated with a succession of two hardgrounds in the Helvetic Alps separating the Upper Öhrli Limestone Member and the base of the Vitznau Formation and the deposition of condensed layers rich in ammonites in the Jura Mountains (Fontanil fauna, Blanc *et al.*, 1992) and the Pyrenees (Marnes de Francazal; Peybernès & Combes, 1994).

Differences in the pattern of sedimentary deposits are, however, observed between the Helvetic and the north-western Tethyan platforms during the early Valanginian. In the Jura Mountains, Provence, and the Pyrenees, photozoan carbonate facies are observed with the Upper Member of the Chambotte Formation, the Calcaire Blanc Supérieur Formation, and the Calcaires Graveleux à Pfenderines, respectively. Conversely, these oligotrophic deposits are not seen in the Helvetic sections. This is related to an enhanced subsidence phase recorded in the Helvetic domain compared to the north-western Tethyan areas (Stampfli *et al.*, 2002), which may be associated with the local influence of nutrient-rich waters onto the northern Tethyan shelf.

Stratigraphically above the lower Valanginian photozoan carbonates of the Jura Mountains, Provence and the Pyrenees, a second major transgression is recorded by the deposition of deeper marly facies, locally associated with hardgrounds and condensed layers containing ammonites (from the *Busnardoides campylotoxus* zone) in Provence (Marnes Grises; Virgone, 1997) and the Pyrenees (Peybernès & Combes, 1994). The onset of condensation may be correlated with the condensed phosphatic layer of the Büls Bed in the Helvetic Alps, which is dated from the late *Tirnovella pertransiens* – early *Busnardoides campylotoxus* ammonite zones (Kuhn, 1996). In the Pyrenees, heterozoan carbonate deposits are described on top of this condensed layer (Calcaires Jaunes à Bryozoaires). The latter formation is attributed to the Hauterivian based on brachiopod stratigraphy (Peybernès & Combes, 1994) but the 'low stratigraphic interest of micropalaeontological descriptors' mentioned by the authors suggests that this age attribution may need to be reviewed. In the Helvetic and the Jura regions, shallow-water heterozoan carbonates (sequence VI and VII, and Bourget Formation,

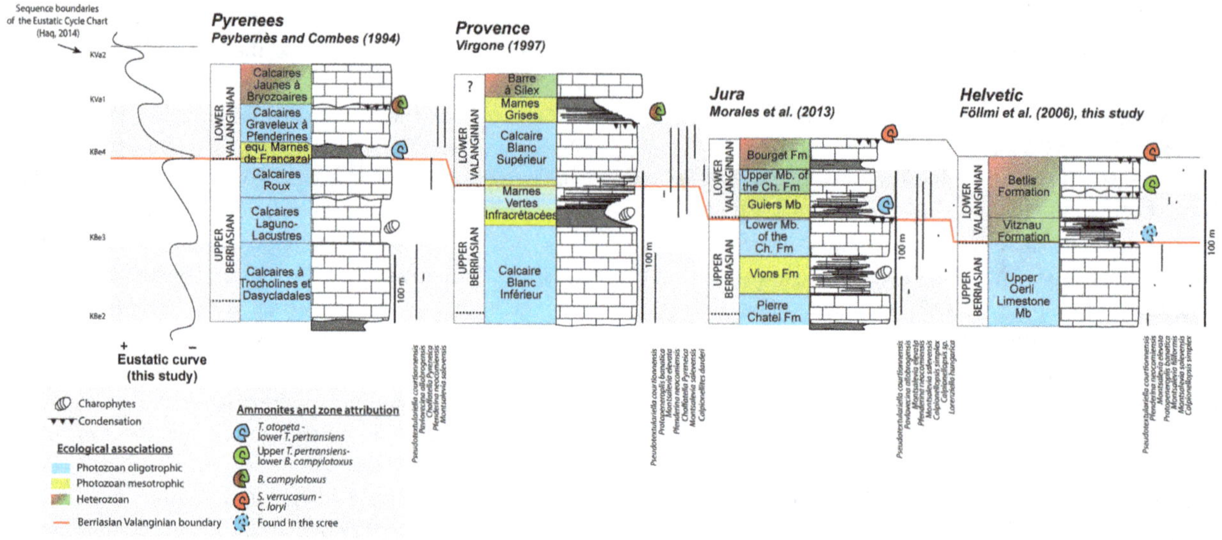

Fig. 16. Comparison of the Pyrenean, Provence, Jura and Helvetic carbonate platform deposits during the late Berriasian and early Valanginian. Lithological logs represent synthetic successions as described by Peybernès & Combes (1994), Virgone (1997), Morales *et al.* (2013), Föllmi *et al.* (2006) and this study. A correlation based on benthic foraminifera, calpionellids and ammonites is proposed, which permits variations in sea-level to be traced during this time interval. A major transgression started from the latest Berriasian onwards; however, TST are not fully recorded in shallow-water series. The Berriasian-Valanginian boundary is therefore placed at the base of the Vitznau Formation (Helvetic Platform) in this scheme, and similarly for time equivalent deposits of the Jura Mountains, Provence and the Pyrenean Platforms. A succession of three sea-level rises is highlighted from the Berriasian-Valanginian boundary to the late Valanginian, which led to the occurrence of three phosphate-rich condensed layers in the Helvetic area.

respectively) appear below the highly condensed sediments of the Marnes à *Astieria* (Jura Mountains) and the Gemsmättli-Pygurus complex.

Sequence boundaries identified in this contribution have been correlated with global sequence boundaries of Haq (2014, Fig. 16). Following calpionellid determinations, SB III would correspond to KBe3 (early zone D), SB V to KBe4 (latest zone D); SB VI (mid zone E) to KVa1 (earliest zone E), and SB VII to KVa2 (Fig. 16). SB I (zone C) may correspond to KBe2 (at the boundary between zones B and C). Interestingly, SB III, SB V and SB VI are associated with biostratigraphic boundaries of foraminiferal faunas (discontinuities d1, d2 and d3 of Darsac, 1983), which might confirm their importance (medium cycle boundaries of Haq, 2014).

Effects of tectonics on palaeogeography

The biostratigraphy of the outer shelf section at Dräckloch indicates the occurrence of an important hiatus related to the absence of the late Berriasian sequences III and IV (Fig. 15). This hiatus is linked with an emersion; its duration corresponds to an important part of the *S. boissieri* ammonite zone, that is, of the late Berriasian. A relatively similar succession is observed in the Helvetic platform section of Lämmerenplatten (Pasquier, 1995), consisting of nearly identical environments (close to or at the barrier) but with a less well-constrained temporal framework. During the Early Cretaceous, major fault zones affected the Helvetic region (Funk, 1985; Detraz *et al.*, 1987) and an important subsidence phase is recorded from the Oxfordian to the Hauterivian (Funk, 1985; Stampfli *et al.*, 2002). Thus, the different stratigraphic records of the three studied successions are probably linked to a phase of tectonic activity during the late Berriasian – early Valanginian, which is likely related to extensive movements affecting the northern Tethyan margin. The generally high subsidence rates documented in the Helvetic plateau was commonly linked to the opening of the Alpine Tethys and of the North Atlantic (Stampfli *et al.*, 2002).

The Dräckloch section, which is close to the platform margin, was probably located near the top of a tilted block, whereas the Säntis section, which is composed of more lagoonal deposits, is thought to have been located within an intrashelf depression formed by the tilting process (Fig. 17). The topographic effect of tilting blocks may also be evident in the Säntis section, where parasequences involving high amplitude relative sea-level variations were observed in the sedimentary succession and microfacies of the Upper Öhrli Limestone Member. The slope section of Vitznau records a FSST, which may also be linked to the topographic effects of block tilting.

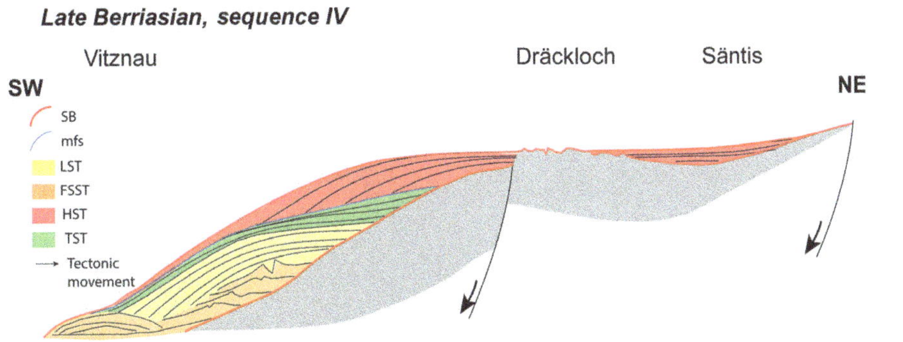

Fig. 17. Schematic representation of the depositional sequences during the latest Berriasian and the earliest Valanginian.

Palaeoenvironmental and palaeoclimate changes

In the late early and early late Berriasian, the increase in P and detrital input observed in sequence I (upper part of Lower Öhrli Limestone Member and Upper Öhrli Marl Member) are linked on the Helvetic platform to a period of transgression (Fig. 18). Outer shelf deposits typically show an increase in detrital minerals basinwards (Fig. 18). The following sequences do not show major changes in these parameters, which is different from the Jura areas, where an increase in quartz and P is observed in the Vions Formation. The sections examined in the Helvetic area were more distant from the continental coast, which may explain these different records.

An increase in P and detrital minerals is also recorded in the Vitznau Formation, which again corresponds to a sea-level rise of probably wider importance (Fig. 16). Platform carbonate production changed from photozoan to heterozoan assemblages in the Helvetic region. Less diagenetically altered, age equivalent sections in the Jura Mountains and Vocontian Basin (La Chambotte and Montclus; Morales *et al.*, 2013) indicate that this interval also corresponds to a maximum in humidity, indicated by high kaolinite contents in both platform and basinal environments. Thus, with an important transgression and a highly hydrolysing climate, biotas were subjected to increasing stress leading to the disappearance of oligotrophic organisms. The presence of phosphates and reworked pelagic sediments on top of the Upper Öhrli Limestone Member at Dräckloch, associated with a series of superimposed hardgrounds in all three sections indicates a platform drowning phase associated with this important transgressive phase. Following the drowning phase, a mesotrophic fauna including bryozoans, crinoids and prolifering *Gryphaea*, brachiopods, ostracods and serpulids, was installed in the distal part of the platform.

With the Betlis Formation (sequences VI and VII), platform carbonate production recovered in heterozoan mode. Excess nutrients are also indicated by the presence of chert nodules, which are related to a higher proportion of filter-feeding siliceous sponges, and the presence of a phosphatic bed separating the Betlis Formation into two members in more distal sections (Kuhn, 1996). This phosphatic bed indicates a second drowning phase of the carbonate platform. The Jura Platform (where a similar heterozoan facies is observed) and the Vocontian Basin sections provide evidence, however, of a decrease in humidity during this period (Morales *et al.*, 2013).

At the top of sequence VII, strong detrital input is recorded. The top of sequence VII is poorly documented in the Jura Mountains (Hennig, 2003) and clay mineral analyses were not performed. Nevertheless, the abundance of millimetre-sized quartz grains (Fig. 8C), sometimes containing a ferruginous coating, indicates an important phase of erosion on the continent. This is correlated with an increase in P contents, which suggest increased nutrient input to the ocean. This phase of enhanced weathering, combined with a sea-level rise was responsible for the third, long-lasting and most important drowning phase of the carbonate platform in the Helvetic area during the Weissert episode.

The early Valanginian negative CIE: a prelude to the Weissert Event?

The comparison of the $\delta^{13}C$ records of the Säntis, Dräckloch and Vitznau sections highlight similar trends, which can be correlated with the Vocontian Basin (section of Montclus, Morales *et al.*, 2013), and which are in adequacy with the sequence stratigraphic interpretation (Fig. 15). A general increase in $\delta^{13}C$ values is observed during the Berriasian, whereas a decrease in $\delta^{13}C$ values is highlighted during the early Valanginian, finally followed by the global positive $\delta^{13}C$ shift characterizing the Weissert episode. On the Helvetic platform, which is marked by emersion and condensation phases, this results in a significant negative shift (of 0·7‰) at the boundary between the Upper Öhrli Limestone Member and the

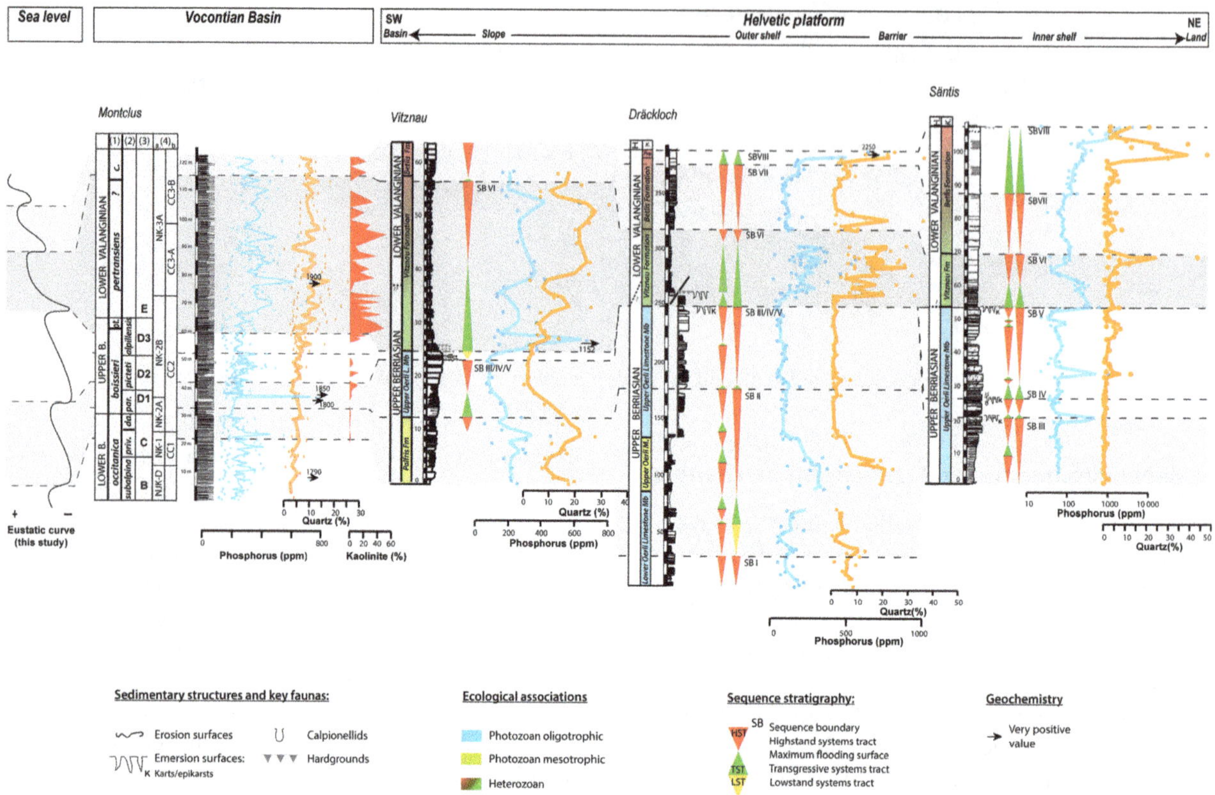

Fig. 18. Stratigraphic variation in phosphorus and quartz contents along a proximal-distal transect (see Fig. 15 for correlation), comparison with the Montclus section (Vocontian Basin, Morales *et al.*, 2013), and correlation with a recent global sea-level reconstruction (1st and 2nd order variations according to Haq, 2014). The phosphorus and quartz contents are expressed in ppm and per cent of the bulk-rock, respectively. The major relative sea-level rise occurring during the latest Berriasian –earliest Valanginian is accompanied by enhanced phosphorus and quartz contents and an ecological change towards heterozoan associations.

Vitznau Formation. By comparison with the $\delta^{13}C$ record of the Vocontian Trough, where the amplitude of variations in $\delta^{13}C$ values should be reduced compared to the platform record (Morales *et al.*, 2013), the duration of the hiatus occurring through the Berriasian-Valanginian boundary may be close to 800 kyr (counting Milankovitch cycles). Thus, our results show that in addition to biostratigraphy and sequence stratigraphy, the general trends of the $\delta^{13}C$ records may be used to correlate the sections within this Helvetic transect. The negative excursion is mainly attributed to a change in carbonate production from photozoan to heterozoan carbonate production on the platform (Föllmi *et al.*, 2006; Morales *et al.*, 2013).

In the Helvetic area, this change in carbonate production coincides with a major relative sea-level rise linked to a transgression and local tectonics, a maximum in humidity, and associated higher nutrient levels. Similarly, a negative $\delta^{13}C$ shift and a change towards heterozoan carbonate deposits are observed in the Jura Mountains (Bourget Formation, Morales *et al.*, 2013), but may not be entirely synchronic with the Helvetic record. This may be explained by different tectonic contexts. A likely similar succession is observed in the Pyrenees, where the heterozoan Calcaires Jaunes à Bryozoaires might be correlated with the upper part of the Betlis and the Bourget formations (of the Helvetic and Jura platforms, respectively). In the Provence area, shallow-water water heterozoan deposits are scarcely observed (with the exception of the Olioulles succession; Virgone, 1997; Schroeder *et al.*, 2000; Masse *et al.*, 2009; Bonin *et al.*, 2012).

Higher nutrient levels control the settlement of heterozoan carbonates, characterized by suspension-feeding organisms (James, 1997). Enhanced humidity and runoff, however, are probably not the main factors driving nutrient fluxes, as the Bourget Formation (Jura Platform) is depleted in kaolinite (relative to smectite and chlorite; Darsac, 1983; Adatte, 1988). Instead, nutrient-rich currents associated with a major transgression may have played a role. The sea-level rise may have favoured water-mass exchanges between the Boreal and Tethyan oceans (van de Schootbrugge *et al.*, 2003). The opening of the

Polish gateway was probably initiated during the early Valanginian, as testified by the deposition of open marine sediments belonging to the *Tirnovella pertransiens* ammonite zone on top of a Tithonian karstified limestone in the Polish Basin (Kutek *et al.*, 1989; Morales *et al.*, 2015), and the migration of Boreal calcareous nanofossils and ammonites to the Tethyan realm (Bulot, 1996).

These nutrient-rich water masses may result from an upwelling system explained by the inundation of continental margins, which would have triggered enhanced evaporation and wind velocities, resulting in stronger westerlies (Poulsen *et al.*, 1998; Godet, 2013). There is, however, no strong evidence for cooler water during the *Tirnovella pertransiens*-lower *Busnardoides campylotoxus* interval. Nutrients may also come from the intense weathering of landmasses located north-eastwards, which may have been transported westwards by the circum-Tethyan current. A very humid climate is indeed known from the Polish Basin (Morales *et al.*, 2015). The existence of these currents might also explain why the Helvetic platform turned to a heterozoan carbonate production earlier than other platforms located further to the west, and why the consistent association of phosphate and platform drowning in the Helvetic region (Kuhn, 1996) is not known in the Jura, Provence and Pyrenees platforms.

In this context, the third and major drowning phase of the Valanginian (corresponding to the Gemsmättli-Pygurus Beds in the Helvetic area) is related to a phase of intense weathering, as witnessed by the high quartz contents in the Gemsmättli-Pygurus complex, linked with a peak in humidity (higher kaolinite contents in the Vocontian Basin, Duchamp-Alphonse *et al.*, 2011; Föllmi, 2012), and combined with a sea-level rise. This drowning phase affected a platform that was already weakened by high nutrient levels resulting in two incipient drowning phases and previous rapid relative sea-level changes.

A general increase in volcanic activity, eventually related to the eruption of the Paranà-Etendeka continental flood basalts is viewed as the main trigger of environmental changes, leading to an increase in pCO_2 levels and profound climate modifications. The increase in magmatic activity may have been linked to enhanced oceanic crust production that triggered eustatic sea-level rise. Martinez *et al.* (2015) recently re-calibrated the Valanginian astronomic time scale and suggested that the main volcanic pulse of the Paranà Etendeka may coincide with the onset of the positive $\delta^{13}C$ excursion. Their calibration also shows a rather good correlation between the initiation of the volcanism, and the early Valanginian decrease in $\delta^{13}C$ values recorded in the Vocontian Basin. Since it is not clear yet if the latter is of global significance, the decrease in stable carbon-isotope values is rather attributed to a change of carbonate production than to volcanic activity.

CONCLUSIONS

The installation, growth and demise of the Berriasian-Valanginian carbonate platform have been documented using a transect of sections across the Helvetic Alps. A more accurate age control is proposed for the studied sections of the Säntis, Dräckloch and Vitznau locations, based on benthic foraminifera, calpionellid and ammonite biostratigraphy, chemostratigraphy, and sequence stratigraphy. This integrated stratigraphic approach provides a base for detailed correlation of the three studied sections, and enables the comparison with equivalent records from the Vocontian Basin, the Jura Mountains, Provence and the Pyrenean platforms. Thereby, this contribution provides a more balanced view of shallow-water ecological changes to global perturbations, by highlighting the superimposed influence of regional tectonic and palaeogeographic factors. The Helvetic sedimentological record is different from those of other northern and north-western Tethyan platforms, which is interpreted as reflecting the combined effect of palaeoceanographic differences, with the Helvetic platform being more exposed to the northern nutrient-carrying Tethyan current, and of enhanced subsidence in the Helvetic domain. Block tilting is the most likely mechanism to explain a major hiatus present in the upper part of the Upper Öhrli Limestone Member in the Dräckloch section, encompassing a significant part of the late Berriasian.

A major sea-level rise is documented, which started in the latest Berriasian and progressively flooded the platform. This sea-level rise, combined with a strongly subsiding setting and enhanced humidity led to the disappearance of photozoan faunas and provoked a first major drowning phase in the Helvetic domain. During the early Valanginian, suspension feeders dominated shallow-water organisms in the northern and north-western Tethyan area, as a response to higher nutrient rates in the ocean. The nearby continents, however, underwent a decrease in humidity and runoff. The turn to heterozoan carbonate production is interpreted as the consequence of palaeoceanographic changes, which may be related to the establishment of upwelling currents. During the late early and early late Valanginian, phases of enhanced detrital input linked with strong continental weathering, and sea-level rises were responsible for a second and a third, widespread demise of the already weakened carbonate platform: near the boundary of the *pertransiens – campylotoxus* zone and during the *verrucosum* zone, respectively. In the Helvetic domain, the latter shallow-water carbonate crisis is evidenced by a quartz-rich phosphatic and glauconitic crust on top of a hardground (Gemsmättli Bed) or a quartz and phosphate-rich, highly bioturbated interval (Pygurus Member). These condensed horizons document the almost complete disappearance of shallow-water calcifying organisms for more than 3 Myr. The renewed development of a carbonate platform in the Helvetic region is only recorded from the middle early Hauterivian onwards.

ACKNOWLEDGEMENTS

The authors are thankful to Pilar Ramirez de Arellano, Cyril Baudon, Aurélie Bonin, Morgan Peel, and Lucie Bonvallet for fieldwork assistance. We also acknowledge Antoine Pictet and Luc Bulot for the preparation and determination of the ammonite and Tiffany Monier for assistance in the laboratory. We thank Ian Jarvis and an anonymous reviewer for their constructive and thoughtful reviews. The detailed and very constructive comments of associated editor Adrian Immenhauser were very helpful in the revision of the manuscript. We also thank editor Peter Swart for his assistance. Financial support from the Swiss National Science Foundation (project 200020_126455) is appreciatively acknowledged.

References

Adatte, T. (1988) *Etude sédimentologique, minéralogique, micropaléontologique et stratigraphique du Berriasien – Valanginien du Jura central.* Université de Neuchâtel, Neuchâtel, 481 pp.

Adatte, T., Stinnesbeck, W. and **Keller, G.** (1996) Lithostratigraphic and mineralogic correlations of near K/T boundary sediments in northeastern Mexico: implications for origin and nature of deposition. In: *The Cretaceous-Tertiary Event and Other Catastrophes in Earth History* (Eds G. Ryder, D. Fastovsky and S. Gartner), **307**, pp. 211–226. Geological Society of America Special Paper, Boulder, Colorado.

Arnaud-Vanneau, A. and **Arnaud, H.** (2005) Carbonate facies and microfacies of the Lower Cretaceous carbonate platforms. In: *The Hauterivian-Lower Aptian Sequence Stratigraphy from Jura Platform to Vocontian Basin: A Multidisciplinary Approach* (Eds T. Adatte, A. Arnaud-Vanneau, H. Arnaud, M.-C. Blanc-Aletru, S. Bodin, E. Carrio-Schaffhauser, K.B. Föllmi, A. Godet, M.C. Raddadi and J. Vermeulen), *Géol. Alpine*, **7**, 39–68.Série Spéciale, Colloques et Excursions, Grenoble.

Arnaud-Vanneau, A. and **Darsac, C.** (1984) Caractères et évolution des peuplements de foraminifères benthiques dans les principaux biotopes des plates-formes carbonatées du Crétacé Inférieur des alpes du nord (France). *Geobios*, **17**, 19–27.

Blanc, E. (1996) *Transect plate-forme – bassin dans les séries carbonatées du Berriasien supérieur et du Valanginien inférieur (domaines Jurassiens et nord-Vocontien). Chronostratigraphie et transferts de sédiments.* Joseph Fournier, Grenoble, 312 pp.

Blanc, E., Arnaud-Vanneau, A., Arnaud, H., Bulot, L.G., Gidon, M., Thieuloy, J.-P. and **Remane, J.** (1992) Les couches du passage Berriasien-Valanginien dans le sectieur du Fontanil (Isère, France). *Géol. Alpine*, **68**, 3–12.

Blau, J. and **Grün, B.** (1997) Late Jurassic/Early Cretaceous revised calpionellid zonal and subzonal division and correlation with ammonite and absolute time scales. *Mineralia Slovaca*, **29**, 297–300.

Bodin, S., Godet, A., Föllmi, K.B., Vermeulen, J., Arnaud, H., Strasser, A., Fiet, N. and **Adatte, T.** (2006) The late Hauterivian Faraoni oceanic anoxic event in the western Tethys: evidence from phosphorus burial rates. *Palaeogeogr. Palaeoclimatol. Palaeoecol.*, **235**, 245–264.

Boisseau, T. (1987) *La plate-forme jurassienne et sa bordure subalpine au Berriasien – Valanginian (Chartreuse-Vercors). Analyse et corrélations avec les séries de bassin.* Université scientifique Technologique et Médicale de Grenoble, Grenoble, 413 pp.

Bonin, A., Vennin, E., Pucéat, E., Guiraud, M., Arnaud-Vanneau, A., Adatte, T., Pittet, B. and **Mattioli, E.** (2012) Community replacement of neritic carbonate organisms during the late Valanginian platform demise: a new record from the Provence Platform. *Palaeogeogr. Palaeoclimatol. Palaeoecol.*, **365–366**, 57–80.

Bujtor, L. (2013) Valanginian perisphinctid ammonites from the Kisújbánya Basin (Eastern Mecsek Mts., Hungary). *Cretac. Res.*, **41**, 1–16.

Bulot, L.G. (1996) The Valanginian Stage. *Bull. Inst. R. Sci. Nat. Belg.*, **66**, 11–18.

Burger, H. (1985) *Palfris-Formation, Öhrli-Formation und Vitznau-Mergel (Basale kreide des helvetikums zwischen reuss und rhein). Stratigraphische, fazielle, mineralogische und paläogeographische Untersuchungen.* ETH Zürich, Zürich, 237 pp.

Burger, H. (1986) Fazielle Entwicklung und paläogeographische Rekonstruktion des helvetischen Schelfs während der untersten Kreide in der Zentral- und Ostschweiz. *Eclogae Geol. Helv.*, **79**, 561–615.

Catuneanu, O., Abreu, V., Bhattacharya, J.P., Blum, M.D., Dalrymple, R.W., Eriksson, P.G., Fielding, C.R., Fisher, W.L., Galloway, W.E., Gibling, M.R., Giles, K.A., Holbrook, J.M., Jordan, R., Kendall, C.G.S.C., Macurda, B., Martinsen, O.J., Miall, A.D., Neal, J.E., Nummedal, D., Pomar, L., Posamentier, H.W., Pratt, B.R., Sarg, J.F., Shanley, K.W., Steel, R.J., Strasser, A., Tucker, M.E. and **Winker, C.** (2009) Towards the standardization of sequence stratigraphy. *Earth Sci. Rev.*, **92**, 1–33.

Choquette, P.W. and **James, N.P.** (1987) Diagenesis in Limestones – 3. The Deep Burial Environment. *Geosci. Canada*, **14**, 3–35.

Darsac, C. (1983) *La plate-forme berriaso-valanginienne du Jura méridional aux massifs subalpins (Ain, Savoie), Institut de Géologie*. Université de Grenoble 1, Grenoble, 319 pp.

Dercourt, J., Gaetani, M., Vrielynck, B., Barrier, E., Biju-Duval, B., Brunet, M.F., Cadet, J.P., Crasquin, S. and **Sandulescu, M.** (2000) *Atlas Peri-Tethys, Palaeogeographical Maps, 24 Maps and Explanatory Notes*. CCGM/CGMW, Paris, 269 pp.

Detraz, H., Charollais, J. and **Remane, J.** (1987) Le Jurassique supérieur – Valanginian des chaînes subalpines septentrionales (massif des Bornes et de Platé, Haute-Savoie; Alpes occidentales): analyse des resédimentations, architecture du bassin et influences des bordures. *Eclogae Geol. Helv.*, **80**, 69–108.

Duchamp-Alphonse, S., Gardin, S., Fiet, N., Bartolini, A., Blamart, D. and **Pagel, M.** (2007) Fertilization of the northwestern Tethys (Vocontian basin, SE France) during the Valanginian carbon isotope perturbation: evidence from calcareous nannofossils and trace element data. *Palaeogeogr. Palaeoclimatol. Palaeoecol.*, **243**, 132–151.

Duchamp-Alphonse, S., Fiet, N., Adatte, T. and **Pagel, M.** (2011) Climate and sea-level variations along the northwestern Tethyan margin during the Valanginian C-isotope excursion: mineralogical evidence from the Vocontian Basin (SE France). *Palaeogeogr. Palaeoclimatol. Palaeoecol.*, **302**, 243–254.

Erba, E., Bartolini, A. and **Larson, R.L.** (2004) Valanginian Weissert oceanic anoxic event. *Geology*, **32**, 149–152.

Ferreri, V., Weissert, H., D'Argenio, B. and **Buonocunto, F.P.** (1997) Carbon isotope stratigraphy: a tool for basin to carbonate platform correlation. *Terra Nova*, **9**, 57–61.

Föllmi, K.B. (1995) 160 m.y. record of marine sedimentary phosphorus burial: coupling of climate and continental weathering under greenhouse and icehouse conditions. *Geology*, **23**, 859–862.

Föllmi, K.B. (2012) Early Cretaceous life, climate and anoxia. *Cretac. Res.*, **35**, 230–257.

Föllmi, K.B., Weissert, H., Bisping, M. and **Funk, H.** (1994) Phosphogenesis, carbon-isotope stratigraphy, and carbonate-platform evolution along the Lower Cretaceous northern Tethyan margin. *Geol. Soc. Am. Bull.*, **106**, 729–746.

Föllmi, K.B., Godet, A., Bodin, S. and **Linder, P.** (2006) Interactions between environmental change and shallow water carbonate buildup along the northern Tethyan margin and their impact on the Early Cretaceous carbon isotope record. *Paleoceanography*, **21**, PA4211.

Föllmi, K.B., Bodin, S., Godet, A., Linder, P. and **van de Schootbrugge, B.** (2007) Unlocking paleo-environmental information from Early Cretaceous shelf sediments in the Helvetic Alps: stratigraphy is the key!. *Swiss J. Geosci.*, **100**, 349–369.

Funk, H. (1985) Mesozoische Subsidenzgeschichte im Helvetischen Schelf der Ostschweiz. *Eclogae Geol. Helv.*, **78**, 249–272.

Funk, H., Föllmi, K.B. and **Mohr, H.M.** (1993) Evolution of the Tithonian-Aptian carbonate platform along the northern Tethyan margin, eastern Helvetic Alps. In: *Cretaceous Carbonate Platforms* (Eds J.A.T. Simo, R.W. Scott and J.-P. Masse), pp. 387–408. American Association of Petroleum Geologists, Tulsa.

Godet, A. (2013) Drowning unconformities: palaeoenvironmental significance and involvement of global processes. *Sed. Geol.*, **293**, 45–66.

Gréselle, B. and **Pittet, B.** (2010) Sea-level reconstructions from the Peri-Vocontian Zone (South-east France) point to Valanginian glacio-eustasy. *Sedimentology*, **57**, 1640–1684.

Haldimann, P.A. (1977) *Sedimentologische Entwicklung der Schichten an einer Zyklengrenze der helvetischen Unterkreide*. Unpublished PhD Thesis. ETH Zürich, Zürich, 182 pp.

Hallam, A., Grose, J.A. and **Ruffell, A.H.** (1991) Palaeoclimatic significance of changes in clay mineralogy across the Jurassic-Cretaceous boundary in England and France. *Palaeogeogr. Palaeoclimatol. Palaeoecol.*, **81**, 173–187.

Haq, B.U. (2014) Cretaceous eustasy revisited. *Global Planet. Change*, **113**, 44–58.

Haq, B.U., Hardenbol, J. and **Vail, P.R.** (1987) Chronology of fluctuating sea levels since the Triassic. *Science*, **235**, 1156–1167.

Hardenbol, J., Thierry, J., Farley, M.B., Jacquin, T., De Graciansky, P.-C. and **Vail, P.R.** (1998) Mesozoic and Cenozoic sequence chronostratigraphic framework of European basins. In: *Mesozoic and Cenozoic Sequence Stratigraphy of European Basins* (Eds P.-C. de Graciansky, J. Hardenbol, T. Jacquin and P.R. Vail), *SEPM Spec. Publ.*, **60**, 3–13.

Hay, W.W. (2008) Evolving ideas about the Cretaceous climate and ocean circulation. *Cretac. Res.*, **29**, 725–753.

Hennig, S. (2003) *Geochemical and Sedimentological Evidence for Environmental Changes in the Valanginian (Early Cretaceous) of the Tethys Region*. ETH Zurich, Zurich, 189 pp.

Huck, S., Heimhofer, U., Rameil, N., Bodin, S. and **Immenhauser, A.** (2011) Strontium and carbon-isotope chronostratigraphy of Barremian-Aptian shoal-water carbonates: Northern Tethyan platform drowning predates OAE 1a. *Earth Planet. Sci. Lett.*, **304**, 547–558.

Ischi, H. (1978) *Das Berriasien-Valanginien in der Wildhorn-Drusberg-Decke zwischen Thuner- und Vierwaldstättersee*. University of Bern, Bern, 142 pp.

James, N.P. (1997) The cool-water carbonate depositional realm. *SEPM Spec. Publ.*, **56**, 1–22.

Kübler, B. (1983) *Dosage quantitatif des minéraux majeurs des roches sédimentaires par diffraction X*. Cahiers de l'Institut de Géologie, Université de Neuchâtel, Suisse, AX1.1 and 1.2, 12 pp.

Kübler, B. (1987) *Cristallinité de l'illite: méthodes normalisées de préparation, méthode normalisée de mesure, méthode* *automatique normalisée de mesure*. Cahiers de l'Institut de Géologie, Université de Neuchâtel, Suisse, ADX 2, 13 pp.

Kuhn, O. (1996) *Der Einfluss von Verwitterung auf die Paläozeanographie zu Beginn des Kreide-Treibhausklimas (Valanginian und Hauterivian) in der West-Tethys*. ETH Zürich, Zurich, 380 pp.

Kutek, J., Marcinowski, R. and **Wiedmann, J.** (1989) The Wawal Section, Central Poland – an important Link between Boreal and Tethyan Valanginian. In: *Cretaceous of the Western Tethys* (Ed. J. Wiedmann), pp. 717–754. Proceedings of the 3rd International Cretaceous Symposium, Tübingen.

Lini, A., Weissert, H. and **Erba, E.** (1992) The Valanginian carbon isotope event: a first episode of greenhouse climate conditions during the Cretaceous. *Terra Nova*, **4**, 374–384.

Martinez, M., Deconinck, J.-F., Pellenard, P., Riquier, L., Company, M., Reboulet, S. and **Moiroud, M.** (2015) Astrochronology of the Valanginian-Hauterivian stages (Early Cretaceous): chronological relationships between the Paraná-Etendeka large igneous province and the Weissert and the Faraoni events. *Global Planet. Change*, **131**, 158–173.

Masse, J.-P., Villeneuve, M., Leonforte, E. and **Nizou, J.** (2009) Block tilting of the North Provence early Cretaceous carbonate margin: stratigraphic, sedimentologic and tectonic data. *Bull. Soc. Géol. France*, **180**, 105–115.

Mohr, H.M. (1992) *Der helvetische Schelf der Ostschweiz am Übergang vom späten Jura zur frühen Kreide*. ETH Zurich, Zurich, 221 pp.

Morales, C., Gardin, S., Schnyder, J., Spangenberg, J., Arnaud-Vanneau, A., Arnaud, H., Adatte, T. and **Föllmi, K.B.** (2013) Berriasian and early Valanginian environmental change along a transect from the Jura Platform to the Vocontian Basin. *Sedimentology*, **60**, 36–63.

Morales, C., Kujau, A., Heimhofer, U., Mutterlose, J., Spangenberg, J.E., Adatte, T., Ploch, I. and **Föllmi, K.B.** (2015) Palaeoclimate and palaeoenvironmental changes through the onset of the Valanginian carbon–isotope excursion: evidence from the Polish Basin. *Palaeogeogr. Palaeoclimatol. Palaeoecol.*, **426**, 183–198.

Pantic, N.K. and **Burger, H.** (1981) Palynologische Untersuchungen in der untersten Kreide des östlichen Helvetikums. *Eclogae Geol. Helv.*, **74**, 661–672.

Pasquier, J.-B. (1995) *Sédimentologie, stratigraphie séquentielle et cyclostratigraphie de la marge Nord-Téthysienne au Berriasien en Suisse occidentale*. Université de Fribourg, Fribourg, 274 pp.

Peybernès, B. and **Combes, P.J.** (1994) Stratigraphie séquentielle du Crétacé basal (intervalle Berriasien-Hauterivien) des Pyrénées centrales et orientales franco-espagnoles. *Cretac. Res.*, **15**, 535–546.

Poulsen, C.J., Seidov, D., Barron, E.J. and **Peterson, W.H.** (1998) The impact of paleogeographic evolution on the surface oceanic. *Paleoceanography*, **13**, 546–559.

Remane, J. (1963) Les calpionelles dans les couches de passage Jurassique – Cretacé de la fosse Vocontienne. *Trav. Lab. Géol. Grenoble*, **39**, 39–82.

Remane, J. (1985) Calpionellids. In: *Plankton Stratigraphy* (Eds H.M. Bolli, J.B. Saunders and K. Perch-Nielsen), **1**, 555–572. Cambridge University Press, Cambridge.

Remane, J., Bakalova-Ivanova, D., Borza, K., Knauer, J., Nagy, I. and **Pop, G.R.** (1986) Agreement on the subdivision of the standard calpionellid zones defined at the IInd planktonic conference, Roma 1970. *Acta Geol. Hung.*, **29**, 5–14.

Remane, J., Haq, B.U. and **Boersma, A.** (1998) Calpionellids. In: *Introduction to Marine Micropaleontology* (Eds B.U. Haq and A. Boersma), 2nd edn, pp. 161–170. Elsevier Science B.V., Amsterdam.

Schlager, W. (1981) The paradox of drowned reefs and carbonate platforms. *Geol. Soc. Am. Bull.*, **92**, 197–211.

Schnyder, J., Gorin, G., Soussi, M., Baudin, F. and **Deconinck, J.-F.** (2005) Enregistrement de la variation climatique au passage Jurassique/Crétacé sur la marge sud de la Téthys: minéralogie des argiles et palynofaciès de la coupe du Jebel Meloussi (Tunisie centrale, formation Sidi Kralif). *Bull. Soc. Géol. France*, **176**, 171–182.

van de Schootbrugge, B., Kuhn, O., Adatte, T., Steinmann, P. and **Föllmi, K.** (2003) Decoupling of P- and Corg-burial following Early Cretaceous (Valanginian-Hauterivian) platform drowning along the NW Tethyan margin. *Palaeogeogr. Palaeoclimatol. Palaeoecol.*, **199**, 315–331.

Schroeder, R., Clavel, B., Conrad, M.A., Zaninetti, L., Busnardo, R., Charollais, J. and **Cherchi, A.** (2000) Correlations biostratigraphiques entre la coupe d'Organyā (Pyrénées Catalanes, NE de l'Espagne) et le Sud-Est de la France pour l'intervalle Valanginien-Aptien. *Treb. Mus. Geol. Barcelona*, **9**, 5–41.

Stampfli, G. and **Hochard, C.** (2009) Plate tectonics of the Alpine realm. In: *Ancient Orogens and Modern Analogues* (Eds J.B. Murphy, J.D. Keppie and A.J. Hynes), **327**, 89–111. Geological Society, London, Special Publications, London.

Stampfli, G., Borel, G.D., Marchant, R. and **Mosar, J.** (2002) Western Alps geological constraints on western Tethyan reconstructions. In: *Reconstruction of the Evolution of the Alpine-Himalayan Orogen* (Eds G. Rosenbaum and G.S. Lister), *J. Virt. Exp.*, **7**, 75–104.

Swart, P.K. and **Eberli, G.** (2005) The nature of the $\delta^{13}C$ of periplatform sediments: implications for stratigraphy and the global carbon cycle. *Sed. Geol.*, **175**, 115–129.

Vail, P.R., Colin, J.-P., Jan Du Chêne, R., Kuchly, J., Mediavilla, F. and **Trifilieff, V.** (1987) La stratigraphie séquentielle et son application aux corrélations chronostratigrahiques dans le Jurassique du bassin de Paris. *Bull. Soc. Géol. France*, **8**, 1301–1321.

Virgone, A. (1997) *Stratigraphie, sédimentologie et dynamique d'une plate-forme carbonatée: le Berriasien supérieur-*

Valanginien basal de Basse Provence Occidentale (S. E. France). Université de Provence -Aix-Marseille I, Marseille, 223 pp.

Virgone, A. and **Masse, J.P.** (1996) Les dépôts carbonates de tempêtes du « Faisceau bioclastique du Mont-Rose » (Valanginien inférieur-Marseille, Sud-Est de la France). *Geol. Alpine*, **72**, 127–143.

Weissert, H., Lini, A., Föllmi, K.B. and **Kuhn, O.** (1998) Correlation of Early Cretaceous carbon isotope stratigraphy and platform drowning events: a possible link?. *Palaeogeogr. Palaeoclimatol. Palaeoecol.*, **137**, 189–203.

Weissert, H., Joachimski, M. and **Sarnthein, M.** (2008) Chemostratigraphy. *Newsl. Stratigr.*, **42**, 145–179.

Wippich, M.G.E. (2003) Valanginian (Early Cretaceous) ammonite faunas from the western High Atlas, Morocco, and the recognition of western Mediterranean, Äòstandard, Äô zones. *Cretac. Res.*, **24**, 357–374.

Wyssling, G.W. (1986) Der frühkretazische helvetische Schelf im Vorarlberg und Allgäu. *Jb. Geol. Bundesanst.*, **129**, 161–265.

Supporting Information

Appendix S1. Stratigraphic variation in carbon and oxygen stable isotopes along a proximal-distal transect (bulk-rock measurements in ‰ VPDB).

Appendix S2. Raw data of bulk-rock $\delta^{13}C$ and $\delta^{18}O$ analyses performed on the Säntis section.

Appendix S3. Raw data of bulk-rock $\delta^{13}C$ and $\delta^{18}O$ analyses performed on the Dräckloch section.

Appendix S4. Raw data of bulk-rock $\delta^{13}C$ and $\delta^{18}O$ analyses performed on the Vitznau section.

Appendix S5. Raw data of bulk-rock mineralogical analyses performed on the Säntis section.

Appendix S6. Raw data of bulk-rock mineralogical analyses performed on the Dräckloch section.

Appendix S7. Raw data of bulk-rock mineralogical analyses performed on the Vitznau section.

10

Chemostratigraphy of the Upper Albian to mid-Turonian Natih Formation (Oman) – how authigenic carbonate changes a global pattern

STEPHAN WOHLWEND*, MALCOLM HART† and HELMUT WEISSERT*

*Geological Institute, ETH Zurich, Sonneggstrasse 5, 8092 Zurich, Switzerland (E-mail: st.wohlwend@gmx.ch)
†School of Geography, Earth & Environmental Sciences, Plymouth University, Drake Circus, Plymouth PL4 8AA, UK

Keywords
Arabian platform, C-isotope chemostratigraphy, Cretaceous, Natih Formation, OAE2, Oman Mountains.

ABSTRACT

The Oman Mountains preserve a Cretaceous continental margin transect with the proximal Arabian carbonate shelf and the adjacent deep Hawasina Basin. Today, the sediments from the Arabian Platform outcrop in the Oman Mountains (Jabal Akdhar and Saih Hatat) and in the Adam Foothills. The western part of the Adam Foothills provides insight into the evolution of a Late Cretaceous intra-platform basin with organic-rich sediments in the central part of this basin. The aims of this study are (a) to establish a biostratigraphy and chemostratigraphy of the Natih Formation and (b) to reconstruct depositional conditions of organic-rich sediments in an intra-platform basin during Cenomanian–Turonian times. The hypothesis that local black shale formation is an expression of global perturbations of the global carbon cycle will be tested. Reconstruction of the depositional history of the Arabian Platform and its intra-platform basin within a global palaeoclimatic framework requires an accurate time frame. The Upper Albian to mid-Turonian biostratigraphy of the Natih Formation has resulted in controversial age models that will be integrated into a solid chemostratigraphic framework with additional biostratigraphic data. A major positive $\delta^{13}C$ excursion (+4·6‰) has been identified as of Middle Cenomanian age, which is confirmed by an ammonite datum. A second positive $\delta^{13}C$ excursion (+4·5‰) following a major negative excursion (−1·0‰) confirms the existence of the Cenomanian/Turonian Boundary Event. The accurate chemostratigraphy and biostratigraphy confirms that major source rocks in the Mishrif-Natih Basin precede OAE2. Low $\delta^{13}C$ values measured in the sediments of the Natih B member are considered a consequence of diagenetic alteration. Elevated organic carbon contents and argillaceous sediments alternating with limestones resulted in diagenetic conditions favouring formation of authigenic calcite depleted in C-13.

INTRODUCTION

Oceanic Anoxic Events (OAE's) were originally defined by Schlanger & Jenkyns (1976) as episodes in Earth history marked by widespread dysoxic to anoxic conditions in the world oceans, with global deposition of sediments enriched in organic carbon. During these events, organic-rich sediments were deposited in deeper basins under fully anoxic conditions and in shallower marginal settings within an expanded oxygen minimum zone (Schlanger & Jenkyns, 1976; Arthur & Schlanger, 1979; Jenkyns, 1980, 2010; Trabucho Alexandre *et al.*, 2010). Oceanic Anoxic Events, and especially OAE2 at the Cenomanian/Turonian boundary, can easily be defined and correlated using C-isotope geochemistry as a stratigraphic tool (Scholle & Arthur, 1980; Jenkyns, 1980; Jarvis *et al.*, 1988, 2006; Weissert, 1989). The Mid-Cenomanian Event I (MCE I) and the onset of OAE2 are characterized by important positive excursions in the $\delta^{13}C$ carbonate bulk-rock record (Schlanger *et al.*, 1987; Erbacher *et al.*, 1996; Jarvis *et al.*, 2006; Voigt *et al.*, 2008).

Neritic carbonates, formed on a carbonate platform and within an intra-platform basin of Albian to Turonian age are exposed in the Adam Foothills of the Oman Mountains. This succession provides the opportunity to investigate the evolution of an equatorial carbonate platform during a time of multiple perturbations of the global carbon cycle. The sediments accumulated on the platform and in the intra-platform Mishrif-Natih Basin, also called Rub'Al Khali Basin by Ziegler (2001), define the Natih Formation. Age control is given by biostratigraphy. However, numerous biostratigraphic investigations have resulted in controversial age models (Simmons & Hart, 1987; Simmons, 1994; Witt & Goekdag, 1994; Van Buchem *et al.*, 1996, 2002, 2011; Sharland *et al.*, 2001; Grélaud *et al.*, 2006; Schroeder *et al.*, 2010). Homewood *et al.* (2008) have summarized the available disputed biostratigraphic and sequence stratigraphic information.

This study aims at using the Natih Formation as an archive of the global C-isotope record. A high-resolution C-isotope record will provide data for an improved stratigraphic resolution of an Arabian platform succession during the time interval of interest. A robust stratigraphy offers the opportunity to test whether or not organic carbon-rich sediments accumulated within the intra-platform area of the Rub'Al Khali Basin are an expression of global carbon cycle perturbations. Of major importance is the accurate chemostratigraphic correlation of the uppermost Albian to lower Turonian Adam Foothills sections with the more distal Fahud Field successions because these are the closest outcrop analogues to the source rocks in the oil-producing subsurface of Oman.

REGIONAL GEOLOGICAL SETTING

Arabian Platform

The eastern part of the Arabian Plate was covered by an extensive shallow-water carbonate platform during Early to Late Cretaceous time (Fig. 1). The subsurface information from these shallow-water carbonate successions in Interior Oman (Hughes-Clarke, 1988; Vahrenkamp, 2010) and from the outcrop equivalents in the Adam Foothills and the Oman Mountains (Simmons & Hart, 1987; Van Buchem *et al.*, 2002; Homewood *et al.*, 2008) show that the carbonate platform can be divided into two larger sedimentary groups (Fig. 2). The Kahmah Group (Glennie *et al.*, 1974) overlies Jurassic and older strata (Rabu, 1987; Pratt & Smewing, 1990, 1993), its top is karstified and it contains evidence of erosion (Immenhauser & Rameil, 2011) related to a major relative sea-level fall (Maurer *et al.*, 2013) in the latest Aptian. The Kahmah Group is covered by shales and fine-grained clastics of the Nahr Umr Formation corresponding to the base of the Wasia Group (Immenhauser *et al.*, 1999). Neritic limestones of the Natih Formation were deposited on top of Nahr Umr clastics. These carbonates record a 50 to 60 km northward progradation of the Arabian carbonate platform (Droste & van Steenwinkel, 2004). The Natih Formation and equivalent stratigraphic units such as the Mishrif Formation in the United Arab Emirates (Burchette, 1993) contain several organic-rich intervals. The end of the Natih Formation and, hence, of the Wasia Group is again marked by a large regional unconformity, the Base Aruma unconformity (Fig. 2). This change in the sedimentary system occurred during mid-Turonian times as a result of the profound erosion following flexure of the continental lithosphere associated with beginning obduction of the Oman ophiolites (Robertson, 1987). The overlying Muti Formation represents infill of the foreland basin starting in the Late Turonian to the Campanian (Robertson, 1987).

Natih Formation

The focus of this study is the Natih Formation, which was deposited on a very extensive, low-relief carbonate platform during the Late Albian to mid-Turonian. Assigned age is based on benthic foraminifera (Simmons & Hart, 1987; Smith *et al.*, 1990; Piuz & Meister, 2013; Piuz *et al.*, 2014), on rudists (Philip *et al.*, 1995) and ammonites (Kennedy & Simmons, 1991; Van Buchem *et al.*, 2005; Bulot, as summarized in Homewood *et al.*, 2008; Meister & Piuz, 2013, 2015). Carbon-isotope chemostratigraphy published by Vahrenkamp (2013) and low-resolution C-isotope stratigraphy by Wagner (1990) supports the biostratigraphically constrained age models.

The carbonate platform covered the Arabian Platform between the Arabian shield to the south and the Tethys margin with the adjacent Hawasina Basin in the north (Philip *et al.*, 2000; Droste & van Steenwinkel, 2004). The Natih Formation was subdivided, using subsurface information from Fahud (Type section: Well Fahud North-3), into seven informal members, labelled by the letters 'a' to 'g' from the top to the base (Hughes-Clarke, 1988). However, following precedents set by previous authors (Homewood *et al.*, 2008), capital letters ('A' to 'G') will be used here. Subdivision is based on variations in the clay-carbonate ratio observed in cores and on gamma-ray data interpretation. Seismic data from Interior Oman (Droste & van Steenwinkel, 2004; Grélaud *et al.*, 2006) and a high-frequency synthetic seismic model from the Adam Foothills Transect (Schwab *et al.*, 2005) reveal that the platform system of the Natih and Mishrif formations was not completely flat. Instead, the platform consisted of several prograding carbonate platform units, separated by intra-platform basin deposits. These intra-shelf-basins show a typical pattern that starts with a regional flat platform top, which diversifies into a shallow-water rudist

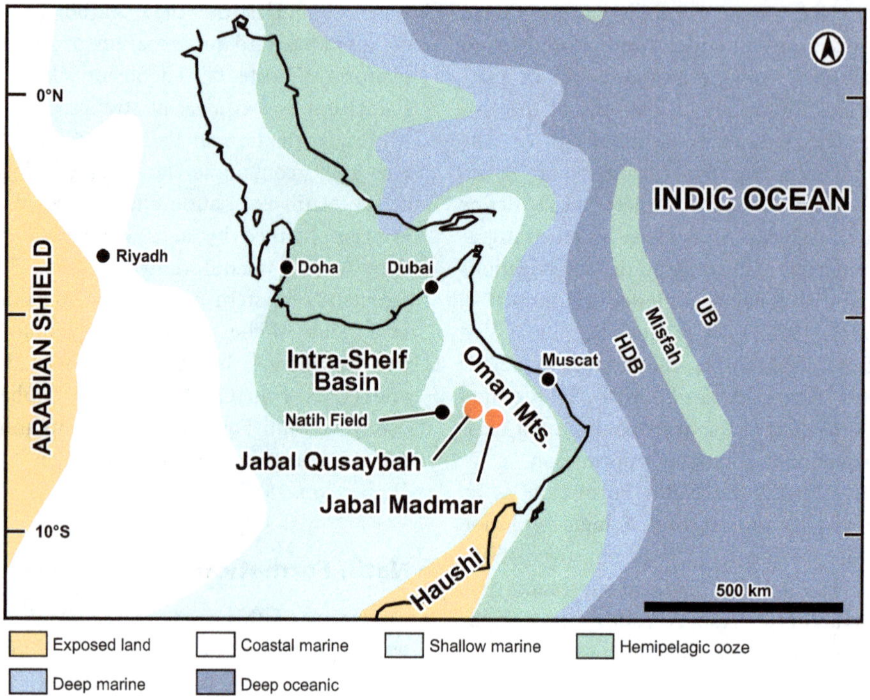

Fig. 1. Schematic palaeogeographical map of the Arabian Peninsula and the Indic Ocean during the deposition of the Natih Formation (late Albian-early Turonian); HDB: Hamrat Duru Basin, UB: Umar Basin; modified after Murris (1980); Philip *et al.* (2000) and Razin *et al.* (2010).

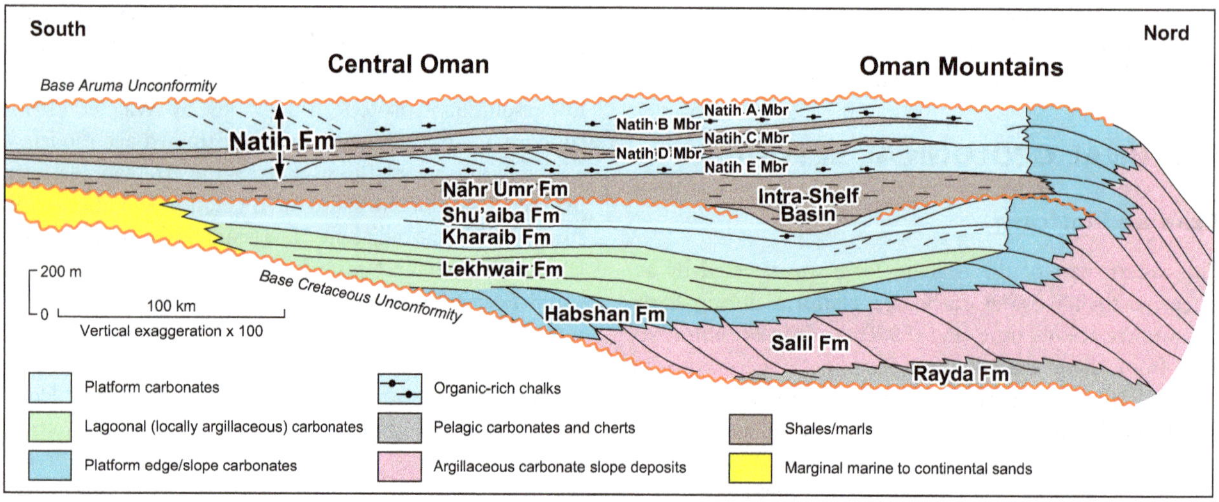

Fig. 2. Geological cross-section through the North Oman Cretaceous Carbonate platform, modified from Droste & van Steenwinkel (2004) and Homewood *et al.* (2008).

barrier platform and a lagoonal part with locally organic-rich sediment infill. The deeper basins were filled with argillaceous carbonates sometimes with a high content of organic carbon, as it is documented in the Natih E and B members (Van Buchem *et al.*, 2002). These argillaceous limestones and marlstones today are hydrocarbon source rocks, which produced the giant Natih and Fahud oil-fields in Oman (Terken, 1999; Droste, 2014).

In the Adam Foothills and also the Oman Mountains several high-resolution sequence stratigraphic reconstructions, created from extensive outcrop studies, serve for better understanding of the depositional system (Van Buchem *et al.*, 1996, 2002; Grélaud *et al.*, 2006; Homewood *et al.*, 2008). In these studies, four fully developed third-order sequences (sequence I to IV) were described. Sequence I includes the lower Natih Formation (Natih G,

F & E members), sequence II the middle Natih Formation (Natih D & C members), sequence III the upper Natih Formation (Natih B & lower A members) and sequence IV the upper Natih A member. In addition to these 4 third-order sequences, 34 higher frequency sequences have been documented (Grélaud *et al.*, 2006; Homewood *et al.*, 2008). Sequences I and III represent the evolution from a carbonate-clay ramp at the base upward into a carbonate ramp (Droste & van Steenwinkel, 2004). During the transgressive parts of these sequences, organic-rich sediments were accumulated in the intra-shelf basin (Van Buchem *et al.*, 1996, 2002). The composition of the organic matter is confirmed as autochthonous Type II, marine algal material by hydrogen index (HI) values between 400 and 650 mgHC/g Total Organic Carbon (TOC), and palynofacies analysis (Van Buchem *et al.*, 2002, 2005). The top of each third-order sequence (top Natih E, top Natih C, and top Natih A) corresponds to a phase of platform emersion due to a relative sea-level fall. In the upper part of the Natih E member, and even more pronounced in the upper part of the Natih A member, a network of channels has been observed on seismic maps and in outcrops in the Adam Foothill and the Oman Mountains (Van Buchem *et al.*, 1996, 2002; Grélaud *et al.*, 2006; Droste, 2010). These channels are more than 2 km wide and reach over 30 km in length. The channels drained into the intra-shelf basins and into seaways between the platforms (Grélaud *et al.*, 2006, 2010).

LOCATIONS AND METHODS

Lithofacies, inorganic and organic C-isotope geochemistry, as well as data on inorganic and organic carbon content, were used to correlate the studied successions. Thin sections were prepared in order to get an overview of the different microfacies and possible depositional settings. The Natih B to A members were investigated by 12 thin sections for diagnostic planktonic and benthic microfossils. The thin sections were covered with acryl and analysed under transmitted light. All geochemical measurements were performed at the isotope laboratory of ETH Zurich (Geological Institute).

Two expanded sections at Jabal Madmar (Wadi H, 112 m and Wadi P, 99 m) and one at Jabal Qusaybah (110·5 m) were sampled at intervals of less than 1 m, sometimes with decimetre spacing (see Fig. 3 for locations). The Jabal Qusaybah section is a composite section starting with the lowermost part (0 to 68 m) at Location 11 (Homewood *et al.*, 2008), followed by the middle part which exposes the most continuous section of the cherty mudstone (68 to 80 m) in the eastern wadi (22°32′39″N, 57°05′08″E). The uppermost part, consisting of the base of the bioclastic pack- to grainstone up to the top of the Natih Formation (80 to 110·5 m), is exposed in the western wadi (22°32′44″N, 57°04′39″E). Two additional shorter sections (Jabal Salakh, 43 m and Wadi Nakhr, 13 m) were sampled for regional correlation of important intervals.

A total of 658 specimens of limestones, argillaceous limestones and carbonate mudstones were sampled for stable carbon and oxygen-isotope analysis. Samples were drilled with a micro-drill in order to avoid diagenetic calcite from veins. Approximately 140 µg of powder was reacted with 100% phosphoric acid at 71°C in a Finnigan GasBench II carbonate device connected to a Thermo-Fisher Delta V PLUS mass spectrometer. The instrument is calibrated with international [National Bureaux Standards (NBS) 19] and internal standards (MS2 ‘Carrara marble’; $\delta^{13}C_{carb}$ = 2·16, $\delta^{18}O$ = −1·85‰). The reproducibility of the measurements based on replicated standards was ±0·05‰ for $\delta^{13}C_{carb}$ and ±0·06‰ for $\delta^{18}O$. The isotope values are reported in the conventional delta notation with respect to Vienna Pee Dee Belemnite (VPDB).

The isotopic composition of organic carbon ($\delta^{13}C_{org}$) was measured on bulk-rock samples from 40 samples of carbonate mudstones and organic-rich black limestones. More than 5 g of the sample material was treated with HCl acid (3 M) over a 24 h period. The samples were treated a second time with HCl to completely remove any carbonate minerals. A few 100 µg of decarbonated powder were weighed in tin capsules and measured using a Thermo-Fisher Flash-EA coupled to a Delta V PLUS mass spectrometer. The reproducibility of the measurements based on replicated standards was <±0·1‰ for $\delta^{13}C_{org}$. The instrument is calibrated with the international standards NBS 22 and IAEA-CH-6 and the isotope values are reported in the conventional delta notation with respect to VPDB.

Total inorganic carbon (TIC) as well as total carbon (TC) in samples was determined using a UIC CM 5012 CO_2 Coulomat. For each analysis, *ca* 10 mg of powdered sediment was used. The TIC contribution was determined by dissolution of the sample in 2N perchloric acid, whereas TC was obtained by sample combustion in a 950°C furnace. The difference between TC and TIC provided the TOC. For a reference sample 100% Na_2CO_3 was used.

RESULTS

Lower Natih Formation at Jabal Madmar (Wadi H)

The carbon-isotope data from the lower Natih Formation (Natih F-E members, Fig. 4) vary between 0·8‰ and 3·6‰ (Fig. 5). The $\delta^{13}C$ profiles show significant variation of up to 2‰ from one sample to the next in the Natih E member (Fig. 5). The systematic scatter around the average (5-point moving average) is usually very small with a few larger outliers of up to 1‰. The oxygen-isotope data vary between −6·1‰ and −2·4‰ (Fig. 5) with most data confined to a narrow range between −5·5‰ and −3·5‰. The only outliers (93 to 95 m, Fig. 5) show a clear offset of around 2‰ to more positive values.

The TOC results reflect the organic-rich intervals in the Natih E member (Fig. 4B and Homewood *et al.*, 2008). In the sequence I-3 (Natih E member, Fig. 5), the weight % of the TOC increases contemporaneously with the deepening trend towards maximum flooding of the fourth-order sequence from around 0·5% to a maximum 7·4%. In the following sequence I-4, the values increase to a maximum of 2·6% towards the maximum flooding surface, as defined by Homewood *et al.* (2008).

Middle Natih Formation at Jabal Madmar (Wadi P)

The carbon-isotope data from the middle Natih Formation (Natih D-B members, Fig. 6) vary between 0·6‰ and 4·6‰ (Fig. 7). The $\delta^{13}C$ profile (Fig. 7) shows two prominent positive peaks in the limestone layers at the top of the Natih D member. The highest values of the first peak are at the base of the limestone bed (D2 in Fig. 6B; 12 m in Fig. 7) and the values decrease upwards towards the marly interval separating the two limestone beds (D2-D1 in Fig. 6B; 12 to 23 m in Fig. 7). The second peak occurs within the marly interval itself. The $\delta^{13}C$-jump to 1·5‰ higher values at the boundary between the Natih D and C members indicates a sedimentary gap corresponding to sub-aerial exposure during a sea-level lowstand. Carbon-isotope values of the Natih C member record a smooth decreasing trend. In the lower Natih B member (Sequence III-1, Fig. 7) a distinct C-isotope excursion to low values near +1‰, was identified in the Wadi P section. The top of the Natih B member is represented by regular bedding and bed thickness ranging from 15 to 20 cm.

The oxygen-isotope data vary between −6·0‰ and −3·4‰ and show a similar pattern to the $\delta^{13}C$-data with clear positive jumps following the individual fourth-order sequence boundaries (top of II-1, II-2 and III-1, Fig. 7). The $\delta^{18}O$ values show an overall trend to more positive values (increase by +1‰) starting at 22 m and ending at the top of the section.

In the uppermost limestone bed of the Natih C member (42·5 m in Fig. 7), an ammonite was found loose within Natih C gravel. The ammonite belongs to the Acanthoceratidae and can be compared to several species of *Calycoceras*, which indicates a Middle Cenomanian age.

Upper Natih Formation at Jabal Qusaybah

The Natih B member outcropping at Jabal Qusaybah most closely corresponds to the Natih B member described from the subsurface in the Natih and Fahud oil fields. Location 11, as it is described in Homewood *et al.* (2008), shows around 30 m of alternating organic-rich carbonate mudstones and limestones with a bedding thickness of 5 to 10 cm (Fig. 8B). For carbon and oxygen measurements, only the limestone layers were sampled in the Natih B member. C-isotope values range from −1·3 to 4·5‰ while oxygen-isotope values fluctuate between −7 and −3‰. The TOC of these dark to black limestone beds varies around 1 to 4 wt%. The argillaceous layers show much higher TOC values of up to 8 wt% (10 m in Fig. 9), which are similar to concentrations measured in the Natih oil field (up to 13·7 wt%; Al Balushi *et al.*,

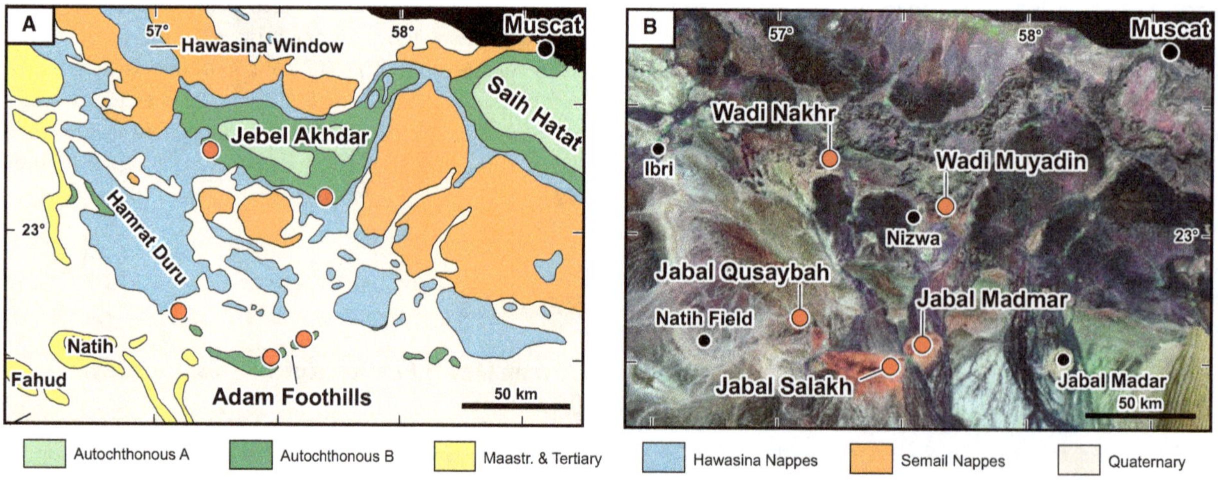

Fig. 3. (A) Simplified geological map after Glennie *et al.* (1974) and Pillevuit *et al.* (1997). (B) Satellite image (Al Balushi *et al.*, 2011).

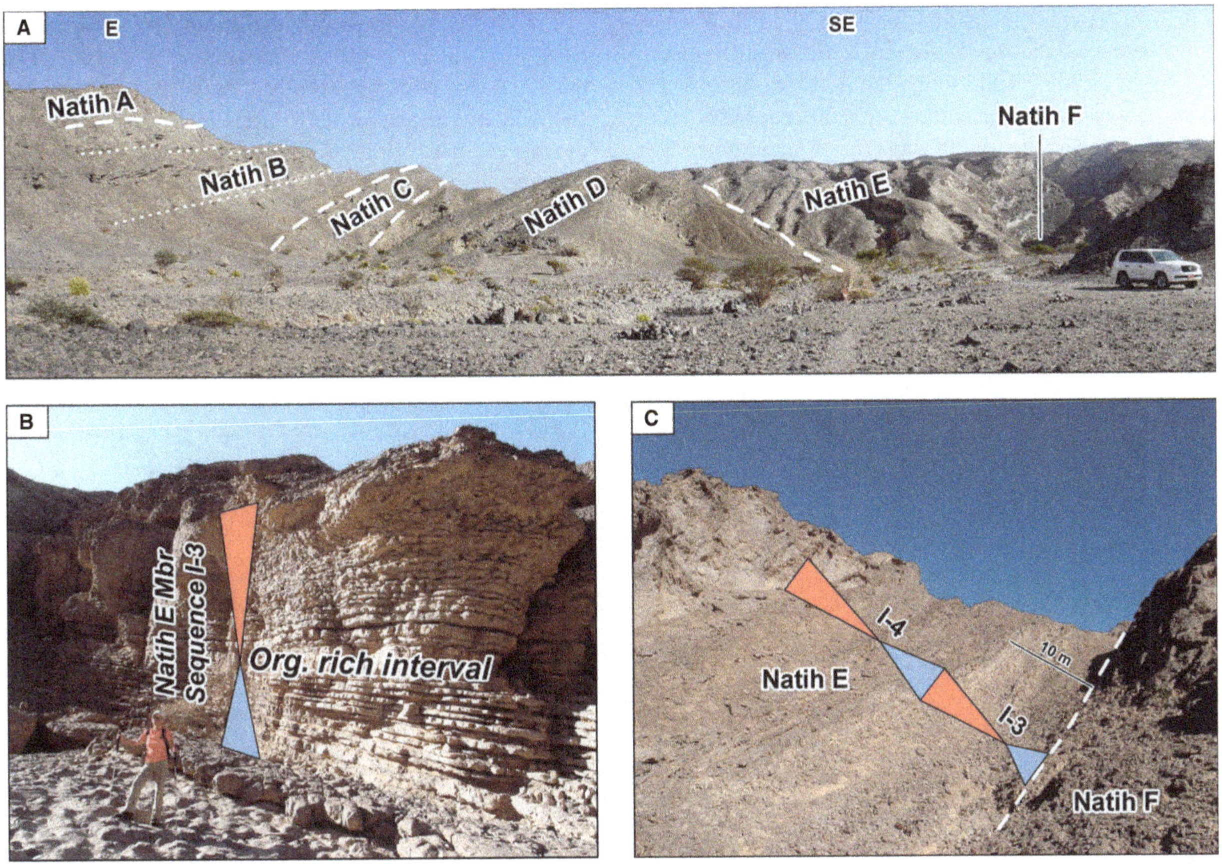

Fig. 4. Detailed pictures of important sections in the lower Natih Fm. Subdivision into members of the Natih Fm after Homewood *et al.* (2008). For location see Fig. 3 and Table 1. (A) Panoramic pictures from the outcrops at Jabal Madmar, Wadi H. (B) Sequence I-3 with the organic-rich interval in the Wadi H at Jabal Madmar. (C) Sequence I-3 & I-4 from the lower Natih E mbr at Jabal Salakh.

Table 1. Locations of the different sampled Natih members and their coordinates

Natih members	Locations	Coordinates (WGS 84)	Homewood *et al.* (2008)
Natih F to D member	Wadi H, Jabal Madmar	22°26′40″N, 57°35′28″E	Location 1
Natih F to E member	Jabal Salakh	22°21′01″N, 57°26′35″E	
Natih D to B member	Wadi P, Jabal Madmar	22°26′35″N, 57°33′56″E	Location 4
Natih B to A member	Jabal Qusaybah	22°32′36″N, 57°04′47″E	Location 11
Natih B member	Wadi Nakhr	23°09′08″N, 57°12′20″E	Location 8

2011). Oysters are the predominant macrofossils in this organic-rich interval of the Natih B member. The oxygen-isotope data show a clear outlier at 21 m above the base of the section with three values around −3‰ (Fig. 9), which is around 2‰ more positive than the overall trend trough the Natih B member. At 42 m above the base of the Jabal Qusaybah section (Fig. 9) organic-rich argillaceous sediments disappear and the TOC decreases to around 0·5 wt%. This significant lithological change reflects the boundary between the Natih B and A members (Fig. 8C). The lowermost 14 m of the Natih A member are represented by a 30 to 40 cm thick bedded, light grey to white limestone. Afterwards the bedding thickness decreases and even more obvious is a colour change to more grey and light brown limestones (Fig. 8D). The carbon-isotope data from the micritic limestones of the Natih A member show a continuous increase in the δ^{13}C-values up to a peak near +4·5‰ (Fig. 8E and 58·2 m in Fig. 9).

From 68 m above the base of the section and upwards, chert nodules are common in the carbonate mudstone. At Location 11, the section has been truncated by the overlying pack- to grainstone (80 to 87 m). The bioclastic pack- to grainstone or calcarenite is described by Homewood *et al.* (2008) as a forced regressive prograding wedge at the surface boundary III-6/7 to III-8. It can be correlated with a calcarenite described from the Fahud area where this also marks the top of sequence III. Detailed section logging in multiple wadis along the northern flank of the Jabal Qusaybah indicates that erosion extended 10 m downward into the underlying cherty limestone. The $\delta^{18}O$ values of this calcarenite body show a clear offset of +1·5 to +2‰ (Fig. 9).

The overlying interval between 87 to 100 m [sequence III-8(7)] is absent in Fahud (Homewood *et al.*, 2008) and therefore can only be described at Jabal Qusaybah. At 100 m above the base of the section, a small fault cuts the top and part of the overlying sequence IV. This is also reflected in an abrupt shift to more positive $\delta^{13}C$ and $\delta^{18}O$ values. Within the uppermost 12 m of the Natih A

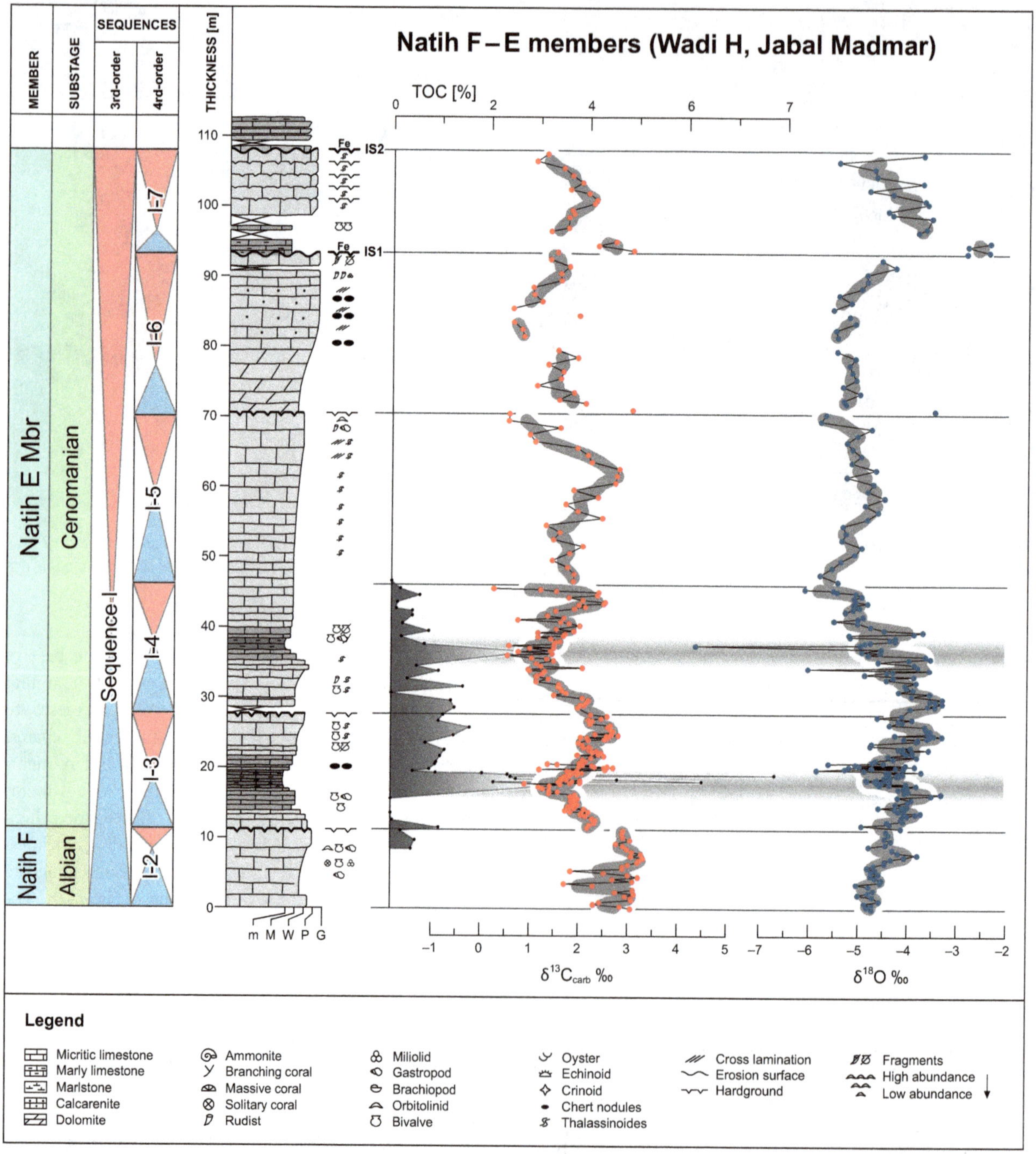

Fig. 5. Natih F and E mbrs from Jabal Madmar, Wadi H (for location see Figs 3 and 4). Sequence stratigraphic data after Van Buchem *et al.* (2002); Grélaud *et al.* (2006) and Homewood *et al.* (2008).

member (sequence V), δ^{13}C-values start to decrease. This trend terminates the δ^{13}C plateau with numbers around 3‰ measured between 60 to 102 m above section base.

In addition to the $\delta^{13}C_{carb}$ data, the $\delta^{13}C_{org}$ composition of 40 samples covering the Natih B member and the lower Natih A member (0 to 70 m), have been analysed (Fig. 9). The $\delta^{13}C_{org}$ data vary around −27‰ (±0·5‰) until 56·1 m above the base of the section. Similar to the $\delta^{13}C_{carb}$ values the $\delta^{13}C_{org}$ also increase by around 2‰ reaching the most positive values (−24·9‰) with the sample at 58·2 m above section base. Afterwards, the values decrease again to −27‰ at the 66 m level.

Micro- and macrofossils in the Natih A member

The microfossil assemblage of the lower Natih A member at Jabal Qusaybah yields a Late Cenomanian to Early Turonian age (Fig. 10). The thin sections QA-4·0 at 46 m (Fig. 10) show the typically Cenomanian taxa *Chrysalidina* sp. and *Nezzazatinella* sp. The occurrence of *Pseudolitinella* sp. (perhaps *P. reicheli*) would indicate a mid-Upper Cenomanian age at 50·8 m above section base. The thin sections between 54·1 to 61·3 m (QA-12·1 – QA 19·3) show a typical Cenomanian/Turonian interval assemblage. In particular, the occurrence of *Marssonella oxycona* is often quite common around the C/T boundary where it is one of the few benthic taxa to survive the OAE2. In this interval, at 56·4 m above the base of the section, *Lingulogavelinella* sp. (also known as *Berthelina* sp.) is quite common. In the same thin sections, *Praeglobotruncana* sp., juvenile *P. stephani*, *Hedbergella amabilisa* and *H. similis* can also be found and clearly confirm the C/T boundary interval. The final occurrence (FO) of *H. praehelvetica* at 61·3 m above the base of the section is an indicator of an earliest Turonian age.

The benthic and planktonic foraminifera between 101 to 110 m (QA-59 – QA-68) are clearly Turonian in age. The FO of *Marginotruncana* sp. cf. *sigali* indicates a mid-Turonian age in the thin-section QA-62. The uppermost

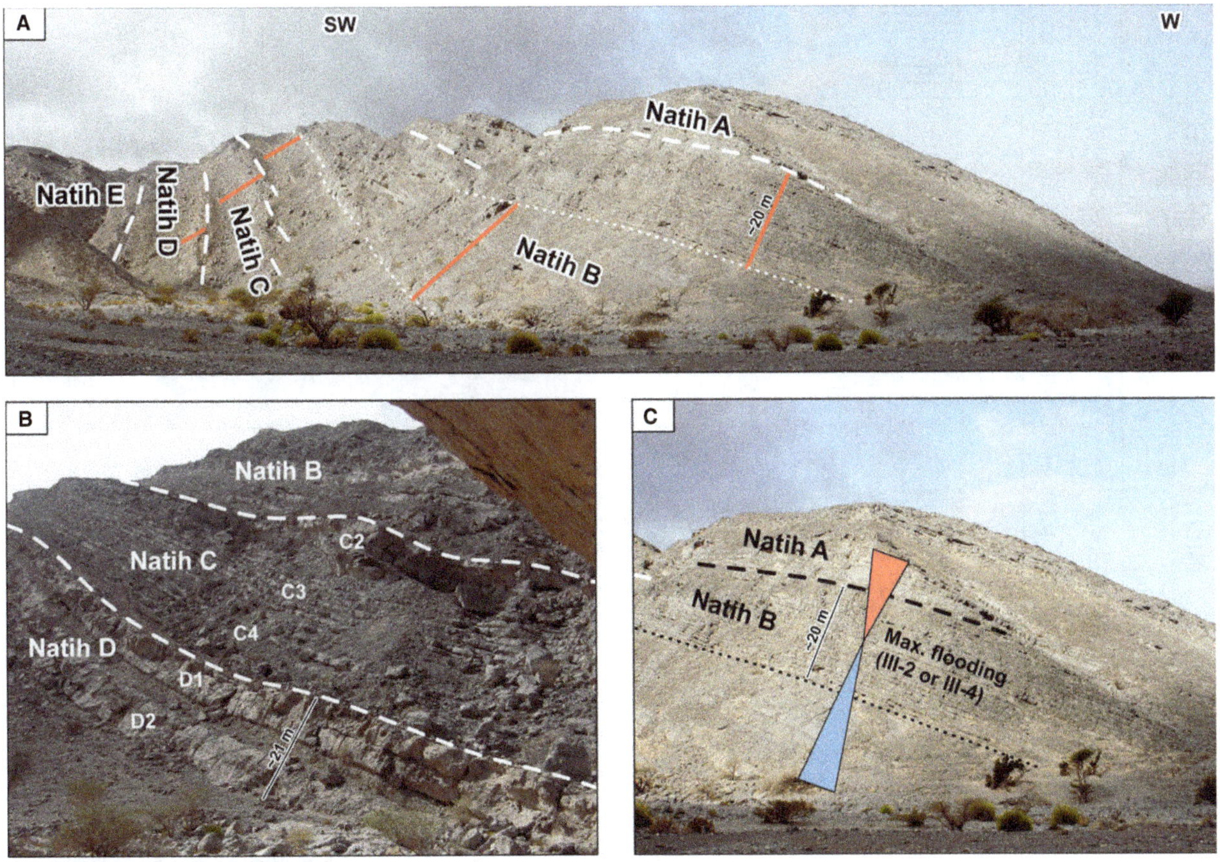

Fig. 6. Detailed pictures of important sections in the middle-upper Natih Fm. Subdivision into members of the Natih Fm after Homewood *et al.* (2008). For location see Fig. 3 and Table 1. (A) Panoramic pictures from the outcrops at Jabal Madmar, Wadi P. The red line indicates the approximate profile for the chemostratigraphy. (B) Detailed photograph of the Natih D-B with subunits (after Homewood *et al.*, 2008) in the Wadi P. (C) Detailed photograph of the Natih B-A transition in the Wadi P. Natih B interval where maximum flooding correlates to the one at Jabal Qusaybah.

two thin sections indicate quite shallow-water and further, the dominance of Miliolidae at 110 m indicates a slightly hypersaline environment (Murray, 1991).

At 59·9 m above the base of the section, just above the most positive C-isotope values, two ammonite specimens have been found *in situ*. While identification has to be taken with caution because of the rather poor preservation, for the first specimen two interpretations are possible. Either it belongs to the subfamily of Euomphaloceratinae ((?) *Kamerunoceras* sp. or (?) *Paramammites* sp.) or to the family of Vascoceratidae. Most likely it is a (?) Euomphaloceratinae with an Early Turonian age. The second ammonite seems to be a (?) *Watinoceras* sp. which would also indicate an Early Turonian age.

At the southern flank of Jabal Salakh (22°20′35″N, 57°15′58″E) 5 additional ammonites have been found *ex situ* in the middle Natih A member. Two specimens can be determined as *Vascoceras* sp. and one as a *Hoplitoides* sp. The two remaining specimens may be attributed to *Hoplitoides* sp. but the poor preservation does not allow a precise determination. The *Vascoceras* sp. and the *Hoplitoides* sp. confirm the Early Turonian age for the middle part of the Natih A member.

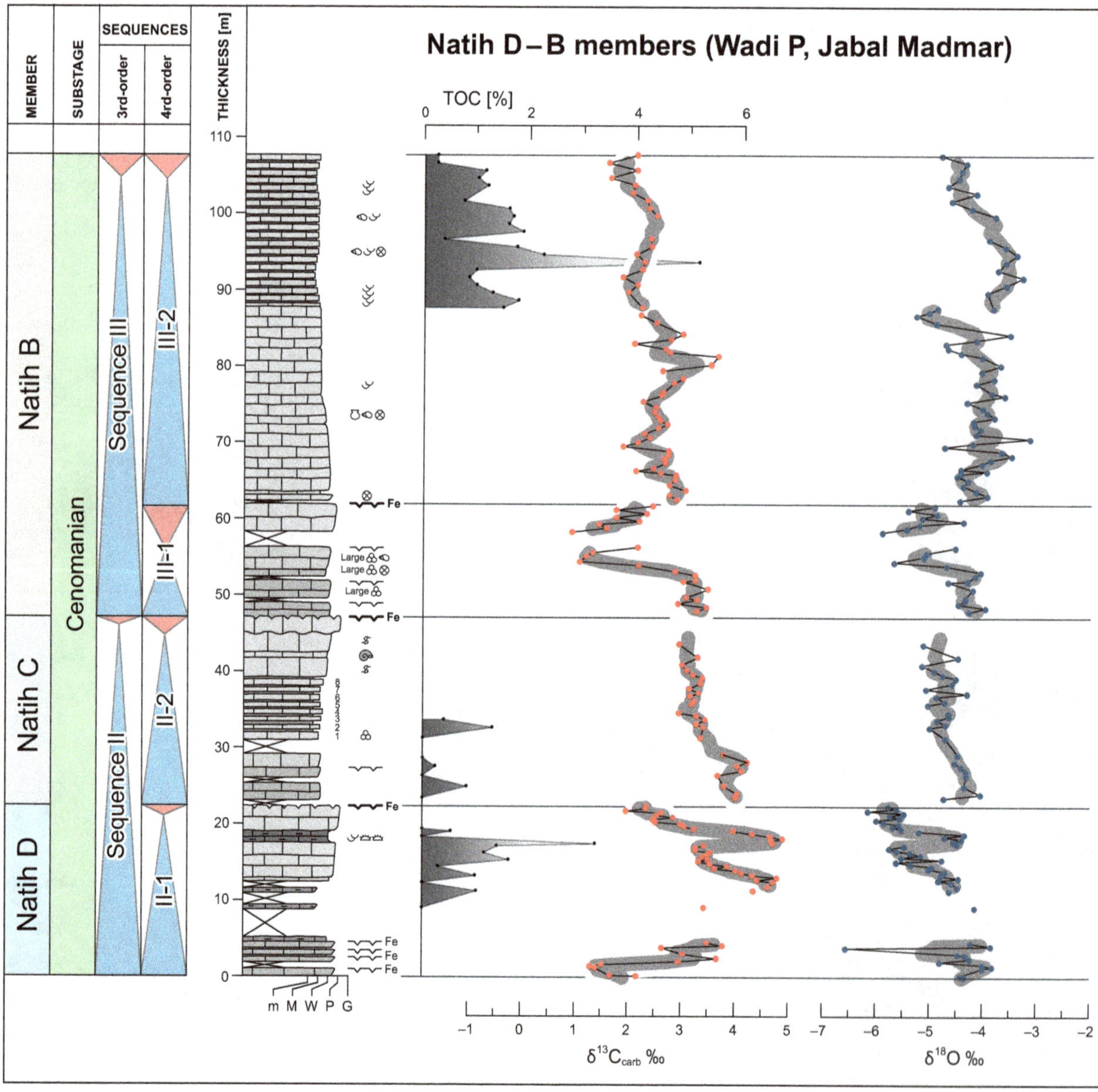

Fig. 7. Natih D, C and B mbrs from Jabal Madmar, Wadi P (for location see Figs 3 and 6). Sequence stratigraphic data after Van Buchem *et al.* (2002); Grélaud *et al.* (2006) and Homewood *et al.* (2008). Legend in Fig. 5.

INTERPRETATION AND DISCUSSION

Correlation of the two organic-rich intervals (Base Natih E and Natih B members)

In the Adam Foothills several workers proposed sequence stratigraphic correlations between the outcropping autochthonous Natih Formation and the subsurface records in the same area (Van Buchem *et al.*, 1996, 2002; Grélaud *et al.*, 2006; Homewood *et al.*, 2008) based on cycles and stacking patterns. Notional, abstract surfaces at turnarounds and at minimum accommodation are used as 'timelines' on cross sections. Carbon-isotope chemostratigraphy is a powerful tool to test whether or not these surfaces are isochronous, because the variation in the δ^{13}C-signal measured in the bulk-rock carbonates is considered to be a mirror of the isotopic composition of the marine (dissolved inorganic carbon) DIC and therefore, it is regarded as a global signal. Two stratigraphic intervals were measured at several locations: the lower Natih E member (Sequence I-3 and I-4; Fig. 11) and the Natih B member (Fig. 12).

The δ^{13}C-values start at 3‰ at Jabal Madmar in the top of the Natih F member followed by two negative excursions of around 1·5‰ amplitude in the lower Natih E member (Fig. 11). The chemostratigraphic data from both studied Natih E sections show similar variations, whereas the absolute δ^{13}C-values are shifted by around 1‰ towards lower values at Jabal Salakh. Both negative carbon-isotope intervals correlate with elevated organic carbon contents (up to 6 wt%, Fig. 11). The organic carbon content is coupled to the fourth-order cycles becoming enriched during maximal flooding (sequences I-3 and I-4). The C-isotope values fluctuating parallel with the fourth -order cycles may record changes in the global carbon cycle driven by orbital change (long eccentricity) as documented for the Albian in pelagic sediments by Giorgioni *et al.* (2012).

The Natih B member, as a hydrocarbon source rock in Interior Oman shows interesting chemostratigraphic features. Wagner (1990) and Vahrenkamp (2013) described a negative $\delta^{13}C_{carb}$ excursion over most of the Natih B member. The chemostratigraphic transect shown in Fig. 12, shows that sediments of the Natih B member outcropping west of Adam (Jabal Madmar) and the south-western part of the Oman Mountains belonged to an intra-shelf basin (Van Buchem *et al.*, 2002 and Razin & Grélaud in Homewood *et al.*, 2008) and that they are enriched in organic carbon. In the Fahud oil field, the source rock is of marine Type I/II origin with TOC contents of up to 15 wt% (Terken, 1999). In the Natih oil field, the TOC content of the source rock is up to 13·7 wt% (Al Balushi *et al.*, 2011). The TOC content at Jabal Qusaybah reaches values up to 8 wt% (Fig. 9) in the carbonate mudstones alternating with limestones. This basin-ramp structure is also reflected in the carbon-isotope measurements along the west-east Adam Foothills section (Fig. 12). Most positive δ^{13}C-values are measured at the 'marginal' intra-shelf basin at Jabal Madmar, while values in the 'central' part of the basin are lower by up to 2‰. The three measurements with less negative δ^{18}O-values (around −3‰) around 21 m above the base of the Jabal Qusaybah section indicate a transgression level. This level is also identical to the subunit boundary between B1 and B2 from the Fahud Field 'F' section (Fig. 12, Vahrenkamp, 2013).

Negative C-isotope excursions: diagenesis or global C-isotope signal?

A plot of all available carbon-isotope data (Fig. 13) from the Natih Formation (Wagner, 1990; Zhao *et al.*, 2012; Vahrenkamp, 2013; Arndt *et al.*, 2014 and this study) points at differences in absolute values. Large variations in amplitude can also be observed. As already shown in Figs 11 and 12 there are intervals in the Natih Formation marked by distinctly lower C-isotope values. The trend to lower δ^{13}C-values can be explained either by an early or late diagenetic overprint, by isotopic offset of the local water mass or by a global change in the δ^{13}C-values. Of importance is the observation that the low C-isotope values of the Natih B member exist in a succession of alternating organic carbon-rich carbonate mudstones and limestones, formed in the intra-platform basin. Whether this 'negative' C-isotope excursion in the upper Cenomanian serves as a regional or global stratigraphic marker or if it is the result of diagenesis is discussed in the following paragraph.

The negative $\delta^{13}C_{carb}$-values measured in the Natih B member show remarkable agreement with the sediments enriched in TOC in the central intra-shelf basin. However, elevated TOC contents are also measured in the basal part of the Natih A member (around 1wt%, Fig. 9), whereas the δ^{13}C-curve returns to more positive values in this part of Natih A. This indicates that the low C-isotope values of Natih B cannot be exclusively explained as result of diagenetic overprint due to elevated organic carbon content. Rather the negative δ^{13}C – shift is linked to the distinct lithology of Natih B, characterized by alternating limestones and organic carbon-enriched carbonate mudstones. Diagenesis of this succession is controlled by both argillaceous sediments and by interbedded neritic carbonates. Elevated clay content in the intra-basin mudstones acted as a decelerator or inhibitor of early marine/meteoric diagenesis (Westphal, 2006). Argillaceous sediments enriched in organic carbon experienced sulphate reduction and anaerobic oxidation of methane during early

Fig. 8. Detailed pictures of important sections in the upper Natih Fm. Subdivision into members of the Natih Fm after Homewood *et al.* (2008). For location see Fig. 3 and Table 1. (A) Panoramic pictures from the outcrops at Jabal Qusaybah. The red line indicates the approximate profile for the chemostratigraphy. (B) Cyclic bedding of marls (LA) and limestones (LB) in the Natih B mbr at Jabal Qusaybah [Lithofacies LA & LB after Al Balushi *et al.* (2011)] (C) Lithological change between the Natih B and A mbr at Jabal Qusaybah. (D) Subdivision of the lower Natih A mbr in the eastern wadi of Location 11. White to grey colour change can be correlated over the entire western Adam Foothills outcrop. Approximate position of the Cenomanian/Turonian Boundary Event (CTBE). (E) CTBE at Location 11 with exact position from the peak (δ^{13}C-value: 4·5‰).

diagenesis. Both processes resulted in alkaline pore fluids (Meister *et al.*, 2007; Schrag *et al.*, 2013). Authigenic carbonate depleted in C-13 was precipitated from supersaturated pore fluids.

The Natih B succession with argillaceous sediments enriched in organic carbon and with interbedded neritic carbonates can, therefore, be considered an example of diagenetic alteration due to the formation of authigenic calcite cement. Similar explanations of low C-isotope values have been proposed by Wagner (1990) and also Vahrenkamp (2013). The coupling of low C-isotope values with mudstones also explains why the δ^{13}C-values measured in Natih B are more negative (up to $-1{\cdot}5‰$) in the deepest part of the intra-shelf basin, where argillaceous sediments rich in organic carbon reach maximum abundance, compared to sediments of the basin margins. At Jabal Qusaybah, the δ^{13}C-values jump by $1{\cdot}5‰$ to higher values at the Natih B to A member transition (Fig. 8C). This lithological boundary marks the end of the organic-rich argillaceous sediments of the Natih B member (Fig. 9). The change in lithology is synchronous in all measured sections (Fig. 12) and could be linked to the global sea-level rise in the latest Cenomanian (Voigt *et al.*, 2006; Haq, 2014). This global sea-level rise probably reconnected the restricted intra-shelf basins with the open ocean. Better oxygenation and reduced input of clay minerals at a time of high sea-level resulted in a change of sedimentary facies and coupled change in diagenetic environment.

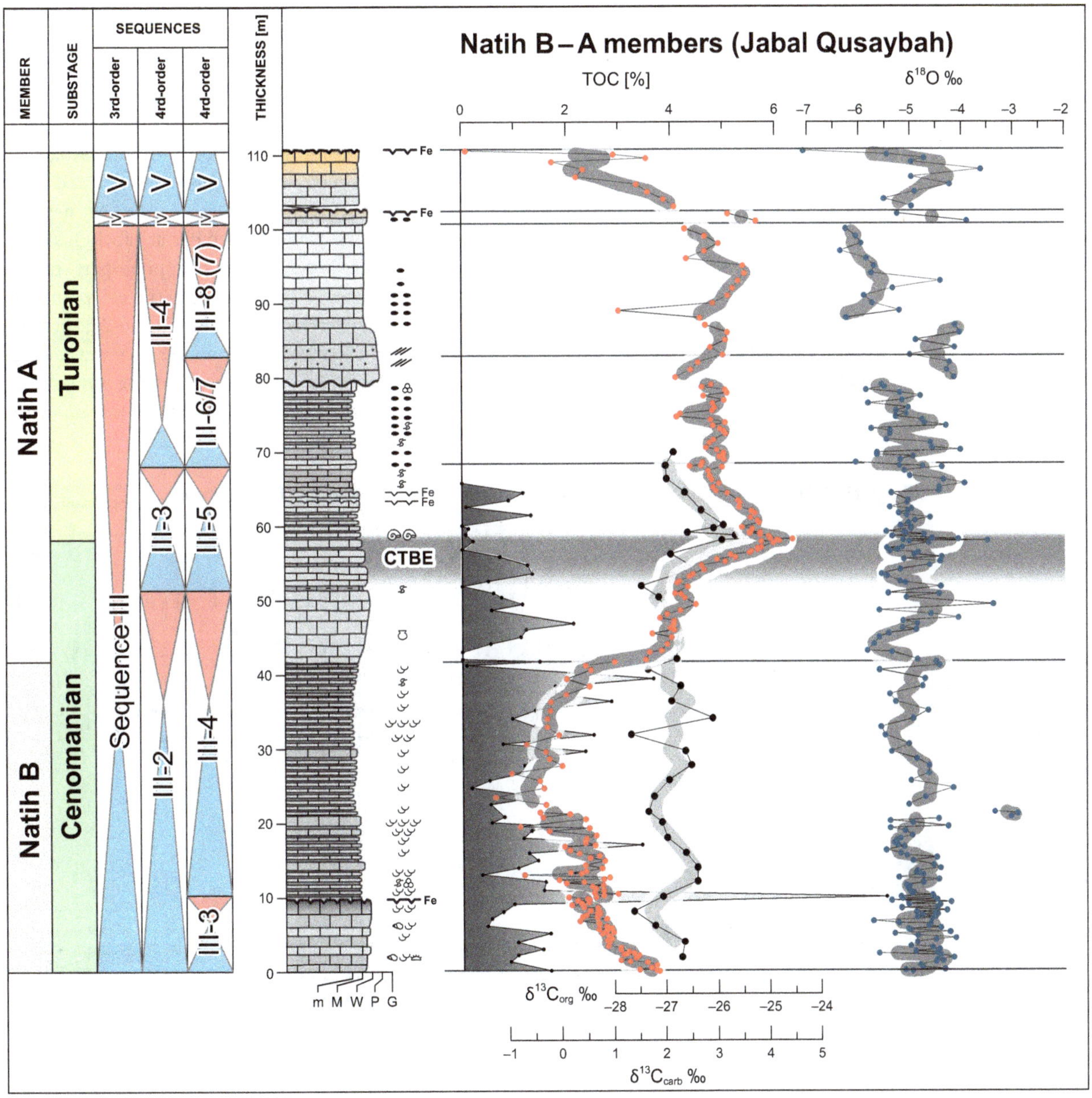

Fig. 9. Natih B and A members from Jabal Qusaybah (for location see Figs 3 and 8). Fourth-order sequence stratigraphic data: left Van Buchem *et al.* (2002) and right Homewood *et al.* (2008). Red points represent $\delta^{13}C_{carb}$ and black ones the $\delta^{13}C_{org}$ data. Legend in Fig. 5.

Low $\delta^{13}C$-values in upper Cenomanian carbonates have not only been identified in the interior of Oman and in certain areas of the Adam Foothills but similar trends can also be found in the High Zagros, SW Iran (Razin *et al.*, 2010), in Central Jordan (Morsi & Wendler, 2010), NE Egypt (El-Sabbagh *et al.*, 2011) and also in parts of the Southern Apennines, Italy (Parente *et al.*, 2007). The sediments from the High Zagros and probably also from Central Jordan may be linked to large intra-shelf basins on the Arabian Platform (Fig. 1). All of these successions may contain unusual amounts of isotopically depleted authigenic calcite. The hypothesis, that not only enrichment of sediments with organic carbon but also composition of the sediments controlled this peculiar diagenesis, will need further testing.

In contrast, the $\delta^{13}C_{org}$-data do not show any significant variation during the negative $\delta^{13}C_{carb}$ interval in the Natih B member (Figs 9 and 13). The $\delta^{13}C_{org}$-values vary around $-27‰$ ($\pm0{\cdot}5‰$) throughout the negative $\delta^{13}C_{carb}$ excursion. Similar to the $\delta^{13}C_{carb}$ values, the $\delta^{13}C_{org}$ also increase by around 2‰ reaching the most positive values ($-24{\cdot}9‰$) 58·2 m above the base of the section. Afterwards the values decrease again to $-27‰$ around 66 m above section base. Therefore, the $\delta^{13}C_{carb}$ and $\delta^{13}C_{org}$ data do not show the same trend. Based on the fact the $\delta^{13}C_{org}$-values correlate well with the global carbon-isotope trend, the negative $\delta^{13}C_{carb}$ excursion in the Natih B member is interpreted as being diagenetic in origin.

An additional factor which may have contributed to the difference between basinal and more coastal sediments is the original mineralogy of the sediments. The $\delta^{13}C_{carb}$-values from shallow-water carbonates are one to several permil higher relative to samples from the open marine pelagic carbonates. This variation in absolute $\delta^{13}C$-values can be explained by a varying amount of aragonite in the bulk rock. Aragonite is enriched in ^{13}C by about 0·9‰ relative to calcite (Lécuyer *et al.*, 2012). Sediments with a larger amount of aragonite therefore show heavier $\delta^{13}C$-values (Swart & Eberli, 2005).

Composite chemostratigraphic section of the Natih Formation

In combining results from different locations in the Adam Foothills (Wadi H and P at Jabal Madmar and Jabal Qusaybah) a composite section for the Late Albian to mid-Turonian (see Fig. 14) has been constructed. The composite reference section shows clearly the negative $\delta^{13}C$-excursion in the Natih B member at Jabal Qusaybah (blue points in Fig. 14). The overlapping part of the Natih B member at Jabal Madmar (green points in Fig. 14) is characterized by $\delta^{13}C$-values up to 2‰ heavier, with the negative excursion representing a central part of the intra-platform basin. The more positive values with lower TOC values were measured along the margin of the basin in carbonate-dominated sediments.

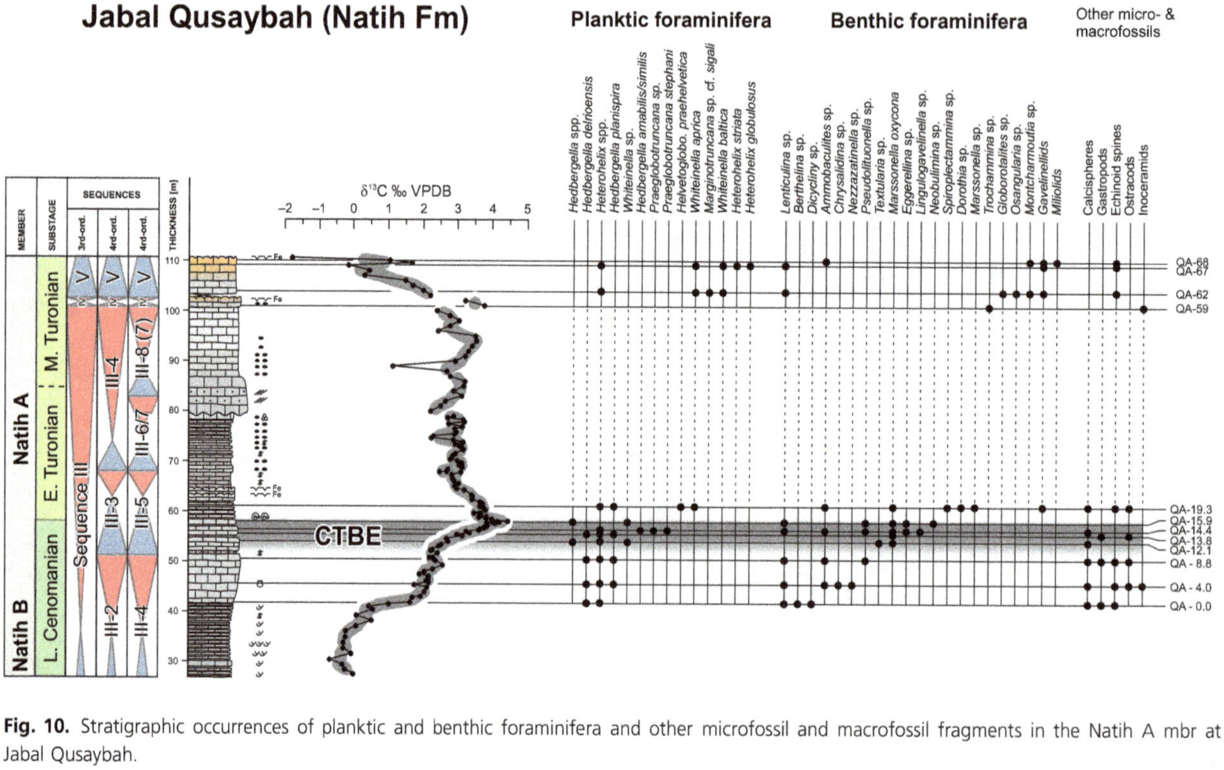

Fig. 10. Stratigraphic occurrences of planktic and benthic foraminifera and other microfossil and macrofossil fragments in the Natih A mbr at Jabal Qusaybah.

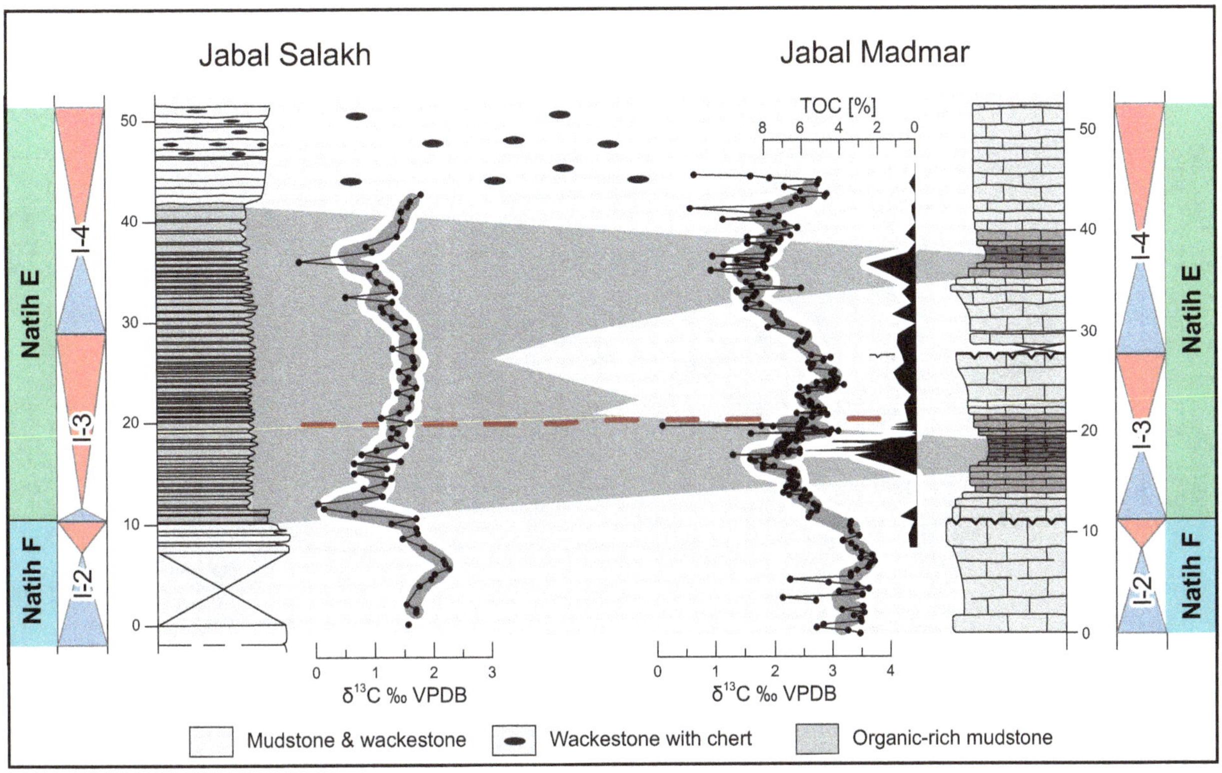

Fig. 11. Correlation from the fourth-order sequences (I-2, I-3 & I-4) from Jabal Madmar and Jabal Salakh. Correlation of the organic-rich mudstone from the intra-shelf basin (Jabal Salakh) and the more proximal section at Jabal Madmar (Homewood *et al.*, 2008). Sequence Stratigraphy after Grélaud *et al.* (2006) and Homewood *et al.* (2008).

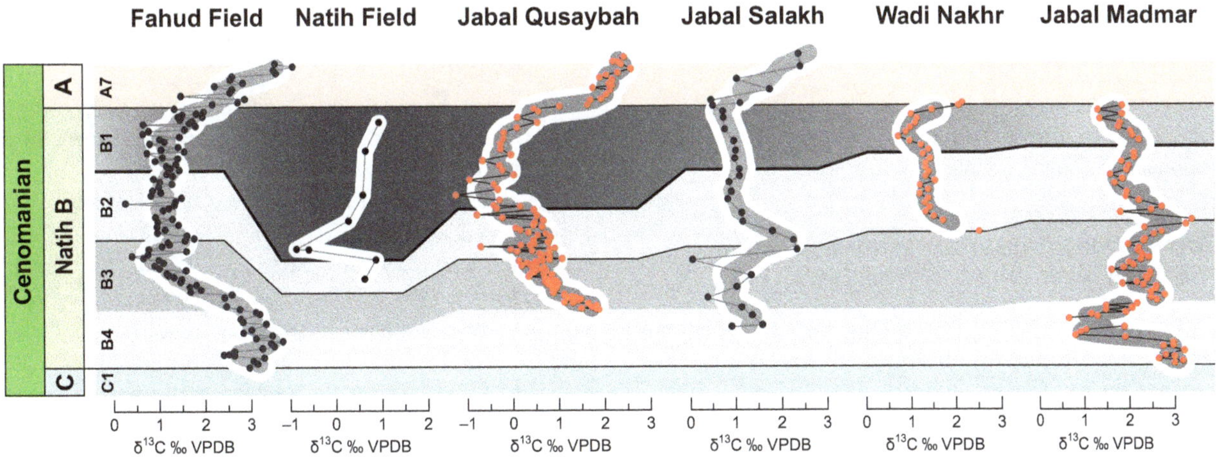

Fig. 12. Several chemostratigraphic results from the Natih B mbr. Fahud Field and Jabal Salakh by Vahrenkamp (2013); Natih Field by Al Balushi *et al.* (2011). Lithostratigraphic subunits (C1, B4-B1 & A7) are correlated with the Fahud Field subdivision (Vahrenkamp, 2013). Datum based at the base of the Natih A mbr.

The C-isotope stratigraphy of the Natih Formation confirms that organic carbon-enriched sediments deposited in the Cenomanian Mishrif-Natih-Basin do not coincide with OAE2. The formation of organic-rich sediments in this intra-platform basin was controlled by regional factors, including intra-platform basin palaeoceanography. Most important was that the sea-level controlled the degree of restriction of the Mishrif-Natih Basin. Correlation of sea-level history with C-isotope stratigraphy indicates that late Cenomanian sea-level rise ended a period of restriction

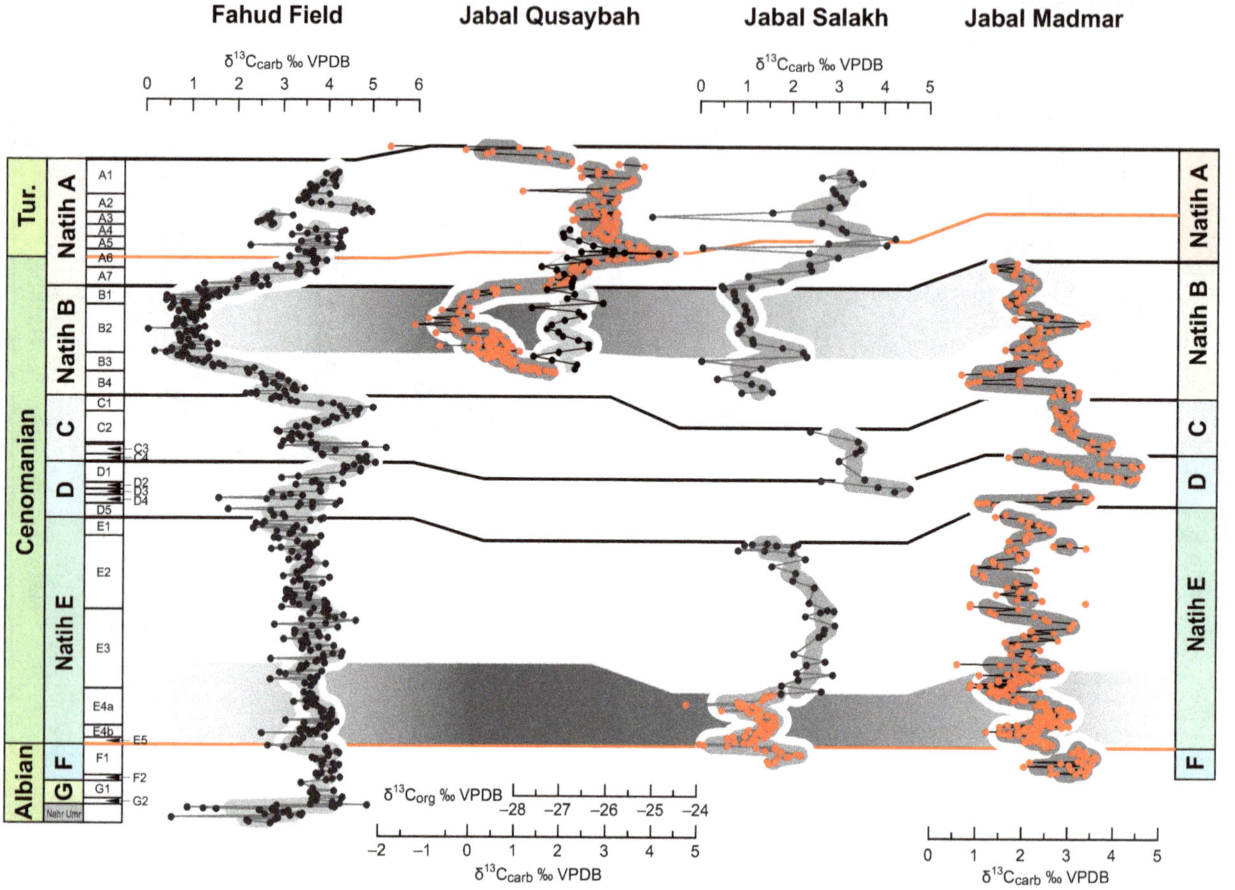

Fig. 13. Field 'F' (Vahrenkamp, 2013) and upper part Jabal Salakh [Razin, 2011 in Vahrenkamp (2013)]; black lines mark sequence stratigraphic boundaries after (Grélaud *et al.*, 2006; Homewood *et al.*, 2008; ; Van Buchem *et al.*, 1996, 2002); red lines are time lines.

with elevated C_{org} burial rates in the basin. The end of C_{org}-rich argillaceous sediments coincides with the end of the regionally limited carbonate C-isotope excursion.

Correlation of Interior Oman and Adam Foothills with global reference curves

A comparison of all carbon-isotope data from Interior Oman and Adam Foothills compared to two reference curves (English Chalk, Jarvis *et al.*, 2006 and Gubbio, Stoll & Schrag, 2000) is shown in Fig. 15. The Oman sections comprise several chemostratigraphic anchor points for a precise timing:

The *Albian/Cenomanian boundary*, defined by Kennedy *et al.* (2004), can be placed between peak b and c of the Albian/Cenomanian Boundary Event (Jarvis *et al.*, 2006). The δ^{13}C-values decrease afterwards in the lower Cenomanian (Mitchell *et al.*, 1996). The same C-isotope positive excursion can be seen in the Natih G and F members in the Fahud section (Vahrenkamp, 2013) and also at the Jabal Madmar (Fig. 13). Therefore, the Albian/Cenomanian boundary was adapted from Vahrenkamp (2013) and it coincides with the top of the sequence I-2 (Homewood *et al.*, 2008).

The cycles identified in the lower Natih E member may be correlated with the Lower Cenomanian Events (LCE I-III; Mitchell *et al.*, 1996). Mitchell *et al.* (1996) already concluded that the LCE's (I-III) are clear evidence of a link between short-term fluctuations in sea-level and significant δ^{13}C excursions. The chemostratigraphic data show the same link on the Arabian Platform and therefore the signal could be a global signal. The amplitude of the LCE's are much larger in the shallow-water sections compared to the more open marine sections (Mitchell *et al.*, 1996).

The *lower/middle Cenomanian boundary* can be placed at the base of the onset of the MCE I (Mitchell *et al.*, 1996; Jarvis *et al.*, 2006). Within the uppermost part of the sequence I-7 the δ^{13}C-values already show a trend to lighter values. Afterwards, within the lower Natih D member, the δ^{13}C-values show a clear trend to more positive values, which culminate in the MCE I (see discussion

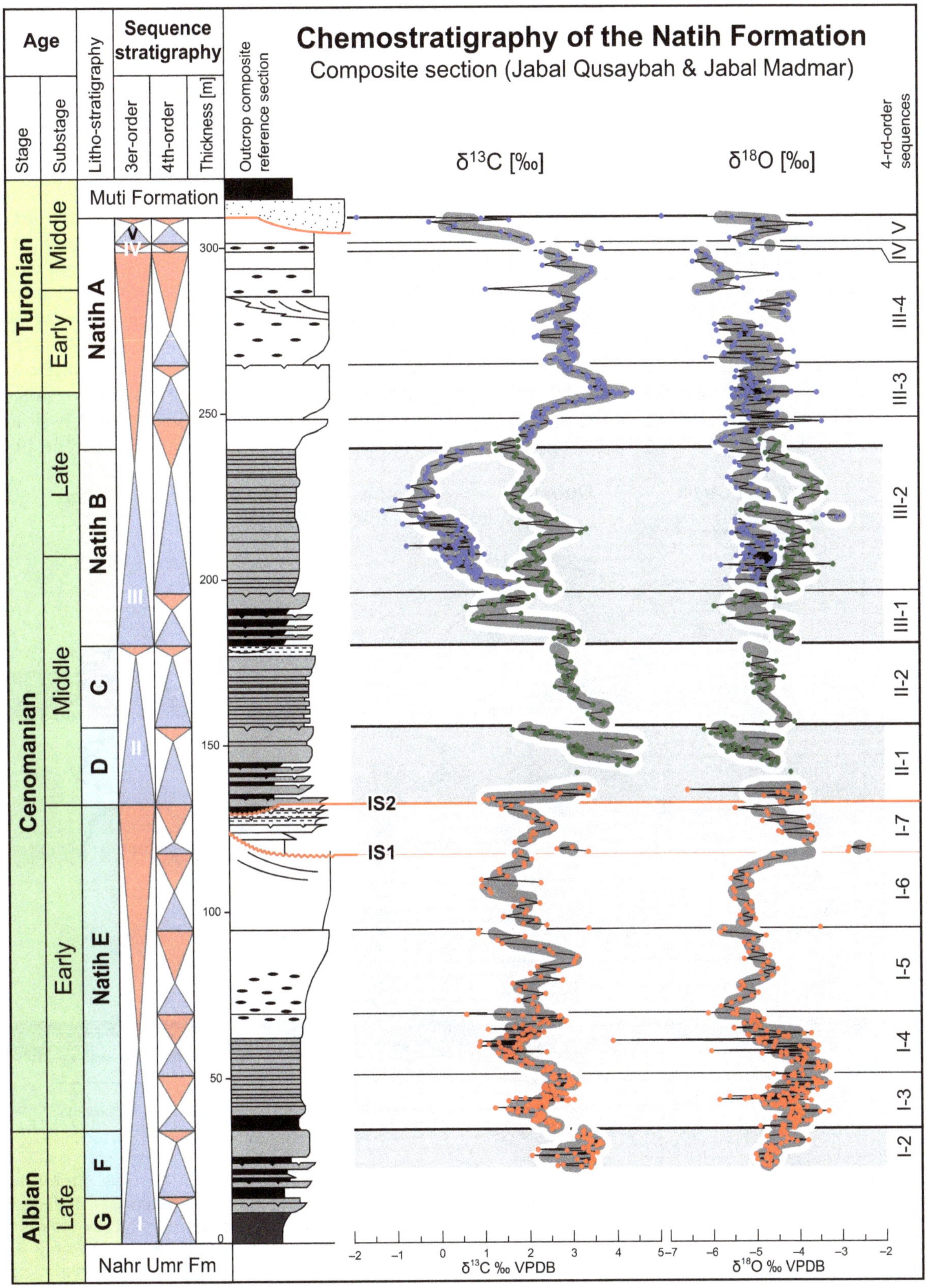

Fig. 14. Composite chemostratigraphic section from the Adam Foothills. Outcrop composite reference section modified after Grélaud *et al.* (2010); red: Wadi H, Jabal Madmar; green: Wadi P, Jabal Madmar and blue: Jabal Qusaybah.

below). Therefore, the lower/middle Cenomanian boundary coincides with the boundary at the Natih E/D boundary, which is a major emersion surface with a hiatus at Jabal Madmar. A stratigraphic lower position in the uppermost part of the Natih E member, as it is discussed by Piuz *et al.* (2014), cannot be excluded.

The first positive carbon-isotope excursion (+4·6‰) coincides with the two massive limestone beds (D2 and D1 in Fig. 6B) at the top of the Natih D member in the sequence II-1 (Homewood *et al.*, 2008). The positive peak is also observed at the Jabal Salakh section (Razin, 2011 in Vahrenkamp (2013); Fig. 13). The correlation with the Fahud area (Vahrenkamp, 2013) contains some uncertainties. The δ^{13}C-trend still increases in the subunit D1 of the Natih D member compared to the Jabal Salakh and Madmar section (Fig. 13). Therefore, the subunit D1 and D2 do not have exactly the same age in the Fahud area compared to the Adam Foothills. Nevertheless, the positive peak can be correlated with the *Middle Cenomanian Event I* (MCE I) (Paul *et al.*, 1994; Mitchell *et al.*, 1996) (Fig. 15). The MCE I can be divided into two smaller peaks MCE Ia & Ib (Mitchell *et al.*, 1996). Whether these two peaks are resolved in the two peaks in the Jabal Madmar at the base of the D2 subunit and just above in the marly interval or as proposed by Vahrenkamp (2013) in the peaks at the top of Natih D and C members remains debatable. It seems that the negative excursion between the two peaks at Fahud (subunit C2-C4; Vahrenkamp, 2013) is absent at Jabal Salakh and Madmar due to emersion at the top of Natih D member.

The *middle/upper Cenomanian boundary* can be placed at the small positive excursion of the Jukes-Browne Isotope Event which occurs in the upper beds of the *Acanthoceras jukesbrownei* and lowest *Calycoceras guerangeri* zones (Jarvis *et al.*, 2006). This isotope event is identified in the middle Natih B member (Fig. 15). The isotope

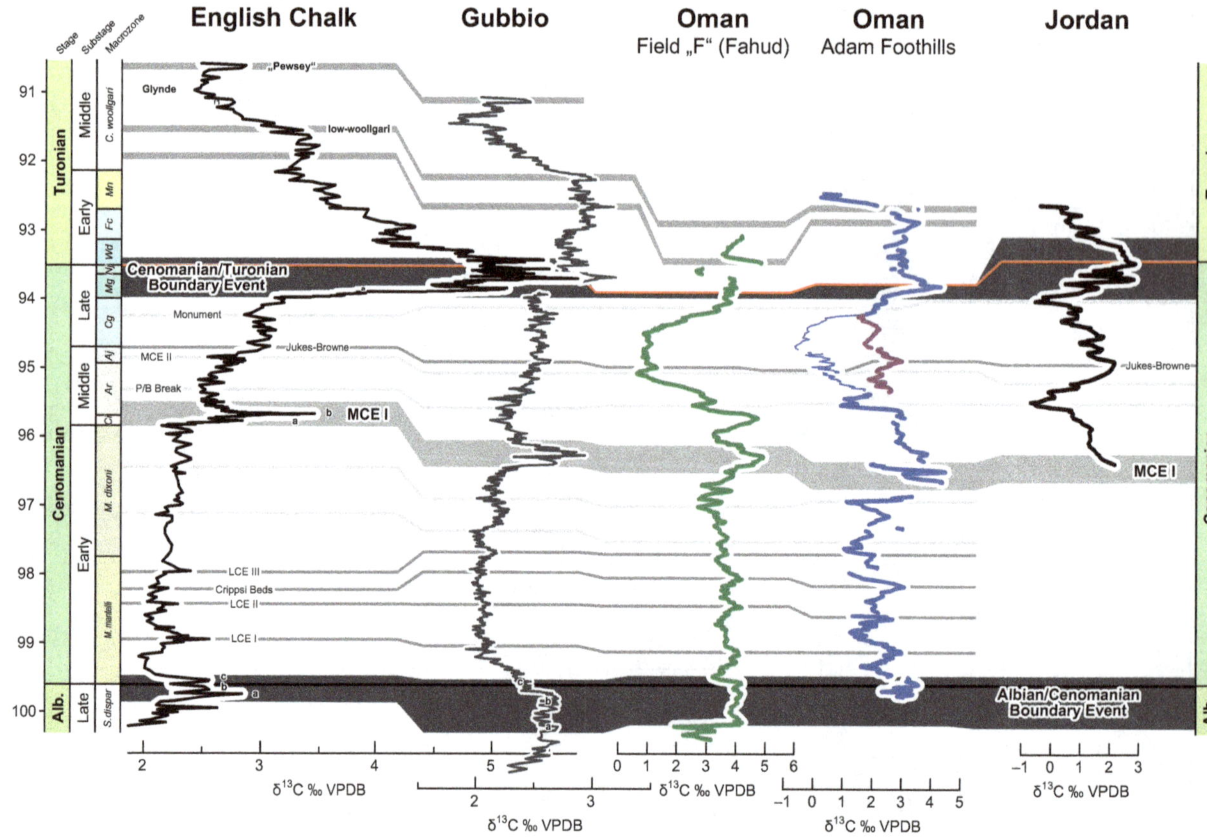

Fig. 15. Correlation of Late Albian to mid-Turonian δ^{13}C events between England, Italy and the Arabian Platform. The English Chalk reference curve in black (Jarvis *et al.*, 2006) is an unsmoothed composite curve plotted against age; Italian profile from Gubbio in grey (Stoll & Schrag, 2000); Oman Field 'F' in green (Vahrenkamp, 2013); Oman Adam Foothills in blue and purple (blue with negative excursion in the Bab basin and purple on the platform) and Jordan (Morsi & Wendler, 2010). Sections are anchored by the onset of the CTBE (increase into the peak 'a'). Correlation of major (grey bands) and minor (grey lines) positive and negative excursion (Jarvis *et al.*, 2006). Ci = *Cunningtoniceras inerme*; Ar = *Acanthoceras rhotomagense*; Aj = *A. jukesbrownei*; Cg = *Calycoceras guerangeri*; Mg = *Metoicoceras geslinianum*; Nj = *Neocardioceras juddii*; Wd = *Watinoceras devonense*; Fc = *Fagesia catinus*; Mn = *Mammites nodosoides*; LCE I-III = Lower Cenomanian Event I-III; MCE I = Mid-Cenomanian Event I.

excursion is more pronounced in the Jabal Madmar section but the correlation by lithology and also chemostratigraphy works also at Jabal Qusaybah and at Fahud (Vahrenkamp, 2013). The position of this substage boundary is supported by ammonites at Jabal Salakh (Meister & Piuz, 2015).

The positive C-isotope excursion (+4·5‰) in the lower part of the Natih A member at Jabal Qusaybah can also be seen in the Jabal Salakh section [P. Razin in Vahrenkamp (2013)] but it is absent in the Fahud area, probably due to an emersion and erosion phase at the end of sequence III in the earliest Turonian (Homewood *et al.*, 2008). The pronounced positive C-isotope excursion (4·5‰) was interpreted as the expression of OAE2 (Fig. 15) and therefore, the *Cenomanian/Turonian boundary* is shifted downwards compared to Vahrenkamp (2013). The position of this boundary is again supported by ammonites at Jabal Qusaybah (Meister & Piuz, 2015). These authors place the Cenomanian/Turonian boundary in the lower part of the 'grey limestone' (see Fig. 8D). Compared to other expanded CTBE-sections (Paul *et al.*, 1999; Jarvis *et al.*, 2006) the section must have been truncated just above the first peak of the CTBE.

The negative $\delta^{13}C$-shift in the Natih A member (sequence V) at the platform top (Jabal Qusaybah) can be correlated with the decreasing trend in the middle *mid-Turonian* observed on a global scale (Fig. 15). The $\delta^{13}C$-plateau ends with a distinct peak which can be seen in the higher values of sequence IV. This peak can be correlated with the Tu10 (Low-woollgari) Event (Voigt *et al.*, 2007). The end of the Natih Formation and, hence, of the Wasia Group, is again marked by a large regional unconformity, the Base Aruma unconformity (Fig. 2). This change in the sedimentary system occurred during the mid-Turonian as a result of the profound erosion due to the flexure of the continental lithosphere associated with beginning obduction of the Oman ophiolites (Robertson, 1987; Van Buchem *et al.*, 2002).

CONCLUSION

This study provides a stratigraphic model for the uppermost Albian to mid-Turonian Natih Formation through integration of stable chemostratigraphy with biostratigraphy and sequence stratigraphy. Two major positive C-isotope excursions can be correlated with global C-isotope anomalies in the Cenomanian and lower Turonian. The lower one, in the top of the Natih D member, corresponds to the Middle Cenomanian Event I (MCE I) and the upper one, in the lower Natih A member, with the Cenomanian/Turonian Boundary Event (CTBE) and Oceanic Anoxic Event 2 (OAE2). The CTBE seems to be truncated just above the first peak of the CTBE, a conclusion confirmed by ammonites and planktic foraminifera. The chemostratigraphic data at Jabal Qusaybah suggest a mid-Turonian age for the uppermost Natih Formation. Therefore, although tectonically controlled, the top of the Natih Formation coincided with a major global short-term sea-level drop in the mid-Turonian. This transition is indicated by a change in the benthic foraminifera to an assemblage dominated by Miliolidae, reflecting elevated salinity in shallow water.

The petroleum source rocks in the lower Natih E member and the Natih B member were not formed during one of the major Oceanic Anoxic Events (OAE1d, MCE and OAE2). The organic-rich interval in the lower Natih E member has an earliest Cenomanian age while the one forming part of the Natih B member clearly pre-dates the OAE2. Both source rock successions were formed in a restricted intra-shelf basin covering wide parts the Arabian Platform. The Natih B member of the intra-platform basin preserves a C-isotope excursion towards low values. This negative excursion is not seen in $\delta^{13}C_{org}$-values. Therefore, the negative isotopic excursion does not represent a change in the global marine carbon pool of the Late Cenomanian. Peculiar conditions during diagenesis of argillaceous intra-platform basin sediments enriched in organic carbon and alternating with neritic limestones may explain this excursion to negative C-isotope values. Early diagenetic sulphate reduction and anaerobic oxidation of methane both contributed to elevated pore-water alkalinity and to precipitation of authigenic carbonate depleted in C-13. Both, argillaceous sediments and neritic carbonates contain a C-isotope signature indicative of unusual diagenetic conditions and not changes in global DIC.

ACKNOWLEDGEMENTS

We thank Irene Meier, Reto Grischott, Maria Isabelle Millán, Stefan Huck, Ricardo Celestino and Bas den Brock for their help in the Oman Mountains during field campaign 2011 to 2014. We also thank Stewart Bishop, Maria Coray-Strasser and Madalina Jaggi (ETH) for laboratory assistance and Frowin Pirovino and Remy Lüchinger (ETH) for thin-section preparation. A special thank goes to Christian Meister and André Piuz in Geneva. The interesting exchange of information about the Natih Formation and, in addition, the determination of several ammonite specimens allowed a better integration of our chemostratigraphic data into a general stratigraphic framework. This study was supported by a Swiss National Science Foundation grant (200020_132775 & 200020_149168) and by ETH Zurich. Furthermore, we thank the two anonymous journal reviewers, for their critical comments and constructive contributions.

References

Al Balushi, S., Macquaker, J., Hollis, C. and **Marshall, J.D.** (2011) Influence of oxic diagenesis on source potential and lithofacies cyclicity: insight from Cenomanian Natih-B Member intra-shelf basinal carbonates, Oman. *Petrol. Geosci.*, **17**, 243–261.

Arndt, M., Virgo, S., Cox, S.F. and **Urai, J.L.** (2014) Changes in fluid pathways in a calcite vein mesh (Natih Formation, Oman Mountains): insights from stable isotopes. *Geofluids*, **14**, 391–418.

Arthur, M.A. and **Schlanger, S.O.** (1979) Cretaceous oceanic anoxic events as causal factors in development of reef-reservoired giant oil fields. *AAPG Bull.*, **63**, 870–885.

Burchette, T.P. (1993) Mishrif Formation (Cenomanian-Turonian), southern Arabian Gulf: carbonate platform growth along a cratonic basin margin. In: *Cretaceous Carbonate Platforms* (Eds J.A.T. Simo, R.W. Scott and J.-P. Masse), *AAPG Mem.*, **56**, 185–199.

Droste, H. (2010) High-resolution seismic stratigraphy of the Shu'aiba and Natih formations in the Sultanate of Oman: implications for Cretaceous epeiric carbonate platform systems. *Geol. Soc. London. Spec. Publ.*, **329**, 145–162.

Droste, H. (2014) Petroleum geology of the Sultanate of Oman. In: *Petroleum Systems of the Tethyan Region* (Eds L. Marlow, C. Kendall and L. Yose), *AAPG Mem.*, **106**, 713–755.

Droste, H. and **van Steenwinkel, M.** (2004) Stratal geometries and patterns of platform carbonates: the cretaceous of Oman. In: *Seismic Imaging of Carbonate Reservoir and System* (Eds G. Eberli, J.L. Masaferro and J.F.R. Sarg), *AAPG Mem.*, **81**, 185–206.

El-Sabbagh, A., Tantawy, A.A., Keller, G., Khozyem, H., Spangenberg, J., Adatte, T. and **Gertsch, B.** (2011) Stratigraphy of the Cenomanian-Turonian Oceanic Anoxic Event OAE2 in shallow shelf sequences of NE Egypt. *Cretaceous Res.*, **32**, 705–722.

Erbacher, J., Thurow, J. and **Littke, R.** (1996) Evolution patterns of radiolaria and organic matter variations: a new approach to identify sea-level changes in mid-Cretaceous pelagic environments. *Geology*, **24**, 499–502.

Giorgioni, M., Weissert, H., Bernasconi, S.M., Hochuli, P.A., Coccioni, R. and **Keller, C.E.** (2012) Orbital control on carbon cycle and oceanography in the mid-Cretaceous greenhouse. *Paleoceanography*, **27**, 1–12.

Glennie, K.W., Boeuf, M.G.A., Hughes-Clarke, M.W., Moody-Stuart, M., Pilaar, W. and **Reinhardt, B.M.** (1974) Geology of the Oman Mountains. *K. Ned. Aardrijksk. Genoot. Tijdschr.*, **50**, 1–423.

Grélaud, C., Razin, P., Homewood, P.W. and **Schwab, A.M.** (2006) Development of incisions on a periodically emergent carbonate platform (Natih Formation, Late Cretaceous, Oman). *J. Sed. Res.*, **76**, 647–669.

Grélaud, C., Razin, P. and **Homewood, P.** (2010) Channelized systems in an inner carbonate platform setting: differentiation between incisions and tidal channels (Natih Formation, Late Cretaceous, Oman). *Geol. Soc. London. Spec. Publ.*, **329**, 163–186.

Haq, B.U. (2014) Cretaceous eustasy revisited. *Global Planet. Change*, **113**, 44–58.

Homewood, P., Razin, P., Grélaud, C., Droste, H., Vahrenkamp, V., Mettraux, M. and **Mattner, J.** (2008) Outcrop sedimentology of the Natih Formation, northern Oman: a field guide to selected outcrops in the Adam foothills and Al Jabal al Akhdar areas. *GeoArabia*, **13**, 39–120.

Hughes-Clarke, M.W. (1988) Stratigraphy and rock unit nomenclature in the oil-producing area of interior Oman. *J. Petrol. Geol.*, **11**, 5–60.

Immenhauser, A. and **Rameil, N.** (2011) Interpretation of ancient epikarst features in carbonate successions – a note of caution. *Sed. Geol.*, **239**, 1–9.

Immenhauser, A., Schlager, W., Burns, S.J., Scott, R.W., Geel, T., Lehmann, J., Van der Gaast, S. and **Bolder-Schrijver, L.J.A.** (1999) Late Aptian to late Albian sea-level fluctuations constrained by geochemical and biological evidence (Nahr Umr formation, Oman). *J. Sed. Res.*, **69**, 434–446.

Jarvis, I., Carson, G.A., Cooper, M.K.E., Hart, M.B., Leary, P.N., Tocher, B.A., Horne, D. and **Rosenfeld, A.** (1988) Microfossil assemblages and the Cenomanian-Turonian (Late Cretaceous) oceanic anoxic event. *Cretaceous Res.*, **9**, 3–103.

Jarvis, I., Gale, A.S., Jenkyns, H.C. and **Pearce, M.A.** (2006) Secular variation in the Late Cretaceous carbon isotopes: a new $d^{13}C$ carbonate reference curve for the Cenomanian-Campanian (99.6-70.6 Ma). *Geol. Mag.*, **143**, 561–608.

Jenkyns, H.C. (1980) Cretaceous anoxic events: from continents to oceans. *J. Geol. Soc. London*, **137**, 171–188.

Jenkyns, H.C. (2010) Geochemistry of oceanic anoxic events. *Geochem. Geophys. Geosyst.*, **11**, 1–30.

Kennedy, W.J. and **Simmons, M.D.** (1991) Mid-Cretaceous ammonites and associated microfossils from the Central Oman Mountains. *Newsl. Stratigr.*, **25**, 127–154.

Kennedy, W.J., Gale, A.S., Lees, J.A. and **Caron, M.** (2004) The global boundary stratotype section and point (GSSP) for the base of the Cenomanian Stage, Mont Risou, Hautes-Alpes, France. *Episodes*, **27**, 21–32.

Lécuyer, C., Hutzler, A., Amiot, R., Daux, V., Grosheny, D., Otero, O., Martineaua, F., Fourela, F., Baltera, V. and **Reynard, B.** (2012) Carbon and oxygen isotope fractionations between aragonite and calcite of shells from modern molluscs. *Chem. Geol.*, **332**, 92–101.

Maurer, F., van Buchem, F.S., Eberli, G.P., Pierson, B.J., Raven, M.J., Larsen, P.H., Al-Husseini, M.I. and **Vincent,**

B. (2013) Late Aptian long-lived glacio-eustatic lowstand recorded on the Arabian Plate. *Terra Nova*, **25**, 87–94.

Meister, C. and **Piuz, A.** (2013) Late Cenomanian-Early Turonian ammonites of the southern Tethys margin from Morocco to Oman: biostratigraphy, paleobiogeography and morphology. *Cretaceous Res.*, **44**, 83–103.

Meister, C. and **Piuz, A.** (2015) Cretaceous ammonites from the Sultanate of Oman (Adam Foothills). *GeoArabia*, **20**, 17–74.

Meister, P., Mckenzie, J.A., Vasconcelos, C., Bernasconi, S., Frank, M., Gutjahr, M. and **Schrag, D.P.** (2007) Dolomite formation in the dynamic deep biosphere: results from the Peru Margin. *Sedimentology*, **54**, 1007–1032.

Mitchell, S.F., Paul, C.R.C. and **Gale, A.S.** (1996) Carbon isotopes and sequence stratigraphy. *Geol. Soc. London. Spec. Publ.*, **104**, 11–24.

Morsi, A.M.M. and **Wendler, J.E.** (2010) Biostratigraphy, palaeoecology and palaeogeography of the Middle Cenomanian-Early Turonian Levant Platform in Central Jordan based on ostracods. *Geol. Soc. London. Spec. Publ.*, **341**, 187–210.

Murray, J.W. (1991) *Ecology and Palaeoecology of Benthic Foraminifera*. Routledge, London and New York, 397 pp.

Murris, R. (1980) Middle East: stratigraphic evolution and oil habitat. *AAPG Bull.*, **64**, 597–618.

Parente, M., Frijia, G. and **Di Lucia, M.** (2007) Carbon-isotope stratigraphy of Cenomanian-Turonian platform carbonates from the southern Apennines (Italy): a chemostratigraphic approach to the problem of correlation between shallow-water and deep-water successions. *J. Geol. Soc. London*, **164**, 609–620.

Paul, C.R.C., Mitchell, S.F., Marshall, J.D., Leafy, P.N., Gale, A.S., Duane, A.M. and **Ditchfield, P.W.** (1994) Palaeoceanographic events in the middle Cenomanian of Northwest Europe. *Cretaceous Res.*, **15**, 707–738.

Paul, C.R.C., Lamolda, M.A., Mitchell, S.F., Vaziri, M.R., Gorostidi, A. and **Marshall, J.D.** (1999) The Cenomanian-Turonian boundary at Eastbourne (Sussex, UK): a proposed European reference section. *Palaeogeogr. Palaeoclimatol. Palaeoecol.*, **150**, 83–121.

Philip, J., Borgomano, J. and **Al-Maskiry, S.** (1995) Cenomanian-Early Turonian carbonate platform of Northern Oman: stratigraphy and palaeo-environments. *Palaeogeogr. Palaeoclimatol. Palaeoecol.*, **119**, 77–92.

Philip, J., Floquet, M., Platel, J.P., Bergerat, F., Sandulescu, M., Baraboshkin, E., Amon, E., Guiraud, R., Vaslet, D., Le Nindre, Y., Ziegler, M., Poisson, A. and **Bouaziz, S.** (2000) Late Cenomanian. In: *Atlas Peri-Tethys Palaeogeographical Maps* (Eds J. Dercourt, M. Gaetani, B. Vrielynck, E.B. Barrier, B. Biju-Duval, M.-F. Brunet, J.P. Cadet, S. Crasquin and M. Sandulescu) CCGM/CGMW, Map 14, Paris.

Pillevuit, A., Marcoux, J., Stampfli, G. and **Baud, A.** (1997) The Oman exotics: a key to the understanding of the Neotethyan geodynamic evolution. *Geodin. Acta*, **10**, 209–238.

Piuz, A. and **Meister, C.** (2013) Cenomanian rotaliids (Foraminiferida) from Oman and Morocco. *Swiss J. Palaeontol.*, **132**, 81–97.

Piuz, A., Meister, C. and **Vicedo, V.** (2014) New Alveolinoidea (Foraminifera) from the Cenomanian of Oman. *Cretaceous Res.*, **50**, 344–360.

Pratt, B.R. and **Smewing, J.D.** (1990) Jurassic and Early Cretaceous platform margin configuration and evolution, central Oman Mountains. *Geol. Soc. London. Spec. Publ.*, **49**, 69–88.

Pratt, B.R. and **Smewing, J.D.** (1993) Early Cretaceous platform-margin configuration and evolution in the central Oman Mountains, Arabian Peninsula. *AAPG Bull.*, **77**, 225–244.

Rabu, D. (1987) Géologie de l'Autochtone des Montagnes d'Oman: la fenêtre du Jabal Akhdar. La semelle métamorphique de la Nappe ophiolitique de Samail dans les parties orientale et centrale des Montagnes d'Oman: une revue. Thèse de Doctorat d'Etat. Université Pierre et Marie Curie, Paris 6. *Bull. Bur. Rech. Géol. Min.*, **130**, 582.

Razin, P., Taati, F. and **van Buchem, F.S.P.** (2010) Sequence stratigraphy of Cenomanian-Turonian carbonate platform margins (Sarvak Formation) in the High Zagros, SW Iran: an outcrop reference model for the Arabian Plate. *Geol. Soc. London. Spec. Publ.*, **329**, 187–218.

Robertson, A.H.F. (1987) Upper Cretaceous Muti Formation: transition of a Mesozoic nate platform to a foreland basin in the Oman Mountains. *Sedimentology*, **34**, 1123–1142.

Schlanger, S.O. and **Jenkyns, H.C.** (1976) Cretaceous oceanic anoxic event: causes and consequences. *Geol. Mijnbouw*, **55**, 179–188.

Schlanger, S.O., Arthur, M.A., Jenkyns, H.C. and **Scholle, P.A.** (1987) The Cenomanian–Turonian oceanic anoxic event: I. Stratigraphy and distribution of organic carbon-rich beds and the marine $d^{13}C$ excursion. In: *Marine Petroleum Source Rocks* (Eds J. Brooks and A. Fleet), *Geol. Soc. London. Spec. Publ.*, **26**, 371–399.

Scholle, P.A. and **Arthur, M.A.** (1980) Carbon isotope fluctuations in Cretaceous pelagic limestone: potential stratigraphic and petroleum exploration tool. *AAPG Bull.*, **64**, 67–87.

Schrag, D.P., Higgins, J.A., Macdonald, F.A. and **Johnston, D.T.** (2013) Authigenic carbonate and the history of the global carbon cycle. *Science*, **339**, 540–543.

Schroeder, R., Van Buchem, F., Cherchi, A., Baghbani, D., Vincent, B., Immenhauser, A. and **Granier, B.** (2010) Revised orbitolinid biostratigraphic zonation for the Barremian-Aptian of the eastern Arabian Plate and implications for regional stratigraphic correlations. *GeoArabia Spec. Publ.*, **4**, 49–96.

Schwab, A.M., Homewood, P.W., Van Buchem, F.S.P. and **Razin, P.** (2005) Seismic forward model of a Natih Formation outcrop: the Adam Foothills Transect (northern Oman). *GeoArabia*, **10**, 17–44.

Sharland, P.R., Archer, R., Casey, D.M., Davies, R.B., Hall, S.H., Heward, A.P., Horbury, A.D. and **Simmons, M.D.**

(2001) Arabian Plate Sequence Stratigraphy, Bahrain, Gulf PetroLink. *GeoArabia Spec. Publ.*, **2**, 371.

Simmons, M.D. (1994) Micropalaeontological biozonation of the Kahmah Group (Early Cretaceous), Central Oman Mountains. In: *Micropalaeontology and Hydrocarbon Exploration in the Middle East* (Ed. M.D. Simmons), pp. 177–220. British Micropalaeontological Society Publication Series, Chapman & Hall, London.

Simmons, M.D. and **Hart, M.B.** (1987) The biostratigraphy and microfacies of the Early to mid-Cretaceous carbonates of Wadi Mi'aidin, Central Oman Mountains. In: *Micropaleontology of Carbonate Environments* (Ed M.B. Hart), pp. 176–207. The British Micropalaeontological Society, Ellis Horwood Ltd, Chichester.

Smith, A.B., Simmons, M.D. and **Racey, A.** (1990) Cenomanian echinoids, larger foraminifera and calcareous algae from the Natih Formation, central Oman Mountains. *Cretaceous Res.*, **11**, 29–70.

Stoll, H.M. and **Schrag, D.P.** (2000) High-resolution stable isotope records from the Upper Cretaceous rocks of Italy and Spain: glacial episodes in a greenhouse planet? *Geol. Soc. Am. Bull.*, **112**, 308–319.

Swart, P.K. and **Eberli, G.** (2005) The nature of the $\delta^{13}C$ of periplatform sediments: implications for stratigraphy and the global carbon cycle. *Sed. Geol.*, **175**, 115–129.

Terken, J.M.J. (1999) The Natih Petroleum System of North Oman. *GeoArabia*, **4**, 157–180.

Trabucho Alexandre, J., Tuenter, E., Henstra, G.A., van der Zwan, K.J., van de Wal, R.S., Dijkstra, H.A. and **de Boer, P.L.** (2010) The mid-Cretaceous North Atlantic nutrient trap: black shales and OAEs. *Paleoceanography*, **25**, 1–14.

Vahrenkamp, V.C. (2010) Chemostratigraphy of the Lower Cretaceous Shu'aiba Formation: a $\delta^{13}C$ reference profile for the Aptian Stage from the southern Neo-TethysOcean. *GeoArabia Spec. Publ.*, **4**, 107–137.

Vahrenkamp, V.C. (2013) Carbon-isotope signatures of Albian to Cenomanian (Cretaceous) shelf carbonates of the Natih Formation, Sultanate of Oman. *GeoArabia*, **8**, 65–82.

Van Buchem, F.S., Razin, P., Homewood, P.W., Philip, J.M., Eberli, G.P., Platel, J.P., Roger, J., Eschard, R., Desaubliaux, G.M.J., Boisseau, T., Leduc, J.P., Labourdette, R. and **Cantaloube, S.** (1996) High resolution sequence stratigraphy of the Natih formation (Cenomanian/ Turonian) in northern Oman: distribution of source rocks and reservoir facies. *GeoArabia*, **1**, 65–91.

Van Buchem, F.S., Razin, P., Homewood, P.W., Oterdoom, W.H. and **Philip, J.** (2002) Stratigraphic organization of carbonate ramps and organic-rich intrashelf basins: Natih Formation (middle Cretaceous) of northern Oman. *AAPG Bull.*, **86**, 21–53.

Van Buchem, F., Huc, A.Y., Pradier, B. and **Stefani, M.M.** (2005) Stratigraphic patterns in carbonate source-rock distribution: second-order to fourth-order control and sediment flux. *SEPM Spec. Publ.*, **82**, 191–223.

Van Buchem, F., Simmons, M., Droste, H. and **Davies, R.** (2011) Late Aptian to Turonian stratigraphy of the eastern Arabian Plate–depositional sequences and lithostratigraphic nomenclature. *Petrol. Geosci.*, **17**, 211–222.

Voigt, S., Gale, A.S. and **Voigt, T.** (2006) Sea-level change, carbon cycling and palaeoclimate during the Late Cenomanian of northwest Europe; an integrated palaeoenvironmental analysis. *Cretaceous Res.*, **27**, 836–858.

Voigt, S., Aurag, A., Leis, F. and **Kaplan, U.** (2007) Late Cenomanian to Middle Turonian high-resolution carbon isotope stratigraphy: new data from the Münsterland Cretaceous Basin, Germany. *Earth Planet. Sci. Lett.*, **235**, 196–210.

Voigt, S., Erbacher, J., Mutterlose, J., Weiss, W., Westerhold, T., Wiese, F., Wilmsen, M. and **Wonik, T.** (2008) The Cenomanian-Turonian of the Wunstorf section (North Germany): global stratigraphic reference section and new orbital time scale for Oceanic Anoxic Event 2. *Newsl. Stratigr.*, **43**, 65–89.

Wagner, P.D. (1990) Geochemical stratigraphy and porosity controls in Cretaceous carbonates near the Oman Mountains. *Geol. Soc. London. Spec. Publ.*, **49**, 127–137.

Weissert, H. (1989) C-Isotope stratigraphy, a monitor of paleoenvironmental change: a case study from the early cretaceous. *Surv. Geophys.*, **10**, 1–61.

Westphal, H. (2006) Limestone-marl alternations as environmental archives and the role of early diagenesis: a critical review. *Int. J. Earth Sci.*, **95**, 947–961.

Witt, W. and **Goekdag, H.** (1994) Orbitolinid biostratigraphy of the Shuaiba Formation (Aptian), Oman-implications for reservoir development. In: *Micropaleontology and Hydrocarbon Exploration in the Middle East* (Ed. M.D. Simmons), pp. 221–242. Chapman & Hall, London.

Zhao, Y., Lokier, S.W. and **Steuber, T.** (2012) A new stable carbon isotope record for the Cenomanian of the Arabian Plate. *EGU General Assembly Conf. Abstr.*, **14**, 3301.

Ziegler, M.A. (2001) Late Permian to Holocene Paleofacies Evolution of the Arabian Plate and its Hydrocarbon Occurrences. *GeoArabia*, **6**, 445–504.

Permissions

All chapters in this book were first published in TDR, by John Wiley & Sons Ltd.; hereby published with permission under the Creative Commons Attribution License or equivalent. Every chapter published in this book has been scrutinized by our experts. Their significance has been extensively debated. The topics covered herein carry significant findings which will fuel the growth of the discipline. They may even be implemented as practical applications or may be referred to as a beginning point for another development.

The contributors of this book come from diverse backgrounds, making this book a truly international effort. This book will bring forth new frontiers with its revolutionizing research information and detailed analysis of the nascent developments around the world.

We would like to thank all the contributing authors for lending their expertise to make the book truly unique. They have played a crucial role in the development of this book. Without their invaluable contributions this book wouldn't have been possible. They have made vital efforts to compile up to date information on the varied aspects of this subject to make this book a valuable addition to the collection of many professionals and students.

This book was conceptualized with the vision of imparting up-to-date information and advanced data in this field. To ensure the same, a matchless editorial board was set up. Every individual on the board went through rigorous rounds of assessment to prove their worth. After which they invested a large part of their time researching and compiling the most relevant data for our readers.

The editorial board has been involved in producing this book since its inception. They have spent rigorous hours researching and exploring the diverse topics which have resulted in the successful publishing of this book. They have passed on their knowledge of decades through this book. To expedite this challenging task, the publisher supported the team at every step. A small team of assistant editors was also appointed to further simplify the editing procedure and attain best results for the readers.

Apart from the editorial board, the designing team has also invested a significant amount of their time in understanding the subject and creating the most relevant covers. They scrutinized every image to scout for the most suitable representation of the subject and create an appropriate cover for the book.

The publishing team has been an ardent support to the editorial, designing and production team. Their endless efforts to recruit the best for this project, has resulted in the accomplishment of this book. They are a veteran in the field of academics and their pool of knowledge is as vast as their experience in printing. Their expertise and guidance has proved useful at every step. Their uncompromising quality standards have made this book an exceptional effort. Their encouragement from time to time has been an inspiration for everyone.

The publisher and the editorial board hope that this book will prove to be a valuable piece of knowledge for researchers, students, practitioners and scholars across the globe.

List of Contributors

Frank J. G. Van Den Belt and Poppe L. De Boer
Department of Earth Sciences, University of Utrecht, 3508 TA, Utrecht, The Netherlands

Frank Van Bergen
Tno/Geological Survey of the Netherlands, Princetonlaan 6, 3508 TA, Utrecht, The Netherlands
Nexen Petroleum UK Ltd., Prospect House, 97 Oxford Road, Uxbridge UB8 1LU, UK

Julien Moreau, Myriam Boussaha, Lars Nielsen and Nicolas Thibault
Department of Geosciences and Natural Resource Management, University of Copenhagen, Øster Voldgade 10, 1350 Copenhagen, Denmark

Clemens V. Ullmann
Camborne School of Mines, University of Exeter, Penryn Campus, Penryn TR10 9FE, UK

Lars Stemmerik
Natural History Museum, University of Cop-enhagen, Øster Voldgade 10, 1350 Copenhagen, Denmark

André Strasser
Department of Geosciences, University of Fribourg, Chemin du Musée 6, 1700, Fribourg, Switzerland

Ian Jarvis
Department of Geography and Geology, Kingston University London, Kingston upon Thames, KT1 2EE, UK

João Trabucho-Alexandre
Department of Earth Sciences, Durham University, Durham, DH1 3LE, UK
Institute of Earth Sciences, Utrecht University, Budapestlaan, 43584 CD, Utrecht, Netherlands

Darren R. Gröcke
Department of Earth Sciences, Durham University, Durham, DH1 3LE, UK

David Uličný and Jiří Laurin
Institute of Geophysics, Academy of Sciences of the Czech Republic, 141 31, Prague, Czech Republic

Muriel Pacton
Geological Institute, ETH-Zürich, Zurich, Switzerland
Laboratoire de Géologie de Lyon, Université Lyon 1, France

Vincent Martinuzzi
Department of Earth Sciences, University of Geneva, Geneva, Switzerland
Geneva Petroleum, Geneva, Switzerland

Gabriela Cusminsky
Departamento de Ecología Crub Unc-Inibioma Conicet, Quintral 1250, 8400, Bariloche, Argentina

Beatrice Burdin
Centre technologique des microstructures, Université Lyon 1, Lyon, France

Kurt Barmettler
Institute of Biogeochemistry and Pollutant Dynamics, ETH-Zürich, Zurich, Switzerland

Crisogono Vasconcelos
Geological Institute, ETH-Zürich, Zurich, Switzerland

Lucas Kämpf
GFZ German Research Centre for Geosciences, Telegrafenberg, Potsdam, 14473, Germany
TU Dresden, Faculty of Environmental Sciences, Institute for Soil Science and Site Ecology, Pienner Strasse 19, Tharandt, 01737, Germany

Philip Mueller, Heiko Thoss, Andreas Güntner, Birgit Plessen, Rudolf Naumann, Achim Brauer and Bruno Merz
GFZ German Research Centre for Geosciences, Telegrafenberg, Potsdam, 14473, Germany

Hannes Höllerer
Research Institute for Limnology Mondsee, University of Innsbruck, Mondseestrasse 9, Mondsee, 5310, Austria

Maria C. Marcano and Kyger C Lohmann
Department of Earth and Environmental Sciences, University of Michigan, 2534 C. C. Little Building, 1100 North University Ave., Ann Arbor, MI 48109-1005, USA

Tracy D. Frank
Department of Earth and Atmospheric Sciences, University of Nebraska-Lincoln, 214 Bessey Hall, Lincoln, NE 68588-0340, USA

Samuel B. Mukasa
College of Engineering and Physical Sciences, University of New Hampshire, Kingsbury Hall W289, 33 Academic Way, Durham, NH 03824-3591, USA

Marco Taviani
Istituto di Scienze Marine (ISMAR) - CNR, Via Gobetti 101, 40129, Bologna Italy
Biology Department, Woods Hole Oceanographic Institution, 266 Woods Hole Road, Woods Hole, MA 02543, USA

Camille Thomas
Department of Earth Sciences, University of Geneva, rue des Maraichers 13, Geneva CH 1205, Switzerland
UMR 42 CARRTEL, Alpine Research Center on Lake Food Webs, University of Savoie Mont-Blanc, 73376 Le Bourget du Lac, France

Yael Ebert
Institute of Earth Sciences, Hebrew University of Jerusalem, Edmond J. Safra Campus, Givat Ram, Jerusalem, IL 91904, USA

Yael Kiro
Lamont-Doherty Earth Observatory, Columbia University, 61 Rt. 9W, Palisades, NY 10964, USA

Mordechai Stein
Geological Survey of Israel, 30 Malkhe Israel St., Jerusalem, IL 95501, USA

Daniel Ariztegui and Gabriel Hunger
Department of Earth Sciences, University of Geneva, rue des Maraichers 13, Geneva CH 1205, Switzerland

Chloe Morales
Institute of Earth Sciences, University of Lausanne, Géopolis, 1015, Lausanne, Switzerland
Marine Palynology Group, Institute of Earth Sciences, University of Utrecht, Van Unnikgebouw, Heidelberglaan

Jorge E. Spangenberg, Thierry Adatte and Karl B. Föllmi
Institute of Earth Surface Dynamics, University of Lausanne, Géopolis, 1015, Lausanne, Switzerland

Annie Arnaud-Vanneau
Association Dolomieu, 6 Chemin des Grenouilles, 38700, La Tronche, France

Stephan Wohlwend and Helmut Weissert
Geological Institute, ETH Zurich, Sonneggstrasse 5, 8092 Zurich, Switzerland

Malcolm Hart
School of Geography, Earth and Environmental Sciences, Plymouth University, Drake Circus, Plymouth PL4 8AA, UK

Index

www.ingramcontent.com/pod-product-compliance
Lightning Source LLC
Chambersburg PA
CBHW082101190326
41458CB00010B/3538
* 9 7 8 1 6 4 1 1 6 0 7 5 9 *